This report contains the collectiv⋯⋯⋯⋯ f
experts and does not necessarily represent the decisions or the stated
policy of the United Nations Environment Programme, the
International Labour Organization, or the World Health
Organization.

Environmental Health Criteria 224

ARSENIC AND ARSENIC COMPOUNDS

Second edition

The first and second drafts of this monograph were prepared, under
the coordination of Dr J. Ng, by the authors A. Gomez-Caminero,
P. Howe, M. Hughes, E. Kenyon, D.R. Lewis, M. Moore, J. Ng, and
by A. Aitio and G. Becking.

Published under the joint sponsorship of the United
Nations Environment Programme, the International
Labour Organization, and the World Health
Organization, and produced within the framework of the
Inter-Organization Programme for the Sound
Management of Chemicals.

World Health Organization
Geneva, 2001

The **International Programme on Chemical Safety (IPCS)**, established in 1980, is a joint venture of the United Nations Environment Programme (UNEP), the International Labour Organization (ILO), and the World Health Organization (WHO). The overall objectives of the IPCS are to establish the scientific basis for assessment of the risk to human health and the environment from exposure to chemicals, through international peer-review processes, as a prerequisite for the promotion of chemical safety, and to provide technical assistance in strengthening national capacities for the sound management of chemicals.

The **Inter-Organization Programme for the Sound Management of Chemicals (IOMC)** was established in 1995 by UNEP, ILO, the Food and Agriculture Organization of the United Nations, WHO, the United Nations Industrial Development Organization, the United Nations Institute for Training and Research, and the Organisation for Economic Co-operation and Development (Participating Organizations), following recommendations made by the 1992 UN Conference on Environment and Development to strengthen cooperation and increase coordination in the field of chemical safety. The purpose of the IOMC is to promote coordination of the policies and activities pursued by the Participating Organizations, jointly or separately, to achieve the sound management of chemicals in relation to human health and the environment.

WHO Library Cataloguing-in-Publication Data

Arsenic and arsenic compounds.

(Environmental health criteria ; 224)

1.Arsenic – toxicity 2.Arsenicals – toxicity 3.Environmental exposure
I.International Programme on Chemical Safety II. WHO Task Group on
Environmental Health Criteria for Arsenic and Arsenic Compounds III.Series

ISBN 92 4 157224 8   (NLM Classification: QV 294)
ISSN 0250-863X

Computer typesetting by I. Xavier Lourduraj, Chennai, India

Printed in Finland
2001/13945 – Vammala – 5000

CONTENTS

ENVIRONMENTAL HEALTH CRITERIA FOR
ARSENIC AND ARSENIC COMPOUNDS

NOTE TO READERS OF THE CRITERIA MONOGRAPHS

Every effort has been made to present information in the criteria monographs as accurately as possible without unduly delaying their publication. In the interest of all users of the Environmental Health Criteria monographs, readers are requested to communicate any errors that may have occurred to the Director of the International Programme on Chemical Safety, World Health Organization, Geneva, Switzerland, in order that they may be included in corrigenda.

* * *

A detailed data profile and a legal file can be obtained from the International Register of Potentially Toxic Chemicals, Case postale 356, 1219 Châtelaine, Geneva, Switzerland (telephone no. + 41 22 - 9799111, fax no. + 41 22 - 7973460, E-mail irptc@unep.ch).

* * *

This publication was made possible by grant number 5 U01 ES02617-15 from the National Institute of Environmental Health Sciences, National Institutes of Health, USA, and by financial support from the European Commission.

The Commonwealth Department of Health and Aged Care, Australia, contributed financially to the preparation of this Environmental Health Criteria monograph. The Task Group meeting was arranged by the National Research Centre for Environmental Toxicology, Australia.

Environmental Health Criteria

PREAMBLE

Objectives

In 1973 the WHO Environmental Health Criteria Programme was initiated with the following objectives:

(i)   to assess information on the relationship between exposure to environmental pollutants and human health, and to provide guidelines for setting exposure limits;

(ii)   to identify new or potential pollutants;

(iii) to identify gaps in knowledge concerning the health effects of pollutants;

(iv)   to promote the harmonization of toxicological and epidemiological methods in order to have internationally comparable results.

The first Environmental Health Criteria (EHC) monograph, on mercury, was published in 1976 and since that time an ever-increasing number of assessments of chemicals and of physical effects have been produced. In addition, many EHC monographs have been devoted to evaluating toxicological methodology, e.g. for genetic, neurotoxic, teratogenic and nephrotoxic effects. Other publications have been concerned with epidemiological guidelines, evaluation of short-term tests for carcinogens, biomarkers, effects on the elderly and so forth.

Since its inauguration the EHC Programme has widened its scope, and the importance of environmental effects, in addition to health effects, has been increasingly emphasized in the total evaluation of chemicals.

The original impetus for the Programme came from World Health Assembly resolutions and the recommendations of the 1972 UN Conference on the Human Environment. Subsequently the work became an integral part of the International Programme on Chemical Safety (IPCS), a cooperative programme of UNEP, ILO and WHO.

In this manner, with the strong support of the new partners, the importance of occupational health and environmental effects was fully recognized. The EHC monographs have become widely established, used and recognized throughout the world.

The recommendations of the 1992 UN Conference on Environment and Development and the subsequent establishment of the Intergovernmental Forum on Chemical Safety with the priorities for action in the six programme areas of Chapter 19, Agenda 21, all lend further weight to the need for EHC assessments of the risks of chemicals.

Scope

The criteria monographs are intended to provide critical reviews on the effect on human health and the environment of chemicals and of combinations of chemicals and physical and biological agents. As such, they include and review studies that are of direct relevance for the evaluation. However, they do not describe *every* study carried out. Worldwide data are used and are quoted from original studies, not from abstracts or reviews. Both published and unpublished reports are considered and it is incumbent on the authors to assess all the articles cited in the references. Preference is always given to published data. Unpublished data are used only when relevant published data are absent or when they are pivotal to the risk assessment. A detailed policy statement is available that describes the procedures used for unpublished proprietary data so that this information can be used in the evaluation without compromising its confidential nature (WHO (1990) Revised Guidelines for the Preparation of Environmental Health Criteria Monographs. PCS/90.69, Geneva, World Health Organization).

In the evaluation of human health risks, sound human data, whenever available, are preferred to animal data. Animal and *in vitro* studies provide support and are used mainly to supply evidence missing from human studies. It is mandatory that research on human subjects is conducted in full accord with ethical principles, including the provisions of the Helsinki Declaration.

The EHC monographs are intended to assist national and international authorities in making risk assessments and subsequent risk management decisions. They represent a thorough evaluation of risks and are not, in any sense, recommendations for regulation or standard setting. These latter are the exclusive purview of national and regional governments.

Content

The layout of EHC monographs for chemicals is outlined below.

* Summary – a review of the salient facts and the risk evaluation of the chemical
* Identity – physical and chemical properties, analytical methods
* Sources of exposure
* Environmental transport, distribution and transformation
* Environmental levels and human exposure
* Kinetics and metabolism in laboratory animals and humans
* Effects on laboratory mammals and *in vitro* test systems
* Effects on humans
* Effects on other organisms in the laboratory and field
* Evaluation of human health risks and effects on the environment
* Conclusions and recommendations for protection of human health and the environment
* Further research
* Previous evaluations by international bodies, e.g. IARC, JECFA, JMPR

Selection of chemicals

Since the inception of the EHC Programme, the IPCS has organized meetings of scientists to establish lists of priority chemicals for subsequent evaluation. Such meetings have been held in Ispra, Italy, 1980; Oxford, United Kingdom, 1984; Berlin, Germany, 1987; and North Carolina, USA, 1995. The selection of chemicals has been based on the following criteria: the existence of scientific evidence that the substance presents a hazard to human health and/or the environment; the possible use, persistence, accumulation or degradation of the substance shows that there may be significant human or environmental exposure; the size and nature of populations at risk (both human and other species) and risks for environment; international concern, i.e. the substance is of major interest to several countries; adequate data on the hazards are available.

If an EHC monograph is proposed for a chemical not on the priority list, the IPCS Secretariat consults with the Cooperating Organizations and all the Participating Institutions before embarking on the preparation of the monograph.

Procedures

The order of procedures that result in the publication of an EHC monograph is shown in the flow chart on p. xvii. A designated staff member of IPCS, responsible for the scientific quality of the document, serves as Responsible Officer (RO). The IPCS Editor is responsible for layout and language. The first draft, prepared by consultants or, more usually, staff from an IPCS Participating Institution, is based initially on data provided from the International Register of Potentially Toxic Chemicals, and reference data bases such as Medline and Toxline.

The draft document, when received by the RO, may require an initial review by a small panel of experts to determine its scientific quality and objectivity. Once the RO finds the document acceptable as a first draft, it is distributed, in its unedited form, to well over 150 EHC contact points throughout the world who are asked to comment on its completeness and accuracy and, where necessary, provide additional material. The contact points, usually designated by governments, may be Participating Institutions, IPCS Focal Points, or individual scientists known for their particular expertise. Generally some four months are allowed before the comments are considered by the RO and author(s). A second draft incorporating comments received and approved by the Director, IPCS, is then distributed to Task Group members, who carry out the peer review, at least six weeks before their meeting.

The Task Group members serve as individual scientists, not as representatives of any organization, government or industry. Their function is to evaluate the accuracy, significance and relevance of the information in the document and to assess the health and environmental risks from exposure to the chemical. A summary and recommendations for further research and improved safety aspects are also required. The composition of the Task Group is dictated by the range of expertise required for the subject of the meeting and by the need for a balanced geographical distribution.

EHC PREPARATION FLOW CHART

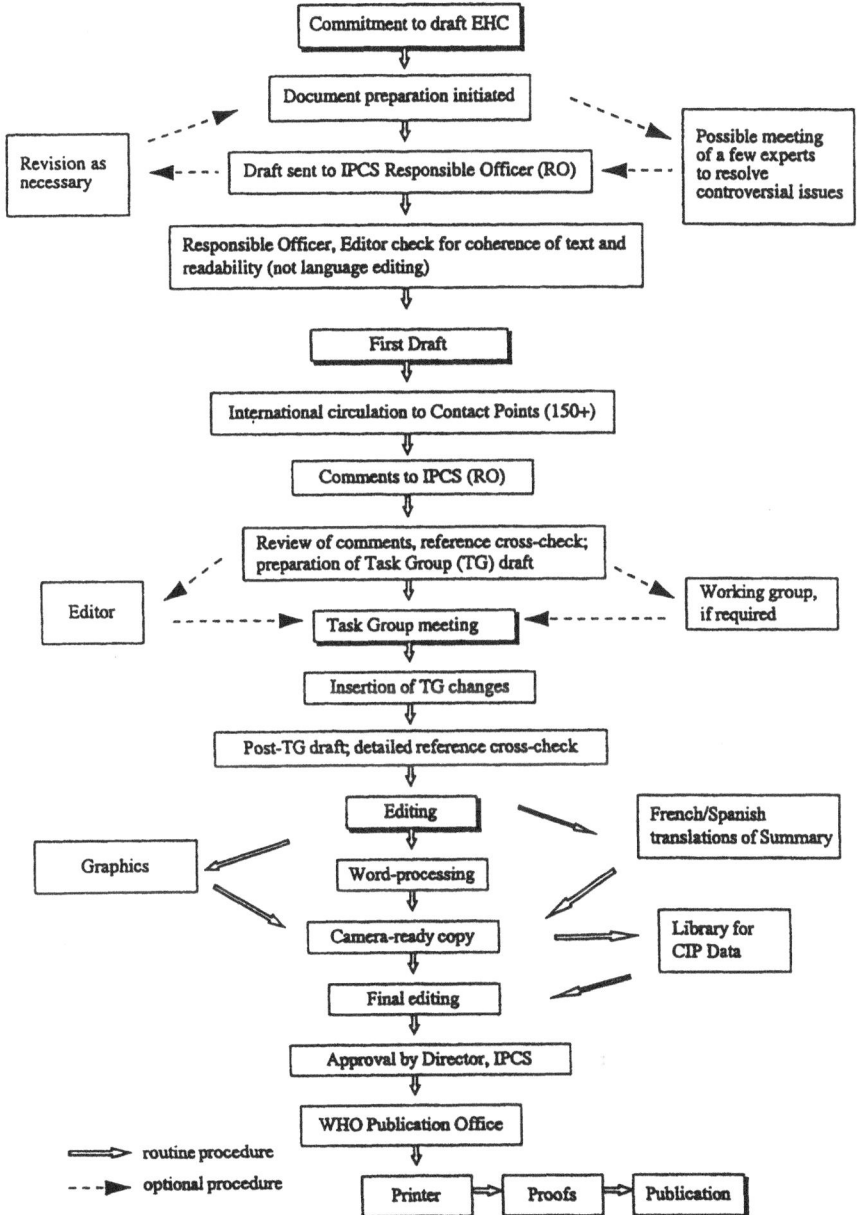

Commitment to draft EHC

⇩

Document preparation initiated

⇩

Draft sent to IPCS Responsible Officer (RO)

Revision as necessary

Possible meeting of a few experts to resolve controversial issues

⇩

Responsible Officer, Editor check for coherence of text and readability (not language editing)

⇩

First Draft

⇩

International circulation to Contact Points (150+)

⇩

Comments to IPCS (RO)

⇩

Review of comments, reference cross-check; preparation of Task Group (TG) draft

Editor

Working group, if required

⇩

Task Group meeting

⇩

Insertion of TG changes

⇩

Post-TG draft; detailed reference cross-check

⇩

Editing

French/Spanish translations of Summary

Graphics

⇩

Word-processing

⇩

Camera-ready copy

Library for CIP Data

⇩

Final editing

⇩

Approval by Director, IPCS

⇩

WHO Publication Office

⇒ routine procedure
- - -▶ optional procedure

⇩

Printer ⇨ Proofs ⇨ Publication

The three cooperating organizations of the IPCS recognize the important role played by nongovernmental organizations. Representatives from relevant national and international associations may be invited to join the Task Group as observers. Although observers may provide a valuable contribution to the process, they can only speak at the invitation of the Chairperson. Observers do not participate in the final evaluation of the chemical; this is the sole responsibility of the Task Group members. When the Task Group considers it to be appropriate, it may meet *in camera*.

All individuals who as authors, consultants or advisers participate in the preparation of the EHC monograph must, in addition to serving in their personal capacity as scientists, inform the RO if at any time a conflict of interest, whether actual or potential, could be perceived in their work. They are required to sign a conflict of interest statement. Such a procedure ensures the transparency and probity of the process.

When the Task Group has completed its review and the RO is satisfied as to the scientific correctness and completeness of the document, it then goes for language editing, reference checking and preparation of camera-ready copy. After approval by the Director, IPCS, the monograph is submitted to the WHO Office of Publications for printing. At this time a copy of the final draft is sent to the Chairperson and Rapporteur of the Task Group to check for any errors.

It is accepted that the following criteria should initiate the updating of an EHC monograph: new data are available that would substantially change the evaluation; there is public concern for health or environmental effects of the agent because of greater exposure; an appreciable time period has elapsed since the last evaluation.

All Participating Institutions are informed, through the EHC progress report, of the authors and institutions proposed for the drafting of the documents. A comprehensive file of all comments received on drafts of each EHC monograph is maintained and is available on request. The Chairpersons of Task Groups are briefed before each meeting on their role and responsibility in ensuring that these rules are followed.

WHO TASK GROUP ON ENVIRONMENTAL HEALTH CRITERIA FOR ARSENIC AND ARSENIC COMPOUNDS

Members

Dr C. Abernathy, Office of Water/Office of Science and Technology, Health and Ecological Criteria Division, US Environmental Protection Agency, Washington, D.C., USA (*Chairperson*)

Dr D. Chakraborti, School of Environmental Studies, Jadavpur University, Calcutta, India

Professor J.S. Edmonds, Department of Chemistry, De Montfort University, Leicester, United Kingdom

Dr H. Gibb, US Environmental Protection Agency, National Center for Environmental Assessment, Washington DC, USA

Dr P. Hoet, Industrial and Occupational Medicine Unit, Catholic University of Louvain, Brussels, Belgium

Dr C. Hopenhayn-Rich, Department of Preventive Medicine and Environmental Health, University of Kentucky, Lexington, KY, USA

Mr P.D. Howe, Centre for Ecology and Hydrology, Monks Wood Experimental Station, Abbots Ripton, Huntingdon, Cambridgeshire, United Kingdom

Dr L. Järup, Department of Epidemiology and Public Health, Imperial College School of Medicine, London, United Kingdom

Dr A.A. Meharg, Department of Plant and Soil Science, Aberdeen, United Kingdom

Professor M.R. Moore, Director, Queensland Health Scientific Services and National Research Centre for Environmental Toxicology, Queensland, Australia (*Vice-Chairperson*)

Dr J. C. Ng, National Research Centre for Environmental Toxicology, Brisbane, Australia

Dr A. Nishikawa, Division of Pathology, National Institute of Health Sciences, Tokyo, Japan

Dr L. Pyy, Director of the Deptartment, Oulu Regional Institute of Occupational Health, Oulu, Finland

Dr M. Sim, Unit of Occupational and Environmental Health, Department of Epidemiology and Preventive Medicine, Monash University, Victoria, Australia

Dr J. Stauber, CSIRO Energy Technology, Lucas Heights Science and Technology Centre, Bangor, NSW, Australia

Professor M. Vahter, Institute of Environmental Medicine, Karolinska Institute, Stockholm, Sweden

Observers/Representatives

Dr P. Imray, Scientific Adviser, Environmental Health Branch, Queensland Health, Brisbane, Australia

Dr L. Tomaska, Canberra, Australia (representing the Australia New Zealand Food Authority)

Mr D. Hughes, MIM Holdings Limited, Brisbane, Australia (representing the Mining Industry)

Secretariat

Dr A. Aitio, International Programme on Chemical Safety, World Health Organization, Geneva, Switzerland

Dr G. Becking, Kingston, Ontario, Canada (*Adviser to the Secretariat*)

Dr K. Buckett, Director DHAC, Public Health Division, Canberra, Australia

Mr P. Callan, Assistant Director, National Health and Medical Research Council, Canberra, Australia

Dr M.F. Hughes, NHEERL/ET/PKB, US Environmental Protection Agency, Research Triangle Park, NC, USA

Dr E.M. Kenyon, NHEERL/ET/PKB, US Environmental Protection Agency, Research Triangle Park, NC, USA

Dr D.R. Lewis, Human Studies Division, NHEERL, US Environmental Protection Agency, Research Triangle Park, NC, USA

Dr M. Younes, International Programme on Chemical Safety, World Health Organization, Geneva, Switzerland

WHO TASK GROUP ON ENVIRONMENTAL HEALTH CRITERIA FOR ARSENIC AND ARSENIC COMPOUNDS

The first and second drafts of this monograph were prepared, under the coordination of Dr J. Ng, by the authors A. Gomez-Caminero, P. Howe, M. Hughes, E. Kenyon, D.R. Lewis, M. Moore, J. Ng, and by A. Aitio and G. Becking. The group of authors met at National Health and Environmental Effects Research Laboratory, US. EPA, North Carolina, on 20–22 July 1998.

A WHO Task Group on Environmental Health Criteria for Arsenic and Arsenic Compounds met at the National Research Centre for Environmental Toxicology, Brisbane, Australia, on 15–19 November 1999. The group reviewed the draft and the peer review comments, revised the draft and made an evaluation of the risks for human health and environment from exposure to arsenic and arsenic compounds.

After the meeting, and based on the peer reviewer comments and Task Group advice, Drs Gibb, Hopenhayn-Rich, Järup, Sim, and Aitio revised and updated the section on Effects on Human Health. This section was then sent for review to a selected group of experts.

The document was revised on the basis of the peer review comments received, these revisions were verified, and the document was finalized by a Review Board, consisting of Drs D. Anderson, H. Gibb, L. Järup, M. Sim and A. Aitio, in TNO BIBRA, Carshalton, UK. The document was finally approved by the Task Group in a mail ballot.

The cut-off date for the literature searches for the document was the Task Group meeting, i.e. November 1999, with the exception of the section on effects on human health, for which the last literature searches were performed in November 2000.

Peer review comments at the first stage international review were received from:

Dr J. Ahlers, Umwelt Bundes Amt, Germany
Dr R. Benson, Region VIII, Environmental Protection Agency, USA
Professor GB Bliss, N.N. Petrov's Research Institute of Oncology,
    Russian Federation
Dr M. Bolger, Food and Drug Administration, USA
Professor M. Cíkrt, Centre of Industrial Hygiene and OccupationaL
    Diseases, Czech Republique
Professor I. Dési, Albert Szent-Györgyi University, Hungary
Professor J Duffus, The Edinburgh Centre for Toxicology, UK
Dr P Edwards, Department of Health, UK
Dr H Falk, Agency for Toxic Substances and Disease Registry, USA
Dr H. Gibb, Environmental Protection Agency, USA
Dr N. Kurzeja European Environmental Bureau, Germany
Dr I. Mangelsdorf, Fraunhofer Institute, Germany
Dr TG Rossman, NYU School of Medicine
Professor H Taskinen, Finnish Institute of Occuational Health
Mr S Tsuda, Ministry of Halth and Welfare, Japan
Dr G. Ungváry, József Fodor National Center for Public Health,
    Hungary
Professor M. Vahter, Karolinska Institute, Sweden,
Bureau of Chemical Safety, Canada
Elf Atochem North America, USA
Environmental Protection Agency Office of Research and
    Development, USA
Eurometaux
Finnish Institute of Occupational Health, Finland

Comments on the revised section on effects on human health were received from members of the Task Group, and from:

Dr D Anderson, TNO BIBRA International, UK
Dr Michael Bates, Kenepuru Science Centre, New Zealand
Dr R. Calderon, National Health and Environmental Effects
    Research Laboratory, US EPA
Professor PE Enterline, University of Pittsburgh, USA
Dr A. Gomez-Caminero, National Health and Environmental Effects
    Research Laboratory, US. EPA

Dr J Lubin, National Cancer Institute, USA
Professor AH Smith, University of California, USA

Dr A. Aitio of the IPCS central unit was responsible for the scientific aspects of the monograph, and Kathleen Lyle for the technical editing.

The efforts of all, especially Queensland Health and the Natinal Research Centre for Environmental Toxicology, Australia, who helped in the preparation and finalization of the monograph are gratefully acknowledged.

ABBREVIATIONS

AAS	atomic absorption spectrometry
ABI	ankle–brachial index
AFS	atomic fluorescence spectrometry
AgDDTC	silver diethyldithiocarbamate
ALA	aminolaevulinic acid
ASV	anodic stripping voltammetry
ATPase	adenosine triphosphatase
AUC	area under the curve
BAL	dimercaprol
BCF	bioconcentration factor
BFD	blackfoot disease
BFD-endemic area	Geographic area in south-western Taiwan, where arsenic-contaminated artesian well water has been used as drinking water, and where BFD is endemic; the area has been also called the "arseniasis" area, or "hyperendemic" area. In this document it is called BFD-endemic area, to differentiate it from other areas e.g. in Taiwan, where high arsenic concentrations in drinking water have been reported
BMI	body mass index
BSO	L-buthionine-(RS)-sulfoximine
CA	chromosome aberrations
CAS	Chemical Abstract Service
CCA	copper chrome arsenate
CCGG	cytosine-cytosine-guanine-guanine
cDNA	complementary DNA
CE	capillary electrophoresis
CI	confidence interval; unless otherwise stated, the 95% CI is given. Accordingly, the term statistically significant in this documents denotes significance at 95% level
CVD	cardiovascular disease

DBDTC	sodium dibenzyldithiocarbamate
DD	duplicate diet study
DMA	dimethylarsinic acid
DMA^{3+}	dimethylarsinous acid
DMA-TGM	dimethylarsinic acid thioglycolic acid methyl ester
DMSA	dimercaptosuccinic acid
DPSCV	differential pulse cathodic stripping voltammetry
DSA	disodium arsenate heptahydrate
Eh	redox potential
EMG	electromyography
ETAAS	electrothermal atomic absorption spectrometry
FAAS	flame atomic absorption spectrometry
FAFS	flame atomic fluorescence spectrometry
FR	frequency ratio
GC	gas chromatography
GM-CSF	granulocyte macrophage-colony stimulating factor
GSH	glutathione
GSSG	oxidized glutathione
GTP	guanosine triphosphate
HFC	high frequency cell
HGAAS	hydride generation atomic absorption spectrometry
HMDE	hanging mercury drop electrode
HPLC	high pressure liquid chromatography
HPRT	hypoxanthine phosphoribosyltransferase
HSDB	Hazardous Substances Data Bank
ICP-AES	inductively coupled plasma atomic emission spectrometry
ICP-MS	inductively coupled plasma mass spectrometry
Ig	immunoglobulin
IHD	ischaemic heart disease
LC	liquid chromatography
LC_{50}	median lethal concentration
MB	market basket survey
MLC	micellar liquid chromatography

MMA	monomethylarsonic acid
MMA^{3+}	monomethylarsonous acid
MMA-TGM	monomethylarsonic acid thioglycolic acid methyl ester
MN	Micronucleus/i
mRNA	messenger RNA
MSMA	monosodium methanearsonate
MTHFR	5,10-methylene-tetradrofolate reductase
NAA	neutron activation analysis
NaFDDC	sodium (bistrifluoroethyl) dithiocarbamate
NER	nucleotide excision repair
4-NQO	4-nitroquinoline oxide
OR	odds ratio
PAD	periodate-oxidized adenosine
PIXES	particle-induced X-ray emission spectrometry
POR	prevalence odds ratio
PVD	peripheral vascular disease
RI	replication index
RPLC	reversed phase liquid chromatography
RTECS	Registry of Toxic Effects of Chemicals
SAH	*S*-adenosylhomocysteine
SAM	*S*-adenosyl methionine
SCE	sister chromatid exchange
SD	standard deviation
SE	standard error of mean
SEM	scanning electron microscopy
SFC	supercritical fluid chromatography
SFE	supercritical fluid extraction
SIR	standardized incidence ratio
SMR	standardized mortality ratio
SRBC	sheep red blood cell
TDT	toluene-3,4-dithiol
TGF	transforming growth factor
TGM	thioglycolic acid methylester

TMA	trimethylarsine
TMAO	trimethylarsine oxide
TWA	time-weighted average
UN	United Nations
UV	ultraviolet
XAFS	X-ray absorption fine structure spectroscopy
XRF	X-ray fluorescence

1. SUMMARY

1.1 Properties and analytical procedures

Arsenic is a metalloid widely distributed in the earth's crust and present at an average concentration of 2 mg/kg. It occurs in trace quantities in all rock, soil, water and air. Arsenic can exist in four valency states: –3, 0, +3 and +5. Under reducing conditions, arsenite (As(III)) is the dominant form; arsenate (As(V)) is generally the stable form in oxygenated environments. Elemental arsenic is not soluble in water. Arsenic salts exhibit a wide range of solubilities depending on pH and the ionic environment.

There is a variety of instrumental techniques for the determination of arsenic. These include AAS, AFS, ICP-AES, ICP-MS and voltammetry. Some of these (e.g. ICP-MS) can serve as element-specific detectors when coupled to chromatographic separation techniques (e.g. HPLC and GC). These so-called "hyphenated" methods are used for determining individual arsenic species. Additional sensitivity for a limited range of arsenic compounds can often be achieved by the use of hydride generation techniques. A test kit based on the colour reaction of arsine with mercuric bromide is currently used for groundwater testing in Bangladesh and has a detection limit of 50–100 µg/litre under field conditions.

1.2 Sources and occurrence of arsenic in the environment

Arsenic is present in more than 200 mineral species, the most common of which is arsenopyrite.

It has been estimated that about one-third of the atmospheric flux of arsenic is of natural origin. Volcanic action is the most important natural source of arsenic, followed by low-temperature volatilization.

Inorganic arsenic of geological origin is found in groundwater used as drinking-water in several parts of the world, for example Bangladesh.

1

Organic arsenic compounds such as arsenobetaine, arseno-choline, tetramethylarsonium salts, arsenosugars and arsenic-containing lipids are mainly found in marine organisms although some of these compounds have also been found in terrestrial species.

Elemental arsenic is produced by reduction of arsenic trioxide (As_2O_3) with charcoal. As_2O_3 is produced as a by-product of metal smelting operations. It has been estimated that 70% of the world arsenic production is used in timber treatment as copper chrome arsenate (CCA), 22% in agricultural chemicals, and the remainder in glass, pharmaceuticals and non-ferrous alloys.

Mining, smelting of non-ferrous metals and burning of fossil fuels are the major industrial processes that contribute to anthro-pogenic arsenic contamination of air, water and soil. Historically, use of arsenic-containing pesticides has left large tracts of agricultural land contaminated. The use of arsenic in the preservation of timber has also led to contamination of the environment.

1.3 Environmental transport and distribution

Arsenic is emitted into the atmosphere by high-temperature processes such as coal-fired power generation plants, burning vegetation and volcanism. Natural low-temperature biomethylation and reduction to arsines also releases arsenic into the atmosphere. Arsenic is released into the atmosphere primarily as As_2O_3 and exists mainly adsorbed on particulate matter. These particles are dispersed by the wind and are returned to the earth by wet or dry deposition. Arsines released from microbial sources in soils or sediments undergo oxidation in the air, reconverting the arsenic to non-volatile forms, which settle back to the ground. Dissolved forms of arsenic in the water column include arsenate, arsenite, methylarsonic acid (MMA) and dimethylarsinic acid (DMA). In well-oxygenated water and sediments, nearly all arsenic is present in the thermodynamically more stable pentavalent state (arsenate). Some arsenite and arsenate species can interchange oxidation state depending on redox potential (Eh), pH and biological processes. Some arsenic species have an affinity for clay mineral surfaces and organic matter and this can affect their environmental behaviour. There is potential for arsenic release when there is fluctuation in Eh, pH, soluble arsenic concentration and sediment organic content. Weathered rock and soil may be transported by wind or water erosion. Many arsenic compounds tend to adsorb to soils, and leaching usually results in transportation over only short distances in soil.

Three major modes of arsenic biotransformation have been found to occur in the environment: redox transformation between arsenite and arsenate, the reduction and methylation of arsenic, and the biosynthesis of organoarsenic compounds. There is biogeochemical cycling of compounds formed from these processes.

1.4 Environmental levels and human exposure

Mean total arsenic concentrations in air from remote and rural areas range from 0.02 to 4 ng/m³. Mean total arsenic concentrations in urban areas range from 3 to about 200 ng/m³; much higher concentrations (> 1000 ng/m³) have been measured in the vicinity of industrial sources, although in some areas this is decreasing because of pollution abatement measures. Concentrations of arsenic in open ocean seawater are typically 1–2 µg/litre. Arsenic is widely distributed in surface freshwaters, and concentrations in rivers and lakes are generally below 10 µg/litre, although individual samples may range up to 5 mg/litre near anthropogenic sources. Arsenic levels in groundwater average about 1–2 µg/litre except in areas with volcanic rock and sulfide mineral deposits where arsenic levels can range up to 3 mg/litre. Mean sediment arsenic concentrations range from 5 to 3000 mg/kg, with the higher levels occurring in areas of contamination. Background concentrations in soil range from 1 to 40 mg/kg, with mean values often around 5 mg/kg. Naturally elevated levels of arsenic in soils may be associated with geological substrata such as sulfide ores. Anthropogenically contaminated soils can have concentrations of arsenic up to several grams per 100 ml.

Marine organisms normally contain arsenic residues ranging from < 1 to more than 100 mg/kg, predominantly as organic arsenic species such as arsenosugars (macroalgae) and arsenobetaine (invertebrates and fish). Bioaccumulation of organic arsenic compounds, after their biogenesis from inorganic forms, occurs in aquatic organisms. Bioconcentration factors (BCFs) in freshwater invertebrates and fish for arsenic compounds are lower than for marine organisms. Biomagnification in aquatic food chains has not been observed. Background arsenic concentrations in freshwater and terrestrial biota are usually less than 1 mg/kg (fresh weight). Terrestrial plants may accumulate arsenic by root uptake from the soil or by adsorption of airborne arsenic deposited on the leaves. Arsenic levels are higher in biota collected near anthropogenic sources or in areas with geothermal activity. Some species accumulate substantial levels, with mean concentrations of up to 3000 mg/kg at arsenical mine sites.

Non-occupational human exposure to arsenic in the environment is primarily through the ingestion of food and water. Of these, food is generally the principal contributor to the daily intake of total arsenic. In some areas arsenic in drinking-water is a significant source of exposure to inorganic arsenic. In these cases, arsenic in drinking-water often constitutes the principal contributor to the daily arsenic intake. Contaminated soils such as mine tailings are also a potential source of arsenic exposure. The daily intake of total arsenic from food and beverages is generally between 20 and 300 µg/day. Limited data indicate that approximately 25% of the arsenic present in food is inorganic, but this depends highly on the type of food ingested. Inorganic arsenic levels in fish and shellfish are low (< 1%). Foodstuffs such as meat, poultry, dairy products and cereals have higher levels of inorganic arsenic. Pulmonary exposure may contribute up to approximately 10 µg/day in a smoker and about 1 µg/day in a non-smoker, and more in polluted areas. The concentration of metabolites of inorganic arsenic in urine (inorganic arsenic, MMA and DMA) reflects the absorbed dose of inorganic arsenic on an individual level. Generally, it ranges from 5 to 20 µg As/litre, but may even exceed 1000 µg/litre.

In workplaces with up-to-date occupational hygiene practices, exposure generally does not exceed 10 µg/m^3 (8-h time-weighted average [TWA]). However, in some places workroom atmospheric arsenic concentrations as high as several milligrams per cubic metre have been reported.

1.5 Kinetics and metabolism

Absorption of arsenic in inhaled airborne particles is highly dependent on the solubility and the size of particles. Both pentavalent and trivalent soluble arsenic compounds are rapidly and extensively absorbed from the gastrointestinal tract. In many species arsenic metabolism is characterized by two main types of reactions: (1) reduction reactions of pentavalent to trivalent arsenic, and (2) oxidative methylation reactions in which trivalent forms of arsenic are sequentially methylated to form mono-, di- and trimethylated products using S-adenosyl methionine (SAM) as the methyl donor and glutathione (GSH) as an essential co-factor. Methylation of inorganic arsenic facilitates the excretion of inorganic arsenic from the body, as the end-products MMA and DMA are readily excreted in urine. There are major qualitative and quantitative interspecies differences in methylation, to the extent that some species exhibit minimal or no arsenic methylation (e.g. marmoset monkey, guinea-pig, chimpanzee). However, in humans and most common laboratory animals, inorganic arsenic is extensively methylated and the metabolites are excreted primarily in the urine. Factors such as dose, age, gender and smoking contribute only minimally to the large inter-individual variation in arsenic methylation observed in humans. However, lower methylation efficiency in children has been observed in only one study out of three. Studies in humans suggest the existence of a wide difference in the activity of methyl-transferases, and the existence of polymorphism has been hypothesized. Animal and human studies suggest that arsenic methylation may be inhibited at high acute exposures. The metabolism and disposition of inorganic arsenic may be influenced by its valence state, particularly at high dose levels. Studies in laboratory animals indicate that administration of trivalent inorganic arsenic such as As_2O_3 and arsenite initially results in higher levels in most tissues than does the administration of pentavalent arsenic. However, the trivalent form is more extensively methylated, leading to similar long-term excretion. Ingested organoarsenicals such as MMA, DMA and arsenobetaine are much less extensively metabolized and more rapidly eliminated in urine than inorganic arsenic in both laboratory animals and humans.

Levels of arsenic or its metabolites in blood, hair, nails and urine are used as biomarkers of arsenic exposure. Blood arsenic is a useful biomarker only in the case of acute arsenic poisoning or stable chronic high-level exposure. Arsenic is rapidly cleared from blood, and speciation of its chemical forms in blood is difficult. Arsenic in hair and nails can be indicators of past arsenic exposure, provided care is taken to prevent external arsenic contamination of the samples. Arsenic in hair may also be used to estimate relative length of time since an acute exposure. Speciated metabolites in urine expressed either as inorganic arsenic or as the sum of metabolites (inorganic arsenic + MMA + DMA) provide the best quantitative estimate of recently absorbed dose of arsenic. However, consumption of certain seafood, mainly seaweed and some bivalves, may confound estimation of inorganic arsenic exposure because of metabolism of arsenosugars to DMA in the body or the presence of DMA in the seafood. Such food should be avoided for 2–3 days before urine sampling for monitoring of exposure to inorganic arsenic.

1.6 Effects on laboratory animals and *in vitro* systems

Both inorganic and organic forms of arsenic may cause adverse effects in laboratory animals. The effects induced by arsenic range from acute lethality to chronic effects such as cancer. The degree of toxicity of arsenic is basically dependent on the form (e.g. inorganic or organic) and the oxidation state of the arsenical. It is generally considered that inorganic arsenicals are more toxic than organic arsenicals, and within these two classes, the trivalent forms are more toxic than the pentavalent forms, at least at high doses. Several different organ systems are affected by arsenic, including skin, respiratory, cardiovascular, immune, genitourinary, reproductive, gastrointestinal and nervous systems.

Several animal carcinogenicity studies on arsenic have been carried out, but limitations such as high dose levels, limited time of exposure and limited number of animals make these inconclusive. However, a recently reported animal model may be a useful tool for future carcinogenicity studies. In that study, female C57B1/6J mice exposed to arsenic in drinking-water containing 500 µg As(V)/litre over 2 years was associated with increased incidence in tumours involving mainly lung, liver, gastrointestinal tract and skin. Inorganic arsenic does not induce point mutations. However, arsenic can produce chromosomal aberrations *in vitro*, affect methylation and repair of DNA, induce cell proliferation, transform cells and promote tumours. One study has indicated that DMA may cause cancer of the urinary bladder in male rats at high doses.

1.7 Effects on human health

Soluble inorganic arsenic is acutely toxic, and ingestion of large doses leads to gastrointestinal symptoms, disturbances of cardio-vascular and nervous system functions, and eventually death. In survivors, bone marrow depression, haemolysis, hepatomegaly, melanosis, polyneuropathy and encephalopathy may be observed.

Long-term exposure to arsenic in drinking-water is causally related to increased risks of cancer in the skin, lungs, bladder and kidney, as well as other skin changes such as hyperkeratosis and pigmentation changes. These effects have been demonstrated in many studies using different study designs. Exposure–response relationships and high risks have been observed for each of these end-points. The effects have been most thoroughly studied in Taiwan but there is considerable evidence from studies on populations in other countries as well. Increased risks of lung and bladder cancer and of arsenic-associated skin lesions have been reported to be associated with ingestion of drinking-water at concentrations ≤50 µg arsenic/litre.

Occupational exposure to arsenic, primarily by inhalation, is causally associated with lung cancer. Exposure–response relationships and high risks have been observed. Increased risks have been observed at cumulative exposure levels ≥ 0.75 (mg/m^3) · year (e.g. 15 years of exposure to a workroom air concentration of 50 µg/m^3). Tobacco smoking has been investigated in two of the three main smelter cohorts and was not found to be the cause of the increased lung cancer risk attributed to arsenic; however, it was found to be interactive with arsenic in increasing the lung cancer risk.

Even with some negative findings, the overall weight of evidence indicates that arsenic can cause clastogenic damage in different cell types with different end-points in exposed individuals and in cancer patients. For point mutations, the results are largely negative.

Chronic arsenic exposure in Taiwan has been shown to cause blackfoot disease (BFD), a severe form of peripheral vascular disease (PVD) which leads to gangrenous changes. This disease has

7

not been documented in other parts of the world, and the findings in Taiwan may depend upon other contributing factors. However, there is good evidence from studies in several countries that arsenic exposure causes other forms of PVD.

Conclusions on the causality of the relationship between arsenic exposure and other health effects are less clear-cut. The evidence is strongest for hypertension and cardiovascular disease, suggestive for diabetes and reproductive effects and weak for cerebrovascular disease, long-term neurological effects, and cancer at sites other than lung, bladder, kidney and skin.

1.8 Effects on other organisms in the environment

Aquatic and terrestrial biota show a wide range of sensitivities to different arsenic species. Their sensitivity is modified by biological and abiotic factors. In general, inorganic arsenicals are more toxic than organoarsenicals and arsenite is more toxic than arsenate. The mode of toxicity and mechanism of uptake of arsenate by organisms differ considerably. This may explain why there are interspecies differences in organism response to arsenate and arsenite. The primary mechanism of arsenite toxicity is considered to result from its binding to protein sulfhydryl groups. Arsenate is known to affect oxidative phosphorylation by competition with phosphate. In environments where phosphate concentrations are high, arsenate toxicity to biota is generally reduced. As arsenate is a phosphate analogue, organisms living in elevated arsenate environments must acquire the nutrient phosphorous yet avoid arsenic toxicity.

Arsenic compounds cause acute and chronic effects in individuals, populations and communities at concentrations ranging from a few micrograms to milligrams per litre, depending on species, time of exposure and end-points measured. These effects include lethality, inhibition of growth, photosynthesis and reproduction, and behavioural effects. Arsenic-contaminated environments are characterized by limited species abundance and diversity. If levels of arsenate are high enough, only species which exhibit resistance may be present.

2. PROPERTIES AND ANALYTICAL PROCEDURES

2.1 Identity

Elemental arsenic (As) is a member of Group 15 of the periodic table, with nitrogen, phosphorus, antimony and bismuth. It has an atomic number of 33 and an atomic mass of 74.91. The Chemical Abstract Service (CAS), National Institute for Occupational Safety and Health Registry of Toxic Effects of Chemicals (RTECS), Hazardous Substances Data Bank (HSDB), European Commission, and UN transport class numbers are 7440-38-2, HSB 509, CG 05235 000, 033-001-00-X and UN 1558, respectively.

This monograph deals with arsenic and inorganic and organic arsenic compounds, except arsine (AsH_3), for which a Concise International Chemical Assessment Document (CICAD) is being prepared.

2.2 Chemical and physical properties of arsenic compounds

Arsenic is a metalloid widely distributed in the earth's crust. It can exist in four valency states; −3, 0, +3, and +5. In strongly reducing environments, elemental arsenic and arsine (−3) can exist. Under moderately reducing conditions, arsenite (+3) may be the dominant form, but arsenate (+5) is generally the stable oxidation state in oxygenated environments.

Arsenic and its compounds occur in crystalline, powder, amorphous or vitreous forms. They usually occur in trace quantities in all rock, soil, water and air. However, concentrations may be higher in certain areas as a result of weathering and anthropogenic activities including metal mining and smelting, fossil fuel combustion and pesticide use.

Arsenical salts exhibit a range of aqueous solubilities depending on the pH and the ionic environment.

There are many arsenic compounds of environmental importance. Representative marine arsenic-containing compounds, of which some are found in terrestrial systems, are shown in Table 1; their molecular structures are shown Fig. 1. Other arsenic compounds discussed in the text are listed in Table 2.

Table 1. Naturally occurring inorganic and organic As species (see Fig. 1 for structures [1]–[22])

CAS No.	Name	Synonyms	Structure
	arsenate		[1]
	arsenite		[2]
124-58-3	methylarsonic acid	monomethylarsonic acid, MMA	[3]
75-60-5	dimethylarsinic acid	cacodylic acid, DMA	[4]
4964-14-1	trimethylarsine oxide		[5]
27742-38-7	tetramethylarsonium ion		[6]
64436-13-1	arsenobetaine		[7]
39895-81-3	arsenocholine		[8]
	dimethylarsinoylribosides		[9]–[19]
	trialkylarsonioribosides		[20], [21]
	dimethylarsinoylribitol sulfate		[22]

Speciation determines how arsenic compounds interact with their environment. For example, the behaviour of arsenate and arsenite in soil differs considerably. Movement in environmental matrices is a strong function of speciation and soil type. In a non-absorbing sandy loam, arsenite is 5–8 times more mobile than arsenate (Gulens et al., 1979). Soil pH also influences arsenic mobility. At a pH of 5.8 arsenate is slightly more mobile than arsenite, but when pH changes from acidic to neutral to basic, arsenite increasingly tends to become the more mobile species, though mobility of both arsenite and arsenate increases with increasing pH (Gulens et al., 1979). In strongly adsorbing soils, transport rate and speciation are influenced by organic carbon

Fig 1. Structures of naturally occurring inorganic and organic arsenic species

$$\begin{array}{cccc}
\text{structure 1} & \text{structure 2} & CH_3-\overset{O}{\underset{OH}{As}}-OH & CH_3-\overset{O}{\underset{CH_3}{As}}-OH \\
1 & 2 & 3 & 4
\end{array}$$

$$(CH_3)_3As{=}O \quad (CH_3)_4As^+ \quad (CH_3)_3As^+CH_2COO^- \quad (CH_3)As^+CH_2CH_2OH$$

$$\quad 5 \qquad\qquad 6 \qquad\qquad\qquad 7 \qquad\qquad\qquad 8$$

9 – 19

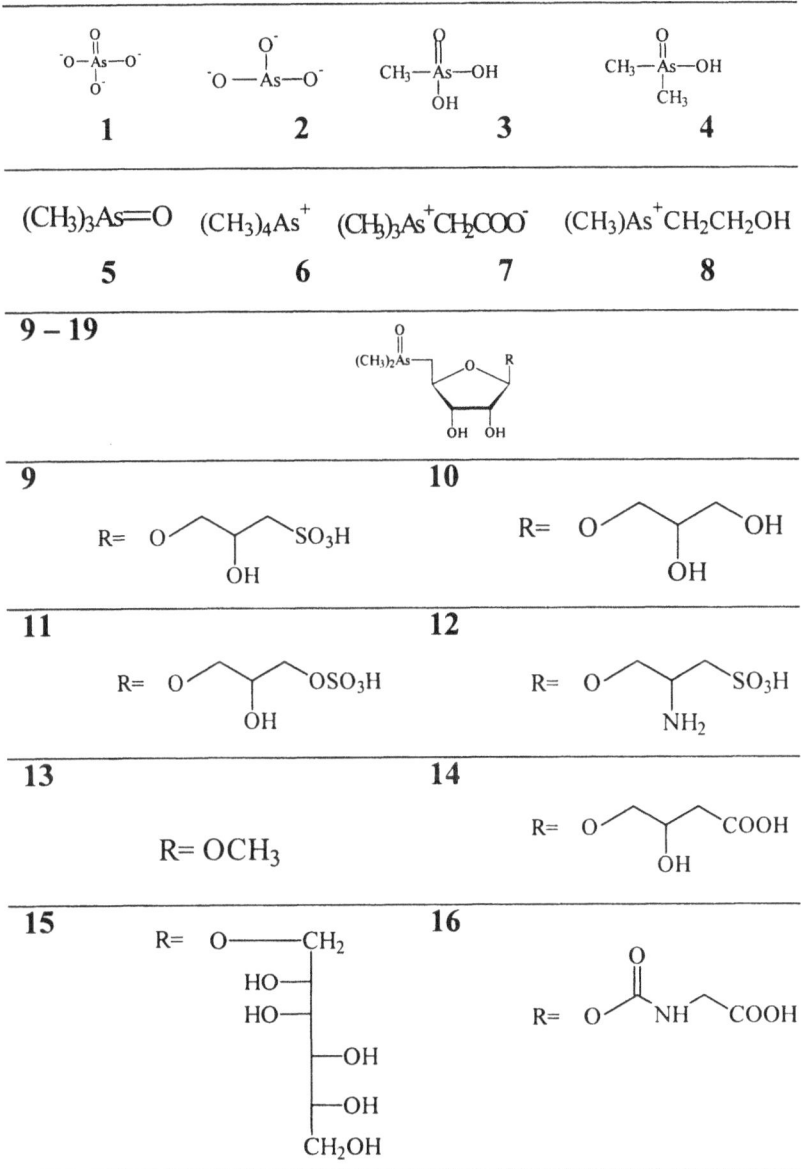

9

10

R= (structure: $O{-}CH_2CH(OH)CH_2{-}SO_3H$)

R= (structure: $O{-}CH_2CH(OH)CH_2{-}OH$)

11

12

R= (structure: $O{-}CH_2CH(OH)CH_2{-}OSO_3H$)

R= (structure: $O{-}CH_2CH(NH_2)CH_2{-}SO_3H$)

13

14

R= OCH_3

R= (structure: $O{-}CH_2CH(OH)CH_2{-}COOH$)

15

16

R= O————CH_2
HO—
HO—
—OH
—OH
CH_2OH

R= (structure: $O{-}C({=}O){-}NH{-}CH_2{-}COOH$)

Fig. 1. (contd.)

17

R=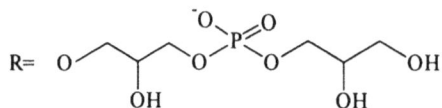

18

R=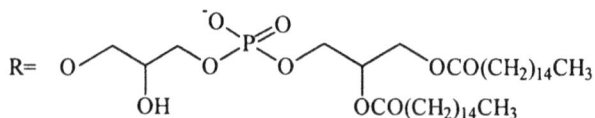

19

R=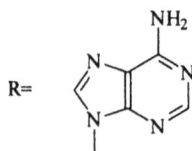

20

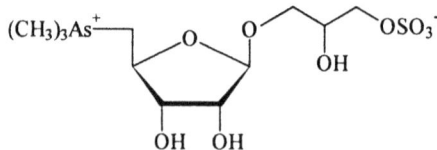

21

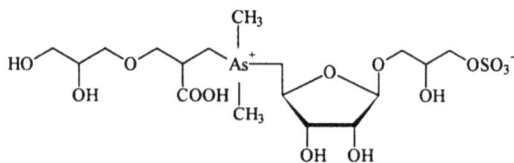

22

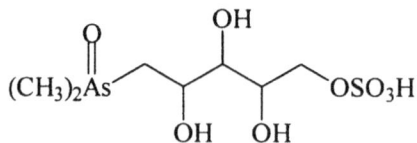

Table 2. Other As compounds of environmental significance referred to in the text

CAS No.	Name	Synonyms	Formula
	Inorganic As, trivalent		
1327-53-3	As(III) oxide	As trioxide, arsenous oxide, white As	As_2O_3 (or As_4O_6)
13768-07-05	arsenenous acid	arsenous acid	$HAsO_2$
7784-34-1	As(III) chloride	As trichloride, arsenous trichloride	$AsCl_3$
1303-33-9	As(III) sulfide	As trisulfide orpiment, auripigment	As_2S_3
	Inorganic As, pentavalent		
1303-28-2	As(V) oxide	As pentoxide	As_2O_5
7778-39-4	arsenic acid	*ortho*-arsenic acid	H_3AsO_4
10102-53-1	arsenic acid	*meta*-arsenic acid	$HAsO_3$
	arsenates, salts of *ortho*-arsenic acid		$H_2AsO_4^-$, $HAsO_4^{2-}$, AsO_4^{3-}
	Organic As		
593-52-2	methylarsine		CH_3AsH_2
593-57-7	dimethylarsine		$(CH_3)_2AsH$
593-88-4	trimethylarsine		$(CH_3)_3As$

Table 2 (contd.)

CAS No.	Name	Synonyms	Formula
98-50-0	(4-aminophenyl)-arsonic acid	arsanilic acid, p-aminobenzene-arsonic acid	H_2N—⬡—$AsO(OH)_2$
139-93-5	4,4-arsenobis(2-aminophenol) dihydrochloride	arsphenamine, salvarsan	$HCl.H_2N$—⬡(OH)—$As=As$—⬡($NH_2.HCl$)(OH)
121-59-5	[4-[aminocarbonyl-amino]phenyl] arsonic acid	carbarsone, N-carbamoylarsanilic acid	NH_2CONH—⬡—$AsO(OH)_2$
554-72-3	[4-[2-amino-2-oxoethyl)amino]-phenyl] arsonic acid	tryparsamide	NH_2COCH_2NH—⬡—$AsO(OH)_2$
121-19-7	3-nitro-4-hydroxy-phenylarsonic acid		O_2N, HO—⬡—$AsO(OH)_2$
98-72-6	4-nitrophenylarsonic acid	p-nitrophenylarsonic acid	O_2N—⬡—$As(OH)_2$
	dialkylchloroarsine		R_2AsCl
	alkyldichloroarsine		$RasCl_2$

14

content and microbial population. Both arsenite and arsenate are transported at a slower rate in strongly adsorbing soils than in sandy soils.

Under oxidizing and aerated conditions, the predominant form of arsenic in water and soil is arsenate. Under reducing and waterlogged conditions (< 200 mV), arsenites should be the predominant arsenic compounds. The rate of conversion is dependent on the Eh and pH of the soil as well as on other physical, chemical and biological factors.

In brief, at moderate or high Eh, arsenic can be stabilized as a series of pentavalent (arsenate) oxyanions, H_3AsO_4, $H_2AsO_4^-$, $HAsO_4^{2-}$ and AsO_4^{3-}. However, under most reducing (acid and mildly alkaline) conditions, arsenite predominates. A pH and Eh diagram is shown in Fig. 2.

2.3 Analytical procedures

Historically, colorimetric and gravimetric methods have been used for the determination of arsenic. However, these methods are either semi-quantitative or lack sensitivity. In recent years, atomic absorption spectrometry (AAS) has become the method of choice, as it offers the possibility of selectivity and sensitivity in the detection of a wide range of metals and non-metals including arsenic. Popular methods for generating atoms for AAS are flame and electro-thermally heated graphite furnaces. However, a commonly used technique for the measurement of arsenic is the highly sensitive hydride generation atomic absorption spectrometric method (HGAAS). However, although it is suitable for total arsenic determination after appropriate digestion the technique is only routinely used to speciate a limited number of compounds – arsenite, arsenate, MMA, DMA, trimethylarsine oxide (TMAO).

Hydride generation followed by cryogenic trapping and AAS detection is a relatively inexpensive technique for the speciation of inorganic arsenic and its methylated metabolites (Ng et al., 1998a), although more expensive hyphenated techniques may also be used.

A number of other approaches have been reported for speciation of arsenic. Inductively coupled plasma-mass spectrometry (ICP-MS)

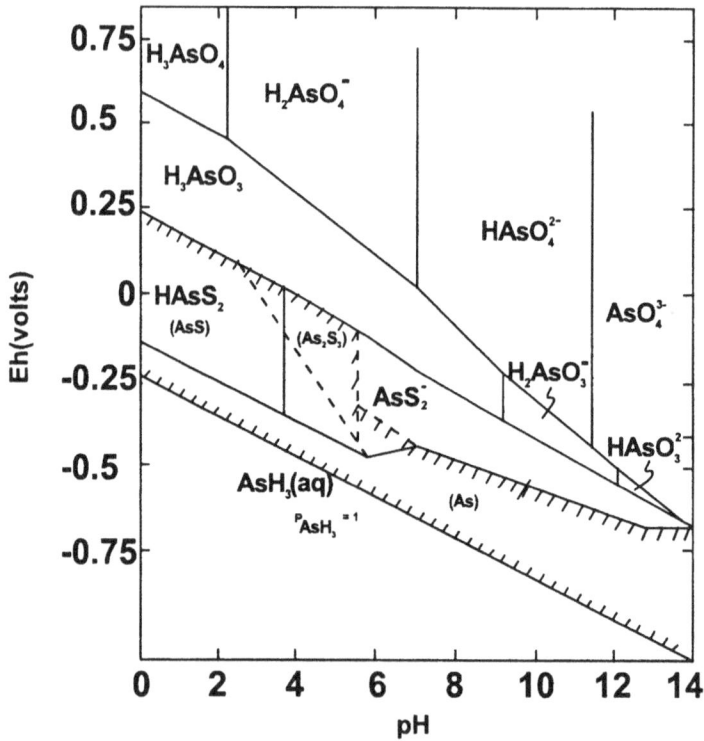

Fig. 2. The Eh-pH diagram for As at 25 °C and one atmosphere with total arsenic 10^{-5} mol/litre and total sulfur 10^{-3} mol/litre. Solid species are enclosed in parentheses in cross-hatched area, which indicates solubility less than $10^{-5.3}$ mol/litre (From: Ferguson & Gavis, 1972).

offers very high sensitivity for the determination of arsenic, and coupled with HPLC enables equally sensitive estimation of a wide variety of arsenic species.

2.4 Sample preparation and treatment

2.4.1 *Sampling and collection*

Care must be taken to avoid contamination and prevent speciation changes during sample collection and storage. Plastic containers should be acid washed and traces of oxidizing and reducing agents avoided to preserve the oxidation state of arsenic compounds. Freezing samples to $-80\ °C$ has also been recommended (Crecelius, 1986). Concentrated hydrochloric acid (1 ml to 100 ml urine) has been added to urine to prevent bacterial growth (Concha et al., 1998a).

For particulates in air and aerosols sampling, various types of filter have been employed including polytetrafluoroethylene (Rabano et al., 1989), cellulose ester (Yager et al., 1997), glass microfibre (Beceiro-Gonzalez et al.,1997) and filter paper (Tripathi et al., 1997).

2.4.2 *Oxidative digestion*

Acid digestion (George & Roscoe, 1951) and dry ashing (George et al., 1973) are the two basic methods which have been widely employed for oxidative digestion of samples before analysis. In more recent years, microwave-assisted digestion has been used (Le et al., 1994b; Thomas et al., 1997). For analysis of biological soft tissues by ICP techniques, a simple partial digestion in a closed vessel at low temperature and pressure is often sufficient for the sample preparation and pretreatment step.

2.4.3 *Extraction*

For speciation of arsenic, solvent extraction is often required before analysis. For example, arsenite and arsenate in soil can be speciated after a hydrochloric acid and chloroform extraction procedure (Chappell et al., 1995; Ng et al., 1998b). Water has been used for the extraction of soluble arsenic compounds from soil with the aid of ultrasonic treatment (Hansen et al., 1992). Forms of arsenic compounds can also be separated by sequential extractions based on procedures described by Tessier et al. (1979). Aqueous methanol has been widely used for the extraction of organic arsenic species (Edmonds & Francesconi, 1981a; Shiomi et al., 1988a;

Shibata et al., 1996; Kuehnelt et al., 1997). Yu & Wai (1991) and Laintz et al. (1992) described the use of sodium bis(trifluoroethyl) dithiocarbamate (NaFDDC) as a selective chelation reagent of arsenic followed by either a gas chromatograph (GC) detection or supercritical fluid chromatography (SFC) detection. The former gave a limit of detection of 10 µg As/litre in water and the latter gave similar sensitivity after 100–1000-fold preconcentration of the chelate complex in organic solvent.

2.4.4 Supercritical fluid extraction

There are very few publications on the use of supercritical fluid extraction (SFE) for the determination of arsenic. Wenclawiak & Krah (1995) reported a procedure for the measurement of arsenic species using SFE followed by GC or SFC detection. The authors described a rapid extraction of organic and inorganic arsenic species from spiked sand and soil samples by SFE with on-line derivatization using thioglycollic acid methylester (TGM) under supercritical conditions. The TGM derivatives are thermally stable, which makes them amenable to GC–SFC determination. The extracts were chromatographed without further clean-up steps. The limits of detection were 1 ng As/µl and 3 ng As/µl injection for DMA-TGM and MMA-TGM respectively.

2.5 Macro-measurement

Most procedures for the separation and determination of arsenic are based on distillation and hydrogen sulfide precipitation methods. Beard & Lyerly (1961) reported a gravimetric method for the measurement of arsenic following extraction of arsenic as $AsCl_3$ by benzene in strong hydrochloric acid. The recovery was close to 100% when 20 mg was spiked into an aqueous solution.

Vogel (1954) described the historic Marsh test, a qualitative method based on the generation of arsine (AsH_3) by the addition of Zn granules to sulfuric acid. If the gas is mixed with hydrogen, and conducted through a heated glass tube, it decomposes into hydrogen and metallic arsenic which is deposited as a brownish-black "mirror" just beyond the heated part of the tube.

2.6 Colorimetric methods

George & Roscoe (1951) reported a spectroscopic emission measurement of the blue complex formed by the reaction of ammonium molybdate and hydrazine sulfate with arsenic in various biological materials. The sensitivity was about 0.01 µg.

George et al. (1973) carried out a collaborative study for a colorimetric measurement of arsenic in poultry and swine tissues using silver diethyldithiocarbamate (AgDDTC) as the complexing agent. The sensitivity was 0.1 mg/kg in tissues. Dhar et al. (1997) reported a detection limit of 0.04 mg/litre with 95% confidence limit using AgDDTC in chloroform with hexamethylenetetramine.

Gutzeit's test (Vogel, 1954) is based on the generation of arsine from arsenic compounds by the addition of zinc granules to concentrated sulfuric acid. The arsine can be detected by means of a strip of filter paper moistened with silver nitrate or mercuric chloride. The arsine reacts with silver nitrate to give a grey spot, and with mercuric chloride to give a yellow to reddish-brown spot. The sensitivity is about 1 µg. A modification of this method, using mercuric bromide, is found in a test kit currently being used in Bangladesh for groundwater testing which has a limit of detection of 50–100 µg/litre under field conditions.

2.7 Methods for total inorganic arsenic

Methods for the analysis of inorganic arsenic based on its conversion to arsenic trichloride or arsenic tribromide by treatment with 6 mol/litre hydrochloric acid or hydrobromic acid have been described. The arsenic trihalide is separated from the remaining organic arsenic either by distillation (Maher, 1983) or by solvent extraction (Brooke & Evans, 1981). The methods have been applied routinely to the measurement of inorganic arsenic in a variety of foodstuffs, including those of marine origin where any inorganic arsenic is a small percentage of the total arsenic present (Flanjak, 1982; Shinagawa et al., 1983).

2.8 Atomic spectrometry

Common flame atomic absorption spectrometric methods are flame AAS (FAAS), electrothermal AAS (ETAAS) and hydride generation AAS (HGAAS). FAAS is relatively less sensitive for the determination of arsenic than ETAAS and HGAAS. Its detection limit is usually in the range of sub-milligram quantities per litre, and therefore it has limited application, especially for biological samples.

ETAAS, referred to also as graphite furnace-AAS (GFAAS), is generally one of the most sensitive atomic spectroscopic methods. Julshamn et al. (1996) reported factors that are known to interfere with the GFAAS determination of arsenic. The study was carried out by four participating laboratories using five marine standard reference materials. A mixture of palladium and magnesium salts has been recommended as a chemical modifier to avoid nickel contamination of the graphite furnace. The use of a pyrolytically coated graphite furnace tube with the L'vov platform improves sensitivity. Larsen (1991) achieved characteristic masses of about 16 pg of arsenic for arsenate, monomethylarsonate, DMA, arsenobetaine, arsenocholine and tetramethylarsonium ion calculated from aqueous standard solutions.

HGAAS is probably the most widely used method for the determination of arsenic in various matrices. Most of the reported errors in the determination of arsenic by HGAAS with $NaBH_4$ can be attributed to variation in the production of the hydride and its transport into the atomizer. The reaction and atomization of arsine have been reviewed and discussed by Welz et al. (1990). The addition of a solution of L-cysteine to a sample before hydride generation eliminates interference by a number of transition metals in the generation of arsine from arsenite and arsenate (Boampong et al., 1988), and improves responses of arsine generated from MMA and DMA in the presence of arsenite and arsenate (Le et al., 1994a).

Holak & Specchio (1991) described the determination of total arsenic, arsenite and arsenate in foods by HGAAS after a chloroform extraction procedure. The recovery was > 80%. Similar methods (Chappell et al., 1995; Ng et al., 1998a) have been developed for arsenic speciation in soils. Ybanez et al. (1992) described a HGAAS determination of arsenic in dry ashed mussel products and reported a detection limit of 0.017 µg As/g with a precision of 3%.

20

HGAAS has been used for arsenic speciation of inorganic arsenic and its urinary metabolites, MMA and DMA, since 1973, when Braman & Foreback (1973) introduced a cold-trapping step into a basic hydride generation system. Since then a number of improvements have been made to this method (Crecelius, 1978; Buchet & Lauwerys, 1981; Van Cleuvenbergen et al., 1988). Ng et al. (1998b) described an optimized procedure for the speciation of arsenic metabolites in the urine of occupationally exposed workers and experimental animals with detection limits of 1, 1.3 and 3 ng per reaction of inorganic arsenic, MMA and DMA (equivalent to 0.25 µg/litre, 0.325 µg/litre, and 0.75 µg/litre respectively), using 4 ml of urine per reaction.

HGAAS has also been widely employed for analysis of arsenic in water (Chen et al., 1994; Chatterjee et al., 1995; Mandal et al., 1996; Dhar et al., 1997; Biswas et al., 1998). Hasegawa et al. (1994) published the first report of trivalent methyl arsenicals, namely monomethylarsonous acid [MMA(III)] and dimethylarsinous acid [DMA(III)], being found and measured in natural waters. Arsenious acid, MMA(III) and DMA(III) were separated from the pentavalent species by solvent extraction using diethylammonium diethyldithio-carbamate (DDDC) and determined by HGAAS after cold trapping and chromatographic separations. The detection limits were 13–17 pmol/litre and 110–180 pmol/litre for the trivalent and pentavalent species respectively.

Atomic fluorescence spectrometry (AFS) has recently been used for the detection of arsenic hydrides in the ultraviolet spectral region because of the small background emission produced by the relatively cool hydrogen diffusion flame (Gomez-Ariza et al., 1998). The use of cold vapour or hydride generation, together with intense light sources, allows very low detection limits to be achieved. For example, arsenic species in seawater have been measured using hydride generation and cold trapping, coupled with AFS detection at 193.7 nm (Featherstone et al., 1998). They found detection limits of 2.3, 0.9, 2.4 and 3.7 ng/litre for arsenite, arsenate, MMA and DMA respectively (in a 5 ml sample), with a precision of 3.5%.

2.9 ICP methodologies

The main advantages of ICP-MS over ICP-AES are lower detection limits (sub-nanogram to sub-picogram) with wide linear range and isotope analysis capability of high precision. The detection limits of ICP-AES are typically in the range of sub-micrograms to sub-nanograms.

ICP-MS is more susceptible to isobaric interferences arising from the plasma. For example, hydrochloric acid and perchloric acid are not desirable for sample preparation, because the chloride ions generated in the plasma combine with the argon gas to form argon chloride (ArCl). This has the same mass as arsenic (75) which could lead to error if not corrected. Therefore, whenever possible, only nitric acid should be used in sample preparation. Careful sample preparation is as important as the final measurement, and special care should be taken to avoid contamination and losses by volatilization, adsorption and precipitation.

2.10 Voltammetry

Voltammetric stripping methods are mostly based on the chemical reduction of As(V) to As(III) before the deposition step, because it has been generally assumed that As(V) is electrochemically inactive. Mercury and gold (or gold-plated) electrodes are most commonly used for the determination of arsenic.

Sadana (1983) used differential pulse cathodic stripping voltammetry (DPCSV) coupled to a hanging mercury drop electrode (HMDE) to determine arsenic in drinking-water in the presence of Cu^{2+} and reported a detection limit of 1 ng/ml and a relative standard deviation of 6.4%. Zima & van den Berg (1994) reported a detection limit of 3 nmol/litre in seawater. DPCSV was employed by Higham & Tomkins (1993) to determine arsenic in canned tuna fish. They evaluated a number of digestion procedures and found the best procedure gave 93–96% recovery. No detection limit was reported.

A gold electrode affords better sensitivity than a mercury electrode. Hua et al. (1987) reported an automated determination of total arsenic in seawater by flow constant-current stripping analysis with a gold film fibre electrode, in which As(V) in the sample was reduced to As(III) with potassium iodide; the detection limit was 0.15 µg/litre. The reduction of As(V) to As(III) can also be achieved by reaction with sulfur dioxide or hydrazinium chloride for use with a gold electrode or HMDE respectively (Esteban et al., 1994).

Huiliang et al. (1988) have shown that As(V) can be reduced to elemental arsenic provided that extremely low reduction potentials are used. They used this method to measure As(V) and total arsenic in seawater and urine. The detection limit was 0.1 µg/litre using constant-current stripping voltammetry on a gold-coated platinum-fibre electrode. Greulach & Henze (1995) developed a cathodic stripping voltammetric method for the determination of As(V) in water and stream sediment on the basis that As(V) can be reduced in perchloric acid solution containing D-mannitol, combined with the accumulation of arsenic by co-precipitation with copper on an HMDE. The detection limit was 4.4 µg/litre.

Pretty et al. (1993) developed an on-line anodic stripping voltammetry (ASV) flow cell coupled to ICP-MS for the determination of arsenic in spiked urine. The detection limit was 130 pg/ml and the recovery was 94–113%.

2.11 Radiochemical methods

Orvini et al. (1974) reported a combustion technique for sample preparation and determination of arsenic, selenium, zinc, cadmium and mercury by neutron activation analysis (NAA) in environmental matrices including a range of standard reference materials. The recoveries were 98–100%. Sharif et al. (1993) described a NAA technique for the determination of arsenic in eight species of marine fishes caught in the bay of Bengal, Bangladesh.

Haddad & Zikovsky (1985) measured several elements including arsenic in air from workroom welding fumes by NAA and reported a detection limit of 0.17 ± 0.07 µg/m^3. Landsberger & Wu (1995) reported the use of NAA to measure arsenic from environmental tobacco smoke in indoor air with a detection limit of 0.2 ng.

Chutke et al. (1994) described a radiochemical solvent extraction procedure for the determination of arsenite using an arsenic-76 tracer. The procedure is based on the complexation of arsenite with toluene-3,4-dithiol (TDT) at pH 2 and subsequent extraction in benzene. This isotopic dilution technique was employed to measure arsenic in a range of standard and certified reference materials. The detection limit was 250 ng with an accuracy of about 4% error and 170 ng with about 12% error.

2.12 X-ray spectroscopy

Particle-induced X-ray emission spectrometry (PIXES) is an analytical technique that entails the bombardment of a sample (target) with charged particles, resulting in the emission of characteristic X-rays of the elements present. PIXES is a multi-elemental technique with a detection limit of approximately 0.1 µg As/g. It has the advantage of using small samples (1 mg or less) and being a non-destructive technique. Applications of PIXES in the environmental field have mostly focused on atmospheric particulate material (aerosol samples) (Maenhaut, 1987).

Castilla et al. (1993) described the determination of arsenite and arsenate by X-ray fluorescence (XRF) spectroscopy in water with a detection limit of 3.1 ng/g. The recovery was $97 \pm 2.1\%$ and $103 \pm 1.4\%$ for arsenite and arsenate respectively. In this method, the water sample was acidified to pH 2 and arsenite co-precipitated with sodium dibenzyldithiocarbamate (DBDTC). Arsenate in the filtrate was then reduced to arsenite with potassium iodide before the co-precipitation step for the XRF measurement.

Although there are a variety of methods to determine the concentration and oxidation states of arsenic in coal and ash, there have been few attempts to determine the mineral forms of arsenic. Huffman et al. (1994) described the use of X-ray absorption fine structure (XAFS) spectroscopy and its capability of providing speciation information at realistic concentrations of 10–100 mg/kg. They identified arsenic present as arsenopyrite in one coal sample and as aluminosilicate slag and calcium orthoarsenate in combustion ashes.

2.13 Hyphenated techniques

Hyphenated techniques is a term referring to the coupling of more than two instrumental systems to form a single technique.

The combination of chromatographic separation with element-specific spectrometric detection has been proved to be particularly useful for the speciation of arsenic compounds at trace levels in environmental samples. Woller et al. (1995) used AFS detection in combination with ultrasonically nebulized liquid chromatography (LC) for on-line speciation of arsenic, but found that the technique had limited sensitivity owing to matrix interferences. More recently, Slejkovec et al. (1998) used LC and purge-and-trap GC interfaced with AFS to separate and quantify six arsenic species with detection limits of 0.5 ng/ml As (100 µl). Gomez-Ariza et al. (1998) coupled anion-exchange HPLC, hydride generation and AFS to achieve detection limits of 0.17, 0.45, 0.30 and 0.38 µg/litre for arsenite, DMA, MMA and arsenate respectively (using a 20 µl loop). Arsenobetaine was also determined by introducing an on-line photo-oxidation step after the chromatographic separation.

Ebdon et al. (1988) described a number of coupled chromatograph–atomic spectrometry methods for arsenic speciation including GC or HPLC with detection by atomic spectrometry, namely FAAS, flame atomic fluorescence spectrometry (FAFS) and ICP-AES. The FAAS system is capable of detection at less than 1 µg/kg (0.22–0.55 ng absolute for different species) when levels permit; HPLC–hydride generation–FAAS is probably the simplest routine method and HPLC–hydride generation–ICP-AES is preferred for multi-elemental analysis. HPLC–ICP-AES has been employed for the speciation of organic arsenic of aquatic origin (Francesconi et al., 1985; Gailer & Irgolic, 1996). Gjerde et al. (1993) described the coupling of microbore columns with direct-injection nebulization to ICP-AES and reported a detection limit of 10 µg/litre (100 pg). Microbore HPLC has the advantage of analysing small sample size using low flow rates (80–100 µl/min) of mobile phases.

Numerous methods (Shum et al., 1992; Larsen et al., 1993; Magnuson et al., 1996; Thomas et al., 1997; Le & Ma, 1997) have been developed for the speciation of arsenic using the separation power of chromatography coupled to the sensitivity of ICP-MS detection. Heitkemper et al. (1989) described an anion-exchange HPLC–ICP-MS method for the speciation of arsenite, arsenate, MMA and DMA in urine with absolute detection limits ranging from 36 to 96 pg (corresponding to 0.7–1.9 µg/litre in a 50 µl injection). Beauchemin et al. (1989) reported detection limits for arsenic species in DORM-1 (a dogfish muscle certified reference material) ranging between 50 and 300 pg using ion pairing and ion exchange HPLC-ICP-MS. Anion exchange is more tolerant because of the higher buffering capacity of the mobile phase. Cation pairing is more suitable for the determination of DMAA and arsenobetaine in biological samples containing high concentrations of salts. Pergantis et al. (1997) analysed and speciated animal feed additives using microbore HPLC–ICP-MS with detection limits ranging from 0.1 to 0.26 pg. Hakala and Pyy (1992) described an ion-pairing HPLC-HGAAS method for speciation of arsenite, arsenate, MMA and DMA in urine with detection limits of 1.0, 1.6, 1.2 and 4.7 µg/litre respectively.

Ding et al. (1995) described the coupling of micellar liquid chromatography (MLC) and ICP-MS for the speciation of arsenite, arsenate, MMA and DMA with detection limits of 90 pg for DMA and 300 pg for the other species. MLC is a type of chromatography that uses surfactants in aqueous solutions, well above their critical micelle concentration, as alternative mobile phases for reversed-phase liquid chromatography (RPLC). MLC extends the analyte candidates to almost all hydrophobic and many hydrophilic compounds providing they can partition to the micelles. Other advantages of MLC over RPLC include simultaneous separation of both ionic and non-ionic compounds, faster analysis times and improved detection sensitivity and selectivity.

Capillary electrophoresis (CE) is a versatile technique for the separation of a variety of analytes ranging from small inorganic ions to large biomolecules such as proteins and nucleic acids. CE-ICP-MS has been described for the speciation of arsenic by Liu et al. (1995) with detection limits of 100 pg arsenite/ml and 20 pg arsenate/ml and Olesik et al. (1995) with a detection limit of 8 µg/litre (1 pg injection).

Although techniques such as HPLC–ICP-MS and MLC–ICP-MS offer the advantages of high sensitivity and selectivity as well as low detection limits, species identification is based on the comparison of chromatographic retention times to those of available standards. When structure information is required, as well as quantification, electrospray HPLC–MS (Siu et al., 1991) and ionspray MS (Corr, 1997) should be considered. Corr & Larsen (1996) reported the use of LC–MS–MS for speciation of arsenic with a detection limit of 2 pg for the tetramethylarsonium cation.

3. SOURCES AND OCCURRENCE OF ARSENIC IN THE ENVIRONMENT

3.1 Natural sources

Arsenic is the main constituent of more than 200 mineral species, of which about 60% are arsenate, 20% sulfide and sulfosalts and the remaining 20% include arsenides, arsenites, oxides and elemental arsenic (Onishi, 1969). The most common of the arsenic minerals is arsenopyrite, FeAsS, and arsenic is found associated with many types of mineral deposits, especially those including sulfide mineralization (Boyle & Jonasson, 1973). The ability of arsenic to bind to sulfur ligands means that it tends to be found associated with sulfide-bearing mineral deposits, either as separate As minerals or as a trace of a minor constituent of the other sulfide minerals. This leads to elevated levels in soils in many mineralized areas where the concentrations of associated arsenic can range from a few milligrams to > 100 mg/kg.

Concentrations of various types of igneous rocks range from < 1 to 15 mg As/kg, with a mean value of 2 mg As/kg. Similar concentrations (< 1–20 mg As/kg) are found in sandstone and limestone. Significantly higher concentrations of up to 900 mg As/kg are found in argillaceous sedimentary rocks including shales, mudstone and slates. Up to 200 mg As/kg can be present in phosphate rocks (O'Neill, 1990).

Concentrations of arsenic in open ocean water are typically 1–2 µg/litre. The concentrations of arsenic in unpolluted surface water and groundwater are typically in the range of 1–10 µg/litre. Elevated concentrations in surface water and groundwater of up to 100–5000 µg/litre can be found in areas of sulfide mineralization (Welch et al., 1988; Fordyce et al., 1995). Elevated concentrations (> 1 mg As/litre) in groundwater of geochemical origins have also been found in Taiwan (Chen et al., 1994), West Bengal, India (Chatterjee et al., 1995; Das et al., 1995, 1996; Mandal et al., 1996) and more recently in most districts of Bangladesh (Dhar et al., 1997; Biswas et al., 1998). Elevated arsenic concentrations were also found in the drinking-water in Chile (Borgono et al., 1977); North Mexico

(Cebrian et al., 1983); and several areas of Argentina (Astolfi et al., 1981; Nicolli et al., 1989; De Sastre et al., 1992). Arsenic-contaminated groundwater was also found in parts of PR China (Xinjiang and Inner Mongolia) and the USA (California, Utah, Nevada, Washington and Alaska) (Valentine, 1994). More recently, arsenic concentrations of < 0.98 mg/litre have been found in wells in south-western Finland (Kurttio et al., 1998). Levels as high as 35 mg As/litre and 25.7 mg As/litre have been reported in areas associated with hydrothermal activity (Kipling, 1977; Tanaka, 1990).

In nature, arsenic-bearing minerals undergo oxidation and release arsenic to water. This could be one explanation for the problems of arsenic in the groundwater of West Bengal and Bangladesh. In these areas the groundwater usage is very high. It has been estimated that there are about 4–10 million tube wells in Bangladesh alone. The excessive withdrawal and lowering of the water table for rice irrigation and other requirements lead to the exposure and subsequent oxidation of arsenic-containing pyrite in the sediment. As the water table recharges after rainfall, arsenic leaches out of the sediment into the aquifer.

However, recent studies seem to favour the reduction of Fe/As oxyhydroxides as the source for arsenic contamination in groundwater (Nickson et al., 1998; BGS, 2000; BGS & DPHE, 2001). Arsenic forms co-precipitates with ferric oxyhydroxide. Burial of the sediment, rich in ferric oxyhydroxide and organic matter, has led to the strongly reducing groundwater conditions. The process has been aided by the high water table and fine-grained surface layers which impede the penetration of air to the aquifer. Microbial oxidation of organic carbon has depleted the dissolved oxygen in the groundwater. The highly reducing nature of the groundwater explains the presence of arsenite (< 50%) in the water. The "pyrite oxidation" hypothesis is therefore unlikely to be a major process, and the "oxyhydroxide reduction" hypothesis (Nickson et al., 1998; Acharyya et al., 1999) is probably the main cause of arsenic contamination in groundwater. Although the oxyhydroxide reduction hypothesis requires further validation, there is no doubt that the source of arsenic in West Bengal and Bangladesh is geological, as none of the explanations for anthropogenic contamination can account for the regional extent of groundwater contamination. During the past 30 years the use of phosphate fertilizers has increased threefold in this region. The widespread withdrawal of groundwater may have mobilized phosphate derived from fertilizers and from the decay of natural organic materials in shallow aquifers. The increase in phosphate concentration could have promoted the growth of sediment biota and the desorption of arsenic from sediments, and the combined microbiological and chemical process might have increased the mobility of arsenic (Acharyya et al., 1999).

Marine organisms naturally accumulate considerable quantities of organic arsenic compounds. In marine animals the bulk of this arsenic is present as arsenobetaine, whereas marine algae contain most of the arsenic as dimethylarsinoylribosides. Humans are therefore exposed to these arsenic compounds through any diet that includes seafoods. This subject is fully discussed in Chapter 4.

Some arsenic compounds are relatively volatile and consequently contribute significant fluxes in the atmosphere. It has been estimated that the atmospheric flux of As is about 73 540 tonnes/year of which 60% is of natural origin and the rest is derived from anthropogenic sources (Chilvers & Peterson, 1987). Volcanic action is the next most important natural source of arsenic after low-temperature volatilization, and on a local scale it will be the dominant atmospheric source.

3.2 Sources of environmental pollution

3.2.1 Industry

It has long been recognized that the smelting of non-ferrous metals and the production of energy from fossil fuel are the two major industrial processes that lead to anthropogenic arsenic contamination of air, water and soil. Other sources of contamination are the manufacture and use of arsenical pesticides and wood preservatives.

Smelting activities generate the largest single anthropogenic input into the atmosphere (Chilvers & Peterson, 1987).

Tailings from metal-mining operations are a significant source of contamination, and can lead to contamination of the surrounding topsoils, and, because of leaching, sometimes the groundwater too. It has been estimated that several billion tons of tailings waste exist in the USA alone (Wewerka et al., 1978). As sulfur is often present in these tailings, exposure to the atmosphere in the presence of water leads to the production of an acid solution that can leach many elements including arsenic.

Elevated concentrations of arsenic in acid sulfate soils in Canada and New Zealand are associated with pyrite (Dudas, 1987). Concentrations of arsenic < 0.5% through lattice substitution of sulfur in this pyrite iron-rich bauxite have been recorded.

In the United Kingdom, the estimated arsenic releases (Hutton & Symon, 1986) were 650 tonnes/year from the non-ferrous metal industry, 9 tonnes/year emission into the atmosphere and 179 tonnes/year to landfill from iron and steel production, and 297 tonnes/year into the atmosphere and 838 tonnes/year to landfill from fossil fuel combustion. In 1996, the estimated total releases of arsenic to the air in the UK were 50 tonnes (DG Environment, 2000).

The working group of the European Union DGV (the directorate with responsibility for the environment) concluded that there were large reductions in the emissions of arsenic to air in several member countries of the European Union in the 1980s and early 1990s. In 1990, the total emissions of arsenic to the air in the member states were estimated to be 575 tonnes, of which 492 tonnes came from stationary combustion (mainly coal and oil combustion) and 77 tonnes from production processes, mainly from the iron and steel industry (35 tonnes) and the non-ferrous metal industry (31 tonnes) (DG Environment, 2000).

Arsenic is present in the rock phosphate used to manufacture fertilizers and detergents. In 1982, the United Kingdom imported 1324×10^3 tonnes of rock phosphate with an estimated arsenic burden of 10.2 tonnes (Hutton & Symon, 1986).

3.2.2 Past agricultural use

In 1983, arsenical pesticides were one of the largest classes of biocontrol agent in the USA (Woolson, 1983). From the 1960s there was a shift, in herbicide use, from inorganic compounds (including lead and calcium arsenate and copper acetoarsenite) to inorganic and organic compounds (arsenic acid, sodium arsenate, MMA and DMA). Use of total arsenical pesticides, excluding wood preservatives, at the time of publication (1983) was estimated at $7-11 \times 10^3$ tonnes As/year. Annual historical applications of lead arsenate to orchards in the USA ranged from 32 to 700 kg As/ha. Residues in orchard soils as high as 2500 mg/kg have been reported, but they are more commonly in the range of 100–200 mg/kg. In Australia between 1900 and 1950 As_2O_3 was widely used for controlling cattle ticks (*Boophilus microplus*), resulting in widespread arsenic contamination (Seddon, 1951).

3.2.3 Sewage sludge

The levels of arsenic in sewage sludge reflect the extent of industrialization of the area served by the local sewage system. Significant quantities may be added by arsenic-contaminated wastewater runoff derived from sources including atmospherically deposited arsenic, residues from pesticide usage, phosphate detergents and industrial effluent, particularly from the metal-processing industry. Levels of 0–188 mg As/kg dry weight have been reported in the United Kingdom (Woolson, 1983). Zhu & Tabatabai (1995) reported levels of 2.4–39.6 mg As/kg with a mean of 9.8 for sewage sludges from waste treatment plants in Iowa, USA.

O'Neil (1990) estimated that in the UK as a whole about 2.5 tonnes As/year is added to the agricultural land by use of sludge, compared to 6.1 tonnes As/year when phosphate fertilizer is used.

3.3 Uses

Arsenic is produced commercially by reduction of As_2O_3 with charcoal. As_2O_3 is produced as a by-product of metal-smelting operations. It is present in flue dust from the roasting of ores, especially those produced in copper smelting. In the 1960s, the pattern of use for As_2O_3 in the USA is believed to have been 77% as pesticides, 18% as glass, 4% as industrial chemicals and 1% as medicine. However, the pattern has changed over the years as the use of arsenic compounds for timber treatment has been increasingly popular since the late 1980s. Worldwide usage in the early 1980s was estimated to be 16 000 tonnes As/year as a herbicide, 12 000 tonnes As/year as a cotton desiccant/defoliant and

16 000 tonnes As/year in wood preservative (Chilvers & Peterson, 1987). By 1990, the estimated end-use of arsenic in the USA was 70% in wood preservatives, 22% in agricultural chemicals, 4% in glass, 2% in non-ferrous alloys and 2% in other uses including semiconductors (US DOI, 1991). Arsenic pentoxide and As_2O_3 are used as additives in alloys, particularly with lead and copper; arsenic and As_2O_3 are used in the manufacturing of low-melting glasses. High-purity arsenic metal and gallium arsenide are used in semiconductor products. Fowler's solution (1% potassium arsenite solution) was used as a medication (Cuzick et al., 1992). As_2O_3 has been used for the treatment of acute promyelocytic leukaemia (Soignet et al., 1998).

Hutton & Symon (1986) reported that about 5000 tonnes/year As_2O_3 is imported to the United Kingdom for conversion to other arsenic compounds. These processes result in an estimated discharge of 87 tonnes As/year in manufacturing sludges on landfilled sites. Currently about 500 tonnes As/year is utilized in copper chrome arsenate (CCA) timber treatment, of which at most 5 tonnes/year is retained in sludges. Small amounts of arsenic are used in the production of glass, and most of the remainder is re-exported.

4. ENVIRONMENTAL TRANSPORT AND DISTRIBUTION

4.1 Transport and distribution between media

4.1.1 Air

Arsenic is primarily emitted into the atmosphere by high-temperature processes such as coal-fired power generation, smelting, burning vegetation and vulcanism. Natural low-temperature biomethylation and microbial reduction also release arsenic into the atmosphere; microorganisms can form volatile methylated derivatives of arsenic under both aerobic and anaerobic conditions, and can reduce arsenic compounds to release arsine gas (Cheng & Focht, 1979; Tamaki & Frankenberger, 1992) (see section 4.2.2). Arsenic is released into the atmosphere primarily as As_2O_3 or, less frequently, as one of several volatile organic compounds. Arsenic released to air exists mainly in the form of particulate matter (Coles et al., 1979). These particles are dispersed by the wind to a varying extent, depending on their size, and the particles are returned to the earth by wet or dry deposition. Arsines that are released from microbial sources in soils or sediments undergo oxidation in the air, reconverting the arsenic to less volatile forms that settle back to the ground (Wood, 1974; Parris & Brinckman, 1976).

Pacyna et al. (1989) studied atmospheric transport of arsenic from various sources in Europe to selected receptor sites in Norway. By modelling long-range transport they were able to calculate a dry deposition velocity for arsenic of 0.4 cm/second. Scudlark & Church (1988) measured arsenic in acid precipitation on the mid-Atlantic coast of the USA during 1985 and 1986. They calculated the total annual arsenic deposition rate to range from 38 to 266 $\mu g/m^2$, with dry deposition estimated to comprise 29–55% of the total. Davidson et al. (1985) calculated the annual dry deposition flux of arsenic to the Olympic National Park, Washington (USA) to range from 76.7 to 208 $\mu g/m^{-2}$. The average annual wet deposition of arsenic at Chesapeake bay (Maryland, USA) was found to be 49 μg As/m^{-2} (Scudlark et al., 1994).

Total atmospheric arsenic emissions from both natural and anthropogenic sources have been estimated to be 31×10^9 g/year, and total atmospheric arsenic removal was estimated to be $30-50 \times 10^9$ g/year. The global tropospheric residence time of arsenic appears to be about 9 days (Walsh et al., 1979). Nakamura et al. (1990) estimated global atmospheric emissions into the atmosphere and deposition of arsenic. Total emissions were estimated at 36×10^9 g/year, with the major source of atmospheric arsenic being anthropogenic emissions; the major natural source of arsenic was volcanic activity. Emissions from anthropogenic sources were estimated at 24×10^9 g/year, representing 64% of total arsenic influxes. Depositions from the atmosphere to the land and the oceans were estimated at 24×10^9 g/year and 9×10^9 g/year respectively. Akeredolu et al. (1994) calculated the total annual transport of arsenic into the Arctic atmosphere at 285 t (285×10^6 g) on the basis of a chemical transport modelling approach previously used for sulfur.

Arsenic in the atmosphere exists primarily adsorbed to particulate matter and mostly to particles < 2 µm in diameter (Coles et al., 1979). Waslenchuk (1978) found that atmospheric arsenate at the continental shelf of the south-eastern USA is associated exclusively with the particulate fraction. Rabano et al. (1989) collected size-fractionated aerosol samples at an urban site during 1987. A greater proportion (75%) of the arsenic was observed in the fine particles (< 2.5 µm). The As(III)/As(V) ratio for both fine and coarse (> 2.5 µm) particles was approximately 1. Similarly, Waldman et al. (1991) reported that 65% of the arsenic in aerosol samples collected at an urban site (China) was associated with fine particles (< 2.5 µm). Kelley et al. (1995) monitored arsenic in aerosol collected from the Kola Peninsula (Russia). They found 68% of arsenic associated with fine particles (< 1 µm), 26% with coarse particles ($1-10$ µm) and 7% with large particles (> 10 µm). The atmospheric residence time of particulate-bound arsenic depends on particle size and meteorological conditions, but a typical value is about 9 days (US EPA, 1982).

4.1.2 Freshwater and sediment

The dissolved forms of arsenic in the water column include arsenate, arsenite, monomethylarsonic acid (MMA) and

35

dimethylarsinic acid (DMA) (Braman & Foreback, 1973). Some As(III) and As(V) species can interchange oxidation states depending on Eh, pH and biological processes (Ferguson & Gavis, 1972). Some arsenic species have an affinity for clay mineral surfaces and organic matter, and this can affect their environmental behaviour. Methylation and demethylation reactions are also important transformations controlling the mobilization and subsequent distribution of arsenicals (Mok & Wai, 1994). Transport and partitioning of arsenic in water depends on the chemical form of the arsenic and on interactions with other materials present. Arsenic may be adsorbed from water on to clays, iron oxides, aluminium hydroxides, manganese compounds and organic material (Callahan et al., 1979; Welch et al., 1988). The distribution and transport of arsenic in sediment is a complex process that depends on water quality, native biota and sediment type. There is a potential for arsenic release when there is fluctuation in Eh, pH, soluble arsenic concentration and sediment organic content (Abdelghani et al., 1981).

Ferguson & Gavis (1972) proposed an arsenic cycle for a stratified lake. In the aerobic epilimnetic water, reduced forms of arsenic tend to be oxidized to arsenate, which co-precipitates with ferric oxyhydroxide. Turbulent dispersion and convection transports some of the arsenate across the thermocline to the oxygen-depleted hypolimnion, where reduction to $HAsO_2$ and AsS_2^- takes place, depending on the sulfur concentration and the Eh. Co-precipitation, adsorption and epitaxial crystal growth cause arsenic to be removed to the sediments, where reduction of ferric iron, arsenate and arsenite result in either solubilization or stabilization as an insoluble sulfide or arsenic metal. Microbial reduction and methylation to arsine solubilize the arsenic (see section 4.2), and diffusion through the sediments or mixing by currents or burrowing organisms (see section 4.1.4) cause arsenic to re-enter the water column.

Aurilio et al. (1994) studied the speciation and fate of arsenic in three lakes of the Aberjona watershed (Massachusetts, USA). Speciation appeared to be controlled by reduction, methylation, and oxidation processes, and by adsorption to and desorption from particles. Biologically mediated reduction, at rates of 0.2–0.5% total arsenic/day, and methylation, at rates of 0.4–0.6% total arsenic/day, occurred in the mixed layers of these lakes. These processes are slow or even absent in the hypolimnion, however, allowing arsenate to accumulate in seasonally anoxic hypolymnetic waters. High micromolar concentrations of arsenic, predominantly arsenite, persisted in the saline, sulfidic monimolimnion of one lake.

Clement & Faust (1981) studied the release of arsenic from contaminated sediments. Anaerobic conditions led to aqueous levels of arsenic, principally as arsenite, about 10 times higher than concentrations reached with aerobic conditions. Under aerobic conditions arsenic in the overlying water comprised 70% arsenate and 20% organic arsenic. The authors found that adsorption–desorption equilibria and the amount of 'available' arsenic present in the sediment greatly influenced the soluble arsenic concentration found in the aqueous phase. In sediment under oxidized conditions arsenic solubility was low and 87% of the arsenic in solution was present as arsenate. On reduction, arsenite became the major arsenic species in solution and solubility increased (Masscheleyn et al., 1991b). Ahmann et al. (1997) identified rapid arsenic mobilization from aquatic sediments in upper Aberjona (Massachusetts, USA) sediment microcosms. The findings suggest that arsenic reduction by microorganisms may contribute to arsenic flux from anoxic sediments in this arsenic-contaminated watershed.

The predominant arsenic species in the water column of lakes is arsenate, as expected in oxidizing environments (Seyler & Martin, 1989). Arsenite is usually present and sometimes dominates in bottom water which contains high concentrations of $Fe(II)$ and low oxygen. Peterson & Carpenter (1983) reported that the arsenate : arsenite concentration ratio was 15 : 1 in the oxic region of the water column and 1 : 12 in the anoxic zone. Seasonal trends reveal higher concentrations of arsenic in summer than in winter. The source of arsenic in the summer is most likely surface sediments that have become anoxic causing a release into the water column of arsenic adsorbed on iron and manganese oxides (Singh et al., 1988; Crecelius et al., 1994).

Pettine et al. (1992) found that arsenate was the predominant arsenic species in the river Po (Italy). The main factors affecting dissolved concentrations included flow and suspended matter concentration and biological activity. The ratio between oxidized and reduced species appears to be significantly influenced by iron and manganese oxides. Abdel-Moati (1990) monitored arsenic in the Nile delta lakes and found arsenate to be the dominant arsenic species (85–95%). Increased arsenite (14–33%) was found near local sewage discharge points. Dimethylarsenic was the dominant organic species, reaching 22% of the total dissolved arsenic.

A temporal study of arsenic speciation in Davis Creek Reservoir, a seasonally anoxic lake in northern California (USA), demonstrated that dimethylarsinic acid increased sufficiently to become the dominant form of dissolved arsenic within the surface photic zone during late summer and early autumn. Methylated forms decreased and arsenate increased when the lake 'turned over' in early December, suggesting a degradation of dimethylarsinic acid (Anderson & Bruland, 1991).

Aggett & O'Brien (1985) report that Lake Ohakuri (New Zealand) becomes stratified during the summer. During this period arsenic released from the sediment accumulates in the hypolimnion until turnover when it is mixed with epilimnetic water. It is estimated that this turnover effect causes a temporary increase in arsenic concentrations of 10–20%. Aggett & Roberts (1986) conclude that arsenate and phosphate are incorporated into Lake Ohakuri sediments by co-precipitation at the time of formation of the hydrous oxides rather than by adsorption on existing surfaces. Aggett & Kriegman (1988) show that in sediment cores from Lake Ohakuri over 90% of arsenic in interstitial waters was present as arsenite, an indication that reduction from arsenate, the predominant form adsorbed from the lake water, was taking place. When conditions at the sediment–water interface became anoxic, arsenite diffused across the interface into the hypolimnion.

Johnson & Thornton (1987) studied the seasonal variation of arsenic in the Carnon river, south-west England (UK). Approximately 85% of the arsenic was found to originate from mine waters. Arsenic is found to a large extent (~80%) in the particulate phase; the authors suggest that sorptive or co-precipitation processes are responsible for the regulation of dissolved concentrations of arsenic in these waters. These processes are largely independent of pH. Adsorption appears to be important in the removal of arsenic from solution, with 80% being removed on entering estuarine waters.

Both adsorption of arsenic on iron-rich oxides on the surface of the sediments and incorporation of arsenic into the sediments by co-precipitation with hydrous iron oxides are factors controlling mobilization of sediment arsenic. The major arsenic species leached was arsenate; release of arsenate was found to be pH dependent and related to the total iron and free iron oxides in the sediments (Mok & Wai, 1989). Arsenate and arsenite differ in adsorption characteristics, and this influences their mobilization and subsequent distribution during water–sediment interactions. The extent of adsorption and remobilization varies with the oxidation state of arsenic, the Eh and the pH of the water. The increase in mobility of arsenate under more reducing conditions is generally attributed to the reduction of Fe^{3+} to Fe(II), with subsequent release of arsenate, and reduction of arsenate to arsenite (Mok & Wai, 1994).

Brannon & Patrick (1987) found that arsenate added to sediment became associated with relatively immobile iron and aluminium compounds. Addition of arsenate to sediments before anaerobic incubation also resulted in accumulation of arsenite and organic arsenic in the interstitial water and exchangeable phases of anaerobic sediments. Seyler & Martin (1989) report that the presence of arsenic in the anoxic zone of a permanently stratified lake was due to adsorption on to iron and manganese.

Sorption of arsenate, MMA and DMA on anaerobic bottom sediments from the Menominee river, Wisconsin (USA) is described by Langmuir isotherms (Holm et al., 1979). Singh et al. (1988) found that the adsorption of arsenite from aqueous solution followed first-order adsorption expression obeying Langmuir's model of adsorption. Similar findings were reported by Yadava et al. (1988) for adsorption of arsenite by china clay, with maximum adsorption at pH 8. Sediment adsorption of arsenate, monosodium methane-arsonate (MSMA) and DMA was positively correlated with clay content (Wauchope & McDowell, 1984). Organic sediments adsorbed arsenic more strongly than sandy sediments (Faust et al., 1987b).

The extent of uptake and the rate of adsorption of arsenate decrease with an increase in temperature from 20 °C to 40 °C. The amount of arsenate adsorbed increases as the pH of the system increases and reaches its maximum at pH 4.2 for haematite and pH 6.2 for feldspar. The removal of arsenate from aqueous solution by adsorption on to geological materials such as haematite and feldspar follows first-order kinetics, and intraparticle diffusion seems to control the mass transfer (Prasad, 1994). The adsorption of arsenate on alumina, haematite , kaolin and quartz was influenced by the charge of the solid surface and the arsenic speciation in solution as determined by pH (Xu et al., 1988).

Adsorption of arsenate by fly ash was significantly greater at pH 4 than at pH 7 or 10 and was found to be almost irreversible. Adsorption fitted both the Freundlich and Langmuir adsorption models (Sen & De, 1987; Diamadopoulos et al., 1993). The partitioning of arsenic between acidic fly ash and leachate is controlled by sorption on iron oxyhydroxide. The leaching of arsenic is mainly controlled by sorption on hydroxylamine-extractable ("amorphous") iron oxyhydroxide; crystalline iron oxides appear to have little influence on the process (Van der Hoek & Comans, 1996). Thanabalasingam & Pickering (1986) found that arsenic sorption by humic acid varies with pH, adsorbate concentration and ash content of the substrate. At fixed pH, the amount of arsenic sorbed conformed to a Langmuir relationship, with calculated capacities in the region of maximum uptake (~pH 5.5) being of the order of 5250–6750 mg/kg for arsenite and 6750–8250 mg/kg for arsenate.

Gupta & Chen (1978) report that arsenic acid and arsenious acid species are effectively adsorbed in the pH range 4–7. Laboratory adsorption experiments indicated that arsenite is less effectively removed than arsenate. Adsorption of MMA and DMA on ferric oxyhydroxide and activated alumina decreased with increasing pH (4–11) (Cox & Ghosh, 1994).

Arsenic in porewater is controlled by the solubility of iron and manganese oxyhydroxides in the oxidized zone and metal sulfides in the reduced zone. Diagenetic sulfides are important sinks for arsenic in reduced, sulfidic sediments. During reduction, oxyhydroxides of iron and manganese dissolve, arsenic sulfides precipitate and arsenic is released to groundwater dominantly as arsenite (Moore et al., 1988). Therefore, mobilization of arsenic is more likely to occur in sediments low in iron and manganese oxyhydroxides, and calcium carbonate (Brannon & Patrick, 1987; Mok & Wai, 1990).

Bright et al. (1994) found that arsenite was the predominant arsenical in sediment porewater throughout a watershed receiving gold-mine effluent; dissolved arsenic in water column samples was mostly arsenate. Arsenic distribution in surficial sediments was controlled partially by the bulk movement of sediments, followed by burial with less-contaminated sediments in the upper reaches of the watershed. Particulate concentrations of arsenic contributed significantly (< 70%) to the total arsenic concentrations in the water column downstream of the gold-mine discharge. Azcue et al. (1994b) found that 66–83% of the arsenic in sediment porewater from a mine-polluted lake was arsenite. The concentration gradient of total dissolved arsenic indicated an upward diffusion of arsenic towards the water column, with the estimated annual fluxes being 0.8–3.8 $\mu g/cm^2$.

Mok & Wai (1990) reported that acid precipitation caused increased release of both arsenate and arsenite from contaminated sediments (pH 2.7). Arsenic release decreased with increasing pH; lowest levels of release were found at pH 8.3 for arsenite and pH 6.3 for arsenate. Release of arsenic increased at more alkaline pH values. Xu et al. (1991) concluded that environmental acidification would increase the leaching of arsenic from sediments to surface waters under reducing conditions as a result of the release of arsenite from iron oxyhydroxide phases, but could also reduce the mobility because of enhanced adsorption under oxidizing conditions. However, a large reduction in pH (to ≤4) would enhance the mobility of arsenic even under oxidizing conditions.

4.1.3 *Estuarine and marine water and sediment*

An arsenic cycle has also been outlined for the estuarine environment. Sanders (1980) found that the major inputs to the marine environment were river runoff and atmospheric deposition. Biological uptake caused changes in arsenic speciation resulting in measurable concentrations of reduced and methylated arsenic species. The overall cycle is similar to the phosphate cycle, but the regeneration time for arsenic is much slower. Arsenic flows into the estuary as arsenate and arsenite from river water and mine adits. There is oxidation of arsenite to arsenate, microbiological reduction of arsenate to arsenite and removal of arsenic by dilution with seawater and subsequent transport out of the estuary. Inorganic arsenic can be adsorbed on to charged particles of iron oxyhydroxides and manganese oxides and deposited as flocculated particles to sediment. There is subsequent release of dissolved arsenite and arsenate following the reduction and dissolution of the iron and manganese carrier phases in the anoxic sediments. Arsenate can be reduced, either microbially or chemically, to arsenite within the anoxic sediment, and arsenic (as arsenate or arsenite) can enter by sediment resuspension (Sanders, 1980; Knox et al., 1984). Studies on the pH dependence of arsenate and arsenite adsorption to soils and sediments and to minerals are not consistent. For example, greater adsorption of arsenate to fly ash occurred at pH4 than at pHs 7 and 10 (Sen & De, 1987; Diamadopoulos et al., 1993), whereas Mok & Wai (1990) reported that absorption increased as pH increased for sediments.

Arsenic entering unpolluted estuaries associated with particulates remains adsorbed, and accumulates in sediment. Remobilization has only a small effect ($< 7\%$) on the dissolved arsenic concentration in the water column. Dissolved arsenic species form complexes with low-molecular-weight dissolved organic matter, and these tend to prevent adsorption and co-precipitation interactions between arsenic and flocculating iron oxyhydroxides and humics (Waslenchuk & Windom, 1978). Langston (1983) reports that more than 80% of arsenic entering Restronguet creek in southwest England (United Kingdom) was retained by the sediment, which consequently acts as a sink for riverine inputs and limits transport of dissolved species to coastal waters. Iron oxyhydroxide scavenging seems to be a predominant factor in the removal of arsenic from the Scheldt estuary (The Netherlands) (van der Sloot et al., 1985). Millward et al. (1997) estimated an annual arsenic budget for the Thames plume (United Kingdom) and found that cycling of arsenic by phytoplankton was the dominant process. Inorganic arsenic was removed from the water column by phytoplankton and recycled during phytoplankton degradation and consumption.

An arsenic budget for Puget sound (Washington, USA) revealed that sediments accumulate less than 30% of the arsenic entering the sound (Crecelius et al., 1975). Carpenter et al. (1978) found that sedimentation processes including adsorption–desorption reactions with natural Puget sound suspended matter remove less than 15% of the dissolved arsenic input, with iron oxyhydroxides dominating what removal does occur. Most of the arsenic entering the sound is removed by advection of surface waters out into the strait of Juan de Fuca . A similar budget for Lake Washington (USA) showed equal inputs of arsenic from the atmosphere and from rivers, and subsequent removal by outflowing water (45%) and by accumulation in the sediments (55%) (Crecelius, 1975).

Riedel (1993) studied the distribution of dissolved and solid arsenic species in contaminated estuarine sediment. Arsenite was the dominant dissolved and solid species in the deeper reduced sediment, and arsenate was dominant in the oxidized surface layer. Arsenite in the interstitial water diffused toward the surface layer, where it was mostly oxidized to arsenate.

Howard et al. (1988) found that the distribution of dissolved inorganic arsenic in an estuary appears to be determined by a combination of secondary inputs arising from old mine drainage and advective transport of arsenic-enriched sediment interstitial waters into the water column. Bioutilization of the element during the warmer months results in the release of dissolved monomethyl-arsenic and dimethylarsenic. Inorganic arsenite and methylated arsenic species can account for up to 41% and 70% of the dissolved arsenic respectively, but only when the water temperature exceeds 12 °C (Howard et al., 1984). Arsenate was the dominant form in a temperate estuary throughout the year except late winter when a dimethylarsenic species was dominant (Riedel, 1993).

Andreae (1978, 1979) monitored seawater samples from the northeast Pacific and southern Californian coast (USA). Methylation and reduction of arsenate to arsenite and methylarsenic acids occur in the photic zone. Arsenic is taken up by planktonic organisms in the surface waters and transported to deeper waters with biogenic debris. At intermediate levels regeneration of arsenate occurs. There was a good correlation between photosynthetic activity and concentration of methylated arsenicals. Andreae & Froelich (1984) and Sadiq (1990) found that arsenate is more abundant in oxic seawaters whereas arsenite is more abundant in anoxic seawaters.

Waslenchuk (1978) found that concentrations of arsenic species in continental shelf waters of the south-eastern USA are controlled mainly by simple mixing of shelf waters and Gulf Stream intrusions.

Riverine and atmospheric arsenic inputs to the shelf waters were relatively insignificant, and uptake of arsenic by biota had only a minor effect on arsenic distribution.

Byrd (1988) studied the seasonal cycle of arsenic on the continental shelf of the South Atlantic. During periods of high winds in the winter and early spring, inorganic arsenic concentrations are reduced to as little as 20% of typical open-ocean concentrations by sorption on to suspended sediments or incorporation into phytoplankton. In the late summer and early autumn arsenic is remobilized and returned to the water column, elevating arsenic concentrations to 50% more than open-ocean concentrations. Belzile (1988) analysed vertical profiles of arsenic in cores from the Laurentian trough in the gulf of St Lawrence. The surface enrichment of solid arsenic and the increase of dissolved arsenic with depth suggested that the mobile portion of arsenic is associated with iron oxyhydroxides. It follows a redox pattern of dissolution in the suboxic zone, upwards diffusion, and precipitation near the sediment–water interface under non-steady-state conditions.

Nereis succinea, a burrowing polychaete, affected distribution and flux of arsenic from sediments by its production of irrigated burrows. These burrows increased both the effective surface area of the sediment and the diffusion of arsenic by a factor of five. Although physical suspension can produce large pulses of materials from contaminated sediments, it is the continuous biological activity that is likely to be more important in the mobilization of arsenic from sediments (Riedel et al., 1987).

Riedel et al. (1989) reported that three species of burrowing invertebrates (*N. succinea*, *Macoma balthica* and *Micura leidyi*) cause a measurable flux of arsenic out of contaminated sediments which was not measurable in the absence of fauna. Arsenic release from sediment was primarily arsenate and arsenite, with trace amounts of methylated arsenic compounds.

4.1.4 Soil

Arsenic from weathered rock and soil may be transported by wind or water erosion. However, because many arsenic compounds tend to adsorb to soils, leaching usually results in transportation over only short distances in soil (Moore et al., 1988; Welch et al., 1988). However, rainwater or snowmelt may leach soluble forms into surface water or groundwater, and soil microorganisms may reduce a small amount to volatile forms (arsines) (Woolson, 1977a; Richardson et al., 1978; Cheng & Focht, 1979; Turpeinen et al., 1999).

Under reducing conditions, arsenite dominates in soil (Deuel & Swoboda, 1972a; Haswell et al., 1985) but elemental arsenic and arsine can also be present (Walsh & Keeney, 1975). Arsenic would be present in well-drained soils as $H_2AsO_4^-$ if the soil was acidic or as $HAsO_4^{2-}$ if the soil was alkaline. Oxidation, reduction, adsorption, dissolution, precipitation and volatilization of arsenic reactions commonly occur in soil (Bhumbla & Keefer, 1994). In the porewater of aerobic soils arsenate is the dominant arsenic species, with small quantities of arsenite and MMA in mineralized areas.

The amount of arsenic sorbed from solution increases as the free iron oxide, magnesium oxide, aluminium oxide or clay content of the soil increases; removal of amorphous iron or aluminium components by treatment with oxalate eliminates or appreciably reduces the arsenic sorption capacity of the soil (Dickens & Hiltbold, 1967; Jacobs et al., 1970a; Galba, 1972; Wauchope, 1975; Livesey & Huang, 1981). Barry et al. (1995) examined the adsorption characteristics of a forest soil profile. The greatest sorption capacity for arsenic occurred at a depth of 30 cm in the profile, in the B2 horizon where there was a predominance of clay and oxyhydroxides of iron and aluminium. Adsorption of arsenic on soil colloids depends on the adsorption capacity and behaviour of these colloids (clay, oxides or hydroxides of aluminium, iron and manganese, calcium carbonates or organic matter). In general, iron oxides/hydroxides are the most commonly involved in adsorption of arsenic in both acidic and alkaline soils (Sadiq, 1997). Manning & Goldberg (1997) studied the adsorption of arsenic in three arid-zone soils. They found that the soil with the highest citrate–dithionite extractable iron and percentage of clay had the highest affinity for arsenite and arsenate and displayed adsorption behaviour similar to that of pure ferric oxide. Adsorption isotherms indicated that arsenate species adsorbed more strongly than arsenite.

The surfaces of aluminium oxides/hydroxides and clay may play a role in arsenic adsorption, but only in acidic soils. Carbonate minerals are expected to adsorb in calcareous soils (Sadiq, 1997), and Goldberg & Glaubig (1988) concluded that carbonates play a major role in arsenate adsorption at pH > 9. Phosphate substantially suppresses arsenate adsorption by soil, with the extent of the suppression varying from soil to soil (Livesey & Huang, 1981). Roy et al. (1986) found that the adsorption of arsenate was significantly reduced by competitive interactions with phosphate in three different soil types (clay, silt loam and ultisol). Darland & Inskeep (1997) found that phosphate effectively competed with arsenate for adsorption sites on sand in batch isotherms as well as in saturated transport studies. The phosphate competition was not, however, sufficient to desorb all of the applied arsenate either in simultaneously applied pulses, or in a column where arsenate was applied before a concentrated pulse of phosphate. Approximately 40% of the applied arsenate remained sorbed to the sand even after the total phosphate loading exceeded the column capacity by more than two orders of magnitude. The authors concluded that rates of arsenate desorption play an important role in transport of arsenate through porous media. Elkhatib et al. (1984) found that arsenite adsorption was not reversible, with only small amounts of sorbed arsenite released during subsequent desorption procedures. No significant correlation was found between arsenic adsorption and soil organic carbon or cation exchange capacity (CEC) (Hayakawa & Watanabe, 1982).

Jones et al. (1997) found that increased mobility of arsenic after liming appears to be consistent with the pH dependence of sorption reactions of arsenic on iron oxide minerals rather than dissolution–precipitation reactions of solid metal arsenates.

Sakata (1987) reports distribution coefficients (K_d) for arsenite for 15 subsurface soils from different sites in Japan with K_d values ranging from 75 to 1200. The distribution coefficient was significantly correlated with the extractable iron content of the soils.

Precipitation is another mechanism of arsenic removal from soil. Thermodynamic calculations showed that in acidic oxic and suboxic soils, iron arsenate may control arsenic solubility, whereas in anoxic soils, sulfides of arsenite may control the concentrations of the dissolved arsenic in soil solutions. In alkaline, acidic, oxic and suboxic soils, precipitation of both iron arsenate and calcium arsenate may limit arsenic concentrations in soil solutions (Sadiq et al., 1983; Sadiq, 1997). Carey et al. (1996) studied the sorption of arsenic in two free-draining sandy soils in New Zealand. They concluded that arsenate sorption occurred primarily through adsorption rather than a precipitation mechanism.

Many soil organisms are capable of converting arsenate and arsenite to several reduced forms, largely methylated arsines which are volatile (see section 4.2). Woolson (1977b) proposed that about 12% of the arsenic applied and present in a soil is lost through volatilization of alkylarsines each year. Woolson & Isensee (1981) report total losses of 14–15% per year from soil treated with sodium arsenite, DMA or MMA. Most of the loss was through volatilisation, although some apparent loss was caused by movement to or mixing with subsoil. Sandberg & Allen (1975) estimated an arsenic loss of 17–35% per year through volatilization. Sanford & Klein (1988) report that arsenic volatilization showed a direct relationship with nutrient levels and microbial growth in soil.

Leaching does not appear to be a significant route of arsenic loss from soil. Arsenic as MMA was applied to three soil types over a 6-year period. Percentage recovery of applied arsenic averaged 67%, 57% and 39% in a fine sandy loam, a silt loam and a sandy loam soil respectively. All of the arsenic recovered in the soils was detected in the ploughed layer (< 30 cm) with no evidence of leaching into deeper zones (Hiltbold et al., 1974). Elfving et al. (1994) monitored the movement of arsenic following the application of lead arsenate to fruit orchards for insect control. The rate of decrease in concentration of arsenic with depth was significantly greater in a sandy soil than in clay, suggesting that downward movement occurred less readily in the former. Peryea & Creger (1994) studied the vertical distribution of arsenic in six contaminated orchard soils. Most of the arsenic was restricted to the upper 40 cm, with maximum arsenic concentrations ranging from 57.8 to 363.8 mg/kg. Absolute soil enrichment with arsenic occurred to depths between 45 and > 120 cm, with arsenic concentrations of 5.3–47.3 mg/kg at 120 cm. The authors state that the deeper movement found in this study compared with many others is due to high loading rates of lead arsenate, coarse soil texture, low organic matter content and use of irrigation. The use of phosphate fertilizers significantly increases the amount of arsenic leached from soil contaminated with lead arsenate pesticide residues (Davenport & Peryea, 1991).

Masscheleyn et al. (1991a) found that at soil Eh levels of 200 and 500 mV arsenic solubility was low and the major part (65–98%) of the arsenic in solution was arsenate. Under moderately reduced soil conditions (at 0 and –100 mV) arsenic solubility was controlled by the dissolution of iron oxyhydroxides. Arsenic was co-precipitated as arsenate with iron oxyhydroxides and released on solubilization. On reduction to –200 mV the soluble arsenic content increased to 13 times what it was at 500 mV.

Richardson et al. (1978) monitored surface runoff of arsenic from a fine montmorillonitic clay after application of arsenic acid for desiccation of cotton (*Gossypium hirsutum*). They calculated that approximately 7% of the amount applied would be transported from the watershed by runoff and erosion, 38% in solution and 62% attached to sediment.

Tammes & de Lint (1969) calculated an average half-life of 6.5 ± 0.4 years for arsenic persistence on two Netherlands soils after application of arsenite.

4.2 Biotransformation

Most environmental transformations of arsenic appear to occur in the soil, in sediments, in plants and animals, and in zones of biological activity in the oceans. Biomethylation and bioreduction are probably the most important environmental transformations of the element, since they can produce organometallic species that are sufficiently stable to be mobile in air and water. However, the biomethylated forms of arsenic are subject to oxidation and bacterial demethylation back to inorganic forms (IPCS, 1981, section 4).

Three major modes of biotransformation of arsenic species have been found to occur in the environment: redox transformation between arsenite and arsenate, the reduction and methylation of arsenic, and the biosynthesis of organoarsenic compounds. There is biogeochemical cycling of compounds formed by these processes (Andreae, 1983).

Arsenic is released into the atmosphere primarily as As_2O_3 or, less frequently, in one of several volatile organic compounds, mainly arsines (US EPA, 1982). Trivalent arsenic and methyl arsines in the atmosphere undergo oxidation to the pentavalent state, and arsenic in the atmosphere is usually a mixture of the trivalent and pentavalent forms (Scudlark & Church, 1988). Photolysis is not considered an important breakdown process for arsenic compounds (Callahan et al., 1979).

Arsenic can undergo a complex series of transformations, including redox reactions, ligand exchange and biotransformation (Callahan et al., 1979; Welch et al., 1988). Factors affecting fate processes in water include the Eh, pH, metal sulfide and sulfide ion concentrations, iron concentrations, temperature, salinity, and distribution and composition of the biota (Callahan et al., 1979; Wakao et al., 1988).

4.2.1 Oxidation and reduction

Oscarson et al. (1980) observed oxidation of arsenite (10 mg/litre) to arsenate in sediments from lakes in Saskatchewan (Canada). The oxidation process was unaffected by flushing nitrogen or air through the system or by the addition of mercuric chloride. The authors therefore concluded that the oxidation was an abiotic process, with microorganisms playing a very minor role in the system. However, Scudlark & Johnson (1982) examined the oxidation of arsenite in seawater at low levels. They found that abiotic oxidation proceeded at a slow and constant rate with rapid oxidation occurring only in the presence of certain aquatic bacteria. The rate of abiotic oxidation, after spiking water with an initial arsenite concentration of 4 µg/litre (53 nmol/litre), was 0.2 µg/litre per day in distilled water and 0.3 µg/litre per day in artificial seawater. Baker et al. (1983a) found no methylated arsenic compounds in sterile lake sediments incubated in the presence of arsenate or arsenite.

Scudlark & Johnson (1982) studied the biological oxidation of arsenite in seawater in Narragansett bay (Rhode Island, USA). They found that oxidation was primarily due to microbial activity. Oxidation obeyed first-order kinetics with a rate constant of 0.06 h^{-1} and half-lives ranging from 8.9 to 12.8 h for initial arsenite concentrations ranging from 7.5 µg/litre to 6.9 mg/litre (0.1–91.8 µmol/litre). Under aerobic conditions the mixed microbial cultures of lake sediments were able to reduce arsenate to arsenite and also to oxidize arsenite to arsenate. However, under anaerobic conditions only reduction was observed (Freeman et al., 1986).

In seawater containing free dissolved oxygen, arsenate is the thermodynamically stable form of the element. Arsenite is present in amounts exceeding those of arsenate only in reduced, oxygen-free porewaters of sediments and in anoxic basins such as the Baltic sea. However, significant amounts of arsenite (up to 10% of total arsenic) are found in the surface and deep waters of the oceans and, conversely, some arsenate is still present in anoxic water (Andreae, 1983). The presence of arsenite in seawater suggests that some reduction of arsenate occurs, and indeed Johnson (1972) demonstrated that bacterial arsenate reduction can take place under laboratory conditions. Matsuto et al. (1984) isolated a cyanobacterium (*Phormidium* sp.) from the coastal marine waters of Suruga bay (Japan) that was capable of readily reducing adsorbed arsenate to arsenite.

Freeman (1985) isolated an *Anabaena oscillaroides*–bacteria assemblage from the arsenic-rich Waikato river (New Zealand) capable of reducing arsenate to arsenite. In continuous culture the cyanophyte–bacteria assemblage could reduce arsenate to arsenite at a rate of 12 ng As/10^6 cells per day. Wakao et al. (1988) detected microbial arsenite oxidation occurring in acid mine waters (pH 2.0–2.4) containing 2–13 mg As/litre. Ahmann et al. (1994) isolated a microorganism from arsenic-contaminated sediment in eastern Massachusetts (USA) which used the reduction of arsenate to arsenite to gain energy for growth. Similarly, Macy et al. (1996) found that an anaerobic bacterium *Chrysiogenes arsenatis* from gold-mine wastewater grew by reducing arsenate to arsenite using acetate as the electron donor and carbon source.

On the basis of both aqueous and solid-phase observations, McGeehan (1996) found that arsenate was reduced to arsenite in flooded soil under batch conditions. Reduction of arsenate to arsenite has also been reported for both freshwater and marine macroalgae (Blasco, 1975; Johnson & Burke, 1978; Andreae & Klumpp, 1979; Wrench & Addison, 1981). Calculations based on the measured rates of reduction indicate that 15–20% of the total arsenic is reduced by phytoplankton during spring and autumn blooms on the continental shelf (Sanders & Windom, 1980).

4.2.2 Methylation

The biomethylation of arsenic was first recognized when arsines were produced from cultures of a fungus, *Scopulariopsis brevicaulis* (Challenger, 1945). Subsequently, the methylation of arsenic by methanogenic bacteria (McBride & Wolfe, 1971) and by reaction with methyl cobalamine (Schrauzer et al., 1972) or L-methionine-methyl-d3 (Cullen et al., 1977) has been demonstrated in laboratory work. Cox & Alexander (1973) showed that cultures of the fungus *Candida humicola* methylate arsenite, arsenate, methylarsonate and DMA to trimethylarsine. Further experiments have shown that growing cells of *C. humicola* can be induced to produce trimethylarsine from arsenate and DMA by preconditioning with DMA (Cullen et al., 1979b). Cullen et al. (1979a) incubated *C. humicola* in the presence of [74]As-arsenate, ^{14}C-methylarsonate or ^{14}C-DMA. They identified arsenite, methylarsonate, DMA and trimethylarsine oxide as intermediates in a biological synthesis of trimethylarsine. However, they tentatively conclude that methylarsonate does not occur as a free intermediate in the arsenate to trimethylarsine pathway.

McBride et al. (1978) reported that dimethylarsine was mainly produced by anaerobic organisms, whereas trimethylarsine resulted from aerobic methylation.

Methylated arsenic compounds were detected in aerobic sediments from various locations in Ontario (Canada) incubated with or without the addition of extraneous arsenic. Two pure bacterial cultures, *Aeromonas* sp. and *Flavobacterium* sp., isolated from lake water, were also found to methylate arsenic compounds in a synthetic medium (Wong et al., 1977).

Baker et al. (1983a) incubated lake sediment in the presence of arsenate or arsenite (7.5 mg As/litre). Methylation occurred over the pH range 3.5–7.5, with analysis revealing the presence of both methyl arsonic acid and dimethylarsinic acid. The amount of arsenic recovered in the methylated species ranged from 0 to 0.4% of the total inorganic arsenic added. Maeda et al. (1988) exposed the cyanobacterium *Phormidium* sp. (isolated from an arsenic-polluted environment) to arsenate (128 mg/kg) and found that 3.2% of the accumulated arsenic had been methylated.

Huysmans & Frankenberger (1991) isolated a *Penicillium* sp. from evaporation pond water capable of methylating and subsequently volatilizing organic arsenic. The conditions optimum for trimethylarsine production were a minimal medium containing 100 mg/litre methylarsonic acid, pH 5–6, a temperature of 20 °C and a phosphate concentration of 0.1–50 mmol/litre.

Reimer & Thompson (1988) found a strong positive correlation between the sum of the methylarsenic compounds and the total dissolved arsenic in marine interstitial waters influenced by mine tailings discharges indicating in situ microbial methylation. Laboratory studies have shown that microorganisms present in both natural marine sediments and sediments contaminated with mine tailings are capable of methylating arsenic under aerobic and anaerobic conditions (Reimer, 1989).

Biomethylation is primarily restricted to the high-salinity regions of estuaries with the presence of methylated arsenic at lower salinities predominantly as a result of the mixing of saline water (containing bioarsenicals) with river water (Howard & Apte, 1989).

Several authors have reported arsenic methylation in macroalgae, particularly in marine organisms (Edmonds & Francesconi, 1977; Andreae & Klumpp, 1979; Wrench & Addison, 1981; Maeda et al., 1987b; Cullen et al., 1994). In fact, most diatoms, dinoflagellates and macroalgae as well as freshwater higher plants, release protein-bound arsenic as a result of sequential methylation and adenosylation (Benson et al., 1988). Baker et al. (1983b) reported that freshwater green algae were capable of methylating sodium arsenite in lake water. Analysis revealed the presence of MMA, DMA and trimethylarsine oxide; however, volatile arsine and methylarsines were not detected. Similarly, Wrench & Addison (1981) identified MMA and DMA as polar arsenic metabolites synthesized by the marine phytoplankton *Dunaliella tertiolecta*. Maeda et al. (1987b) exposed five arsenic-resistant freshwater algae from an arsenic-polluted environment to arsenate. Small amounts of methylated arsenic compounds were detected and these were strongly bound with proteins or polysaccharides. Methylated arsenic compounds were found mainly in the lipid-soluble fractions and the major form was a dimethyl arsenic compound. No methylation occurred in algal cells (*Chlorella vulgaris*) exposed to arsenate under in vitro conditions; however, in vivo a small fraction of the arsenic accumulated was first transformed to methyl and dimethyl arsenic compounds during the early exponential phase and finally transformed to trimethylarsenic species (Maeda et al., 1992b). The marine algae *Ecklonia radiata* and *Polyphysa peniculus* methylated arsenate to produce a dimethylarsenic derivative. It was concluded that methionine or *S*-adenosylmethionine was the source of the methyl groups in this biological alkylation (Edmonds & Francesconi, 1988a; Cullen et al., 1994). *S*-adenosylmethionine is also likely to be the source of adenosyl and ribosyl groups in the arsenosugars.

The organic arsenical arsenobetaine was first identified in the late 1970s (Edmonds & Francesconi, 1981b) and has now been isolated in a variety of marine organisms (Edmonds & Francesconi, 1981b; Norin & Christakopoulos, 1982; Shiomi et al., 1984; Edmonds et al., 1992). Edmonds & Francesconi (1981a) identified arsenosugars isolated from brown kelp (*Ecklonia radiata*) as intermediates in the cycling of arsenic and stated that these compounds could be subsequently metabolized to arsenobetaine. Edmonds et al. (1982) have shown that the simpler arsenosugars in the brown alga are degraded under anaerobic conditions to dimethyloxarsylethanol. The transformation of dimethyloxarsyl-ethanol to arsenobetaine would require both a reduction-methylation step and an oxidation step; these are probably bacterially mediated (Edmonds & Francesconi, 1987a, 1988b). Edmonds & Francesconi (1988b) concluded that arsenobetaine is probably formed by the conversion of arsenate to dimethyl(ribosyl)arsine oxides by algae, and that the microbially mediated transformation to arsenobetaine or its immediate precursors occurs in sediments. Phillips & Depledge (1985, 1986) proposed that phospholipids containing arsenoethanol-amine or arsenocholine moieties may be formed as intermediates in the formation of arsenosugars and arsenobetaine. Edmonds et al. (1992) identified arsenocholine-containing lipids as natural products in the digestive gland of the rock lobster (*Panulirus cygnus*). Phillips & Depledge (1985) concluded that arsenic replaces nitrogen in phospholipid synthesis leading to a large number of arsenic-containing intermediates, which would be either water-soluble or lipid-soluble. Arsenic-containing compounds are catabolized as they pass through the food web, yielding arsenobetaine as a stable end-product.

Inorganic arsenic administered orally to brown trout (*Salmo trutta*) was detected in tissues as organoarsenical species, whereas arsenic administered by injection was taken up as inorganic arsenic and slowly converted to the organic form. It was concluded that biosynthesis of arsenic was occurring in the gastrointestinal tract (Penrose, 1975). Oladimeji et al. (1979) reported that arsenic given as an oral dose to rainbow trout (*Oncorhynchus mykiss*) was rapidly converted to organic forms. The ratio of total organic to inorganic increased with time in all tissues, with the organic arsenic fraction accounting for about 50% after 6 h and over 80% within 24 h. The major organic arsenical appeared to be an arsenobetaine-related compound. Similarly, Penrose et al. (1977) found that sea urchins (*Strongylocentrotus droebachiensis*) were also able to convert inorganic arsenic to an organic form, but to a more limited degree than trout. However, Wrench et al. (1981) concluded that organic arsenic synthesized in the brine shrimp (*Artemia salina*) is methylated by intestinal microflora and not by the filter feeder itself.

Maeda et al. (1990c) found that 85% of arsenic accumulated by the guppy (*Poecilia* sp.) was in the di- and tri-methylated forms. The percentage of organic species was much higher than that found in phytoplankton and zooplankton in the same model ecosystem. Similarly, Maeda et al. (1990a) found that biomethylation of arsenic increased successively with trophic level in another model ecosystem: goldfish (*Carassius* sp.) > zooplankton (*Moina* sp.) > alga (*Chlorella* sp.).

4.2.3 Degradation

4.2.3.1 Abiotic degradation

The rates of photochemical decomposition of arsenite, DMA, MMA and arsenobetaine have been studied in both distilled water and seawater. All species were found to degrade rapidly in aerated distilled water. In deaerated solutions the rate of oxidation of arsenite was almost two orders of magnitude slower. Half-lives for the degradation of DMA, MMA and arsenite were 9.2, 11.5 and 0.9 min respectively for aerated distilled water and 25, 19 and 8 min for deaerated distilled water. In seawater, the rates of photochemical decomposition were slower. For example, in seawater only 20% of DMA was converted to MMA after 300 min with no other products detected, whereas in distilled water DMA was completely degraded within 100 min (Brockbank et al., 1988). This study suggests that UV irradiation is of limited use for the pretreatment of saline samples to convert organoarsenic species to As(V) before analysis. The implications for photochemical decomposition of arsenic species in natural waters is not clear, because sunlight is deficient in the lower-wavelength bands generated by the mercury lamp used in this study. In addition, colloids and suspended particulates in the photic zone may play a significant role in arsenic decomposition in natural waters.

Von Endt et al. (1968) concluded that degradation of MSMA in soil was primarily due to soil microorganisms rather than abiotic factors. In 60-day tests in non-sterile soil 1.7–10% of the ^{14}C-MSMA was degraded, whereas under steam-sterilized conditions only 0.7% was degraded.

4.2.3.2   Biodegradation

The predominant form of arsenic in water is usually arsenate (Callahan et al., 1979; Wakao et al., 1988), but aquatic micro-organisms may reduce the arsenate to arsenite and a variety of methylated arsenicals.

Marine organisms tend to contain much higher levels of arsenic than terrestrial organisms; this is because of the high arsenate/phosphate ratio in oceans, which is a consequence of the very low phosphate concentration. Most of the arsenic accumulated in marine organisms is in a water-soluble form of arsenic, namely arsenobetaine. Hanaoka et al. (1987) incubated marine sediments in the presence of arsenobetaine and demonstrated microbial degradation, with arsenate, arsenite, MMA, DMA and arsenobetaine being identified. Further experiments revealed the formation of trimethylarsine oxide during aerobic incubation of bottom sediments with arsenobetaine as the carbon source (Kaise et al., 1987). Under aerobic conditions, arsenobetaine is converted to its metabolites to a much greater extent than other methylarsenicals. Under anaerobic conditions little or no degradation of arsenobetaine occurred, whereas trimethylarsine oxide and DMA were converted to less methylated compounds (Hanaoka et al., 1990). Degradation of arsenobetaine has also been demonstrated in the water column in the presence of suspended substances (Hanaoka et al., 1992).

Organoarsenical pesticides (e.g. MMA and DMA) applied to soil are metabolized by soil bacteria to alkylarsines, MMA, and arsenate (ATSDR, 1993). The half-time of DMA in soil is about 20 days (ATSDR, 1993).

Cheng & Focht (1979) added arsenate, arsenite, methylarsonate and DMA to three different soil types. Arsine was produced in all three soils from all substrates but methylarsine and dimethylarsine were only produced from methylarsonate and DMA respectively. Both *Pseudomonas* sp. and *Alicaligenes* sp. produced arsine as the sole product when incubated anaerobically in the presence of arsenate or arsenite. The authors concluded that reduction to arsine, not methylation to trimethylarsine, was the primary mechanism for gaseous loss of arsenicals from soil.

Degradation of MSMA by soil microorganisms was studied by Von Endt et al. (1968). In 60-day tests they found that 1.7–10% of the ^{14}C-MSMA was degraded; four soil microorganisms isolated in pure cultures degraded 3–20% of ^{14}C-MSMA to ^{14}CO$_2$ when grown in liquid culture at 10 mg MSMA/litre. Woolson & Kearney (1973) showed that sodium DMA was degraded to arsenate in soil under aerobic conditions but not under anaerobic conditions. Degradation of MSMA has been shown to be associated with soil organic matter oxidation. In a loamy soil, degradation increased with increasing organic matter content (Dickens & Hiltbold, 1967). Akkari et al. (1986) studied the degradation of MSMA in soils at concentrations up to 5 mg As/kg. It was found that degradation followed first-order kinetics. The rate constant was temperature dependent only at soil water contents less than field capacity, and the temperature effect was less under flooded conditions. The differences in degradation rate under aerobic conditions and 20% water content were related to differences in the texture of the three soils. Half-lives for the clay and silty loam soils were 144 and 88 days respectively. Under anaerobic (flooded) soil conditions MSMA degradation occurs by reductive methylation to form arsenite and alkylarsine gases. The half-life values for the two soils indicate significantly faster degradation at 25 and 41 days respectively. The third soil, a sandy loam, produced the slowest degradation rate ($t_{1/2}$ = 178 days) probably because of its low organic matter content which may have supported fewer microorganisms.

The overall percentage of DMA (sodium salt) and MMA mineralized in a silty clay soil after 70 days ranged from 3% to 87% – values much higher than arsenic loss as volatile arsines (0.001–0.4%). Arsenate was the main metabolite from the degradation of both sodium DMA and MMA. The amount of sodium DMA mineralized was linearly related to the concentration of sodium DMA in the soil, indicating that the rate is first order. Mineralization of sodium DMA increased with increasing soil moisture and temperature. It was concluded that the loss of arsenic from some soils to the atmosphere may not be a major pathway and that inorganic arsenic may accumulate in soil from arsenical usage (Gao & Burau, 1997).

4.2.4 Bioaccumulation

Bioconcentration of arsenic under laboratory conditions occurs in aquatic organisms, primarily in algae and lower invertebrates. Bioconcentration factors (BCFs) measured in freshwater invertebrates for several arsenic compounds generally ranged up to 20; bioconcentration factors in fish were < 5; higher concentration factors have been observed in algae. Biomagnification in aquatic food chains does not appear to be significant (Callahan et al., 1979). Terrestrial plants may accumulate arsenic by root uptake from the soil or by adsorption of airborne arsenic deposited on the leaves, some species accumulating substantial levels.

4.2.4.1 Microorganisms

Maeda et al. (1987a) exposed cyanobacteria (*Nostoc* sp.) to arsenate concentrations of 1 and 10 mg As(V)/litre for 32 days with no effect on cell growth. *Nostoc* sp. accumulated 32 and 77 mg As/kg (dry cell) respectively at the two exposure concentrations.

Lindsay & Sanders (1990) report BCFs ranging from 1132 to 3688 for estuarine phytoplankton (*Thalassiosira pseudomonas*, *Skeletonema costatum* and *Dunaliella tertiolecta*) exposed to 25 µg As(V)/litre as arsenate for up to 48 h.

Phytoplankton take up arsenate readily and incorporate a small proportion into the cell. Most of the arsenate is reduced, methylated and released to the surrounding media. Phytoplankton batch cultures exposed to elevated levels of arsenate take up additional arsenic during the log phase of growth. Studies using ^{74}As indicate that the uptake rate varies from 0.15 ng As(V)/10^6 cells per hour in unenriched cultures to 2.3 ng As(V)/10^6 cells per hour in cultures containing 25 µg As(V)/litre. Cultured *Skeletonema costatum* increase their arsenic concentrations approximately 40% from 22 to 29 mg/kg (dry weight) in response to arsenate concentrations of 6–25 µg As(V)/litre (Sanders & Windom, 1980).

Phytoplankton readily incorporated dissolved arsenic, with average arsenic residues increasing from 5.7 to 17.7 mg/kg (dry weight) when cultured for 48–96 h at 25 µg As(V)/litre as arsenate (Sanders et al., 1989). Arsenate added to a freshwater model ecosystem was readily accumulated in plankton with arsenic residues of 37–47 mg/kg (dry weight) at 5 µg As(V)/litre and > 200 mg/kg at 50 µg As(V)/litre after 65-day exposures. Accumulation in other biota was much lower than for phytoplankton (Reuther, 1992).

Giddings & Eddlemon (1977) studied the uptake of radioactively labelled arsenic (added as sodium arsenate at 50 µg As(V)/litre) in model ecosystems (7 and 70 litres) for 5 weeks. Mean BCFs for algae ranged from 370 for sand microcosms to 4300 for lake mud microcosms. Algal arsenic concentrations were significantly greater in the 70-litre microcosms and in the sand microcosms than in the 7-litre and sediment microcosms.

Green algae (*Chlorella vulgaris*) exposed to arsenate concentrations of 7 to 9 mg As(V)/litre accumulated maximum residues of 3.75 g total As/kg (dry mass) within 10 days (Maeda et al., 1992c).

Maeda et al. (1985) found that arsenate uptake increased with an increase in the arsenic exposure concentration with *C. vulgaris* isolated from an arsenic-polluted environment. Maximum BCFs of 200–300 were observed during the log phase. At the highest exposure concentration (10 g As(V)/litre) algae were able to accumulate 50 g As/kg (dry weight). Approximately half of the arsenic taken up was estimated to be adherent to the extraneous coat of the cell with the remainder accumulated by the cell. Arsenate accumulation was affected by the growth phase; arsenic was most actively accumulated when the cell was exposed to arsenic during the early exponential phase (Maeda et al., 1992a).

Accumulation of arsenic (1 mg As(V)/litre as arsenate) by *Dunaliella* sp. was rapid, with equilibrium established within 8 h. Arsenic accumulation was studied at temperatures ranging from 10 °C to 33 °C, pH 4–10, light intensity from 0 to 10 000 lux and sodium chloride concentrations from 1 to 100 g/litre. Maximum arsenic residues under optimum conditions (22 °C; pH 8.2; 5000–10 000 lux and 20 g NaCl/litre) were 2000 mg As/kg. Increased phosphate significantly decreased the uptake of arsenic in the culture (Yamaoka et al., 1988). Yamaoka et al. (1992) found that *D. salina* accumulated more arsenic at nitrogen concentrations of 72 mg/litre than at 4.5 mg/litre.

4.2.4.2   Macroalgae

Fucus vesiculosus accumulated approximately 120 mg As/kg during an 85-day exposure to 7.5 µg As(V)/litre as arsenate. Filamentous algae and planaria accumulated less than 40 mg As/kg (dry weight), and cyanobacteria and various zooplankton accumulated less than 20 mg As/kg (Rosemarin et al., 1985).

Klumpp (1980) studied the effect of a variety of factors on the uptake of labelled arsenic by the seaweed *Fucus spiralis*. Neither pH (pH 7–9) nor salinity (9–36 g/litre) affected the uptake of arsenic; however, uptake at 30 °C was twice that at 16 °C. Arsenate uptake was reduced with increasing phosphate concentration (40–400 µmol/litre).

Lee et al. (1991) grew the aquatic plant *Hydrilla verticillata* in both mine-waste pool water and deionized distilled water contaminated with arsenate (0.4 and 0.8 mg As(V)/litre) for up to 16 days. Accumulation of arsenic reached steady state at 2–6 days in pool water at BCFs of 110–190. In deionized water maximum arsenic accumulation occurred after 8 days at a BCF of around 300. Phosphate ($\geq$ 12 mg/litre) inhibited the uptake of arsenic by *H. verticillata*.

4.2.4.3 Aquatic invertebrates

Sanders et al. (1989) studied the uptake of arsenic from water and from phytoplankton by the copepod *Eurytemora affinis* and the barnacle *Balanus improvisus*. In 24-h tests, *E. affinis* exhibited no uptake of dissolved arsenic; the arsenic content of copepods fed phytoplankton increased to 11.2 mg/kg (dry weight) compared with 8.9 mg/kg in controls. In 22-day tests, *B. improvisus* exposed to dissolved arsenate (55 µg As(V)/litre) in water did not accumulate arsenic, with levels remaining around 0.88 mg/kg; however, levels in shell material increased from 0.3 mg/kg to 2 mg/kg. Barnacles fed arsenic-contaminated phytoplankton (~18 mg/kg) exhibited an increase in total arsenic concentrations from 0.3 mg/kg to 1.7 mg/kg. In further experiments with oysters (*Crassostrea virginica*) no accumulation of arsenic from water was observed in 28-day tests, but tissue concentrations increased significantly from 5.3 mg/kg to 8.2 mg/kg in oysters fed arsenic-contaminated phytoplankton. Zaroogian & Hoffman (1982) reported maximum total arsenic residues in soft tissues of oysters (*Crassostrea virginica*) of 12.6, 12.7 and 14.1 mg/kg (dry weight) at arsenite exposure concentrations of 1.2 (control), 3 and 5 mg As(III)/litre during 112-day exposures. Generally, arsenic body burdens increased with increases in phytoplankton concentration and it appears that food contributes more to arsenic uptake than do seawater arsenic concentrations. No relationship between arsenic uptake and seawater arsenic concentrations was found.

Ünlü & Fowler (1979) exposed mussels (*Mytilus galloprovincialis*) to arsenate ([74]As) concentrations ranging from 20 to 100 µg As(V)/litre at 12 °C and 21 °C. Mean concentration factors after 20 days were low, at respectively 8.8 and 12.1 for the two temperatures; however, mussels did accumulate significantly more arsenic at 21 °C than at 12 °C. Arsenic uptake was inversely related to salinity over the range 31–19 g/litre. Arsenic loss was essentially biphasic, with biological half-times of approximately 3 and 32 days for the fast and slow compartments respectively. The active secretion of arsenic in the byssal threads contributed to the total elimination of the element from the mussels. Similarly, Ünlü (1979) found a biphasal loss of arsenic from crabs (*Carcinus maenas*) during a 43-day depuration period. The elimination of [74]As by the crabs after ingestion of arsenic-contaminated mussels was dependent on the chemical form of the arsenic. After ingestion of mussel containing mostly lipid- and water-soluble arsenic species (undetermined), biological half-times were 3.4 and 19.6 days for the first and second phase of loss. After ingestion of mussel containing mostly arsenite and residual arsenic, half-times were 1.6 and 9.3 days respectively.

Naqvi et al. (1990) exposed red crayfish (*Procambarus clarkii*) to MSMA at concentrations of 0.5, 5 and 50 mg As/litre. Uptake of arsenic was dose-dependent but not time-dependent. Maximum whole-body residues were 1.36, 4.29 and 9 mg As/kg respectively for each of the exposure concentrations during the 8-week uptake period.

Gibbs et al. (1983) reported equilibrium BCFs, based on [74]As, for the cirratulid polychaete *Tharyx marioni* ranging from 4.5 at an exposure concentration of 10 mg As(V)/litre (as arsenate) to 111.6 at 0.01 mg/litre after 7 days. A lower BCF of only 15.9 at 0.01 mg/litre was reported for the polychaete *Caulleriella caputesocis*.

Shrimps exposed to water concentrations ranging from 0.1 to 1.5 mg As(V)/litre (as arsenate) or food (*Chlorella* sp.) containing, 1940 mg total As/kg contained arsenic residues ranging from 18.9 to 31.8 mg/kg (dry weight) (Maeda et al., 1992c).

Fowler & Ünlü (1978) reported BCFs of less than 10 for shrimps exposed to arsenate ([74]As) concentrations of 20–100 μg As(V)/litre for 14 days. Arsenic loss was biphasic with half-lives of 3 and 26 days for the fast and slow compartments respectively. Moults shed during loss contained 2–5% of the shrimp's [74]As body burden.

Lindsay & Sanders (1990) found no bioaccumulation of arsenate directly from the water (25 μg As(V)/litre) or from food for the grass shrimp (*Palaemonetes pugio*). Brine shimp (*Artemia* sp.) grown in elevated arsenic concentrations exhibited small, but significant, increases in arsenic content from an average of 16.8 mg/kg (dry weight) in controls to 17.8 mg/kg at 25 µg As(V)/litre; no accumulation was observed when brine shrimps were fed arsenic-contaminated food.

4.2.4.4   Fish

Barrows et al. (1980) exposed bluegill sunfish (*Lepomis macrochirus*) to 130 µg As(III)/litre of As₂O₃ for 28 days. The maximum BCF was found to be 4, with a half-life in tissues of 1 day. Nichols et al. (1984) found no accumulation of arsenic in a 6-month study on coho salmon (*Oncorhynchus kisutch*) exposed to As₂O₃ concentrations of < 300 µg As(III)/litre. Whole-body residues were below 0.4 mg As/kg (wet weight) and were not dose dependent.

Sorensen (1976) found that green sunfish (*Lepomis cyanellus*) exposed to higher arsenic concentrations of 100, 500 and 1000 mg As(V)/litre (as arsenate) accumulated whole-body arsenic concentrations of 33.4, 541.2 and 581.6 mg/kg (BCFs ranging from 0.3 to 1.1). Green sunfish exposed to 60 mg As(V)/litre for 6 days accumulated mean arsenic residues of 158.7, 47.7, 18.9 and 14.2 mg/kg in the gallbladder (plus bile), liver, spleen and kidney respectively (BCFs ranging from 0.2 to 2.6) (Sorensen et al., 1979).

Cockell & Hilton (1988) fed rainbow trout (*O. mykiss*) on diets containing As₂O₃ (180–1477 mg As/kg diet), disodium arsenate heptahydrate (DSA) (137–1053 mg As/kg diet), DMA (163–1497 mg As/kg diet) or arsanilic acid (193–1503 mg As/kg) for 8 weeks. For each of the arsenicals investigated, carcass arsenic concentration showed a dose–response relationship to dietary arsenic concentration and exposure rate. At lower levels of exposure (137 mg As/kg diet), dietary DSA yielded the highest mean carcass arsenic concentrations (6.9 mg As/kg), but at higher levels, dietary As₂O₃ (1477 mg As/kg diet) yielded the highest mean residues (21.6 mg As/kg). Inorganic arsenicals were accumulated from the diet to a greater degree than the organic forms. In a 16-week study, dietary DSA (8–174 mg As/kg diet) accumulated in the carcass (0.25–5.7 mg As/kg), liver (0.7–34.4 mg As/kg) and kidney (1.1–31.9 mg As/kg) in a dose-related manner (Cockell et al., 1991).

Oral administration of sodium arsenate to estuary catfish (*Cnidoglanis macrocephalus*) and school whiting (*Sillago bassensis*) resulted in an accumulation of trimethylarsine oxide in their tissues (Edmonds & Francesconi, 1987b). Yelloweye mullet (*Aldrichetta forsteri*) fed the organic arsenicals 2-dimethylarsinylethanol, 2-dimethylarsinylacetic acid or 2-dimethyllarsinothioylethanol showed no arsenic accumulation in their tissues; fish fed arsenate-contaminated food showed a small but significant increase in arsenic concentration (muscle tissue = 1 mg As/kg wet weight). However, administering arsenobetaine or arsenocholine in the diet led to muscle concentrations of around 24 mg As/kg (wet weight) (Francesconi et al., 1989).

Oladimeji et al. (1984) fed rainbow trout (*O. mykiss*) on a diet containing 10, 20 or 30 mg As(III)/kg (as sodium arsenite) (equivalent to 0.2, 0.4 and 0.6 mg/kg fish wet weight per day) for up to 8 weeks. Arsenic accumulation was dose related, with residues ranging from 1.28 to 1.52 mg/kg (dry weight) for muscle, 1.55 to 5.21 mg/kg for liver, 0.84 to 1.88 mg/kg for gills and 1.21 to 1.98 mg/kg for skin tissue.

4.2.4.5 Terrestrial plants

Arsenic species can enter into edible tissues of food crops through absorption (i.e. not just surface contamination) (Woolson, 1973; Helgesen & Larsen, 1998). Helgesen & Larsen (1998) demonstrated that bioavailability of arsenic pentoxide to carrots in soil from a wood preservative treatment plant (soil was contaminated with CCA) was 0.47 ± 0.06% of total soil arsenic burden. This study showed that arsenite, arsenate, MMA and DMA were present in carrot tissue, where only arsenite and arsenate were present in soil. In soils dosed with arsenate (0–500 µg/g) at the concentrations which inhibited growth of vegetable crops (green bean, lima bean, spinach, cabbage, tomato and radish), high levels of accumulation when found in the edible parts of radish (76 µg/g) spinach (10 µg/g) and green bean (4.2 µg/g). Arsenic accumulation in Lima bean, cabbage and tomato ranged from 0.7–1.5 µg/g. The studies of Woolson (1973) and Helgesen & Larsen (1998) highlight the potential of movement of arsenic species from soil into agronomic crops.

Uptake of arsenate (10 mg/litre [133 µmol/litre]) by moss (*Hylocomium splendens*) from nutrient solution displayed saturation kinetics at pH 5 that could be described in terms of Michaelis–

Menten parameters with a mean K_m value of 31.4 mg/litre (418 µmol/litre). Phosphate was a competitive inhibitor of arsenate uptake with an inhibition constant (K_m phosphate) of 82 µmol/litre (Wells & Richardson, 1985).

Asher & Reay (1979) studied the uptake of arsenate (1 mg/litre [15 µmol/litre]) from nutrient solution by barley (*Hordeum vulgare*) seedlings. They found that uptake consisted of a rapid initial phase followed by a less rapid 'steady-state' phase, both of which were strongly inhibited by phosphate and positively correlated with temperature.

The marsh plant species *Spartina alterniflora* was grown in sediment treated with ~50 µg As(V)/litre (as arsenate) and accumulated significantly elevated total concentrations of arsenic after 9 days; new and old leaf blades contained mean arsenic concentrations of 6.3 and 5 mg/kg (dry weight) respectively, relative to 1 and 0.4 mg/kg in control plants (Sanders & Osman, 1985).

Meharg & Macnair (1991b) found that non-tolerant genotypes of *Holcus lanatus* accumulated significantly more arsenate than tolerant plants during a 6-h period of growth in 3.75 mg As(V)/litre (0.05 mol/m^3) arsenate. They found that tolerant plants transported a much greater proportion of arsenic to their shoots than non-tolerant plants. Phosphate (0.05 or 0.5 mol/m^3) decreased arsenate uptake in both tolerant and non-tolerant genotypes. Arsenate tolerance involves reduced accumulation of arsenate through suppression of the high-affinity phosphate–arsenate uptake system (Meharg et al., 1994).

Anastasia & Kender (1973) grew lowbush blueberry (*Vaccinium angustifolium*) plants in greenhouse soil at As_2O_3 concentrations ranging from 7.7 (controls) to 84.5 mg As(III)/kg for 17 weeks. Arsenic was accumulated in a dose-dependent manner with arsenic residues of 0.78–15 mg/kg for leaves, 0.27–13.3 mg/kg for stems and 2.4–164.2 mg/kg for roots.

Otte et al. (1990) grew *Urtica dioica* and *Phragmites australis* in soil containing up to 30 mg As/kg added as lead arsenate or sodium DMA. Concentrations of arsenic in shoots and roots of *P. australis* increased significantly only at the highest arsenic concentration in soil with mean values of up to 1 mg/kg (dry weight) in shoots and 44.3 mg/kg in roots, whereas the arsenic content of *U. dioica* increased by a factor of 4 at 5 mg As/kg with plants accumulating mean arsenic concentrations of up to 150 mg/kg in roots at the highest exposure.

Onken & Hossner (1995) grew rice (*Oryza sativa*) in two soil types treated with up to 45 mg As(III) or As(V)/kg (as arsenite or arsenate) for 60 days. The arsenic concentration of rice plants correlated with the mean soil solution arsenate concentration in the clay soil and to the mean soil solution arsenite for the silt loam. The rate of arsenic uptake by plants increased as the rate of plant growth increased.

4.2.4.6   Terrestrial invertebrates

Meharg et al. (1998) exposed earthworms (*Lumbricus terrestris*) to arsenate (40 mg/kg dry weight) for 23 days. There was a steady-state increase in residues for depurated and undepurated worms and by 12 days earthworm residues were equivalent to those of the soil. Arsenic residues were accumulated to three times soil levels by the end of the 23-day exposure in depurated worms; however, undepurated worms did not appear to bioconcentrate arsenic beyond the level of the surrounding soil.

4.2.4.7   Birds

Daghir & Hariri (1977) administered arsanilic acid (used as a feed medication for poultry) to White Leghorn laying hens at 50 and 100 mg/kg for 15 weeks. Maximum concentrations in eggs were reached after 4–5 weeks at 0.13 and 0.24 mg As/kg (dry weight) for the two dose levels respectively. Residual arsenic was negligible 2 weeks after the withdrawal of the drug from the feed.

Proudfoot et al. (1991) found a higher concentration of arsenic in liver and muscle of broilers that were fed arsanilic acid (99 mg/kg diet) compared with controls. Mean arsenic residues of up to 1.5 mg/kg and 0.4 mg/kg were measured for the two tissues respectively. Broilers fed a diet containing 100 or 500 mg/kg arsanilic acid accumulated up to 2.3 and 8 mg As/kg in liver tissue at the two exposure concentrations respectively. Lower levels were accumulated in muscle tissue, with arsenic concentrations of up to 0.15 and 0.67 mg/kg (VanderKop & MacNeil, 1989).

Holcman & Stibilj (1997) fed Rhode Island Red hens on diets containing 7.5, 15 or 30 mg As(III)/kg (as As_2O_3) for 19 days. Eggs were collected on days 8–19 day of the experiment, and arsenic residues were consistent throughout this period. Mean concentrations were respectively 0.2, 0.42 and 0.96 mg As/kg (dry weight) in egg yolk and in 0.06, 0.14 and 0.3 mg As/kg egg white for the three exposure concentrations.

Hoffman et al. (1992) fed mallard on a diet containing 200 mg As(V)/kg (as sodium arsenate) for 4 weeks. Arsenic accumulated in the liver at a concentration of 2.3 mg As/kg (wet weight). Birds maintained on a restricted protein and exposed to the same arsenic-contaminated diet accumulated 5.1 mg As/kg. Stanley et al. (1994) maintained mallards on diets containing 25, 100 or 400 mg As(V)/kg (as sodium arsenate) for 16–18 weeks. Arsenic was accumulated in a dose dependent manner; mean concentrations in adult livers were 0.49–6.6 mg As/kg (dry weight), in duckling livers from 0.65–33 mg/kg and in whole eggs from 0.46–3.6 mg/kg.

5. ENVIRONMENTAL LEVELS AND HUMAN EXPOSURE

5.1 Environmental levels

Arsenic is a natural component of the earth's crust, and found in all environmental media. Concentrations in air in remote locations range from < 1 to 3 ng/m^3, but concentrations in cities may range up to 100 ng/m^3. Concentrations in water are usually < 10 µg/litre, although higher concentrations can occur near natural mineral deposits or anthropogenic sources. Natural levels in soils usually range from 1 to 40 mg/kg, but pesticide application or waste disposal can produce much higher values.

5.1.1 Air

Levels of arsenic in ambient air are summarized in Table 3. Examples are given of mean total arsenic concentrations in remote and rural areas ranging from 0.02 to 4 ng/m^3. Levels of arsenic in outdoor air near to urban and industrial sources are summarized in Table 4. Examples are given of mean total arsenic concentrations in urban areas ranging from 3 to 200 ng/m^3; much higher concentrations (> 1000 ng/m^3) have been measured in the vicinity of industrial sources. Arsenic in ambient air is usually a mixture of arsenite and arsenate, with organic species being of negligible importance except in areas of substantial methylated arsenic pesticide application or biotic activity. Schroeder et al. (1987) reviewed worldwide arsenic concentrations associated with particulate matter. They identified arsenic levels ranging from 0.007 to 1.9 ng/m^3 for remote areas, 1 to 28 ng/m^3 for rural areas and 2 to 2320 ng/m^3 in urban areas. The highest arsenic levels detected in the atmosphere were near non-ferrous-metal smelters.

Typical background levels for arsenic are now 0.2–1.5 ng/m^3 for rural areas, 0.5–3 ng/m^3 for urban areas and < 50 ng/m^3 for industrial sites (DG Environment, 2000).

Table 3. Concentrations of As in ambient air[a]

Location	Sampling period	Particle size and/or species	Concentration (ng/m^3)[b]	Reference
Antarctica	NS		0.019	Brimblecombe (1979)
Antarctic Ocean	1988–1989	As_i	0.05 (0.01–0.2)	Nakamura et al. (1990)
	1988–1989	As_o	0.002 (single sample)	Nakamura et al. (1990)
North Pacific Ocean	1981–1987	As_i	0.1 (0.01–0.95)	Nakamura et al. (1990)
	1981–1987	As_o	0.008 (0.001–0.03)	Nakamura et al. (1990)
North Atlantic Ocean	1989	As_i	0.1 (0.01–0.45)	Nakamura et al. (1990)
	1989	As_o	0.007 (0.001–0.3)	Nakamura et al. (1990)
Baltic Sea	1985		1.1 (0.3–3.7)	Häsänen et al. (1990)
Mid-Atlantic coast, USA	1985–1986		1.05	Scudlark & Church (1988)
Continental shelf waters, south-eastern USA	1975–1976	particulate	1.7 (0.2–4.3)	Waslenchuk (1978)
Northern Chesapeake Bay, USA	1990–1991	< 10 µm	0.66 (0.11–1.96)	Wu et al. (1994)
Rural US sites (National Parks)	1979–1981	0.45 µm	< 1.6–2.3 (range of means)	Davidson et al. (1985)

Table 3 (contd.)

Midwestern USA	1990		1.6 (0.7–2.5)	Burkhard et al. (1994)
Natural geysers, northern California, USA	1989	As(III) As(V)	0.22 & 0.54 (0.06–3.08) 0.46 (0.08–1.3) & 2.29 (0.7–6.54)	Solomon et al. (1993) Solomon et al. (1993)
Bagauda, Nigeria	1976		0.6	Beavington & Cawse (1978)
Pelindaba, South Africa	1976		1.7	Beavington & Cawse (1978)
Chilton, United Kingdom	1976		4.2	Beavington & Cawse (1978)
Rural sites, United Kingdom	1972–1973		1.5–2.5 (range of means)	Peirson et al. (1974)
Rural area near Thessaloniki, Greece	1989–1990		2.7	Misaelides et al. (1993)
Birkenes, Norway	1978–1979 1985–1986	particulate particulate	1.2 (0.02–12) 0.63 (< 0.04–4.6)	Amundsen et al. (1992) Amundsen et al. (1992)

[a] As = inorganic As; As = organic As; NS = not stated
[b] Mean and ranges of total As unless stated otherwise

69

Table 4. Concentrations of As in outdoor air near urban and industrial sources

Location	Distance from source (km)	Sampling period	Particle size and/or species	Concentration (ng/m³)[a]	Reference
Industrial sites, UK	NS	1972–1973	NS	1.2–24 (ng/kg, range of means)	Peirson et al. (1974)
Urban area, Thessaloniki, Greece	NS	1989–1990	0.45 μm	4.1	Misaelides et al. (1993)
Urban area, Yokohama, Japan	NS	1988	0.45 μm; inorganic As	2.5 (1–5.1)	Nakamura et al. (1990)
			0.45 μm; organic As	0.01 (0.001–0.64)	Nakamura et al. (1990)
Los Angeles, USA	NS	1987	< 2.5 μm; As(III)	7.4 (< 1.2–44)	Rabano et al. (1989)
	NS	1987	> 2.5 μm; As(III)	1.8 (< 0.9–4.8)	Rabano et al. (1989)
	NS	1987	< 2.5 μm; As(V)	5.2 (< 0.9–18.7)	Rabano et al. (1989)
	NS	1987	> 2.5 μm; As(V)	2.2 (< 0.8–6.6)	Rabano et al. (1989)
Wuhan City, China	NS	1988	< 2.5 μm	25	Waldman et al. (1991)
	NS	1988	≥ 2.5 μm < 10 μm	13	Waldman et al. (1991)
Calcutta, India	NS	NS	0.45 μm	180 (91–512)	Chakraborti et al. (1992)

Table 4 (contd.)

			particulate		
Kola peninsula, Russia near (Cu–Ni smelter)	NS	1993	particulate	28	Kelley et al. (1995)
Caletones, Chile (near Cu smelter)	< 10	1987–1990	0.4 µm	1483	Romo-Kröger & Llona (1993)
	< 20	1987–1990	0.4 µm	131	Romo-Kröger & Llona (1993)
	< 30	1987–1990	0.4 µm	14	Romo-Kröger & Llona (1993)
	< 10	1987–1990	0.8 µm	29	Romo-Kröger & Llona (1993)
	< 20	1987–1990	0.8 µm	5	Romo-Kröger & Llona (1993)
	< 30	1987–1990	0.8 µm	3.5	Romo-Kröger & Llona (1993)
	13	1991	< 2.5 µm	241[b]	Romo-Kröger et al. (1994)
	13	1991	2.5–10 µm	26	Romo-Kröger et al. (1994)

[a] Mean and ranges of total As unless stated otherwise
[b] Fine particle concentration was 23 ng/m^3 during a strike period at the smelter
NS = not stated

71

5.1.2 Precipitation

Arsenic has been detected in rainwater at mean concentrations of 0.2–0.5 µg/litre (Welch et al., 1988). Peirson et al. (1974) report mean arsenic concentrations in rainfall ranging from < 6 µg/litre for a rural site to 45 µg/litre at a North Sea gas platform. Arsenic concentrations in precipitation at the mid-Atlantic coast of the USA ranged from < 0.005 to 1.1 µg/litre with an average of 0.1 µg/litre (Scudlark & Church, 1988). Andreae (1980) collected rainwater samples from non-urban sites in California (USA) and state parks in Hawaii and found mean arsenic concentrations ranging from 0.013 to 0.032 µg/litre. Samples from a rural site in Washington state (USA) contained a mean concentration of 1.1 µg As/litre; the author states that the site is 154 km north of a large copper smelter. Vermette et al. (1995) monitored arsenic levels in wet deposition at three sites (Colorado, Illinois and Tennessee, USA) and found mean concentrations ranging from 0.09 to 0.16 µg/litre. Reimann et al. (1997) monitored rainwater samples during the summer of, 1994 in eight Arctic catchments. Median arsenic concentrations (0.45 µm) ranged from 0.07 µg/litre at the most remote site to 12.3 µg/litre near a smelter.

Barbaris & Betterton (1996) analysed fresh snowpack samples from high-elevation forests of north-central Arizona (USA) during late winter and early spring 1992–1994. Arsenic concentrations ranged from 0.02 to 0.4 µg/litre with a mean value of 0.14 µg/litre.

5.1.3 Surface water

Levels of arsenic in seawater are summarized in Table 5. Concentrations of arsenic in open ocean seawater are typically 1–2 µg/litre. The dissolved forms of arsenic in seawater include arsenate, arsenite, MMA and DMA, with adsorption on to particulate matter being the physical process most likely to limit dissolved arsenic concentrations (Maher & Butler, 1988). Levels of arsenic in estuarine water are summarized in Table 6. Tremblay & Gobeil (1990) noted that arsenic concentrations increased with increasing salinity (0–31 g/litre) from 0.5 to 1.4 µg/litre (6.6 to 18.9 nmol/litre) in the St Lawrence estuary (Canada) and from 0.1 to 1.4 µg/litre (1.1 to 18.7 nmol/litre) in its tributary Saguenay fjord. Penrose et al. (1975) monitored seawater in Moreton's Harbour, Newfoundland

Table 5. Background concentrations of As in seawater

Location	Sampling period	Sampling details and/or species	Concentration (μg/litre)[a]	Reference
Gulf of Mexico	NS	0.2 μm filtered	0.04	Chakraborti et al. (1986)
Pacific Ocean	NS	no arsenite detected	1.8 (1.6–2.1) 1.2–1.5	Bodewig et al. (1982) Andreae (1978, 1979)
Coastal waters, South Australia	NS	dissolved; particulate As below limit of detection (0.6 ng/litre)	1.3 (1.1–1.6)	Maher (1985a)
Continental shelf waters, south-eastern USA	1975–1976	depth 30 m and 500 m	1.1 and 1.5	Waslenchuk (1978)
Coastal waters, south-east Spain	NS	below surface	1.5 (0.45–3.7)	Navarro et al. (1993)
Baltic Sea	1982–1983	0.45 μm filtered	0.76 (0.45–1.1)	Stoeppler et al. (1986)
Coastal waters, Malaysia	NS	0.45 μm filtered; 66% arsenate; 33% arsenite	0.95 (0.65–1.8)	Yusof et al. (1994)
Bohai Bay, China	1979	39°10′–38°40′N; 117°37′–180°00′E	1.4 (0.56–2.1)	Tan et al. (1983)

[a] Mean and ranges of total As unless stated otherwise
NS = not stated

Table 6. Concentrations of As in estuarine waters

Location	Sampling Period	Sampling details and/or species	Concentration (µg/litre)[a]	Reference
Tamar estuary, UK	1984	Glass fibre filtered	2.7–8.8 (range)	Howard et al. (1988)
Rhone estuary, France	1984–1988	4–24% arsenite; surface	1.3–3.8 (range)	Seyler & Martin (1990)
Gironde estuary, France	1984	4–14% arsenite; surface	0.7–2.5 (range)	Seyler & Martin (1990)
Loire estuary, France	1984	4–25% arsenite; surface	1.5–3.0 (range)	Seyler & Martin (1990)
Schelde estuary, Belgium	1984	2–16% arsenite	1.8–4.9 (range)	Andreae & Andreae (1989)
Huang He river estuary, China	NS	dissolved As, surface water	Total = 3.6 (2.8–4.3) Organic = 2.3 (1.3–2.9) Inorganic. = 1.4 (0.7–2.3) Arsenite = 0.5 (0.3–0.8) Arsenate = 0.8 (0.2–1.4)	Li et al. (1989)

[a] Mean and ranges of total As unless stated otherwise
NS = not stated

near a long-term stibnite mine. Total inorganic arsenic concentrations were 5.3 µg/litre near the mine but declined to normal levels (1–2 µg/litre) within 200 m.

Howard et al. (1988) report that concentrations of methylated arsenic increased with salinity in the Tamar estuary (United Kingdom). Concentrations of monomethylarsenic ranged from 0.02 to 0.46 µg As/litre and dimethylarsenic from 0.02 to 1.27 µg As/litre; these two methylated forms of arsenic were typically 4% and 10% of the total soluble arsenic levels respectively.

Levels of arsenic in surface freshwaters are summarized in Table 7. Surveys of arsenic concentrations in rivers and lakes indicate that most values are below 10 µg/litre, although individual samples may range up to 1 mg/litre (Page, 1981; Smith et al., 1987; Welch et al., 1988). Mean total arsenic concentrations of 2000 µg/litre have been recorded near a pesticide plant, with MMA being the predominant arsenic species (Faust et al., 1983; 1987a). Crearley (1973) measured arsenic in two lakes near a manufacturing plant which had been producing arsenic-based cotton desiccants/defoliants for 30 years. Mean arsenic concentrations of 7900 and 3200 µg/litre were found. During the dry season total dissolved arsenic concentrations (< 0.45 µm) of up to 250 µg/litre were recorded near industrial discharges to the Xiangjiang river (China); however, maximum levels during the rainy season were generally less than 30 µg/litre (Chunguo & Zihui, 1988).

High levels of arsenic have been recorded in thermal waters. Tanaka (1990) found a mean concentration of 570 µg/litre in geothermal waters throughout Japan, with a maximum level of 25.7 mg/litre.

5.1.4 Groundwater

Levels of arsenic in groundwater are summarized in Table 8. Arsenic levels in groundwater average about 1–2 µg/litre, except in areas with volcanic rock and sulfide mineral deposits where arsenic levels can range up to 3400 µg/litre (Page, 1981; Welch et al., 1988; Robertson, 1989). In some mining areas arsenic concentrations of up to 48 mg/litre have been reported (Welch et al., 1988). Korte & Fernando (1991) reported that arsenic levels in arsenic-contaminated

Table 7. Concentrations of As in surface freshwaters

Location	Sampling period	Sampling details and/or As source	Concentration (µg/litre)[a]	Referenc
Brazos river, Texas, USA	NS	0.2 µm filtered, arsenite	0.05	Chakraborti et al. (1986)
Madison river, Montana, USA	NS	geothermal	51	Sonderegger & Ohguchi (1988)
Finfeather lake, Texas, USA	1973	near manufacturing plant for As-based cotton defoliants	7900 (6000–8600)	Crearley (1973)
Municipal lake, Texas, USA	1973	as above	3200 (1700–4400)	Crearley (1973)
Maurice river, NJ, USA	1982–1983	upstream of pesticide plant	3.3 (1.05–4.4)	Faust et al. (1987a)
	1982–1983	0.6 km downstream	2222 (1320–4160)	Faust et al. (1987a)
	1982–1983	4.2 km downstream	266 (118–578)	Faust et al. (1987a)
Union lake, NJ, USA	1982–1983	14–17 km downstream	86.1 (27.1–267)	Faust et al. (1987a)
Bowron lake, British Columbia, Canada	1992	reference lake; no mining activity	0.26 (<0.2–0.42)	Azcue et al. (1994a)
Lake water, British Columbia, Canada	1992	near abandoned gold mine	0.25 (< 0.2–0.3)	Azcue et al. (1994a)
Asososca lake, Nicaragua	1991–1992	volcanic crater; includes surface, intermediate and bottom samples	5.9 (0.85–15.8)	Cruz et al. (1994)

Table 7 (contd.)

Location	Year	Description	Concentration[a]	Reference
Moira lake, Ontario, Canada	1987–1988	past mining activity; 15% particle sorbed	43 (4–94)	Diamond (1995)
Lakes, Northwest Territories, Canada	1975	gold mining activity	700–5500 (range)	Wagemann et al. (1978)
Subarctic lakes, Northwest Territories, Canada	1991	gold mining activity	270 (64–530)	Bright et al. (1996)
Yangtze river (source area), China	NS	filtered water (< 0.45 µm)	3.1 (0.1–28.3)	Zhang & Zhou (1992)
Antofagasta, Chile	1958–1970	Toconce river, Andes mountains	< 800	Borgono et al. (1977)
Mutare river, Zimbabwe	1993	near gold/As mine dumps	13–96 (range of means)	Jonnalagadda & Nenzou (1996b)
Odzi river, Zimbabwe	1993	2.2 km downstream from gold/As mine dumps (after confluence with Mutare river)	1–3 (range of means)	
Xolotlan lake, Nicaragua	NS	volcanic crater; range of means	10.2–30.1 (range of means)	Lacayo et al. (1992)
Waikato river, New Zealand	1993–1994	volcanic source	32.1 (28.4–35.8)	McLaren & Kim (1995)
Lake water, Lapland, Finland	1992	0.1 m below surface	0.17 (median)	Mannio et al. (1995)
Nakhon Si Thammarat province, Thailand	1994	mining activity	217.5 (4.8–583)	Williams et al. (1996)

[a] Mean and ranges of total As unless stated otherwise
NS, not stated

Table 8. Concentrations of As in groundwater

Location	Sampling period	As source	Concentration (µg/litre)[a]	Reference
Hungary	NS	deep groundwater	68 (1–174)	Varsanyi (1989)
South-west Finland	1993–1994	well-waters; natural origin	17–980 (range)	Kurttio et al. (1998)
New Jersey, USA	1977–79	well-waters	1 (median)	Page (1981)
			1160 (maximum)	Page (1981)
Western USA	NS	geochemical environments	48 000 (maximum)	Welch et al. (1988)
South-west USA	1970	alluvial aquifers	16–62 (range of means)	Robertson (1989)
Southern Iowa and western Missouri, USA	NS	natural origin	34–490 (range)	Korte & Fernando (1991)
North-eastern Ohio, USA	NS	natural origin	< 1–100 (range)	Matisoff et al. (1982)
Lagunera region, northern Mexico	NS	well-waters	8–624 (range)	Del Razo et al. (1990)
Cordoba, Argentina			> 100	Astolfi et al. (1981)
Chile			470–770 (range)	De Sastre et al. (1992)

Table 8 (contd.)

Location	Date	Description	Concentration	Reference
Pampa, Cordoba, Argentina	NS	2–15 m, 61°45'–63°W; 32°20'–35°00'S	100–3810 (range)	Nicolli et al. (1989)
Kuitun-Usum, Xinjiang, PR China	1980	well-waters	850 (maximum)	Wang et al. (1993)
Hsinchu, Taiwan	NS	well-waters	< 0.7	Chen et al. (1994)
West Bengal, India	NS	As-rich sediment	193–737 (range of means) 3700 (maximum)	Chatterjee et al. (1995) Chatterjee et al. (1995)
Calcutta, India	1990–1997	near pesticide production plant	< 50–23 080 (range)	Chakraborti et al. (1998)
Bangladesh	1996–1997	well-waters	< 10–> 1000 (range)	Dhar et al. (1997)
Nakhon Si Thammarat Province, Thailand	1994	shallow (alluvial) ground-water; mining activity	503.5 (1.25–5114)	Williams et al. (1996)
	1994	deep groundwater; mining activity	95.2 (1.25–1032)	Williams et al. (1996)

[a] Mean and ranges of total As unless stated otherwise
NS = not stated

water supply wells in southern Iowa and western Missouri (USA) ranged from 34 to 490 µg/litre. The authors state that the arsenic appears to be of natural origin. Similarly, Matisoff et al. (1982) found no evidence for an anthropogenic source contributing to elevated groundwater levels of arsenic (< 1 to 100 µg/litre) in north-eastern Ohio (USA). Arsenic levels in groundwater were found to exceed 10 µg/litre in 5.6–9.5% of samples collected in Germany during the period 1992–1994 (Umweltbundesamt, 1997). Varsanyi (1989) found arsenic concentrations in deep groundwater in Hungary to range from 1 to 174 µg/litre with an average value of 68 µg/litre. High arsenic levels originating from arsenic-rich bedrock were found in drilled wells in south-west Finland, with concentrations ranging from 17 to 980 µg/litre (Kurttio et al., 1998). Del Razo et al. (1990) monitored groundwater in the Lagunera region of northern Mexico. Total arsenic concentrations ranged from 8 to 624 µg/litre with over 50% of samples > 50 µg/litre. The predominant arsenic species in 93% of samples was arsenate, although in 36% of samples 20–50% arsenite was found. Chen et al. (1994) report that arsenic levels in the groundwater of south-west Taiwan contained mean dissolved arsenic levels of 671 µg/litre. Arsenic levels in the well-waters of Hsinchu (Taiwan) were less than 0.7 µg/litre.

Arsenic contamination of groundwater from arsenic-rich sediment has been reported in both India and Bangladesh. Chatterjee et al. (1995) analysed groundwater from six districts of West Bengal (India). Mean total arsenic levels ranged from 193 to 737 µg/litre with a maximum value of 3700 µg/litre. Mean arsenite levels in the groundwater were around 50% of the total arsenic. Mandal et al. (1996) reported that 44% of groundwater samples collected in West Bengal (India) up to January 1996 contained total arsenic levels > 50 µg/litre. Dhar et al. (1997) found that 38% of groundwater samples collected from 27 districts of Bangladesh contained total arsenic levels > 50 µg/litre.

During 1990 and 1991 Chatterjee et al. (1993) sampled groundwater in the vicinity of a chemical plant in Calcutta, India, which had produced the insecticide Paris green (acetocopper arsenite) for 20 years. Groundwater contained total arsenic levels ranging from < 0.05 to 58 mg/litre; the highest total arsenic level included 75% arsenite.

5.1.5 Sediment

Arsenic concentrations in sediments are summarized in Table 9. Sediments in aquatic systems often have higher arsenic concentrations than those of the water (Welch et al., 1988). Most sediment arsenic concentrations reported for rivers, lakes and streams in the USA range from 0.1 to 4000 mg/kg, with higher levels occurring in areas of contamination (Welch et al., 1988). Arsenic concentrations of < 10 000 mg/kg (dry weight) were found in surface sediments near a copper smelter (Crecelius et al., 1975). Sediment arsenic concentrations of < 3500 mg/kg were reported for lakes in the Northwest Territory (Canada) which had received past inputs from gold-mining activity (Wagemann et al., 1978). Mean total arsenic concentrations of 500 mg/kg (dry weight) were measured in sediment near a pesticide plant and at a lake 14–17 km downstream mean concentrations of almost 3000 mg/kg had accumulated (Faust et al., 1987a). Arsenate was the predominant arsenic species, with inorganic arsenic amounting to 70–90% of the total arsenic measured (Faust et al., 1983). Bright et al. (1996) found total arsenic concentrations ranging from 1043 to 3090 mg/kg in the top 10 cm of sediment from subarctic lakes contaminated by gold-mining activity. Total dissolved arsenic levels in porewater ranged from 800 to 5170 μg/litre (0.7% organic arsenic). Ebdon et al. (1987) reported that methylated arsenic species represented 1–4% of the total arsenic in sediment porewater from the Tamar estuary, south-west England (United Kingdom). Similar findings were reported by de Bettencourt (1988) for the Tagus Estuary (Portugal).

Chunguo & Zihui (1988) studied arsenic accumulation in sediment of the Xiangjiang river (China), which receives inputs from a variety of industrial plants. Total arsenic concentrations upstream of industrial inputs were 13.2 mg/kg during the rainy season and 81.4 mg/kg during the dry season. Near to industrial discharges maximum total arsenic concentrations exceeded 1000 mg/kg during the dry season (approximately 70% as iron or aluminium arsenate) but rarely reached 100 mg/kg during the rainy season.

Farmer & Lovell (1986) monitored arsenic concentrations in surface sediments of Loch Lomond (Scotland, UK); no recent significant sources of environmental arsenic contamination were identified. They found natural enrichment of sediment to levels of up

Table 9. Concentrations of As in sediment

Location	Sampling period	Sampling details and/or As source	Concentration (mg/kg dry weight)[a]	Reference
Estuarine/marine				
UK estuaries	1977–1979	100 μm sieved	2–94 (range)	Langston (1980)
Estuaries, south-west England, UK	1978–1979	past mining activity	7–2500 (range)	Langston (1980)
Tamar estuary, UK	1984	inorganic As	29.2	Howard et al. (1988)
Northern Tyrrhenian/eastern Ligurian Seas, Italy	1985–1989	surface sediment	4–88 (range)	Leoni & Sartori (1996)
Bohai bay, China	1979	39°00'–38°40'N; 117°37'–180°00E	12.8 (9.9–16.4)	Tan et al. (1983)
Eastern Mississippi bight, USA	1987–1989	surface sediment	7.5 (< 1–16)	Presley et al. (1992)
Commencement bay, Washington, USA	1981	surface sediment; industrial inputs	12–288 (range of means)	Schults et al. (1987)

Table 9 (contd.)

Bothnian sea, Sweden/Finland	1991–1993	surface sediment; open sea basin (water depth > 60 m)	61	Leivuori & Niemistö (1995)
Bothnian bay, Sweden/Finland	1991–1993	surface sediment; open sea basin (water depth >60 m); industrial inputs	278	Leivuori & Niemistö (1995)
Moreton's Harbour, Newfoundland	1972–1974	< 40 m from stibnite mine	847–2600 (range)	Penrose et al. (1975)
	1972–1974	> 40 m from stibnite mine	9.1–34.4 (range)	Penrose et al. (1975)
Continental shelf, south-east Australia	1972		18 (2–180)	Davies (1974)
Upper Spencer gulf, South Australia	NS	surface sediment; smelting activity	5.8 (0.34–160)	Tiller et al. (1989)
Freshwater				
Clark Fork river, Montana, USA	1991	past mining, milling and smelting activity	4–404 (range)	Brumbaugh et al. (1994)

83

Table 9 (contd.)

Location	Sampling period	Sampling details and/or As source	Concentration (mg/kg dry weight)[a]	Reference
Maurice river, NJ, USA	1982–1983	upstream of pesticide plant	25.3 (4.1–48.5)	Faust et al. (1987a)
	1982–1983	0.6 km downstream	515 (291–809)	Faust et al. (1987a)
	1982–1983	4.2 km downstream	23.5 (16–30.2)	Faust et al. (1987a)
Union lake, NJ, USA	1982–1983	14–17 km downstream	2922 (83.6–23 200)	Faust et al. (1987a)
Bowron lake, British Columbia, Canada	1992	reference lake; no mining activity	19 (16–23)	Azcue et al. (1994a)
Lake water, British Columbia, Canada	1992	past mining (gold)	342 (80–1104)	Azcue et al. (1994a)
Subarctic lakes, Northwest territories, Canada	1991	gold mining activity; 0–10 cm sampling depth	1716 (1043–3090)	Bright et al. (1996)
Lakes, northern Sweden	1988	within 80 km smelter; 0–1 cm sampling depth	584 (9–4169)	Johnson et al. (1992)

[a] Mean and ranges of total As unless stated otherwise
NS = not stated

to 675 mg As/kg compared with typical background concentrations of 15–50 mg/kg.

5.1.6 Sewage sludge

Zhu & Tabatabai (1995) monitored total arsenic levels in sewage sludges from waste treatment plants in Iowa (USA). Concentrations ranged from 2.4 to 39.6 mg/kg with a mean of 9.8 mg/kg.

5.1.7 Soil

Levels of arsenic in soil are summarized in Table 10. Arsenic is found in the earth's crust at an average level of 2 mg/kg. Most natural soils contain low levels of arsenic, but industrial wastes and pesticide applications may increase concentrations. Background concentrations in soil range from 1 to 40 mg/kg, with a mean value of 5 mg/kg (Bowen, 1979; Beyer & Cromartie, 1987). Naturally elevated levels of arsenic in soils may be associated with geological substrata such as sulfide ores. Anthropogenically contaminated soils can have concentrations of arsenic up to several percent (NAS, 1977; Porter & Peterson, 1977). Arsenic concentrations of up to 27 000 mg/kg were reported in soils contaminated with mine or smelter wastes (US EPA, 1982). Chatterjee & Mukherjee (1999) reported arsenic levels of 20 100–35 500 mg/kg in soil around the effluent dumping point of an arsenical pesticide manufacturing plant. Peat may contain considerable quantities of arsenic. Minkkinen & Yliruokanen (1978) found maximum arsenic concentrations in various Finnish peat bogs of between 16 and 340 mg/kg dry peat. However, Shotyk (1996) analysed peat cores from the Jura mountains (Switzerland) and found mean total arsenic concentrations of 3.6 mg/kg at a depth of < 30 cm and 0.16 mg/kg at between 69 and 84 cm. Higher levels of arsenic were found in the mineral sediments underlying the peat bogs, with mean concentrations of 6.4 mg/kg at 170 cm and 15.9 mg/kg at 650 cm. Soil on agricultural land treated with arsenical pesticides may retain substantial amounts of arsenic. Mean total arsenic concentrations of 50–60 mg/kg have been recorded for agricultural soils treated with arsenical pesticides (Takamatsu et al., 1982; Sanok et al., 1995). Walsh & Keeney (1975) reported that arsenic-treated soils contained up to 550 mg As/kg. Stilwell & Gorny (1997) found mean arsenic concentrations ranging

Table 10. Concentrations of As in soil

Location	Sampling period	Soil depth (cm)	Notes	Concentration (mg/kg) (dry weight)[a]	Reference
USA	1961–1975	20	1318 sampling sites	7.2 (< 0.1–97)	Shacklette & Boerngen (1984)
Annapolis valley, USA	NS	NS	non-orchard	TR–7.9 (range)	Bishop & Chisholm (1962)
			orchard soil treated with arsenicals	9.8–124.4 (range)	Bishop & Chisholm (1962)
NY, USA	1992–1993	0–25	orchard soil	1.8–3.0 (range)	Merwin et al. (1994)
	1992–1993	0–25	orchard soil previously treated with lead arsenate	1.6–141 (range)	Merwin et al. (1994)
Manitoba, Canada	1982–1984	surface	peat soil	4 (1–19.6)	Zoltai (1988)
Alberta, Canada	NS	NS	acid sulfate soil; soil horizons E-C	1.5–45 (range)	Dudas (1984)
Upper Austria	NS	surface		6.2 (1–39)	Aichberger & Hofer (1989)
The Netherlands	1976–1977	0–20	agricultural soil	12 (0.1–110)	Wiersma et al. (1986)
Norway	NS	0–60	agricultural soils	2.4 (0.8–17)	Esser (1996)
Southern Norway	1981–1983	3–5	< 50 km from coast	5 (1.4–14.8)	Steinnes et al. (1989)
	1981–1983	3–5	> 100 km from coast	2.2 (1.3–5)	Steinnes et al. (1989)
Poland	1982–1986	0–20	arable soils	2.6 (0.5–15)	Dudka & Markert (1992)
South-east Spain	NS	10–15		16.8 (8.75–34.5)	Navarro et al. (1993)

Table 10 (contd.)

Location	Year	Depth (cm)	Description	Concentration	Reference
Mekong delta, Vietnam	NS	0–140	acid sulfate soil	6–41 (range)	Gustafsson & Tin (1994)
Taiwan	1983	0–15	agricultural soil	5.65 (0.01–16.16)	Chang et al. (1999)
Japan	NS	NS	agricultural soil	9.9	Harako (1986)
			agricultural soil; volcanic region; < 1% of total As was organic	609 (maximum 1400)	Harako (1986)
Nagpur city, India	1992	NS	urban	6.3	Chutke et al. (1995)
South Australia	1974–1979	0–10	uncontaminated	3.9	Merry et al. (1983)
Tasmania	1974–1979	0–10	uncontaminated	0.6	Merry et al. (1983)
South Australia and Tasmania	1974–1979	0–10	orchard soil	29 (< 0.5–115)	Merry et al. (1983)
Long Island, NY, USA	NS	0–18	sandy loam soil	2.3	Sanok et al. (1995)
	NS	0–18	sandy loam soil; potato soils treated with lead arsenate	27.8–51 (range of means)	Sanok et al. (1995)
Japan	1980	0–15	orchard soil treated with arsenicals; < 1% of total As was organic	10.6–61.5 (range of means)[b]	Takamatsu et al. (1982)
	1980	0–15	orchard soil treated with arsenicals; < 1% of total As was organic	0.27–1.9 (range of means)[c]	Takamatsu et al. (1982)
	1980	0–15	paddy soil polluted by mining activity; < 3% of total As was organic	2.5–81.9 (range of means)[b]	Takamatsu et al. (1982)

Table 10 (contd.)

Location	Sampling period	Soil depth (cm)	Notes	Concentration (mg/kg) (dry weight)[a]	Reference
	1980	0–15	paddy soil polluted by mining activity; < 3% of total As was organic	0.43–5.7 (range of means)[c]	Takamatsu et al. (1982)
South-west England, UK	1984	0–15	past mining activity	322 (144–892)	Xu & Thornton (1985)
	NS	NS	contaminated with mine and smelter waste; water soluble As 0.5–2.9 mg/kg	8510–26 530 (range)	Porter & Peterson (1977)
North-west England, UK	NS	surface	control site	5.0	Ismael & Roberts (1992)
	NS	surface	250 m from As refinery	155.9	Ismael & Roberts (1992)
Zimbabwe	NS	0–10	gold/As mine dumps	9530	Jonnalagadda & Nenzou (1996a)
Northern Peru		0–10	near copper mine	143–3052 (range of means)	Bech et al. (1997)
Obuasi, Ghana	1992–1993		0.3 km from gold ore processing plant	48.9	Amonoo-Neizer et al. (1996)
Southern Ontario, Canada	1974	0–5	urban area	9.8 (2.7–41)	Temple et al. (1977
	1974	0–5	<700 m from secondary lead smelter	107 (4.7–2000)	Temple et al. (1977)

Table 10 (contd.)

Toronto, Canada	NS	0–1	near secondary lead smelter	17.9–3007 (range of means)	Dolan et al. (1990)
Utah, USA	NS	surface	1–2 km from copper smelter	75–540 (range of means)	Ball et al. (1983)
	NS	surface	10–25 km from copper smelter	6–150 (range of means)	Ball et al. (1983)
Nakato, Niigata Prefecture, Japan	1994	15	site of factory producing As sulfide (35 years before)	2.4–72.7 (range of means)	Nakadaira et al. (1995)
Australian Capital Territory, Australia	NS	surface and subsurface	urban area, former site of arsenical pesticide plunge sheep dip for tick control (1946-1960)	32–1597	Ng et al. (1998b)
New South Wales, Australia	NS	NS	urban area, former arsenical pesticide plunge cattle dip sites	730–2100	Ng & Moore (1996)
	NS	NS	copper chrome arsenate contaminated site	52–138	Ng & Moore (1996)
Queensland, Australia	NS	0–12.5	site of tannery (1891–1972)	80 (< 1–435)	Sadler et al. (1994)
	NS	25–72.5	site of tannery (1891–1972)	121 (< 1–1010)	Sadler et al. (1994)

[a] Mean and ranges of total As unless stated otherwise
[c] Arsenite
[b] Arsenate
NS = not stated; TR = trace

89

from 9 to 139 mg/kg (dry weight) in soil (upper 5 cm) below decking treated with copper chrome arsenate (CCA).

Uptake and effects of arsenic on organisms are related to bioavailable arsenic rather than total arsenic. Xu & Thornton (1985) measured mean total arsenic levels of 300 mg/kg in garden soils (south-west England, UK) at sites of past mining activity; however, water-soluble and acid-fluoride extractable arsenic represented < 1% and < 2% of total arsenic respectively. Kavanagh et al. (1997) report total arsenic concentrations ranging from 174 to 477 mg/kg for agricultural soil and from 1200 to 22 290 mg/kg for mine waste in the Tamar valley, south-west England. The proportion of water-extractable arsenic in agricultural topsoils ranged from 0.05 to 0.3% and in mine wastes from 0.02 to 1.2%. Similarly, McLaren et al. (1998) found total arsenic concentrations of 37–3540 mg/kg (dry weight) in surface soil (0–10 cm) contaminated by cattle dip (sodium arsenite) compared with water-extractable arsenic concentrations in the same samples ranging from 0.2 to 22.4 mg/kg . The highest total arsenic level recorded was 14 800 mg/kg (water-extractable arsenic = 1.2 mg/kg) at a depth of 40–45 cm. Ng et al. (1998b) measured total arsenic concentrations of 32–1597 mg/kg in soil which had been contaminated 30 years previously with arsenical pesticides. Chemical speciation showed that arsenite ranged from 0.32–56% of total arsenic. In a rat model, the absolute bioavailability of these contaminated soils relative to arsenite and arsenate ranged from 1.02 to 9.87% and 0.26 to 2.98% respectively.

Doyle & Otte (1997) found that the presence of vegetation and burrowing organisms significantly increased the concentration and accumulation of arsenic in salt-marsh soils.

Chutke et al. (1995) analysed dust samples collected in Nagpur city (India) during 1992. Mean arsenic levels for residential/commercial areas, industrial areas and highways were respectively 10.2, 18 and 17 mg/kg (dry weight). Stone & Marsalek (1996) collected samples of sediment from road surfaces in an urban area of Ontario (Canada) during 1991 and found total arsenic concentrations ranging from 1 to 33 mg/kg with a mean of 3.4 mg/kg.

5.1.8   Biota

Background arsenic concentrations in living organisms are usually less than 1 mg/kg (fresh weight) in freshwater and terrestrial biota. The levels are higher in biota collected from mine waste sites, arsenic-treated areas, near smelters and mining areas, near areas with geothermal activity and near manufacturing sites of arsenical defoliants and pesticides (Eisler, 1988). Marine organisms, however, can normally contain arsenic residues ranging from 1–2 mg/kg to more than 100 mg/kg (Lunde, 1977; Maher & Butler, 1988; Phillips, 1990). Neff (1997) reviewed levels of total arsenic in marine organisms and calculated geometric means ranging from < 1 mg/kg for marine mammals to 50 mg/kg for snails. An overall geometric mean for a wide variety of marine biota was calculated to be 11 mg/kg (dry weight). There is a substantial number of publications on the levels of arsenic in biota, and the following examples have been chosen to provide an overview.

5.1.8.1   Freshwater

Freshwater plants in uncontaminated environments tend to contain arsenic concentrations < 10 mg/kg (Reay, 1972; Outridge & Noller, 1991). Reay (1972) reported a considerable accumulation of arsenic in freshwater plants in the Waikato river (New Zealand). The elevated arsenic concentrations in the water (30–70 µg/litre) arising from geothermal activity gave rise to concentrations of < 971 mg As/kg in aquatic plants. Arsenic concentrations of < 1200 mg/kg (dry weight) were reported by Mudroch & Capobianco (1979) for aquatic macrophytes growing in an area of the Lake Ontario drainage basin (Canada) contaminated with mine effluent. Wagemann et al. (1978) reported arsenic concentrations ranging from 150 to 3700 mg/kg for macrophytes in lakes (Northwest Territory, Canada) which had received past inputs from gold-mining activity. During 1983, Tanner & Clayton (1990) analysed macrophytes from Lake Rotoroa (New Zealand), which had been sprayed with sodium arsenite herbicide in 1959. Arsenic concentrations ranged from 540 to 780 mg/kg (dry weight) in surficial sediments and from 193 to 1200 mg/kg in macrophytes. Similar levels of arsenic accumulation to that seen in aquatic plants has been observed for zooplankton (700–2400 mg/kg) in lakes receiving mine drainage water (Wagemann et al., 1978).

Freshwater bivalves have been used to measure arsenic in several biomonitoring programmes. Leland & Scudder (1990) monitored freshwater bivalves (*Corbicula fluminea*) in the San Joaquin valley (California, USA), an area influenced by high levels of elements in irrigation wastewater. Mean concentrations of arsenic in bivalves ranged from 5.3 to 13.9 mg/kg (dry weight). A highly significant relationship was observed between arsenic residues and the HNO_3-extractable arsenic : iron ratio of suspended matter. Similarly, Johns & Luoma (1990) found mean arsenic levels ranging from 5.4 to 11.5 mg/kg (dry weight) for the same species for the Sacramento/San Joaquin river delta (California, USA). Arsenic levels in mussels from the St Lawrence river (Canada) ranged from 2.8 to 8.6 mg/kg (dry weight) (Metcalfe-Smith, 1994).

Freshwater fish have not been shown to accumulate arsenic to the same degree as lower aquatic organisms. Arsenic residues in freshwater fish have been monitored in the USA over a period of approximately 10 years. The geometric means (mg/kg wet weight), with the range in parentheses, of total arsenic concentrations were 0.27 (0.05–2.92) during 1976–1977, 0.14 (0.05–1.69) during 1980–1981 and 0.14 (0.27–1.5) during 1984 (May & McKinney, 1981; Lowe et al., 1985; Schmitt & Brumbaugh, 1990). Mean total arsenic residues in freshwater fish near a copper smelter (Sweden) ranged from 0.05 to 0.24 mg/kg (wet weight) compared with 0.06 to 0.09 mg/kg for a control lake (Norin et al., 1985). Takatsu & Uchiumi (1998) analysed fish from a naturally acidified volcanic lake (Lake Usoriko, Japan) with low phosphate levels (< 0.02 mg/litre). Mean arsenic levels were 0.28 and 0.27 mg/kg (wet weight) for gills and bone respectively and 6.1 mg/kg for eye tissue. Arsenic residues have also been measured in fish from the San Joaquin valley area of California (USA) exposed to agricultural subsurface drainage water. Mean arsenic concentrations ranged from 0.18 to 0.44 mg/kg (dry weight) (maximum value 0.97 mg/kg) for bluegill sunfish (*Lepomis macrochirus*) and from 0.23 to 0.39 mg/kg (maximum value 1.5 mg/kg) for common carp (*Cyprinus carpio*) (Saiki & May, 1988). Mean arsenic concentrations for striped bass (*Morone saxatilis*) from the same area ranged from 0.23 to 0.65 mg/kg (dry weight) compared with mean values of 1.23–1.44 mg/kg for bass from San Francisco bay (California, USA) (Saiki & Palawski, 1990).

Clark et al. (1998) found a mean arsenic concentration of 6.87 mg/kg (wet weight) in tadpoles of the cricket frog (*Acris crepitans*) collected in 1994 downstream from Finfeather Lake (Texas, USA); the lake was contaminated during 53 years (1940–1993) of industrial production of arsenic-based cotton desiccants/defoliants.

5.1.8.2 Marine

Marine biota tend to accumulate much higher levels of arsenic than freshwater species (see section 4.2.3.2). Very little information is available on arsenic levels in natural phytoplankton populations. Benson & Summons (1981) reported 9 mg/kg total arsenic in a mixed marine phytoplankton population near Cape Ferguson (Queensland, Australia). Sanders (1979a) found that mean total arsenic concentrations in marine macroalgae ranged from 1.4 mg/kg (*Rhodophyceae*) to 10.3 mg/kg (*Phaeophyceae*). The absolute concentration of inorganic arsenic was not significantly different between groups, suggesting that the variation is due to metabolic differences between algal classes rather than to differences in the environmental concentration of arsenic. Mean total arsenic concentrations in macroalgae collected in the South Atlantic ranged from 5.3 to 70.2 mg/kg, with inorganic arsenic residues ranging from 0.2 to 2.0 mg/kg (Muse et al., 1989). Lai et al. (1998) report seasonal changes in arsenic speciation in the brown alga *Fucus gardneri* in Vancouver (Canada). During the summer algae contain 9 mg As/kg with most (79–98%) being extractable, whereas during the winter months residues range from 16 to 22 mg/kg with extraction efficiencies of 5.8–49%. Klumpp & Peterson (1979) found mean arsenic concentrations ranging from 83.7 to 141.4 mg/kg (dry weight) (maximum 189.3 mg/kg) for macroalgae in Restronguet creek, south-west England (UK) (an estuary influenced by past mining activity). Penrose et al. (1975) monitored marine biota near the site of a disused stibnite mine (pre–1916). Mean arsenic residues were 17.2 mg/kg for macroalgae and 3.8–11.5 mg/kg for invertebrates near the mine site compared with 9.8–12.1 mg/kg for macroalgae and 1.6–4.0 mg/kg for invertebrates at a control site.

Stronkhorst (1992) reported mean arsenic concentrations in mussels of 1 mg/kg (wet weight) for two Dutch estuaries. Similar levels (mean values ranging from 1.1 to 2.7 mg/kg) were reported in clams and oysters collected from U.S. coastal waters in use for shellfish production during 1985 and 1986 (Capar & Yess, 1996). Shellfish from the Arabian Gulf contained mean arsenic concentrations ranging from 3 to 15.8 mg/kg (wet weight) (Attar et al., 1992; Madany et al., 1996). Molluscs sampled in Restronguet creek contained arsenic concentrations ranging from 35 to 64 mg/kg (dry weight) (Klumpp & Peterson, 1979). Benson & Summons (1981) found that arsenic was accumulated to substantial levels in the kidney of molluscs from the Great Barrier Reef (Australia) with residues ranging from 481 to 1025 mg/kg (dry weight).

Langston (1980) found that the highest arsenic concentrations in estuarine benthic organisms (< 190 mg/kg [dry weight]) were found at sites where high arsenic : iron ratios exist in the sediment. Concentrations of arsenic in estuarine organisms correlated more significantly with the arsenic : iron ratio in sediments than arsenic levels alone.

Arsenic residues in marine fish appear to show substantial variation. Hellou et al. (1992) found mean arsenic residues of 3.2 mg/kg (dry weight) (1.6–4.2 mg/kg) in Atlantic tuna (*Thunnus thynnus*). Engman & Jorhem (1998) reported arsenic residues in marine fish muscle ranging from 0.59 to 17 mg/kg (fresh weight) with a mean value of 4.5 mg/kg. The mean value was 60 times greater than that found for freshwater fish in the same study. Several studies of marine fish from the Arabian Gulf have shown that in general mean arsenic concentrations range from < 1 to < 10 mg/kg in muscle (Tariq et al., 1991; Attar et al., 1992; Madany et al., 1996). However, higher concentrations have been reported, for example, Attar et al. (1992) found mean muscle concentrations of up to 32.3 mg As/kg (wet weight) in black-banded bream (*Acanthopagrus bifasciatus*). Bohn (1975) reported mean arsenic concentrations for marine fish from West Greenland ranging from 21.9 to 240 mg/kg (dry weight).

Maher (1983, 1988) analysed a variety of marine biota and found mean total arsenic concentrations (dry weight) ranging from 2.7 mg/kg (fish muscle) to 114 mg/kg (macroalgae: *Cystophora moniliformis*). Mean inorganic arsenic concentrations were low (0.02–3.6 mg/kg) in all marine organisms with organic arsenic representing 70–98% of the total arsenic. Arsenobetaine was the most abundant arsenic species found in marine invertebrates and fish muscle tissue (Edmonds & Francesconi, 1981c; Shiomi et al., 1984; Maher, 1985b; Matsuto et al., 1986).

Mean arsenic concentrations in the liver and muscle tissue of marine mammals were found to be generally less than 1 mg/kg (Julshamn et al., 1987; Muir et al., 1988; Skaare et al., 1990; Miles et al., 1992; Varanasi et al., 1994).

5.1.8.3   Terrestrial

The arsenic content of plants grown on soils that had never been treated with arsenic-containing pesticides varied from 0.02 to about 5 mg/kg (dry weight). Plants grown on arsenic-contaminated soils may, however, contain considerably higher levels, especially in the roots. Plants growing on arsenical mine wastes (south-west England, UK) contained mean arsenic levels ranging from 350 to 2040 mg/kg (dry weight); a maximum concentration of 6640 mg/kg was reported for *Jasione montana* (Porter & Peterson, 1975). Benson et al. (1981) reported mean arsenic concentrations of 1480 and 1070 mg/kg (dry weight) for the grasses *Agrostis stolonifera* and *A. tenuis* growing on arsenical mine waste. De Koe (1994) found arsenic concentrations of up to 1800 and 1900 mg/kg in senescent shoots and roots respectively of grass species growing on gold-mine spoils (north-east Portugal). Jonnalagadda & Nenzou (1997) report arsenic concentrations in couch grass (*Cynodon dactylon*) growing on or near gold/arsenic mine dumps (Zimbabwe) ranging from 200 to 1660 mg/kg (dry weight) in stems and from 1020 to 10 880 mg/kg in roots. Mean concentrations of arsenic in the leaves of plants growing near a copper mine (northern Peru) ranged from 111 to 1651 mg/kg (dry weight) (Bech et al., 1997). Temple et al. (1977) found mean arsenic levels of 5.8 mg/kg in grass samples and 7.4 mg/kg in tree and shrub foliage from within 700 m of a secondary lead smelter; samples collected at a control site contained < 1 mg/kg.

Grass growing on plots which had been previously treated (7–11 years before) with lead arsenate contained mean arsenic residues of 1.5 mg/kg, compared with 0.9 mg/kg in grass from untreated sites. After a further 2 years mean arsenic concentrations were 0.88 and 0.56 mg/kg for treated and untreated sites respectively (Chisholm & MacPhee, 1972). Merry et al. (1986) reports that pasture plants growing at sites formerly used as orchards (soil concentration 80 mg As/kg) contained less than 2.5 mg As/kg (dry weight).

Biomonitoring studies at six background sites in Norway found mean arsenic concentrations in moss (*Hylocomium splendens*) ranging from 0.1 to 2.2 mg/kg (Berg et al., 1995a); an overall mean of 0.36 (< 0.03–3.2) mg/kg was reported by Berg et al. (1995b). Similarly, Glooschenko & Arafat (1988) sampled sphagnum moss (*Sphagnum fuscum*) throughout northern Canada. A mean background concentration of 0.66 mg As/kg (dry weight) was found, with elevated levels (> 3 mg/kg; maximum 31 mg/kg) in the vicinity of mining and smelting areas. Lichen biomonitoring of arsenic in a geothermal area of central Italy revealed a mean concentration of 1.19 mg/kg (0.19–3.55 mg/kg) (dry weight) (Loppi & Bargagli, 1996).

Monitoring of conifer needles has been carried out at sites remote from pollution sources, with mean arsenic concentrations ranging from 5 to 58 μg/kg for Norway spruce (*Picea abies*) and from 2 to 8 μg/kg for balsam fir (*Abies balsamea*) (Lin et al., 1995; Wyttenbach et al., 1997). However, much higher concentrations (0.46–3.1 mg/kg) have been reported for leaves from loblolly pine trees (*Pinus taeda*) growing on land affected by coal-pile leachate (Carlson & Carlson, 1994). Dmuchowski & Bytnerowicz (1995) monitored Scots pine (*Pinus sylvestris*) needles at three sites in Poland during 1983–1985. Mean arsenic concentrations were 0.54 mg/kg (dry weight) in a primeval forest (eastern Poland), 0.88 mg/kg near the city of Warsaw and 1.5 mg/kg at a polluted site in Silesia. Mankovska (1986) analysed pine needles (*Pinus silvestris*) from the vicinity of a smelter and found arsenic concentrations ranging from ~15 to ~22 mg/kg within 1000 m of the smelter (soil concentrations ranged from 30 to > 120 mg/kg).

Byrne & Tusek-Znidaric (1983) found arsenic concentrations ranging from 34 to 182 mg/kg (dry weight) in caps and stalks of the common mushroom (*Laccaria amethystina*) from rural sites in Slovenia; soil arsenic concentrations ranged from 3.2 to 27 mg/kg.

Beyer & Cromartie (1987) analysed earthworms from a diverse variety of sites in Maryland, Pennsylvania and Virginia (USA). Arsenic concentrations ranged from trace levels to 0.8 mg/kg (dry weight) at uncontaminated sites, mining sites and industrial sites.

However, a single earthworm sample at a mining site contained 10 mg/kg (soil concentration 20 mg/kg) although all other samples from mining sites contained only trace amounts of arsenic. Total arsenic concentrations ranging from 3.2 to 17.9 mg/kg (dry weight) were found in earthworms sampled from six sites in Austria. There was no correlation between the total arsenic concentrations in the earthworms and the soil. The major arsenic compounds detected in the earthworms were arsenous acid and arsenic acid; arsenobetaine, dimethylarsinic acid and two dimethylarsinoylribosides were also detected (Geiszinger et al., 1998).

Arsenic residues in birds tend to be low (< 1 mg/kg) with little accumulation even at sites with higher environmental concentrations (Martin & Nickerson, 1973; Blus et al., 1977; White et al., 1980; Ohlendorf et al., 1991; Pain et al., 1992; Vermeer & Thompson, 1992; Custer & Hohman, 1994; Guitart et al., 1994; Hothem & Welsh, 1994). Of 18 osprey (*Pandion haliaetus*) livers analysed by Wiemayer et al. (1980), 14 contained less than 1.5 mg/kg (wet weight); arsenic concentrations in the other four birds ranged from 2 to 16.7 mg/kg. The bird with the highest concentration was in a weak condition with very low fat reserves. Erry et al. (1999) analysed tissue samples from raptors in south-west England, an area with naturally and anthropogenically (through mining) elevated arsenic levels and compared the results with birds from another geographical area. Mean arsenic residues of 0.278, 0.346 and 0.187 mg/kg (dry weight) in the kidney, liver and muscle of kestrels (*Falcio tinnunculus*) were approximately three times higher in south-west England than in south-west Scotland. However, in another two raptors (sparrowhawk *Accipiter nisus* and barn owl *Tyto alba*) arsenic levels were not elevated in south-west England. The authors suggested that the difference could be attributed to differences in both diet and arsenic metabolism. Vermeer & Thompson (1992) analysed livers from birds collected in the vicinity of a copper mine; mean arsenic concentrations ranged from 0.08 to 3.23 mg/kg (wet weight). Goede (1985) found mean arsenic concentrations ranging from 0.5 to 3.2 mg/kg in the feather shafts of wading birds (Waddenzee, Netherlands); liver concentrations ranged from 4 to 14 mg As/kg (dry weight).

Elfving et al. (1979) analysed small mammals (voles and mice) from apple orchards which had received lead and calcium arsenate applications for many years. Arsenic concentrations in the soil ranged from 31 to 94 mg/kg (dry weight) and in the small mammals from 0.05 to 0.96 mg/kg (whole-body). Arsenic concentrations at a control site were 2.4 mg/kg in soil and < 0.03 mg/kg in small mammals. Ismael & Roberts (1992) monitored arsenic residues in vegetation and small mammals near an arsenic refinery. Mean arsenic levels in vegetation were 0.2 and 37.3 mg/kg for a control site and 250 m from the arsenic refinery respectively. Mean whole-body arsenic residues in four species of small mammal ranged from 0.4 to 3.2 mg/kg (fresh weight) at the control site and from 0.4 to 2.4 mg/kg near to the refinery. Significantly higher levels were found at the control site for three of the four species; the common shrew *Sorex araneus*, a carnivorous species, accumulated the highest levels of arsenic at both sites.

Arsenic was not detected (detection limit 5 µg/kg) in kidney tissue of mink (*Mustela vison*) collected in Georgia, North Carolina and South Carolina (USA) (Osowski et al., 1995). Langlois & Langis (1995) did not detect arsenic in muscle tissue (detection limit 50 µg/kg) of hares or martens in northern Quebec (Canada). Norstrom et al. (1986) found a mean arsenic concentration of 0.07 mg/kg (dry weight) in livers of polar bears (*Ursus maritimus*) in the Canadian Arctic. Norheim et al. (1992) reported mean arsenic concentrations of 0.06 and 0.04 mg/kg (wet weight) in the livers of adult and juvenile polar bears respectively at Svalbard (Norway).

5.2 General population exposure

Arsenic is widely distributed and human exposure is inevitable. Exposure of the general population to the various species of arsenic (inorganic and organic) will vary according to local geochemistry and the level of anthropogenic activity and can occur through the intranasal, oral and dermal routes.

5.2.1 Air

Arsenic in ambient air is associated with particulate matter and is predominantly a mixture of arsenite and arsenate. Organic species are of negligible significance except in areas where there has been substantial use of methylated arsenic pesticides or in areas with high biotic activity (ATSDR, 1993). As discussed in section 5.1.1 (see Tables 3 and 4), arsenic concentrations associated with particulate matter vary world wide as follows: $0.007-1.9$ ng/m^3 in remote areas; $1-28$ ng/m^3 in rural locations, and $2-2320$ ng/m^3 in urban environments (Schroeder et al., 1987). The highest concentrations are found near non-ferrous-metal smelters.

5.2.2 Food and beverages

Arsenic has been found in all foodstuffs analysed. Although most monitoring data is given as the concentration of total arsenic, arsenic in foods is a mixture of inorganic species and organo-arsenicals including arsenobetaine. The actual total arsenic concentrations in foodstuffs from various countries will vary widely depending on the food type, growing conditions (type of soil, water, geochemical activity, use of arsenical pesticides) and processing techniques.

From monitoring studies in the USA (Gunderson, 1995, Yost et al., 1998; US NRC, 1999), in the United Kingdom (UK MAFF, 1997), Canada (Dabeka et al., 1993) and Australia (ANZFA, 1994), by far the highest concentrations of total arsenic is found in seafood. Meats and cereals have higher concentrations than vegetables, fruit and dairy products. On the basis of limited data, it has been estimated that the percentage of inorganic arsenic is about 75% in meats, 65% in poultry, 75% in dairy products, and 65% in cereals (US EPA, 1988; Yost et al., 1998). Tao & Bolger (1998) estimated an inorganic arsenic intake for US men and women aged 60–65 years of 13 and 10 µg respectively. Other age groups had lower estimated daily intakes of inorganic arsenic, varying from 1.3 µg for infants to 9.9 µg for men aged 25–30 years. Additional samples, and a wider range of foodstuffs, need to be analysed in various countries before a definite conclusion can be reached on the normal range of inorganic arsenic in foods. In fruits, and vegetables and seafood the organic species predominate, with inorganic arsenic contributing 10%, 5% and 0–10% respectively. On the basis of these preliminary data it has been estimated that approximately 25% of the daily intake of dietary arsenic is inorganic (US EPA, 1988, Yost et al., 1998). A report from the Netherlands (Vaessen & van Ooik, 1989) estimated that inorganic arsenic in seafood was 0.1 to 41% of the total. Edmonds & Francesconi (1993) reviewed all data on inorganic arsenic in seafoods (excluding algae) available at that time and concluded that inorganic arsenic represented less than 1% of total arsenic at low arsenic burdens and fell to about 0.5% of total arsenic at concentrations of about 20 mg/kg. Mohri et al. (1990) estimated that the customary Japanese diet contained 5.7% inorganic arsenic with an intake ranging from 27 to 376 µg total arsenic/day.

Concentrations of arsenic in various food groups found in Canada are given in Table 11. Analysis of various beverages from Denmark found 3–11 µg/litre (Pedersen et al., 1994). Very few data were found on the concentration of arsenic in human breast milk. One study of 10 lactating women by Concha et al. (1998c) found a range of 0.83–7.6 µg/kg fresh weight (median 2.3 µg/kg) in breast milk from women consuming > 200 µg arsenic per day from drinking-water. Thus breast-feeding provided 1–2 µg As/day, compared to 100–200 µg As/day from formula mixed with the arsenic-rich water.

Table 11. Total As concentrations in various food groups from Canada[a]

Food category	Sample size	Mean	Range
		(µg As/kg wet weight)	
Milk and dairy products	89	3.8	< 0.4–26
Meat and poultry	124	24.3	< 1.3–536
Fish and shellfish	40	1662	77.0–4830
Soups	28	4.2	< 0.2–11
Bakery goods and cereals	177	24.5	< 0.1–365
Vegetables	262	7.0	< 0.1–84
Fruit and fruit juices	176	4.5	< 0.1–37
Fats and oils	21	19.0	< 1.0–57
Sugar and candies	49	10.9	1.4–105
Beverages[b]	45	3.0	0.4–9
Miscellaneous[c]	33	12.5	< 0.8–41

[a]   Data from: Dabeka et al. (1993).
[b]   Includes: coffee, tea, soft drinks, wine and canned and bottled beer.
[c]   Includes: bran muffins, muffins with and without raisins, gelatine desserts, raisins, baked beans, weiners, and raw and canned beets.

Examples of mean total daily intakes of arsenic from food and beverages in different countries are given in Table 12. The variation in dietary intake of total arsenic in adults reflects in large part the variability in the consumption patterns of arsenic-rich food groups (fish/shellfish and meats) confirming the need to consider such regional variations in arsenic intake when assessing human health effects for arsenic.

Table 12. Estimated average dietary intake of As in various countries

Country	Sampling method [a]	Total As intake (µg/day)	Reference
Australia	MB		ANZFA (1994)
	adult male	73	
	adult female	53	
	2-year old	17	
Brazil	DD (students)	19	Fávaro et al. (1994)
	S. Catarina 1 region	53	
	Manaus region	140–159	
		16–17	
Canada	TD		Dabeka et al. (1993)
	5 cities, adult males	59	
	5 cities, 1–4 years	15	
Croatia	MB	12	Sapunar-Postruznik et al. (1996)
Japan	DD (adult-male and female)	182	Mohri et al. (1990)
Spain	TD (Basque region, adults)	291	Urieta et al. (1996)
UK	TD (adults)	63	UK MAFF (1997)
USA	MB		Yost et al. (1998)
	adults	53	
	0.5–2 years	28	

[a] DD = duplicate diet study; MB = market basket survey; TD = total diet study; mean concentrations not reported.

The risk of arsenic exposure to populations living in or near arsenic-contaminated environments (i.e. mine-tailing sites, CCA and arsenical pesticide contamination soils), must be considered. In particular, contamination of home-grown vegetables and reared livestock, or wild collected foods must be considered. Helgesen & Larsen (1998) demonstrated that 0.47–0.6% of total soil arsenic (from a CCA plant) was bioavailable to carrot. Woolson (1973) dosed soils with 0–500 mg/kg arsenate and showed that at the arsenate dose that limited growth by 50%, arsenic in edible parts was up to 87 mg/kg (see section 4.2.4.5).

5.2.3 *Drinking-water*

Concentrations of arsenic in fresh surface water and groundwater, potential sources of drinking-water, are given in sections 5.1.3 and 5.1.4. Arsenate is the predominant species, but some groundwaters have been found to contain a high proportion of arsenite (section 5.1.4). Concentrations of methylated species in natural waters are usually less than 0.3 µg/litre (ATSDR, 1993). Unless stated otherwise in this section, monitoring data for drinking-water is reported as total arsenic.

A summary of the monitoring of drinking-water carried out in the USA by the US EPA during 1976–1993 has been published by Borum & Abernathy (1994). Concentrations of arsenic were in the range of < 2.5–28 µg/litre for surface waters and < 5–48 µg/litre for groundwater sources. Detection limits of 2 or 5 µg/litre preclude more accurate estimates of the lower limit of these ranges. On the basis of these data, it was estimated that approximately 2% of the population of the USA is exposed to > 10 µg/litre arsenic in drinking-water. Additional data sources (US EPA, 1993) provide support for this estimate, and have identified areas with higher concentrations of arsenic in drinking-water. In 1978 arsenic was detected in 67% of the 3834 drinking-water samples analysed (detection limit 0.1 µg/litre) with a mean concentration of 2.4 µg/litre (Borum & Abernathy, 1994). In areas with elevated geological concentrations of arsenic (e.g. California and Nevada) mean arsenic concentrations up to 80 µg/litre have been reported, with maximum reported levels of > 1400 µg/litre.

A limited summary of the monitoring data collected in 1985–1988 for total arsenic in drinking-water in six Canadian provinces has been published (NHW/DOE, 1993). Of the 717 samples of surface water, 3.6% were > 5 µg/litre and 5% of the 600 groundwater samples contained arsenic at concentrations > 5 µg/litre.

Although arsenic levels in natural waters are usually low (a few µg/litre), there are several areas in the world where humans consume drinking-water containing > 100 µg As/litre resulting from natural geochemical activity. In the West Bengal region of India it was estimated that over 1 000 000 people consume drinking-water containing > 50 µg/litre (up to 3.7 mg/litre) arising from normal geochemical processes (Das et al., 1995; Chowdhury et al., 1997). In the areas of West Bengal and Bangladesh, 38% of groundwaters sampled in 27 districts were > 50 µg/litre (Dhar et al., 1997). Natural geochemistry resulted in the pre–1970 exposure of about 100 000 people in the south-western coastal region of Taiwan to variable but high (10–1800 µg/litre, mean 500 µg/litre) concentrations of arsenic in drinking-water (Guo et al., 1994). A similar problem was reported in Chile where 100 000 people consumed drinking-water containing 800 µg As/litre between 1959 and 1970, when the concentration was lowered to about 50 µg/litre(Borgono et al., 1977). About 200 000 people in north central Mexico were reported to be exposed to >50 µg/litre arsenic in drinking-water (410 µg/litre in at least one village) (Cebrian et al., 1983).

In major Australian drinking-water systems levels of arsenic range up to 15 µg/litre, but typical concentrations are usually < 5 µg/litre (NHMRC, 1996).

5.2.4 Soil

Although ingestion of arsenic in soil and dust may not be a significant source of arsenic intake in adults, it may be significant for children, particularly in locations near industrial and hazardous waste sites. As described in section 5.1.7 (Table 10), background concentrations of total arsenic in soil are 1–40 mg/kg dry weight with a mean of 5 mg/kg (Beyers & Cromartie, 1987). The comparative bioavailability of arsenic in soil from a CCA-contaminated site and soil contaminated by arsenic solutions used in cattle tick control were reported by Ng & Moore (1996). In a rat model, soil from the cattle dip site had a bioavailability of 8.1 ± 4%; 14.4 ± 7% and 60 ± 3.4% when compared with orally administered sodium arsenite, calcium arsenite and sodium arsenate respectively. For CCA-contaminated soil the corresponding comparative bio-availabilities were 13.0 ± 4.5%; 32.2 ± 11.2% and 38.0 ± 13.2%. Also using a rat model, Ng et al. (1998b) have reported the absolute bioavailability of arsenic in soils containing 32–1597 µg As/kg (0.32–56% arsenite) from a combination of arsenical pesticides and natural geological formations in a residential area. The absolute bioavailability ranged from 1.02 to 9.87% relative to arsenite and from 0.26 to 2.98% relative to arsenate.

Freeman et al. (1993) determined both the absolute and comparative bioavailability of arsenic in soil from a smelter site using male rabbits and monkeys. When compared to the intravenous administration of sodium arsenate, the absolute bioavailability was reported as 25.9% in rabbits and 24.2% in monkeys. When compared to an oral dose (gavage) of sodium arsenate, the comparative bioavailabilities were 67.8% in rabbits and 43.6% in monkeys, in general agreement to findings of Ng et al. (1998b) in rats. Such data on availability of arsenic in soil needs to be considered in assessing human uptake of arsenic from soil (for more details on bioavailability see Table 15).

5.2.5 *Miscellaneous exposures*

Smokers are exposed to arsenic by the inhalation of mainstream cigarette smoke. It has been estimated that someone in the USA smoking 40 cigarettes per day would inhale about 10 µg of arsenic (ATSDR, 1993).

Proprietary herbal asthma medicines have been shown to contain up to 107 mg/g of inorganic arsenic (Chan, 1994).

5.3 Occupational exposures

There is the potential for significant occupational exposure to arsenic in several industries, in particular non-ferrous smelting, electronics, wood preservation, wood joinery shops, arsenic production, glass manufacturing, and the production and application of arsenical pesticides. Exposure is primarily through inhalation of arsenic-containing particulates, but ingestion and dermal exposure may be significant in particular situations. (e.g. preparation of CCA-treated timber). It is extremely rare for workers to be exposed to arsenic alone: the exposure is usually to arsenic in combination with other elements. Data on typical exposure levels of arsenic in the workplace are difficult to obtain and may vary considerably between different locations of the same industry because of the level of occupational hygiene in place and the chemical properties of the materials processed. Also, they are often out of date with regard to the current level of industrial hygiene. Currently, countries which have occupational regulations for arsenic have set the limit for inorganic arsenic between 0.01 and 0.1 mg/m^3 (ILO, 1991; DFG, 1999; Ministerie van Soziale Zaken en Werkgelegenheid, 2000; OSHA, 2000). The following examples are given to illustrate levels found in specific industries in various locations worldwide and provide some information on present and past exposures of workers to arsenic. They should not be considered as representative of all similar industrial sites.

Some workplace exposures dating from before 1980 are summarized in IPCS (1981). For example, in a Swedish copper smelter during the mid-1950s levels of arsenic ranged between 0.06 and 2 mg/m^3, but at the same facility in the 1970s levels of arsenic between 0.002 and 0.23 mg/m^3 in the air breathed by the workers were reported. Several other studies described in IPCS (1981) reported levels of arsenic in non-ferrous metal production to be between 0.001 and 0.3 mg/m^3 depending on the job location and the level of ventilation.

Welch et al. (1982), on the basis of industrial hygiene measurements made from 1943 to 1965, estimated average arsenic concentrations of workers in various departments of a copper smelter in the USA who were employed before 1956. Very high exposures (> 5 mg/m^3) were estimated in the following departments: arsenic roaster (20 mg/m^3), electrostatic precipitator (13 mg/m^3), arsenic refinery (7.5 mg/m^3), and main flue (7 mg/m^3). High exposures (0.5–4.99 mg/m^3) were estimated in these departments: masons' shop (3 mg/m^3), ore roaster (1 mg/m^3), materials crushing (1 mg/m^3) and reverberatory furnaces (0.6 mg/m^3). Medium (0.1–0.49 mg/m^3) or low exposures (< 0.1 mg/m^3) were estimated in the 10 other departments of the smelter in which arsenic measurements were carried out. In another copper smelter in the USA, Pinto et al. (1976) reported an overall mean arsenic concentration of 0.05 mg/m³ (range 0.003–0.3 mg/m³) on the basis of data from 24 workers wearing personal air samplers on 5 consecutive days. For 1973, a more detailed exposure estimation within the 32 departments of the smelter was made on the basis of the individual 24-h urinary excretion of arsenic in 1000 workers (Pinto et al., 1978). The highest average urinary arsenic excretions (μg As/litre of urine) were calculated for the following departments: electrostatic precipitator (526 mg As/litre), arsenic plant (516 μg As/litre), roaster (414 μg As/litre) and boiler room (409 μg As/litre). Eleven other departments had urinary excretion levels between 289 and 201 μg As/litre of urine, and the remaining 17 areas had levels between 180 and 58 μg As/litre, the lowest level calculated for the refined casting department. On the basis of a study by Enterline & Marsh (1982) airborne arsenic levels (as μg/m³) will be about one-third of the urinary excretion concentrations (as μg As/litre of urine). These authors also summarized the airborne arsenic levels in the smelter between 1938 and 1957. Levels varied by department, but were all high. For example, between 1947 and 1953, in a total of 25 samples from the arsenic plant they found airborne arsenic concentrations ranging from 0.8 to 41.4 mg/m³.

Vahter et al. (1986) reported airborne arsenic levels (8 h TWA) of 1–194 μg/m³ in a copper smelter. Daily urinary excretion of total arsenic metabolites ranged from 16 to 328 μg As/g creatinine. Correlation between urinary excretion of arsenic species and 8-h TWAs of arsenic between 0.8–45 μg/m³ in 24 workers in a copper smelter and an As_2O_3 refinery were reported by Hakala & Pyy (1995). The best correlation was obtained between urinary excretion of the sum of arsenite and arsenate species in urine samples taken 8 h after exposure. An exposure to an 8-h TWA of 10 μg/m³ was calculated to lead to an inorganic arsenic concentration of 5 μg/litre in urine. Jakubowski et al. (1998) reported levels of arsenic in a copper smelter between 1 and 746 μg/m³ in the worker's breathing zone (8-hTWA) resulting in daily urinary excretion of 2–850 μg As/g creatinine. On the basis of results from this study and three others, the authors calculated that daily exposure to arsenic concentrations of 10 or 50 μg/m³ corresponded to concentrations of total urinary metabolites of 30 μg/litre and 70 μg/litre (specific gravity 1.024) respectively. This compares to urinary excretions (total As metabolites) of 5–30 µg As/litre in people not excessively exposed via the workplace or from the consumption of seafood (Foa et al., 1984; Vahter et al., 1986). Simonato et al. (1994) reported urinary excretion of 183–205 µg As/g creatinine of arsenic metabolites in a cohort of gold-miners and refinery workers. Using airborne arsenic data for 1952–1991, Ferreccio et al. (1996) categorized workers' exposure to arsenic in various units of a copper mine and smelter complex (in µg As/m[3)] as follows: workshop and administration, 9.8; administrative area, 1.6; mine, 2.3; oxide production, 3.1; sulfur plant, 8.4; smelter, 201.7. In comparison, Offergelt et al. (1992) reported levels of arsenic (TWA) between 6 and 502 µg/m^3 in a sulfuric acid plant. As part of an epidemiological investigation on lung cancer mortality of workers in non-ferrous mines, Liu & Chen (1996) measured airborne arsenic levels in 1978, 1981 and 1988. In chronological order, concentrations of arsenic reported (in mg/m^3) were 0.23 (range 0.004–0.577 in 6 samples); 0.06 (range 0.003–0.166 in 14 samples) and 0.32 (range 0.028–1.442 in 8 samples).

Workers in certain glass-manufacturing industries may be exposed to airborne arsenic through the use of As_2O_3 (IARC, 1993). Workers in the heavy crystal industry in Germany were found to have urinary arsenic concentrations ranging from 3 to 114 µg/g creatinine, with 36% of the cases in 1976 and 18% of cases in 1981 being above the upper normal limit of 25 µg As/g creatinine (Schaller et al., 1982). A study in the USA of 35 crystal glassworkers within the mix-and-melt and batch-house areas indicated the potential for arsenic exposure (Chrostek et al., 1980). Personal air monitoring of 8 workers found airborne arsenic concentrations of 2–11 mg/m^3. The mean urinary arsenic excretion in 18 workers involved in weighing and mixing chemicals in a specialist glass-manufacturing facility was 79.4 µg/g creatinine compared to 4.4 µg/g creatinine in controls (Farmer & Johnson, 1990). In a Belgian glass factory, Roels et al. (1982) measured urinary excretion of arsenic in 10 workers ranging between 10 and 941 µg/g creatinine compared to a range of 7.6 to 59 µg/litre in control workers. The authors concluded that the high urinary arsenic concentrations in the workers were more related to oral intake due to poor hygienic practices than to pulmonary uptake.

Airborne arsenic levels in a wood joinery shop handling treated wood were reported to be 0.043–0.36 mg/m³ (IPCS, 1981). In a more recent study of joinery shops (Nygren et al., 1992), airborne arsenic concentrations between 0.54 and 3.1 µg/m³ were reported. In two workshops machining wood impregnated with CCA, levels of arsenic in personal air samples were reported to be 30–67 µg/m³ in plant A (8 workers) and 10–62 µg/m³ in plant B (8 workers) (Subra et al., 1999).

Workers in coal-powered power plants may also be exposed to arsenic found in the coal, or more likely that found in the fly ash during cleaning. Yager et al. (1997) reported arsenic concentrations (8-h TWA) between 0.17 and 375.2 µg/m³ (mean 48.3) in the breathing zone of maintenance workers in a coal-fired power plant in Slovakia. The urinary excretion of total urinary arsenic metabolites ranged between 2.6 and 50.8 µg As/g creatinine (mean 16.9). The authors estimated a mean urinary excretion of 13.2 µg As/g creatinine, in workers exposed to fly ash, from an 8-h TWA exposure to 10 µg As/m³, suggesting that the bioavailability of arsenic in coal fly ash is approximately one-third that seen in smelters. Concentrations of arsenic in the breathing zone of underground gold-miners in Ontario (Canada) were reported to range between 2.4 and 5.6 µg/m³ (geometric mean) with urinary arsenic concentrations reported to range between 23.5 and 25.9 µmol As/mol creatinine (Kabir & Bilgi, 1993). The median total urinary arsenic concentration of the miners was significantly higher than that of a control group, but no correction was made for differences in dietary habits of the two groups. In a study relating arsenic exposures to lung cancer among tin-miners in Yunnan province (China), Taylor et al. (1989) reported mean concentrations of airborne arsenic to range from 0.42 mg/m³ in 1951 to 0.01 mg/m³ in 1980.

5.4 Total human intake of arsenic from all environmental pathways

For healthy humans who are not occupationally exposed the most significant pathway of exposure to arsenic is through the oral intake of food and beverages. In areas with elevated concentrations of arsenic in drinking-water, this source make a significant contribution to the total intake of inorganic arsenic. For example, a consumption of 1.4 litres of drinking-water containing > 50 µg

As/litre could provide over 70 µg inorganic arsenic compared to an estimated daily intake, based on very preliminary data, of 12–14 µg inorganic arsenic from typical North American diets (Yost et al., 1998).

As shown in Table 12, the total estimated daily dietary intake of arsenic may vary widely, mainly because of wide variations in the consumption of fish and shellfish. Data in Table 12 are for total arsenic intake and do not reflect the possible variation in intake of the more toxic inorganic arsenic species (see sections 5.2.2 and 5.2.3). In areas where drinking-water contains > 50 µg As/litre, water may be the major source of inorganic arsenic. All other routes of intakes of arsenic (intranasal and dermal) are of minor importance in comparison to the oral route (ATSDR, 1993). For example, inhalation would add about 1 µg As/day from airborne particulates and approximately 6 µg As /day may be inhaled from 20 cigarettes.

The most appropriate approach to determining the internal dose of inorganic arsenic in individuals in specific populations is to measure the arsenic species in urine. Concentrations of total urinary arsenic and metabolites of inorganic arsenic (inorganic arsenic, MMA and DMA) provides estimates of the exposure (uptake) to total arsenic and inorganic arsenic, respectively (see section 6.3). Reported concentrations of metabolites of inorganic arsenic in urine with no known exposure to arsenic are generally < 10 µg/litre in European countries (Apel & Stoeppler, 1983; Valkonen et al., 1983; Foa et al., 1984; Vahter & Lind, 1986; Andren et al., 1988; Jensen et al., 1991; Buchet et al., 1996; Trepka et al., 1996; Kristiansen et al., 1997; Kavanagh et al., 1998). Similar or slightly higher concentrations are reported from studies in some parts of the USA (Smith et al., 1977, Morse et al., 1979; Binder et al., 1987; Kalman et al., 1990; Pollisar et al., 1990; Gottlieb et al., 1993; Bates et al., 1995; Lewis et al., 1999) and around 50 µg/litre in Japan (Yamamura et al., 1979; Yamauchi et al., 1992). In West Bengal and Bangladesh arsenic concentrations > 1 mg/litre have frequently been observed (Chatterjee et al., 1995; Das et al., 1995).

The concentration of arsenic metabolites in urine correlates well with the concentration of arsenic in the drinking-water. However, the relationship may vary considerably depending on the amount of water consumed and the amount of water used for preparation of drinks and food. For example, studies from California and Nevada (USA) showed that a water concentration of 400 μg/litre corresponded to about 230 μg/litre in urine (total arsenic) and 100 μg/litre in water corresponded to 75 μg/litre in urine (Valentine et al., 1979). Similarly, people in Alaska drinking water containing about 400 μg/litre had on average 180 μg/litre in urine, and those drinking water containing 50–100 μg/litre had on average 45 μg/litre in urine (Harrington et al., 1978). Thus, urinary arsenic concentration was about half of that in the water. However, people living in areas of northern Argentina with drinking-water containing 200 μg/litre had much higher arsenic concentrations in the urine (metabolites of inorganic arsenic) – on average 250–450 μg/litre (Vahter et al., 1995a; Concha et al., 1998a). The fluid intake of these people consisted mainly of drinking-water or drinks prepared at home from the drinking-water. Also, most of the food consumed was prepared at home using the local drinking-water. In areas in the north-east of Taiwan where drinking-water concentrations were 50–300 μg/litre, people had similar concentrations in the urine; about 140 μg/litre (Chiou et al., 1997a).

Soil (section 5.2.4) may be a significant source of arsenic intake, particularly for children. However, the bioavailability may vary considerably.

Some studies have been conducted for the purpose of evaluating whether there is an increased body burden of arsenic in children living near arsenic-contaminated sites relative either to children from areas of low arsenic exposure or to adults. For example, Binder et al. (1987) reported that total urinary arsenic excretion was significantly increased in children living in a Montana (USA) community with high levels of arsenic in soil (average ~400–700 mg/kg) compared to a community with low arsenic levels in soil (44 mg/kg). In first-morning urine samples taken in July in the high-arsenic community, mean total arsenic in urine averaged 54 μg/litre (53.8 μg/g creatinine) compared to 16.6 μg/litre (17.1 μg/g creatinine) in the low-arsenic community.

Trepka et al. (1996) studied differences in arsenic exposure among children of different age groups in various areas of Germany, as assessed by urinary arsenic excretion. They reported no marked age- or gender-related differences, although urinary arsenic excretion was significantly increased in children from the most polluted area (5.1 µg/litre vs. 4 µg/litre in the control area). However, the authors did not consider this increase to be toxicologically significant. In contrast, Diaz-Barriga et al. (1993) reported increases in urinary arsenic in children living closest to a copper smelter (median soil levels ~500 mg/kg arsenic; range 69–594 µg/g creatinine in urine) compared to children living 7–25 km away (median soil levels ~11–14 mg/kg arsenic). Urinary arsenic excretion (normalized to creatinine) was more than doubled, and arsenic levels in hair were more than 10-fold higher.

6. KINETICS AND METABOLISM IN LABORATORY ANIMALS AND HUMANS

Humans are exposed to many different forms of inorganic and organic arsenic species (arsenicals) in food, water and other media. Study of the kinetics and metabolism of arsenicals in animals and humans can thus be quite complex, as a result of differences in physicochemical properties and bioavailability of the various forms of arsenic. Arsenic metabolism is also characterized by relatively large qualitative and quantitative interspecies differences. Given the relatively large interspecies differences in arsenic metabolism, and that there is considerable information on human metabolism of arsenicals, discussion of animal studies focuses on areas where human data is inadequate or where animal data can serve to aid in the interpretation of toxic effects caused by arsenicals. This chapter covers both inorganic and organic arsenicals.

6.1 Inorganic arsenic

The metabolism and disposition of inorganic arsenic may depend on its valence state, particularly at high doses. The two most common valence states to which humans might be environmentally exposed are the trivalent and pentavalent forms. Since these two forms are readily interconverted, studies cited in this review were evaluated with particular attention to whether methods used were appropriate to ensure that inorganic arsenic was maintained in the intended valence state until the time of administration. Arsenite, but not arsenate, exists mainly in the non-ionized form at physiological pH and relatively low Eh.

6.1.1 Absorption

6.1.1.1 Respiratory deposition and absorption

Human inhalation exposure to inorganic arsenic can occur as a consequence of industrial activity (e.g. smelting of ores) and energy production (e.g. coal-fired power plants), and during cigarette smoking. Arsenic in air exists on particulate matter and thus respiratory absorption of arsenic is a two-stage process, involving deposition of the particles on to airway and lung surfaces, followed by absorption of arsenic from deposited particulates. The extent of deposition of inhaled arsenic will depend largely on the size of the inhaled particulates, and absorption of deposited arsenic is highly dependent on the solubility of the chemical form of arsenic.

a) Animal studies

Quantification of the relative amount of airborne arsenic that is deposited in various parts of the respiratory tract is not possible because there is a lack of such animal inhalation studies. However, intratracheal instillation studies provide information on the extent of absorption of various chemical forms of inorganic arsenic. In general, solubility appears to be the most important physicochemical property determining the extent of lung clearance, although wetting capacity and pulmonary toxicity may also have an important influence. It is important to note that clearance of particulates by the mucociliary escalator may also result in oral exposure.

Pershagen et al. (1982) found that lung concentrations of arsenic in hamsters given weekly intratracheal instillations of As_2O_3, arsenic trisulfide or calcium arsenate differed by a factor of approximately 10-fold after 4 weeks. The much more rapid clearance of As_2O_3 was attributed to its being much more soluble *in vivo* than the other two arsenicals. The authors speculated that the clearance of calcium arsenate was much slower than that of arsenic trisulfide because of its higher wetting capacity, which would result in more calcium arsenate being transported to the alveolar regions of the lung where clearance is slower. The authors also indicated that the pulmonary toxicity of calcium arsenate may have impaired normal clearance mechanisms which would have prolonged lung retention.

Marafante & Vahter (1987) reported that the extent of absorption of inorganic arsenicals from the lungs of hamsters after intratracheal instillation was directly correlated with their *in vivo* solubility as determined by the amount of radiolabelled arsenical retained at an intramuscular injection site. The lung retention of arsenic (2 mg As/kg) 3 days after an intratracheal instillation of sodium arsenite, sodium arsenate, arsenic trisulfide and lead arsenate was respectively 0.06%, 0.02%, 1.3% and 45.5% of the dose. Similar observations have been reported by Buchet et al. (1995), who found that 24 h after a single intratracheal instillation of soluble arsenic salts in hamsters (NaAsO$_2$ and Na$_2$HAsO$_4$) at 50 and 100 µg/kg, the amount of arsenic detected in the lung was not different from that found in the control animals. Buchet et al. (1995) also observed that both lung retention and urinary excretion indicate a prolonged contact of the lung tissue with particulate arsenic rather than soluble arsenic salts. Some 48 h after intratracheal administration of arsenic in the form of fly ash or copper smelter dust, lung retention amounted to 25–35% of the administered dose (50–100 µg As/kg). Rosner & Carter (1987) also reported, based on results of intratracheal instillation studies in hamsters, that the more soluble forms of arsenic, sodium arsenate and sodium arsenite (5 mg/kg dose), had a relative bioavailability 10-fold greater than gallium arsenide (GaAs). Webb et al. (1987) also reported that decreasing the particle size by a factor of 2 increased the *in vivo* dissolution rate and toxicity of GaAs in rats after intratracheal instillation.

b) Human studies

The available human data are insufficient to allow quantitative estimation of regional arsenic deposition in the respiratory tract. Occupational studies in which both the concentration of inorganic arsenic in the breathing zone and the urinary excretion of inorganic arsenic and its metabolites were determined provide information on arsenic absorption. These studies (e.g. Vahter et al., 1986; Yamauchi et al., 1989a; Offergelt et al., 1992; Hakala & Pyy, 1995; Yager et al., 1997) demonstrate that excretion of inorganic arsenic and sometimes total arsenicals and methylated metabolites are significantly increased in workers exposed to arsenic in their breathing zone. This indicates that arsenic is absorbed from the respiratory tract, but does not provide sufficient information for quantitative estimation of arsenic absorption after inhalation because the contribution of oral exposure after mucociliary clearance – and in some instances probably also from diet and drinking-water – cannot be assessed.

Comparison of studies relating occupational arsenic exposure in different industrial settings to urinary arsenic excretion suggests that there are differences in respiratory absorption depending on the form of arsenic. Using equations relating urinary arsenic excretion to air concentrations, Yager et al. (1997) noted that in several studies the predicted urinary arsenic output for workers exposed to 10 µg/m^3 arsenic was more than one-third lower for boiler maintenance workers in a coal-fired power plant than it was for copper-smelter workers. This finding was attributed to the fact that the arsenic in coal fly ash in their study was predominantly in the form of calcium arsenate, whereas in the copper smelter work environment the arsenic was in the form of As$_2$O$_3$. Such an interpretation is consistent with the much greater retention of calcium arsenate than As$_2$O$_3$ in hamster lung that was reported by Pershagen et al. (1982).

6.1.1.2 *Gastrointestinal absorption*

Arsenic can be absorbed from the gastrointestinal tract after ingestion of arsenic-containing food, water, beverages or medicines, or as a result of inhalation and subsequent mucociliary clearance. The bioavailability of ingested inorganic arsenic will vary depending on the matrix in which it is ingested (e.g. food, water, beverages, soil), the solubility of the arsenical compound itself and the presence of other food constituents and nutrients in the gastrointestinal tract.

a) *Animal studies*

Soluble arsenates and arsenites are rapidly and extensively absorbed from the gastrointestinal tract of common laboratory animals after a single oral dose (Table 13). The mouse data of Vahter & Norin (1980) indicate that arsenite may be more extensively absorbed from the gastrointestinal tract than arsenate at lower doses (e.g. 0.4 mg As/kg), whereas the reverse appears to occur at higher doses (e.g. 4.0 mg As/kg). In these same studies (Vahter & Norin, 1980) about the same percentage faecal elimination was observed following the same dose given orally and subcutaneously, indicating nearly complete gastrointestinal absorption (Table 13).

Studies conducted by Odanaka et al. (1980) suggest that much less pentavalent arsenic is absorbed from the gastrointestinal tract after oral administration – 48.5% of dose (5 mg/kg) in urine, compared to the 89% of dose (4 mg/kg) in urine found by Vahter & Norin (1980) (Table 13). This difference may be attributable to the fact that the mice in the study of Vahter & Norin were not fed for at least 2 h before and 48 h after dosing, whereas the mice in the Odanaka et al. studies were not food restricted. Studies by Kenyon et

Table 13. Cumulative 48-h elimination (% of dose) of As in urine and faeces of laboratory animals after oral and parenteral administration of inorganic As

Species	As form	Dose	Route	Urine	Faeces	Total	Reference
Rat	Arsenic acid	5 mg/kg	oral	17.2	33.0	50.2	Odanaka et al. (1980)
		1 mg/kg	i.v.	51.0	0.8	51.8	
Hamster	Arsenic acid	5 mg/kg	oral	43.8	44.1	87.9	Odanaka et al. (1980)
		1 mg/kg	i.v.	83.9	4.0	87.9	
Hamster	As trioxide	4.5 mg/kg	oral	43.5	9.4	52.9	Yamauchi & Yamamura (1985)
Mouse	Arsenic acid	5 mg/kg	oral	48.5	48.8	97.3	Odanaka et al. (1980)
		1 mg/kg	i.v.	86.9	2.6	89.5	
Mouse[a]	Sodium arsenate	0.4 mg As/kg	s.c.	86 ± 3.6	6.4 ± 2.1	92.4	Vahter & Norin (1980)
		0.4 mg As/kg	oral	77 ± 3.6	8.0 ± 1.6	85	
		4.0 mg As/kg	oral	89 ± 3.6	6.1 ± 1.2	95.1	

Table 13 (contd.)

Mouse[a]	Sodium arsenite	0.4 mg As/kg	s.c.	73 ± 5.3	3.8 ± 1.6	76.8	Vahter & Norin (1980)
		0.4 mg As/kg	oral	90 ± 2.4	7.1 ± 2.0	97.1	
		4.0 mg As/kg	oral	65 ± 2.1	9.1 ± 1.9	74.1	
Mouse	Sodium arsenate	0.00012 mgAs/kg	oral	65.0	16.5	81.5	Hughes et al. (1994)
		0.0012 mgAs/kg	oral	68.3	13.5	81.8	
		0.012 mgAs/kg	oral	72.1	10.5	82.6	
		0.12 mgAs/kg	oral	71.0	14.6	85.6	
		1.2 mgAs/kg	oral	68.7	18.2	86.9	
Rabbit	Sodium arsenite	0.050 mg As/kg	i.p.	75.7	9.9	85.6	Marafante et al. (1982)

[a] Data given as mean ± SEM

al. (1997) suggest that feeding a diet lower in fibre or "bulk" to female B6C3F$_1$ mice increased absorption of sodium arsenate by ~10% compared to a standard rodent chow diet, after a single oral dose of 5 mg As/kg.

Yamauchi et al. (1986b) studied the absorption and metabolism of GaAs (a relatively insoluble arsenical compared to sodium arsenite and sodium arsenate) in hamsters after a single oral or intraperitoneal dose. Faecal elimination of total arsenic after 5 days averaged 87.5 ± 13.8%, 79.4 ± 10.6% and 77.9% after oral doses of 10, 100 and 1000 mg/kg GaAs, respectively. However, after a single intraperitoneal dose of 100 mg/kg GaAs only 0.38% of the total arsenic dose was eliminated in faeces after 5 days. During this same time period less than 1% of the dose was eliminated in urine irrespective of the route of administration, indicating that GaAs is minimally absorbed from the gastrointestinal tract. It is noteworthy that a consistently greater amount of DMA was excreted in urine over time after intraperitoneal administration than after oral administration. This indicates that arsenic liberated from GaAs undergoes methylation, and is consistent with results reported in hamsters after intratracheal instillation of GaAs (Rosner & Carter, 1987).

The bioavailability of arsenic from soils has been assessed using various animal models because this can be a significant issue in risk assessment for contaminated industrial sites where there is potential for arsenic exposure via soil ingestion. As summarized in Table 14, these studies indicate that oral bioavailability of arsenic in a soil or dust vehicle is often lower than that of the pure soluble salts typically used in toxicity studies. However, bioavailability is substantially dependent on the soil type. A study by Vahter (1988) showed that although some soil samples from former wood-treatment plants containing 1.1 mg As/g were highly toxic, other soil samples containing 9 mg As/g were without any effect when tested in mice. Similarly, the mean relative bioavailability of arsenic in mining wastes compared to that of sodium arsenate administered to young swine and analysed as arsenic in urine was found to range from 7–52% (US EPA, 1996). Davis et al. (1992) have pointed out that this is due mainly to mineralogical factors which control solubility in the gastrointestinal tract, such as solubility of the arsenic-bearing mineral itself and encapsulation within insoluble matrices (e.g.

Table 14. Oral bioavailability of As from soil, based on studies in laboratory animals

Species	Duration (h)	Intravenous dose	Soil or sample	Soil dose	Bioavailability[a] (%, mean ± SD)	Method	Reference
Beagle dog	120	2 mg As(V)	Netherlands bog ore[b]	6.6–7.0 mg As	8.3 ± 2.0	AUC[c] for urinary excretion	Groen et al. (1994)
New Zealand white rabbit	120	1.95 mg As(V)/kg	smelter impacted soil (Anaconda, Montana, USA)	0.78 mg As/kg 1.95 mg As/kg 3.9 mg As/kg	24 ± 3.2	AUC for urinary excretion, no dose-dependency observed	Freeman et al. (1993)
			sodium arsenate	1.95 mg/kg	50 ± 5.7		
Cynommolus monkey	168	0.62 mg As(V)/kg	soil (Anaconda, Montana, USA)	0.62 mg As/kg	14 (11)	AUC for urinary excretion (AUC for blood in parentheses)	Freeman et al. (1995)
			house dust (same location)	0.26 mg As/kg	19 (10)		
			sodium arsenate	0.62 mg As/kg	68 (91)		

119

Table 14 (contd.)

Species	Duration (h)	Intravenous dose	Soil or sample	Soil dose	Bioavailability[a] (%, mean ± SD)	Method	Reference
Immature swine	144	0.01–0.31 mg As/kg	soil[d]	0.04–0.24 mg As/kg	52	AUC for blood	US EPA (1996)
			slag	0.61–1.52 mg As/kg	28		
			sodium arsenate	0.01–0.11 mg As/kg	68		
Rat	96	0.5 mg As(III)/kg	soil (Watson, Australia)[e]	0.5–5.0 mg As/kg	0.55–2.98 (As(V))[f]	AUC for blood	Ng et al. (1998b)
		0.5 mg As(V)/kg			1.02–9.87(As(III))		

[a] Comparison to intravenous administration
[b] Bog ore is naturally high in As
[c] AUC, area under the curve
[d] Soil or slag from Ruston/North Tacoma Superfund site in Tacoma, Washington (USA)
[e] Soil from site contaminated with Arsenical pesticides (former cattle dip); soils contained 32–1597 mg As/kg soil
[f] Figures are relative to sodium arsenate and sodium arsenite

silica). The comparative bioavailabilities of arsenic in soil from a site contaminated with copper chrome acetate (CCA) and soil contaminated by arsenic solutions used in cattle tick control were reported by Ng & Moore (1996), using a rat model. Soil from the cattle dip site had a bioavailability of $8.1 \pm 4\%$, $14.4 \pm 7\%$ and $60 \pm 3.4\%$ when compared with orally administered sodium arsenite, calcium arsenite and sodium arsenate respectively. For CCA-contaminated soil the corresponding comparative bioavailabilities were $13.0 \pm 4.5\%$, $32.2 \pm 11.2\%$ and $38.0 \pm 13.2\%$. Ng et al. (1998b), also using a rat model, have reported the absolute bioavailability of arsenic in soils containing 32–1597 µg As/kg (0.32–56% arsenite) from a combination of arsenical pesticides and natural geological formations in a residential area. The absolute bioavailability ranged from 1.02 to 9.87% relative to arsenite and from 0.26 to 2.98% relative to arsenate.

Several older studies reviewed in the previous IPCS arsenic document (IPCS, 1981, sections 6.1.1 and 6.2) demonstrated that composition of the diet can alter gastrointestinal absorption of arsenic. Some more recent studies have examined the mechanism of arsenical uptake and interaction with nutrients at the intestinal level. Gonzalez et al. (1995), using isolated perfused rat small intestine, demonstrated that uptake of pentavalent arsenic is carried out by a saturable transport process and that addition of phosphate markedly decreased arsenic absorption, most likely because arsenate and phosphate can share the same transport mechanism. Hunder et al. (1993), using isolated rat jejunal segments, found that increasing concentrations of arsenite (2.5–250 µmol/litre) and arsenate (2.5–2500 µmol/litre) caused a dose-dependent decrease in the intestinal transfer of water, sodium, glucose and leucine, with arsenite being about 5-fold more potent than arsenate.

b) Human studies

In common with studies in experimental animals, controlled ingestion studies in humans indicate that trivalent and pentavalent arsenic are both well absorbed from the gastrointestinal tract (Table 15). For example, Pomroy et al. (1980) reported that healthy male human volunteers excreted $62.3 \pm 4.0\%$ of a 0.06-ng dose of [74]As-arsenic acid (As(V)) in urine over a period of 7 days, whereas only $6.1 \pm 2.8\%$ of the dose was excreted in the faeces. Few other

Table 15. Metabolism and urinary excretion of inorganic and organic arsenicals in humans after experimental administration

Form	No. of subjects	Dose and frequency	Time interval	% dose in urine	% of total urinary metabolites				Reference
					As(V)	As(III)	MMA	DMA	
As acid	6	0.01 µg	5 days	57.9	27.2 (IAs)[a]		20.6	51.0	Tam et al. (1979)
As acid	6	0.06 ng	7 days	62.3	ND	ND	ND	ND	Pomroy et al. (1980)
As trioxide	1	700 µg	3 days	68.2	7.9	31.7	28.2	32.2	Yamauchi & Yamamura (1979)
Sodium arsenite	3	500 µg	4 days	45.1	25 (IAs)		21.3	53.7	Buchet et al. (1981a)
Sodium metaarsenite	1	125 µg × 5 days	14 days	54	16 (IAs)		34	50	Buchet et al. (1981b)
	1	250 µg × 5 days	14 days	73	7 (IAs)		20	73	
	1	500 µg × 5 days	14 days	74	19 (IAs)		21	60	
	1	1000 µg × 5 days	14 days	64	26 (IAs)		32	42	
Sodium MMA	4	500 µg	4 days	78.3	ND	ND	87.4	12.6	Buchet et al. (1981a)
Sodium DMA	4	500 µg	4 days	75.1	ND	ND	ND	100	Buchet et al. (1981a)

Subjects in all studies cited were adult males

[a] IAs, sum of As(III) and As(V)

controlled human ingestion studies have actually reported data on both urine and faecal elimination of arsenic. However, between 45% and 75% of the dose of various trivalent forms of arsenic is excreted in the urine within a few days (Table 15), which suggests that gastrointestinal absorption is both relatively rapid and extensive.

6.1.1.3   Dermal absorption

Wester et al. (1993) studied the percutaneous absorption of arsenic acid (H_3AsO_4) from water and soil both *in vivo* using rhesus monkeys and *in vitro* with human skin. *In vivo*, absorption of arsenic acid from water (loading 5 $\mu l/cm^2$ skin area) was 6.4 ± 3.9% at the low dose (0.024 ng/cm^2) and 2.0 ± 1.2% at the high dose (2.1 $\mu g/cm^2$). Absorption from soil (loading 0.04 g $soil/cm^2$ skin area) *in vivo* was 4.5 ± 3.2% at the low dose (0. 04 ng/cm^2) and 3.2 ± 1.9% at the high dose (0.6 $\mu g/cm^2$). Thus, *in vivo* in the rhesus monkey, percutaneous absorption of arsenic acid is low from either soil or water vehicles and does not differ appreciably at doses more than 10 000-fold apart. Wester et al. (19993) also reported that for human skin, at the low dose, 1.9% was absorbed from water and 0.8% from soil over a 24-h period.

Limited data suggest that the *in vitro* percutaneous absorption of inorganic arsenicals may differ substantially depending on chemical form or species used. Rahman et al. (1994) evaluated the percutaneous absorption of sodium arsenate *in vitro* using clipped full-thickness dorsal skin of $B6C3F_1$ mice. They found that a constant fraction of the applied dose was absorbed over a 24-h period irrespective of dose level (5–500 ng or 0.36–360 mg/kg for soil), but that the vehicle or vehicle volume had significant effects. Using 100 μl water as a vehicle resulted in ~60% of applied dose being absorbed, whereas using a volume of 250 μl water resulted in ~37% absorption, which was about the same percentage absorbed if the chemical was applied in the solid form. Absorption of sodium arsenate from soil was minimal (< 0.3%), which is similar to what was reported for arsenic acid in the studies of Wester et al. (1993).

6.1.1.4   Placental transfer

a)   Animal studies

Both older and more recent studies have documented the ability of trivalent and pentavalent inorganic arsenic to cross the placenta in laboratory animals. Lindgren et al. (1984) reported that in pregnant mice given a single intravenous injection (4 mg As/kg) of sodium arsenate or sodium arsenite, both forms passed through the placenta easily and to approximately the same extent. These investigators also reported that the rate of placental transfer was lower in a marmoset monkey (non-methylating species) injected intravenously with arsenite than in mice, and suggested that this was a consequence of stronger binding in maternal tissues.

Hood et al. (1987) compared the fetal uptake of sodium arsenate after oral (40 mg/kg) or intraperitoneal (20 mg/kg) administration to pregnant CD-1 mice on day 18 of gestation. Arsenic levels peaked later and over 5-fold lower in fetuses of mice dosed orally, most likely reflecting both slower uptake from the gastrointestinal tract and greater opportunity for methylation in the liver before the arsenic reached the systemic circulation. The quantity of dimethylated metabolite present in the fetuses rose over time (to ~80% of total metabolites present for both routes of administration) and remained relatively constant from ~10 h after dosing until the study ended, 24 h after dosing.

Hood et al. (1988) also compared the fetal uptake of sodium arsenite after oral (25 mg/kg) or intraperitoneal (8 mg/kg) administration to mice that were 18 days pregnant. As was the case with arsenate, injected mice achieved both higher fetal and placental levels of arsenic more quickly than did mice dosed orally. Both valence forms followed similar time-course trends after oral administration. However, levels of arsenic in fetuses of dams injected with arsenite reached a plateau 12–24 h after dosing, whereas levels of arsenic in fetuses of dams injected with arsenate peaked at 2–4 h after dosing and then declined quickly. The proportion of arsenic present in fetuses as methylated metabolite increased over time to 88% and 79% after oral and intraperitoneal administration, respectively. A higher fraction of monomethylated arsenic was present in fetuses of dams dosed with arsenite than with arsenate. The authors concluded that much of the arsenic reaching the fetus has already been transformed to the less acutely toxic methylated metabolites.

b)   Human studies

Case reports of arsenic poisoning in pregnant women resulting in death of the fetus accompanied by toxic levels of arsenic in fetal organs and tissues demonstrate that arsenite (As_2O_3) readily passes through the placenta (Lugo et al., 1969; Bollinger et al., 1992). In a more recent study, Concha et al. (1998b) reported that arsenic concentrations were similar in cord blood and maternal blood (~9 µg/litre) of maternal–infant pairs exposed to drinking-water containing high levels of arsenic (~200 µg/litre). Another study of an "unexposed" population in the southern USA found that concentrations of arsenic in cord blood and maternal blood (about 2 µg/litre) were also similar, and suggests that arsenic readily crosses the placenta (Kagey et al., 1977).

6.1.2   Distribution

6.1.2.1   Fate of inorganic arsenic in blood

a)   Animal studies

Inorganic arsenic is rapidly cleared from the blood in most common laboratory animals, including mice, rabbits, and hamsters (Vahter & Norin, 1980; Marafante et al., 1982, 1985; Yamauchi & Yamamura, 1985). The notable exception to this is the rat in which the presence of arsenic is prolonged owing to accumulation in erythrocytes (Vahter, 1981; Marafante et al., 1982; Lerman & Clarkson, 1983). For example, Marafante et al. (1982) reported that levels of arsenic in erythrocytes were 2-fold, 102-fold and 268-fold higher at 1, 16 and 48 h after dosing in rats compared to rabbits that received the same intraperitoneal dose of arsenite; in the plasma of these same animals the rat : rabbit ratio of arsenic never exceeded 1. Lerman & Clarkson (1983) further noted that higher levels of arsenic were achieved much more rapidly in the blood of rats dosed intravenously with arsenite than with arsenate, and that 95% or more of the arsenic in erythrocytes was in the form of DMA by 4 h after dosing. It appears that rat haemoglobin specifically binds DMA, and this greatly increases the biological half-life of inorganic arsenic and DMA in rats (Vahter, 1981; Vahter et al., 1984).

Although clearance of both arsenate and arsenite from blood in other mammalian species is rapid, differences dependent on both valence state and dose have been observed. Vahter & Norin (1980) reported that at a high oral dose of arsenic (4 mg As/kg), arsenite-dosed mice had a higher erythrocyte to plasma ratio (~2–3), whereas in arsenate-dosed mice the ratio was much closer to 1. No such difference was observed at a lower oral dose (0.4 mg As/kg) of arsenate or arsenite. Delnomdedieu et al. (1994b) investigated the *in vitro* uptake of arsenite and arsenate in intact rabbit erythrocytes. They reported that ~76% of arsenite, compared to ~25% of arsenate, was taken up within 0.5 h, and that arsenite subsequently bound with intracellular glutathione (GSH), whereas arsenate entered the phosphate pathway, depleting ATP and increasing inorganic phosphate levels.

Yamauchi & Yamamura (1985) characterized the forms of arsenic present and their distribution over time in whole blood, plasma and erythrocytes in male Syrian golden hamsters given a single oral dose of 4.5 mg/kg As_2O_3 (As(III)). Arsenic levels in whole blood had dropped to control levels by 72 h after dosing, indicating rapid clearance from the blood. In whole blood, inorganic arsenic, MMA and DMA concentrations peaked 1, 12 and 24 h after dosing, respectively; the levels of MMA and DMA achieved relative were 1/3 and 1/10 as high as those of inorganic arsenic. Inorganic arsenic and MMA were found mainly in erythrocytes and DMA occurred chiefly in plasma.

Marafante et al. (1985) also characterized the forms of arsenic present and their distribution over time in blood of rabbits dosed with arsenate. Male New Zealand white rabbits were injected intravenously with 0.4 mg As/kg and blood samples were taken at intervals from 15 min to 6 h after dosing. Within 15 min after dosing, 10% of the arsenic in the plasma was in the form of arsenite, 30% was present as arsenate and 60% was bound to plasma proteins. Arsenate was rapidly cleared from plasma (first-order $t_{1/2}$ ~1 h). Both arsenite and plasma protein-bound arsenic exhibited biphasic kinetics, with half-times of 10 min and 2 h for arsenite and 15 min and 2.5 h for protein-bound arsenic. In studies of giant Flemish rabbits using carrier-free [74]As (arsenate), DeKimpe et al. (1996) reported biphasic blood clearance rates of about 2 h and 58 h in plasma and 8.6 h and 170 h in erythrocytes. Taken together, the findings of Marafante et al. (1985) and DeKimpe et al. (1996) suggest a triphasic, rather than biphasic, clearance in plasma. These findings indicate that binding of arsenic to plasma protein is not strong. The kinetics and concentration of arsenite in erythrocytes were quite similar to that seen in plasma, but arsenate concentrations in erythrocytes were only about 1/10 as high. DMA levels peaked at about 4 h after dosing in both plasma and erythrocytes, but the concentration in erythrocytes was only about 20% of that found in plasma.

b) Human studies

Inorganic arsenic is reported to be rapidly cleared from blood. Results from some older studies reviewed in the previous arsenic IPCS document (IPCS, 1981, section 6.1.2), suggest that the kinetics of arsenic clearance in plasma and erythrocytes are similar, although levels in erythrocytes tended to be approximately 3-fold higher a few hours after exposure (similar to findings in laboratory animals). Mealey et al. (1959) measured the plasma and erythrocyte levels of radioactive arsenic after intravenous injection of labelled arsenite. The rate of decline of arsenic in the erythrocytes was comparable with that in plasma, but the erythrocytes contained about 3 times more arsenic than the plasma 10 h after the injection. The plasma curve showed a three-compartment model (IPCS, 1981, section 6.1.3). The first half-life seemed to be very short, and the bulk of the arsenic was removed from the plasma at a high rate. Some 24 h after dosing, less than 0.1% of the dose remained. The second phase of the curve showed a half-time of about 30 h. The third phase of the curve, beginning about 1 week after the injection, showed a very low rate of disappearance with a half-time of over 200 h.

Zhang et al. (1996a,b, 1997, 1998a,b) have reported on the distribution of arsenical species in serum and arsenic–protein binding in serum of patients with renal disease. The predominant species of arsenic present in serum were DMA (~15–30%) and arsenobetaine (~54–76%), with the remainder being protein-bound and inorganic arsenic; MMA was undetectable (Zhang et al., 1996a, 1997). Zhang et al. (1998a,b) further reported that only inorganic arsenic was bound to serum proteins, and that transferrin is the main carrier protein. It should be noted that since individuals with renal disease tend to accumulate arsenic in serum, these results may not be typical of the general population.

6.1.2.2 *Tissue distribution*

a) Animal studies

Studies in rabbits, rats, mice, hamsters and monkeys demonstrate that arsenic, administered orally or parenterally, in either the trivalent or pentavalent form, is rapidly distributed throughout the body (Lindgren et al., 1982; Marafante et al., 1982; Vahter et al., 1982; Vahter & Marafante, 1985; Yamauchi & Yamamura, 1985). Many of these studies have used radiolabelled arsenic, and it is noteworthy that arsenic-derived radioactivity is generally present in all tissues examined (Lindgren et al., 1982; Marafante et al., 1982; Vahter et al., 1982; Vahter & Marafante, 1985).

Comparative studies of arsenate and arsenite distribution at comparable dose levels provide insights on the influence of valence state on arsenic distribution. Lindgren et al. (1982) studied the distribution of arsenic in male C57BL mice intravenously administered 0.4 mg As/kg as either sodium arsenate or sodium arsenite at 0.5, 6, 24 and 72 h after dosing. Highest concentrations of arsenic-derived radioactivity were present in liver, kidney and gallbladder at 0.5 h after administration of either arsenate or arsenite, reflecting their rapid elimination (see section 6.1.4). Arsenate administration resulted in much lower arsenic concentrations in liver and gallbladder, but higher concentrations in kidney compared to administration of arsenite within 0.5 h after dosing, indicating valence-dependent differences in route of elimination. In general, concentrations of arsenic in organs tended to be higher after administration arsenite than of arsenate, with the notable exception of the skeleton at all time points (Table 16). This latter finding was ascribed to arsenate being a structural analogue of phosphate and substituting for it in the apatite crystal of bone. The greater retention of arsenite in tissues is a consequence of its reactivity and binding with tissue constituents, most notably sulfhydryl groups (Vahter & Marafante, 1983).

Table 16. Comparison of tissue distribution over time in mice given a single intravenous injection of ^{74}As-As (0.4 mg As/kg) as either sodium arsenate or sodium arsenite[a]

Tissue	Valence	Concentration of ^{74}As (ng/g) at specified time			
		0.5 h	6 h	24 h	72 h
Brain	5+	20 ± 3.6	26 ± 2.5	1.8 ± 0.1	0.6 ± 0.1
	3+	21 ± 1.5	41 ± 3.6	3.3 ± 0.3	0.9 ± 0.1
Stomach	5+	165 ± 15	81 ± 10	24 ± 1.6	11 ± 0.7
	3+	418 ± 58	118 ± 13	79 ± 5.2	27 ± 3.6
Duodenum	5+	553 ± 91	77 ± 14	5.6 ± 0.6	2.0 ± 0.5
	3+	1016 ± 96	150 ± 12	14 ± 1.9	4.6 ± 0.6
Small intestine	5+	214 ± 12	53 ± 8.4	3.9 ± 0.9	1.6 ± 0.3
	3+	582 ± 87	124 ± 6.4	9.0 ± 1.1	3.6 ± 0.3
Liver	5+	571 ± 68	77 ± 12	8.1 ± 0.4	3.3 ± 0.2
	3+	1589 ± 222	188 ± 11	29 ± 1.4	12.1 ± 1.0
Gall bladder	5+	1255 ± 298	200 ± 98	< 10	< 10
	3+	5172 ± 3022	422 ± 224	< 10	< 10
Kidney	5+	2355 ± 185	209 ± 33	20 ± 1.5	7.7 ± 0.7
	3+	1603 ± 211	200 ± 15	20 ± 0.9	7.6 ± 0.5
Lung	5+	291 ± 26	131 ± 19	8.0 ± 0.7	2.3 ± 0.1
	3+	540 ± 59	243 ± 37	23 ± 2.9	5.6 ± 0.4
Skin	5+	184 ± 18	46 ± 3.9	16 ± 1.3	9.1 ± 0.9
	3+	205 ± 22	125 ± 5.5	66 ± 7.8	42 ± 4.3
Skeleton	5+	388 ± 52	98 ± 24	41 ± 3.8	17 ± 1.5
	3+	247 ± 34	82 ± 4.5	8.8 ± 1.3	3.6 ± 0.9
Epididymis	5+	127 ± 10	66 ± 11	16 ± 2.3	9.7 ± 1.6
	3+	187 ± 15	151 ± 5.6	61 ± 8.1	36 ± 4.5
Testis	5+	48 ± 4.8	34 ± 4.0	5.7 ± 0.4	0.9 ± 0.3
	3+	47 ± 1.7	60 ± 5.0	11 ± 1.0	1.2 ± 0.4

[a] Table assembled from Lindgren et al. (1982)

Both species-specific and valence-state-dependent differences have been demonstrated in the biliary excretion of arsenic. Studies reviewed in the previous arsenic document (IPCS, 1981, 6.1.2) indicate that excretion of trivalent arsenic into the bile is much more extensive in rats than in rabbits or dogs (Klaassen, 1974). The studies of Lindgren et al. (1982) suggest that arsenite is excreted to a greater extent than arsenate in the bile of mice (Table 16); these authors also attribute the higher concentrations of arsenite-derived radioactivity in the duodenum of mice to greater biliary excretion of arsenite. Excretion of arsenite into the bile of rats is also more rapid and efficient than that of arsenate – 19% vs. 6% of the dose in 2 h (Gyurasics et al., 1991). Mechanistic studies indicate that transport of either arsenate or arsenite into the bile of rats is dependent on GSH, since agents that decrease hepatobiliary transport of GSH (e.g. diethyl maleate) also decrease hepatobiliary transport of arsenic (Alexander & Aaseth, 1985; Gyurasics et al., 1991). It has been demonstrated in recent studies that arsenite as well as other trivalent arsenicals directly form complexes with GSH (Scott et al., 1993; Delnomdedieu et al., 1994a).

Numerous studies reveal that skin, hair, and tissues high in squamous epithelium (e.g. mucosa of the oral cavity, oesophagus, stomach and small intestine) have a strong tendency to accumulate and maintain higher levels of arsenic (e.g. Lindgren et al., 1982; Yamauchi & Yamamura, 1985). This is apparently a function of the binding of arsenic to keratin in these tissues (Lindgren et al., 1982). Autoradiographic studies have also revealed a tendency for arsenic to accumulate in the epididymis, thyroid and lens of the eye of mice (Lindgren et al., 1982).

Arsenic can cross the blood–brain barrier; it is found in brain tissue after oral or parenteral administration of trivalent or pentavalent inorganic arsenic in all species studied (e.g. see Table 16). However, the levels are uniformly low both across time and relative to other tissues, which indicates that arsenic (when administered in the form of sodium salts) does not readily cross the blood–brain barrier or accumulate in brain tissue after acute dosing (Lindgren et al., 1982; Marafante et al., 1982; Vahter et al., 1982; Vahter & Marafante, 1985; Yamauchi & Yamamura, 1985; Itoh et al., 1990).

Relatively few studies have examined the distribution of arsenic metabolites in tissues, owing to limitations in availability of appropriate analytical techniques. Given the vigorous treatment necessary to extract arsenicals from tissues, and the ease with which arsenate and arsenite are interconverted, any reports that distinguish between arsenate and arsenite in tissues should be interpreted with caution. DeKimpe et al. (1996) reported the tissue distribution of arsenic metabolites at 4, 20 and 120 h after intraperitoneal injection of a trace amount of [74]As-arsenate in male Flemish giant rabbits. The analytical methodology used was ion-exchange chromatography separation of ultrafiltrates with radiometric detection. The predominant metabolite present in tissues was DMA, followed by the inorganic arsenic species, with MMA being generally detected in all tissues, although making up a smaller percentage of the total metabolites in most cases. The percentage of total metabolite present as DMA increased steadily in bone marrow, heart, liver, muscle, pancreas, small intestine and spleen, but levelled off or declined in kidney and lung. Marafante et al. (1982) also found that inorganic arsenic was the predominant form of arsenic in rat and rabbit liver and kidney ultrafiltrable fraction (cut-off 25 000 Da) 1 h after intraperitoneal injection of 50 µg As/kg sodium arsenite; the analytical methodology used was ion-exchange chromatographic separation with radiometric detection. However, the fraction of DMA was higher at 16 h after injection in kidney of both rats and rabbits, but only in the liver of rats. The fraction present as MMA was uniformly low, generally less that one-tenth that of inorganic arsenic.

Yamauchi & Yamamura (1985) studied the tissue distribution over time of arsenic metabolites in male Syrian golden hamsters given a single oral dose of 4.5 mg/kg As_2O_3. Hydride generation atomic absorption spectrophotometry (HGAAS) with a cold trap was used to speciate arsenicals in whole tissues after alkaline digestion. The predominant form of arsenic present in all tissues at 1, 6, 12, 24, 72 and 120 h after dosing was inorganic arsenic. Interestingly, concentrations of MMA in tissue were uniformly 2–4-fold greater than DMA at all time points, although much more DMA (22% of the dose) was excreted in urine over 5 days than MMA (2.5% of the dose). Highest concentrations of MMA were achieved in lungs and spleen at 12 h and kidney at 24 h, and highest concentrations of DMA occurred in liver, lung and kidney at 24 h.

The subcellular distribution of total arsenic administered as either sodium arsenate or sodium arsenite has been studied in mice, rats, rabbits and marmoset monkeys (Marafante et al., 1982; Vahter et al., 1982; Vahter & Marafante, 1983; Vahter & Marafante, 1985). In general the subcellular localization and retention of arsenic accounts for its much slower elimination in rats than in other species. In rats arsenic is strongly associated with high-molecular-weight cellular components in liver and kidney, whereas in rabbits it is associated with low-molecular-weight, more readily diffusible cellular components (Marafante et al., 1982). This research group also reported that in the marmoset monkey arsenic, administered as arsenite or arsenate, shows a unique strong tendency to bind with the rough endoplasmic reticulum in the liver that they had not observed in other laboratory animals (Vahter et al., 1982; Vahter & Marafante, 1985).

b)   Human studies

As in experimental animals, postmortem analysis of human tissues reveals that arsenic is widely distributed in the body after either long-term relatively low-level exposure or poisoning (Dang et al., 1983; Gerhardsson et al., 1988; Raie, 1996). Dang et al. (1983) used neutron activation analysis (NAA) to measure total arsenic in various tissues of individuals (age and sex not specified) dying in accidents in the Bombay area (India) (Table 17). Notable results from this study are that arsenic concentrations are quite low in both blood and brain relative to other tissues and that arsenic concentration in any given tissue was quite variable.

Table 17.   As levels in human tissues from accident victims in Bombay area of India[a]

Tissue	No. of samples	As concentration (ng/g wet weight)	
		Range	mean ± SD
Blood	8	3.1–13.8	5.9 ± 3.9
Brain	12	2.5–6.0	3.9 ± 1.0
Liver	19	4.5–27.7	14.5 ± 6.9
Kidney	13	1.6–62.8	12.4 ± 20.7
Lung	13	2.5–81.8	19.9 ± 22.7
Spleen	18	3.6–46.2	15.2 ± 16.6

[a]   Table assembled from Dang et al. (1983)

Yamauchi & Yamamura (1983) analysed by HGAAS the levels of total arsenic and major arsenic metabolites in a variety of human tissues obtained from adult patients (age 36–79) dying of cerebral haemorrhage, pneumonia or cancer in Kawasaki (Japan) (Table 18). No sex-dependent differences in arsenical tissue levels were observed; inorganic arsenic was the predominant form in tissues, followed by DMA. MMA levels were uniformly low and detected only in liver and kidney. It is interesting to note that total arsenic levels were higher than those reported in the Indian study of Dang et al. (1983), and levels in brain tended to be more comparable to levels in other tissues. Inter-individual variation in total tissue arsenic was also quite high, as observed in the Dang study using NAA.

Table 18.   Tissue concentrations of metabolites and total As in normal tissues and organs of adult Japanese people[a]

Tissue/organ	No. of samples	Arsenical concentration (mean) in tissue (ng/g wet weight)[b]			
		Inorganic As	MMA	DMA	Total As
Aorta	16	535	< LD	16.4	551 ± 350
Adrenal gland	19	301	< LD	25.8	327 ± 364
Cerebellum	30	132	< LD	< LD	132 ± 60.2
Cerebrum	30	76.8	< LD	< LD	76.3 ± 43.9
Kidney	24	97.7	3.6	27.6	129 ± 72.3
Liver	23	116	5.9	14.0	129 ± 39.7
Lung	22	96.9	< LD	7.6	104 ± 29.5
Muscle	22	88.1	< LD	18.9	106 ± 32.7
Pancreas	18	139.7	< LD	14.7	154 ± 71.4
Skin	22	149.6	< LD	3.7	153 ± 97.7
Spleen	20	91.4	< LD	12.6	101 ± 49.4

[a]   Table assembled from Yamauchi & Yamamura (1983)
[b]   < LD indicates less than limit of detection, which was 1 ng As/g wet tissue weight

Raie (1996) used NAA to compare tissue arsenic levels in infants (1 day–5 months) and adults from the Glasgow (Scotland, UK) area. Mean levels of arsenic (μg/g dry weight) in liver, lung and spleen in infants vs. adults were 0.0099 vs. 0.048, 0.007 vs. 0.044, and 0.0049 vs. 0.015, respectively. These data suggest that arsenic accumulates in tissues with age, which is consistent with observations in laboratory animals (Marafante et al., 1982).

Studies have been conducted in humans with the goal of determining whether there are differences in tissue accumulation of arsenic (and other metals as well) in differing disease states. Warren et al. (1983) compared trace element levels in brain and other tissues of multiple sclerosis and non-multiple-sclerosis patients and found no significant difference in any tissue arsenic levels. Narang & Datta (1983) have reported that concentrations of arsenic in both liver and brain of patients who died of fulminant hepatitis are high compared to those in patients who died of non-hepatic-related causes. Collecchi et al. (1985) compared the distribution of arsenic and cobalt in cancerous and non-cancerous laryngeal tissue and plasma of patients with and without laryngeal cancer. Malignant tissue had significantly higher levels of arsenic than normal tissue, and plasma arsenic levels were also significantly higher in cancer patients than in controls. Zhang et al. (1996b, 1997) have reported that arsenic levels in serum are significantly elevated in patients with chronic renal disease (~5–6-fold), whether on dialysis or not, and that accumulation and removal of the dominant species of arsenic in serum (DMA and arsenobetaine) was non-selective for dialysed patients.

6.1.3 *Metabolic transformation*

Arsenic metabolism is characterized in many species by two main types of reactions: (1) reduction of pentavalent to trivalent arsenic, and (2) oxidative methylation reactions in which trivalent forms of arsenic are sequentially methylated to form mono-, di- and trimethylated products using *S*-adenosyl methionine (SAM) as the methyl donor and GSH as an essential co-factor (see Fig. 3)[1]. One

---

[1] Although this sequence of methylation reactions is generally considered to be a detoxification reaction, recent publications may indicate that an intermediate, MMA(III), is highly toxic (Styblo and Thomas, 1997; Petrick et al., 2000; Styblo et al., 2000), and is released from the site of methylation as it has been detected in bile and urine (Aposhian et al., 2000; Gregus et al., 2000).

As(V)

| 2GSH
| GSSG

As(III)

Methyltransferase
[unclassified]

SAM

S-adenosylhomo-
cysteine

CH₃ As(V)

$CH_3 As(V)$

2GSH
GSSG

CH₃ As(III)

$CH_3 As(III)$

Methyltransferase
[unclassified]

SAM
S-adenosylhomo-
cysteine

$(CH_3)_2 As(V)$

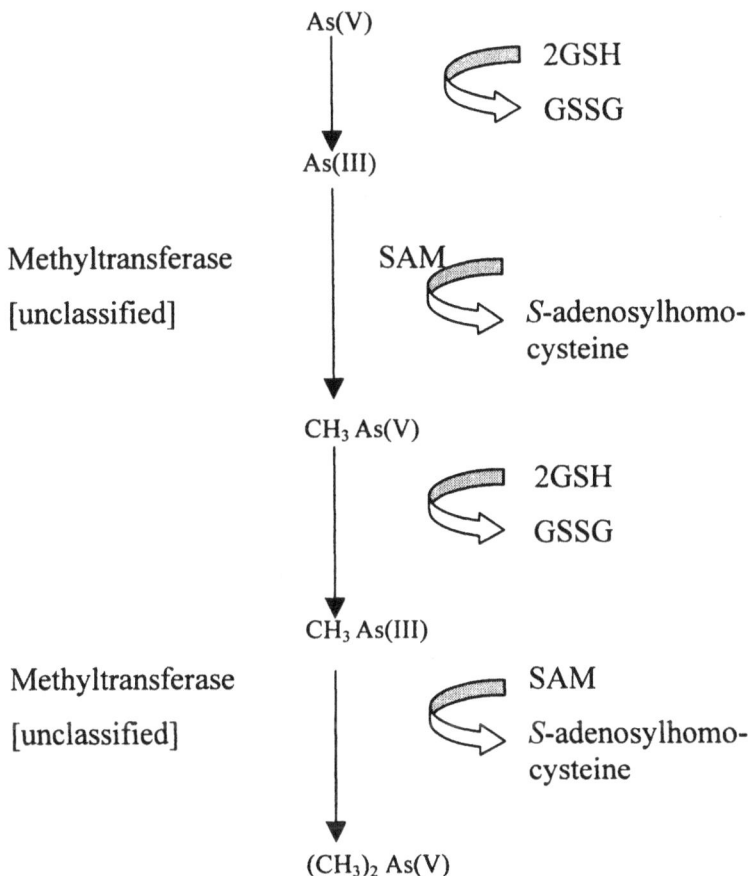

Fig. 3. Arsenic methylation in mammals. Reducing equivalents are supplied by glutathione (GSH) and *S*-adenosyl (methionine (SAM) serves as the methyl donor. Reduction of pentavalent to trivalent forms is required for methylation. Trimethylated forms are produced in small amounts if animals or humans are administered DMA.

unusual feature of arsenic metabolism is that there are extreme qualitative and quantitative interspecies differences in methylation to the extent that some species do not appear to methylate arsenic at all (Styblo et al., 1995; Vahter, 1999).

6.1.3.1   Animal studies

Reduction of pentavalent to trivalent arsenic species is required for methylation of arsenic (Lerman et al., 1983; Cullen et al., 1984a,b; Marafante et al., 1985; Thompson, 1993). Arsenate reduction is known to occur non-enzymatically under conditions of low oxygen tension (i.e. an anaerobic environment such as exists in the gut) or over time at pH 2 or lower (Vahter & Envall, 1983). *In vitro* mechanistic studies have demonstrated that the ubiquitous cellular tripeptide GSH is able to reduce arsenate to arsenite in both aqueous systems (Scott et al., 1993; Delnomdedieu et al., 1994a) and in intact erythrocytes (Delnomdedieu et al., 1994b). Interestingly, bacteria have the capability to enzymatically reduce inorganic arsenate to arsenite (Rosen, 1995; see also section 4.2). It has been hypothesized that mammalian cells also have this capability (Healy et al., 1998), but this has not been conclusively demonstrated.

In vivo reduction of arsenate to arsenite before methylation was demonstrated by Vahter & Envall (1983). The appearance of arsenic metabolites in the urine of catheterized New Zealand white rabbits was followed over a period of 4 h after intravenous injection of 0.04 mg As/kg arsenate. Arsenate (unmetabolized) was excreted in greatest amounts during the first hour and thereafter declined, whereas excretion of arsenite progressively increased and cumulatively amounted to 10% of the administered dose. Metabolism of arsenate to DMA, which requires reduction from pentavalent to trivalent form, also progressively increased over the 4-h time period, but peaked later than arsenite. *In vivo* reduction of arsenate to arsenite has also been demonstrated in marmoset monkeys, which exhibit little to no methylation of arsenic. Arsenate levels in the plasma of marmosets injected intravenously with 0.4 mg As/kg peaked at 0.5 h after injection and thereafter declined, whereas the amount of arsenite increased sharply through 6 h (Vahter & Marafante, 1985).

Arsenic methylation activity is localized in the cytosol and appears to occur sequentially and mainly in the liver. Styblo et al. (1996), using a rat liver cytosol system, found that whether the starting material was arsenate or arsenite, MMA was detected sooner and peaked earlier than DMA. Similar findings have been reported using rat liver slices (Buchet & Lauwerys, 1985). Georis et al.

(1990), also using rat tissue slices, reported that liver, kidney and lung all had the capacity to methylate arsenite, but that the capacity of the liver was clearly greater.

Both *in vivo* and *in vitro* studies have demonstrated that SAM and GSH are essential co-factors in enzymatic arsenic methylation (Buchet & Lauwerys, 1985, 1987, 1988; Marafante et al., 1985; Hirata et al., 1988, 1989; Styblo et al., 1996). For example, Marafante et al. (1985) compared the biotransformation and tissue retention in control and periodate-oxidized adenosine (PAD) treated rabbits dosed intravenously with 0.4 mg As/kg [74]As[arsenate]. PAD depletes the intracellular SAM pool by inhibiting its synthesis. Approximately 35% of the arsenate dose was excreted in urine as DMA in 24 h in control rabbits, compared to ~5% in the PAD-treated rabbits. Tissue retention of arsenate-derived radioactivity was also significantly higher in PAD-treated rabbits than in controls at 24 h in all major organs examined. These data indicate that methylation is an equally important mechanism for expediting the excretion of both arsenate and arsenite from the body. In addition, since DMA was found only in liver, but not other organs, 1 h after [74]As[arsenate] administration, the authors considered this to indicate that the liver is the main site of arsenic methylation.

Healy et al. (1998), using cytosol prepared from liver, lung, kidney and testes of male B6C3F$_1$ mice, reported that all the tissues had the capacity to methylate arsenite. However, the specific activity (defined as pmol [^{3}H] MMA formed per hour/mg protein at 37 °C and reported as mean ± SEM) was greatest in testes (1.45 ± 0.08) followed by kidney (0.70 ± 0.06), liver (0.40 ± 0.06) and lung (0.22 ± 0.01). These findings suggest that although the liver may have the greatest overall arsenite methylation capacity (on the basis of tissue mass), extrahepatic metabolism may also be significant. This would be particularly the case for routes of exposure such as inhalation, where there is opportunity for first-pass metabolism in the lung.

The capacity of the gut microbiota to metabolize arsenic has also been investigated in rats and mice. Rowland & Davies (1981) demonstrated that the low oxygen environment of the intestine itself stimulates the rapid reduction of arsenate to arsenite, and that arsenate reduction was further stimulated by the presence of bile acids and gut contents from male Wistar rats. These authors also reported that methylation apparently did not occur in incubations of rat small-intestinal contents, but production of both MMA and DMA occurred in incubations of caecal contents. Hall et al. (1997) reported that caecal contents from male CD-1 mice incubated under physiological conditions methylated 33% of 0.1 µmol/litre arsenite, but only 8% of 0.1 µmol/litre arsenate over a period of 6 h. After 21 h incubation, 36% and 29% of the applied dose of arsenite and arsenate, respectively were in the form of MMA, and DMA accounted for only approximately 3% of the methylated metabolites present. These data suggest that gut microbial metabolism could contribute significantly to methylation in laboratory animals. However, Vahter & Gustafsson (1980) showed that the methylation of arsenic was similar in germfree and conventional mice with normal intestinal microflora, indicating that methylation of arsenic by intestinal microorganisms contributes little to the overall methylation *in vivo*.

Major qualitative and quantitative interspecies differences in arsenic methylation are apparent in laboratory animals when they are compared on the basis of metabolites excreted in urine (e.g. see Tables 19 and 20). For example, methylated metabolites are virtually undetectable in the urine of marmoset monkeys administered either arsenate or arsenite (Vahter et al., 1982; Vahter & Marafante, 1985) and chimpanzees administered arsenate (Vahter et al., 1995b). Studies using liver cytosol from marmoset and tamarin monkeys (Zakharyan et al., 1996) and guinea-pigs (Healy et al., 1997) also indicate that these species are deficient in methyltransferase activity compared to species such as the rabbit.

It is not known with certainty if enzymatic methylation of arsenic is saturable under *in vivo* exposure conditions in laboratory animals. Vahter (1981) reported a significant dose-dependent decrease in the urinary excretion of DMA in mice administered inorganic arsenic as either arsenate or arsenite (see Tables 19 and 20). These findings are consistent with either saturation or inhibition of methylation. No similar clear-cut trend was seen in the study by Hughes (1994), in which a lower dose was used.

In vitro studies have demonstrated that arsenite can inhibit the formation of DMA from MMA. Styblo et al. (1996), using a rat liver

Table 19. Urinary excretion of As metabolites after a single dose of pentavalent inorganic As

Species	Route	Dose (mg/kg)	Time (h)	% dose in urine	Mean (± SD or SE) % dose excreted in urine as metabolite				Reference
					As(V)	As(III)	MMA	DMA	
Mice	s.c.	0.4	0–48	~96	40.9 ± 3.6 (IAs)[a]		0.4 ± 0.1	44.5 ± 2.0	Vahter (1981)
	oral	0.04	0–48	~94	16.6 ± 0.8 (IAs)		0.8 ± 0.1	76.6 ± 4.9	
		0.4	0–48	~93	20.8 ± 2.3 (IAs)		0.9 ± 0.2	71.1 ± 0.9	
		2.0	0–48	~92	37.9 ± 4.4 (IAs)		1.7 ± 0.5	52.4 ± 5.9	
		4.0	0–48	~84	39.5 ± 2.3 (IAs)		1.3 ± 0.2	43.7 ± 2.7	
Mice[b]	oral	0.00012	0–48	~65	5.2 ± 2.4	0.6 ± 0.4	0.12 ± 0.01	59.1 ± 6.2	Hughes et al. (1994)
		0.0012	0–48	~68	3.2 ± 0.2	0.7 ± 0.3	0.22 ± 0.13	64.2 ± 9.9	
		0.012	0–48	~72	14.6 ± 6.6	0.7 ± 0.4	0.35 ± 0.09	56.4 ± 11.0	
		0.12	0–48	~70	6.0 ± 1.7	1.2 ± 0.2	0.65 ± 0.18	63.1 ± 3.2	
		1.2	0–48	~68	10.2 ± 1.4	6.1 ± 1.7	1.01 ± 0.04	51.4 ± 2.4	
Mice	oral	5.0	0–48	48.5	16.7 (IAs)		1.8	30	Odanaka et al. (1980)
	i.v.	1.0	0–48	86.9	47.4 (IAs)		2.1	37.4	

Table 19 (contd.)

Species	Route	Dose (mg/kg)	Time (h)	% dose in urine	Mean (± SD or SE) % dose excreted in urine as metabolite				Reference
					As(V)	As(III)	MMA	DMA	
Rat	oral	5.0	0–48	17.2	14.1 (IAs)		0.9	2.2	Odanaka et al.
	i.v.	1.0	0–48	51.0	47.6 (IAs)		0.7	2.7	(1980)
Hamster	oral	5.0	0–48	43.8	17.7 (IAs)		4.6	21.5	Odanaka et al.
	i.v.	1.0	0–48	83.9	42.4 (IAs)		1.8	39.7	(1980)
Rabbit	i.v.	0.04	0–72	65.5 ± 3.8	31.5 ± 7.1 (IAs)		–	34.0 ± 3.3	Vahter & Marafante (1983)
Marmoset	i.v.	0.4	0–72	39.3 ± 3.8	~20	~20	–	< 0.1	Vahter & Marafante (1985)

[a] IAs is inorganic As, figures are mean ± SE
[b] Figures are mean ± SD

Table 20. Urinary excretion of As metabolites after a single dose of trivalent inorganic As

Species	Route	Dose (mg/kg)	Time (h)	% dose in urine	Mean (± SE) % dose excreted in urine as metabolite			Reference
					Inorganic As	MMA	DMA	
Mice	s.c.	0.4	0–48	~75	11.9 ± 0.8	0.9 ± 0.2	62.0 ± 3.4	Vahter (1981)
	oral	0.04	0–48	~88	7.9 ± 1.2	1.0 ± 0.1	79.4 ± 0.5	
	oral	0.4	0–48	~91	7.8 ± 1.1	0.7 ± 0.1	82.7 ± 2.7	
	oral	2.0	0–48	~86	37.9 ± 1.1	1.9 ± 0.1	68.8 ± 3.0	
	oral	4.0	0–48	~75	39.5 ± 0.8	1.3 ± 0.1	55.9 ± 2.3	
Rat	oral	0.4	0–48	~6	2.3 ± 0.3	0.2 ± 0.1	3.7 ± 0.3	Vahter (1981)
Rat	i.p.	0.05	0–48	~5.5	~1.6	~0.1	~4.0	Marafante et al. (1982)
Hamster	oral	4.5	0–120	48.5	23.3	2.5	22.1	Yamauchi & Yamamura (1985)
Rabbit	i.v.	0.04	0–72	52.8 ± 4.8	8.1 ± 1.3	---	43.9 ± 4.4	Vahter & Marafante (1983)
Rabbit	i.v.	0.4	0–72	88.5 ± 2.2	17.0 ± 1.4	5.9 ± 0.2	65.5 ± 3.8	Vahter & Marafante (1987)
Marmoset	i.p	0.4	0–96	29.8	~29.8	ND	ND	Vahter et al. (1982)

ND = not detected

cytosol system, found that as the initial concentration of arsenite was increased from 0.1 to 50 µmol/litre, the amount of DMA produced decreased and there was an increased time lag before DMA was detected. Production of MMA from arsenite increased in proportion to the amount of arsenite in the assay system.

On the basis of case reports in the medical literature, it has been theorized that prolonged exposure to arsenic can result in the development of tolerance. This hypothesized tolerance could in theory be a result of increased excretion due to enhanced methylation, or an increase in some other excretory mechanism. Healy et al. (1998) investigated the possibility that arsenite methyltransferase activity is inducible by exposure to arsenic itself. When male B6C3F$_1$ mice received 25 or 2500 µg As/litre arsenate in their drinking-water for 32 or 91 days, there was no increase in arsenite methyltransferase activity in liver, testes, kidney or lung. This is consistent with the finding of Hughes & Thompson (1996) that subchronic exposure of mice to 25 or 2500 µg As/litre arsenate in their drinking-water for 28 days resulted in no increased urinary excretion of methylated metabolites.

Although studies in mice exposed to arsenate do not provide evidence for induction of arsenite methyltransferases, studies in the older literature (Bencko & Symon, 1969; Bencko et al., 1973) of mice exposed to arsenite suggest that there is enhanced tissue clearance upon continuous exposure. Interestingly, enhanced efflux of arsenite has been demonstrated to be a mechanism of resistance to arsenic toxicity in Chinese hamster V79 cells (Wang et al., 1996b). Albores et al. (1992) have reported that the metal-binding protein metallothionein is inducible *in vivo* in rats injected with arsenite, but not with arsenate. Kreppel et al. (1990) have also reported that arsenite is a much more effective inducer of metallothionein in mice *in vivo* than *in vitro*. Healy et al. (1998) have suggested that this may be a factor in older reports of enhanced arsenic clearance in mice dosed with arsenite. However, it should be noted that arsenite does not bind metallothionein (Albores et al., 1992).

The question of whether alterations in nutritional status can influence arsenic methylation has also been investigated in animal models. Vahter & Marafante (1987) examined the effect of low dietary intake of methionine, choline or protein on excretion of methylated metabolites in rabbits given a single intravenous dose of 0.4 mg As/kg as arsenite. Total arsenic excretion in urine was significantly decreased compared to controls in all diet groups. DMA excretion (expressed as percentage of the dose) was also significantly decreased relative to controls (65.5 ± 3.1) by diets low in choline (43.9 ± 1.6), methionine (39.3 ± 1.7) and protein (51.9 ± 4.3).

6.1.3.2   *Human studies*

Controlled ingestion studies indicate that both arsenate and arsenite are extensively methylated in humans, as is also observed in laboratory animals, with DMA being the principle methylated metabolite excreted in human urine (Table 15). A noteworthy difference between humans and laboratory animals is that MMA is excreted in the urine of humans to a greater extent (see Tables 18, 19 and 20). The biological basis for this difference is unknown, but it is consistent with the large interspecies differences observed in arsenic methylation among experimental animals. It is also noteworthy that, on the basis of data summarized from a number of studies of different human populations by Hopenhayn-Rich et al. (1993), the proportion of MMA excreted in human urine is highly variable.

In some studies, ratios of arsenic metabolites in urine (e.g. DMA/inorganic arsenic, MMA/inorganic arsenic or DMA/ MMA have been used to draw conclusions regarding saturation or inhibition of methylation. Such conclusions should be evaluated with caution because of the inherent numerical and statistical properties of ratio data. Two specific problems are that (1) small changes in a metabolite present in small amounts can result in large changes in the ratio, which can exaggerate or distort the magnitude of observed differences and (2) metabolites present at levels near the detection limit may be associated with a higher degree of measurement error, which could also distort the magnitude of differences when used to calculate ratio data. Application of statistical analysis to metabolite ratio data is also complex because of the degree of correlation in metabolite data, particularly when expressed as a percentage of total metabolites excreted. All of these factors require that conclusions based on ratio data be evaluated critically and independently.

Humans acutely intoxicated by high doses of inorganic arsenic show a marked delay in the urinary excretion of DMA (Mahieu et al., 1981; Foa et al., 1984). However, in the case of exposure to arsenic via drinking-water, even at very high arsenic concentrations, the methylation of arsenic seems to be relatively unaffected by the dose. In a case study by Kosnett & Becker (1988), after subacute exposure to drinking-water containing arsenic at a concentration of 25 000 µg/litre, a 36-year-old man yielded a urinary arsenic collection containing about 6000 µg/24 h, 26% as inorganic arsenic and 74% as methylated metabolites. Results from *in vitro* studies using human hepatocytes suggested that the delay in urinary excretion of DMA might occur because the high tissue concentration of arsenite inhibits or saturates the methyltransferase catalysing the second methylation step (Styblo et al., 1999).

The proportion of methylated metabolites in urine can vary considerably. For example, in the literature review performed by Hopenhayn-Rich et al. (1993), the average proportions of MMA and DMA in urine of occupationally and environmentally exposed population groups (range of average total urinary arsenic from 10.2 to 245 µg/litre) ranged from 9 to 20% and 61 to 70%, respectively. Data on the variation in human populations have been comprehensively reviewed (NRC, 1999).

Studies focused on populations highly exposed to arsenic in drinking-water also indicate that methylation patterns are not highly correlated with exposure level, but that there is a high level of inter-individual variability (Warner et al., 1994; Hopenhayn-Rich et al., 1996a). In another study (Hopenhayn-Rich et al., 1996b), methylation patterns in a population of northern Chilean subjects ($n = 73$) were compared (each subject served as their own control) before and after changing from drinking-water containing higher (600 µg/litre) to lower (45 µg/litre) levels of arsenic. There was a small but significant decrease in urinary inorganic arsenic (from 17.8% to 14.1%). The authors note that there was large inter-individual variation in methylation profiles and that factors such as smoking, gender, age, years of residence and ethnicity accounted for only ~20% of the variation observed. They further speculate that much of the observed inter-individual variation might be explained by genetic differences in the activity of methylating enzymes and related co-factors.

Vahter et al. (1995a) reported a unique pattern of urinary methylated metabolite excretion in a population of healthy native Andean women in north-western Argentina consuming an apparently protein-adequate diet. Reported arsenic concentration in the drinking-water of this population was ~200 µg/litre. These women excreted mainly inorganic arsenic (median 25%, range 6.5–42%) and DMA (median 74%, range 54–93%) in their urine and very little MMA (median 2.1%, range of 0.6–8.3%). The authors suggest that this finding indicates the existence of genetic polymorphism in the control of arsenic methyltransferases. They also suggest that the higher urinary DMA excretion in women in the village with the highest arsenic in drinking-water (~200 µg/litre) compared to that of women in the villages with lower arsenic in drinking-water (2.5–31 µg/litre) indicates induction of DMA excretion. It is worthy of note that differences in the activities of other methyltransferases have been explained by the existence of genetic polymorphisms (Weinshilboum, 1992).

In further studies of that Andean Argentinian population, Concha et al. (1998a) reported striking differences in urinary excretion patterns of arsenic metabolites in children compared to adult women. In one village with a predominantly indigenous Indian population consuming drinking-water high in arsenic (~200 µg/litre), children (age 3–15 years) excreted a much higher median percentage of inorganic arsenic in urine (49% vs. 25%) and a much lower median percentage of DMA in urine (47% vs. 74%) compared to adult women (age 20–47 years); this difference was observed even though the median concentrations of arsenic metabolites in urine (sum of inorganic arsenic plus both methylated metabolites) did not differ greatly for the children and the women (323 µg/litre vs. 303 µg/litre). A low median percentage of MMA excreted in urine was also observed both in the women (2.1%) and in the children (3.6%) which is consistent with previously reported results (Vahter et al., 1995a). Another significant finding in these children was that with increasing excretion of total arsenic metabolites in urine, the percentage of inorganic arsenic decreased and the percentage of DMA increased; the authors interpreted this as evidence for induction of arsenic methylation with increasing exposure (Concha et al., 1998b). It should be noted that in the very few studies that have looked at methylation patterns in children, percentages of metabolites excreted in urine are similar to adults (Buchet et al.,

145

1980; Kalman et al., 1990). However, in both these studies arsenic exposure was relatively low, as indicated by total concentration of arsenic metabolites excreted in urine (i.e. < 20 µg/litre).

Data suggestive of gender differences in arsenic metabolism have been reported in studies conducted in Chile and Taiwan (Hopenhayn-Rich et al., 1996a; Hsu et al., 1997). In both of these studies relatively more DMA was excreted by women than men. In this connection it is also of interest to note that Concha et al. (1998b) reported significant increases in the percentage of DMA excreted in urine in Argentinian women during pregnancy which is one possible reason for gender differences reported in some studies.

Inorganic arsenic metabolism is known to be affected by liver disease in humans. Buchet et al. (1984) compared the urinary excretion of inorganic arsenic and its methylated metabolites in normal human subjects and patients with various forms of liver disease after intravenous injection of 7.14 µg As/kg as sodium arsenite. Liver disease had no effect on the total amount of arsenic excreted within 24 h, but dramatically shifted the proportion of MMA and DMA excreted in the urine. The percentage of arsenic excreted as MMA was decreased in liver disease patients compared to controls (6.1 ± 0.7 vs. 12.8 ± 0.7) and the percentage of DMA was increased (40.7 ± 1.9 vs. 24.3 ± 1.6). Geubel et al. (1988) reported similar findings in subjects with cirrhotic liver disease. They further noted that in patients with other non-hepatic disease, the arsenic methylation was unaffected.

6.1.4 Elimination and excretion

6.1.4.1 Animal studies

Urine is the primary route of elimination for both pentavalent and trivalent inorganic arsenicals in most common laboratory animals (Table 14). With the exception of the rat, which exhibits slower overall elimination of arsenic, 50% or more of a single oral dose of arsenic is usually eliminated in urine within 48 h. Urine is also the primary route of elimination in species such as the marmoset which do not methylate arsenic. Vahter et al. (1982) reported that when arsenite was administered intraperitoneally to marmosets at a dose of 0.4 mg As/kg, 29.8% of the dose was eliminated in urine over 4 days, compared to only 4.1% in the faeces. Similarly, when administered an intravenous dose of 0.4 mg As/kg arsenate, marmosets excreted 39.3% of the dose in the urine and only 2.1% in the faeces over 72 h (Vahter & Marafante, 1985).

Comparison of urinary and faecal elimination in mice that have been given the same dose of arsenic by oral and parenteral routes (e.g. Vahter & Norin, 1980) reveals that only ~4–8% of the dose is eliminated in faeces irrespective of route of administration. This suggests that, for both arsenate and arsenite, biliary elimination in mice is quite low (< 3% over 48 h – see Table 13) and that most arsenic appearing in the faeces after oral dosing was unabsorbed from the gastrointestinal tract.

Urinary elimination of arsenate in laboratory animals – at least for mice – does not appear to be capacity-limited or dose-dependent. Hughes et al. (1994) reported that 66–79% of a single oral dose of sodium arsenate was eliminated in the urine in 48 h over a 10 000-fold dose range. Vahter & Norin (1980) reported a significant decrease in both urinary and total excretion of arsenic in mice when administered as arsenite, which is apparently a function of greater arsenite binding in tissues with increasing dose.

6.1.4.2   Human studies

Inorganic arsenic is eliminated primarily via the kidney in humans as well as laboratory animals. Studies in adult human males voluntarily ingesting a known amount of either trivalent or pentavalent arsenic indicate that 45–75% of the dose is excreted in the urine within a few days to a week (Table 13). Relatively few studies in volunteers have included measurement of arsenic in both faeces and urine. However, Pomroy et al. (1980) reported that 6.1% ± 2.8% of a single oral dose of arsenic acid (As(V)) was excreted in the faeces over a period of 7 days, compared to 62.3% ± 4.0% of the dose excreted in urine. It should be noted that Pomroy et al. used radiolabelled arsenate, which enabled distinction between ingested arsenic acid and dietary arsenic. No quantitative data was available that directly addressed the issue of biliary excretion of trivalent or pentavalent arsenic in humans.

Arsenic is excreted by routes other than just urine and faeces, but in general these routes of excretion are quantitatively minor. Studies reported in the previous IPCS arsenic document (IPCS, 1981, section 6.1.3) indicate that arsenic is excreted in sweat to some degree. Owing to its ability to accumulate in keratin-containing tissues, skin, hair and nails could also be considered potential excretory routes for arsenic, although they would in general be quantitatively minor.

Both earlier (IPCS, 1981) and recent studies indicate that arsenic can be excreted in human milk, although the levels are low (Dang et al., 1983; Grandjean et al., 1995; Concha et al., 1998b). For example, in the Bombay area (India) Dang et al. (1983) reported arsenic levels ranging from 0.2 to 1.1 ng/g in breast milk of nursing mothers 1–3 months postpartum. Concha et al. (1998b) found that the average concentration of arsenic in breast milk of was quite low (3.1 µg/litre) even when urinary arsenic excretion was high (230–300 µg/litre) from 3 weeks to 5 months postpartum in a study of Andean women in Argentina consuming drinking-water high in arsenic (~200 µg/litre). Significantly, low-arsenic excretion in breast milk of nursing mothers led to a decrease in urinary arsenic concentration of their infants during the nursing period.

6.1.5 Retention and turnover

6.1.5.1 Animal studies

Lindgren et al. (1982) compared the whole-body retention of arsenate and arsenite administered intravenously as the sodium salts to male C57BL mice at a dose of 0.4 mg As/kg. Retention was higher in arsenite-treated mice than in arsenate-treated mice at all times measured, i.e. 44.4% vs. 20.4% at 6 h after dosing, 14% vs. 3.3% at 24 h and 5.6% vs. 1.7% at 72 h after dosing. Vahter & Norin (1980) earlier reported that, in male CBA mice dosed orally with 0.4 mg As/kg arsenate or arsenite, whole-body retention was similar over the 35-day time course of the experiment. In contrast, and similar to what was observed with intravenously dosed mice in the study by Lindgren et al. (1982), whole-body retention was clearly consistently higher in arsenite-dosed mice than in arsenate-dosed mice when dosed orally with 4 mg As/kg: 35 days after administration the high/low dose retention ratios were 11 for arsenite-dosed and 6 for arsenate-dosed mice.

6.1.5.2 Human studies

Pomroy et al. (1980) studied the whole-body retention of [74]As (6.4 µCi, 0.06 ng As) administered once orally as arsenic acid (As(V)) in healthy male volunteers (age 28–60 years) using whole-body counting for periods of < 103 days. Although the averaged whole-body clearance data for the six subjects in the study were best described by a triexponential model, it should be noted that the inter-individual variation was quite high. It was reported that 65.9% of the dose was cleared with a half-life of 2.09 days, 30.4% with a half-life of 9.5 days and 3.7% with a half-life of 38.4 days. No comparable data for humans was located for trivalent inorganic arsenic.

6.1.6 Reaction with body components

Numerous mechanistic studies have documented basic differences in the interaction of pentavalent and trivalent inorganic arsenic with body components, and this is an important determinant in observed differences in tissue distribution. Pentavalent inorganic arsenic can act as a phosphate analogue. At the molecular level this means that arsenate can compete with phosphate for active transport processes. This is why the addition of phosphate can decrease intestinal uptake (Gonzalez et al., 1995) and renal tubular reabsorption of arsenate (Ginsburg & Lotspeich, 1963). Arsenate can also substitute for phosphate in the hydroxyapatite crystal of bone, which accounts for the higher concentrations of arsenic-derived radioactivity in bone after administration of arsenate compared to arsenite (Lindgren et al., 1982). At the biochemical level, arsenate can uncouple oxidative phosphorylation in mitochondria by substituting for inorganic phosphate in the synthesis of ATP (Gresser, 1981); it can also inhibit glycolysis by competing with phosphate to form the dysfunctional compound 1-arseno-3-phosphoglycerate, rather than 1:3-diphosphoglycerate (Mayes, 1983).

Arsenite reacts readily with vicinal sulfhydryl groups of a variety of essential enzymes and proteins. It is the affinity of arsenite for sulfhydryl groups that accounts for its accumulation in keratin-rich tissues such skin, hair and nails. Arsenite also interacts with the ubiquitous sulfhydryl-containing cellular tripeptide GSH at many different levels in the methylation process. These include, but may not be limited to, reduction of arsenic from pentavalency to trivalency following the addition of a methyl group, and formation of complexes with trivalent arsenicals which may be substrates for methylation (Styblo et al., 1996). See section 7.1.10.1 for further discussion.

6.2 Organic arsenic compounds

The kinetics and metabolism of MMA, DMA, trimethylarsine (TMA) and trimethylarsine oxide (TMAO), as well as arsenobetaine and arsenocholine, are discussed in this section. In general, organoarsenicals are less extensively metabolized than inorganic arsenic and more rapidly eliminated in both laboratory animals and humans.

6.2.1 Absorption

6.2.1.1 Respiratory deposition and absorption

No quantitative data concerning the respiratory deposition and absorption of organoarsenicals are available for humans or laboratory animals. However, increased urinary excretion of arsenic during the work week with a return to baseline levels on weekends in workers spraying the herbicide monosodium methanearsonate indicates that respiratory absorption of organoarsenicals can occur under occupational exposure conditions (Abdelghani et al., 1986).

6.2.1.2 Gastrointestinal absorption

a) Animal studies

Methylated arsenicals are absorbed from the gastrointestinal tract after oral administration to experimental animals. In male Syrian golden hamsters administered a single oral dose of 50 mg/kg MMA, 36.6 and 60.9% of the dose was eliminated in urine and faeces, respectively, within 5 days. When the same dose was administered by intraperitoneal injection, much more was eliminated in urine (82.6%) and much less in faeces (1%) during the same time period (Yamauchi et al., 1988). The authors noted that this indicated that a relatively large fraction of the administered oral dose was unabsorbed from the gastrointestinal tract compared to their previous studies with DMA (Yamauchi & Yamamura, 1984a) and arseno-betaine (Yamauchi et al., 1986a).

Yamauchi & Yamamura (1984a) reported that 48.9% of a single oral dose of 40 mg/kg DMA was eliminated in the urine of hamsters within 5 days (~36% in faeces). Similarly, Marafante et al. (1987) reported that 56.3% of a single oral dose of 40 mg As/kg DMA was eliminated in urine of male Syrian golden hamsters within 48 h (41.2% in faeces). Gastrointestinal absorption of DMA may be more extensive in mice. In this same study 67.6% and 29.2% of the dose was eliminated in the urine and faeces, respectively, of male ICR mice administered the same oral dose of DMA.

For the trimethylated organoarsenicals – TMA and TMAO – absorption from the gastrointestinal tract of male Syrian golden hamsters is extensive (Yamauchi et al., 1989b; 1990). Yamauchi et al. (1990) reported that $76.9 \pm 2.4\%$ of a single oral dose of 10 mg As/kg TMA was eliminated in urine within 48 h but only $0.11 \pm 0.03\%$ in faeces. Similarly, $88.2 \pm 9.57\%$ of a single oral dose of 10 mg As/kg TMAO was eliminated in urine within 48 h but only $0.55 \pm 0.44\%$ in faeces.

Arsenobetaine, sometimes referred to as "fish arsenic" because it is the predominant organoarsenical present in a number of species of fishes and crustacea, undergoes rapid and almost complete absorption from the gastrointestinal tract of laboratory animals. Vahter et al. (1983) reported in that male NMRI mice given an oral dose of [73]As-arsenobetaine (4 mg As/kg), 73% and 95% of the dose was recovered in the urine after 24 and 72 h respectively. Similarly, Yamauchi et al. (1986a) found that male Syrian golden hamsters dosed orally with 36 mg/kg arsenobetaine excreted 70% of the dose in urine within 12 h and 90% within 5 days. Arsenocholine, also found in seafood, is extensively absorbed from the gastrointestinal tract of mice and rats, with ~70% of the administered oral dose (4 mg As/kg) excreted in the urine within 72 h (Marafante et al., 1984).

b) Human studies

Limited experimental studies in human volunteers suggest that both MMA and DMA are absorbed readily and to a similar extent from the gastrointestinal tract. Buchet et al. (1981a) reported that on average 78.3% of an oral dose of 500 µg of MMA and 75.1% of an oral dose of 500 µg DMA were excreted in urine within 4 days.

Studies have been conducted on the metabolism of organoarsenicals ingested in seafood. In one study in which an adult male Japanese volunteer consumed ~10 µg As/kg trimethyl arsenic in prawns (98.8% trimethylarsenic by analysis, presumably in the form of arsenobetaine), ~90% of the ingested arsenic was excreted in urine within 72 h (Yamauchi & Yamamura, 1984b). In another study conducted in human volunteers consuming flounder, in which the predominant form of arsenic is arsenobetaine, an average of 60% or more of the dose was eliminated in urine within 2 days (Freeman et al., 1979). This suggests that arsenobetaine is readily and rapidly absorbed from the gastrointestinal tract.

6.2.1.3 Dermal absorption

No data concerning the dermal absorption of organoarsenicals in humans were located, but both *in vivo* and *in vitro* dermal absorption data have been reported for arsenical herbicides in laboratory animals. Rahman & Hughes (1994), using clipped dorsal skin of $B6C3F_1$ mice, found that a constant fraction of the dose (~ 12.4%) in water vehicle was absorbed during a 24-h period over the entire applied dose range (10–500 µg) for both the monosodium and disodium salts of monomethylarsonate, and that this was unaffected by vehicle volume. Using the same experimental system with DMA, Hughes et al. (1995) again found no significant dose-dependency in absorption over a 24-h period. However, vehicle volume exerted a significant effect on absorption, which ranged from ~7–40% and decreased with increasing volume of water. In both these studies percutaneous absorption of the arsenical herbicides from soil was very low (< 1%).

Shah et al. (1987) studied the *in vivo* percutaneous absorption of MMA (monosodium salt) and DMA (disodium salt) in young (33-day-old) and adult (82-day-old) Fischer 344 rats. Three levels of each compound (MMA [monosodium salt]: 16.4, 98.6 and 496 $\mu g/cm^2$; DMA [disodium salt]: 16.4, 98.6 and 496 $\mu g/cm^2$) were applied in aqueous vehicle, and absorption over 72 h was determined. Although both compounds exhibited similar absorption values within either young or adult animals over the dose range studied, the young animals absorbed significantly less. The total percutaneous absorption (mean of all doses for both compounds) was 15.1 and 3.01% of the recovered dose in old and young rats, respectively.

6.2.1.4 Placental transfer

No human or animal data directly assessing the ability of organoarsenicals to cross the placenta have been located that have appeared since publication of the last arsenic environmental health criteria document (IPCS, 1981). Older studies have demonstrated that dimethylarsenic acid is capable of crossing the placenta of rats (Stevens et al., 1977). Studies in laying hens also indicate that the organoarsenical feed additive Roxarsone (3-nitro-4-hydroxyphenyl-arsonic acid) accumulates in significantly in eggs as the level in the diet is increased (Chiou et al., 1997b).

6.2.2 Distribution

6.2.2.1 Fate of organic arsenic in blood

a) Animal studies

Yamauchi et al. (1988) reported the time-course distribution of MMA and DMA in whole blood and plasma after a single oral dose of 50 mg MMA/kg body weight in hamsters. MMA concentration in blood peaked at 6 h after dosing, and thereafter declined to control values at 120 h. Distribution of MMA was similar between plasma and erythrocytes through 12 h, but then more tended to be associated with blood cells. DMA levels in plasma peaked at 12 h, but there was no significant change in inorganic or trimethylated arsenic in blood of control compared to dosed hamsters.

Yamauchi & Yamamura (1984b) also studied the time-course distribution in the whole blood of DMA and its metabolites after a single oral dose of 50 mg DMA/kg (mean arsenic dose 1440 μg) in hamsters. Total arsenic in blood peaked at 6 h and consisted of 61% DMA, 26.2% TMAO, 11.8% inorganic arsenic and 1.07% MMA. DMA levels had returned to control values by 24–72 h after dosing. Vahter et al. (1984) reported the distribution of total arsenic between plasma and erythrocytes of mice at 0.5 and 6 h after intravenous injection of 0.4 mg As/kg [74]As-DMA. The plasma : erythrocytes ratio of arsenic at was 2.2 at 0.5 h and 1.4 at 6 h, and levels had declined 76-fold in plasma and 50-fold in erythrocytes 6 h after administration.

The trimethyl arsenic compounds, TMA and TMAO, are even more rapidly cleared from the blood of hamsters than are MMA and DMA. Yamauchi et al. (1990) reported that the half-life of TMA in blood was 3.3 h in hamsters administered a single oral dose of 10 mg As/kg. In hamsters administered a single oral dose of 10 mg As/kg TMAO, arsenic levels in whole blood and plasma peaked within 1 h and thereafter declined very rapidly (Yamauchi et al., 1990). In both studies, only trimethylated arsenic levels detected in blood were related to exposure since levels of inorganic arsenic did not differ between exposed and control animals and other methylated arsenicals were not detected (Yamauchi et al., 1989b, 1990).

b)   Human studies

Studies concerning the fate of organoarsenicals in human blood are almost totally lacking. After ingestion of 10 µg/kg of trimethylarsenic (98.8% by analysis, presumably arsenobetaine) in prawns, trimethylarsenic levels were approximate 2.5-fold higher in plasma than in erythrocytes at 2 h after ingestion in the single subject studied. Levels declined thereafter and were at background by 24 h (Yamauchi & Yamamura, 1984b).

6.2.2.2   Tissue distribution

a)   Animal studies

Yamauchi et al. (1988) reported data on the time-course tissue distribution of hamsters given a single oral dose of 50 mg/kg MMA. Peak MMA concentrations were achieved within 6–12 h after dosing and were highest in the kidney, followed by spleen, lung, skin, liver, muscle and brain. MMA itself accumulated in the kidney and declined very slowly. DMA was also detected in several tissues, with highest levels achieved in lung, followed by kidney and liver. Trimethylated arsenic was not detected in any tissues.

Long-term pharmacokinetic studies are generally lacking for organoarsenicals, but Jaghabir et al. (1994) have performed such a study in New Zealand white rabbits administered multiple oral doses of MMA as the monosodium salt (MSMA). The limitation of this study is that only total arsenic was measured. Rabbits were given an oral dose of 5 mg MSMA/kg 4 days a week for 4 weeks with serial sacrifices at 2 weeks and 4 weeks after the start of exposure and then 1 week and 2 weeks after exposure ended. Significant accumulation of arsenic was observed in muscle and fur after 4 weeks of exposure with significant clearance of arsenic from muscle 1 week after exposure ended, but no significant clearance from fur after 2 weeks of no exposure. Levels of arsenic in kidney were significantly higher than liver at both 1 and 2 weeks after the end of exposure, but did not differ greatly during exposure.

Yamauchi & Yamamura (1984a) studied the tissue distribution of DMA and metabolites in hamsters administered a single oral dose of 50 mg/kg DMA. DMA levels were elevated in all tissues examined, including the brain, indicating that DMA passes the blood–brain barrier, though not to a large degree. DMA concentrations peaked at 6 h in all tissues examined except hair, with levels highest in lung, followed by kidney, spleen, liver, skin, muscle and brain. It is notable that the peak DMA concentration in lung was over 4-fold higher than in the next highest organ. DMA concentrations had declined to control levels by 120 h after dosing. TMA concentrations peaked in most tissues at 6 h after DMA dosing; the highest concentration was achieved in lung, which had a 5-fold higher level than the next highest tissue, which was kidney. Interestingly, MMA levels were also elevated in some tissues of DMA-dosed hamsters compared to controls.

The tissue distribution of DMA has also been studied in mice. Vahter et al. (1984) reported that after intravenous administration of [74]As-DMA (0.4 mg As/kg) to male NMRI mice, the highest levels of [74]As-derived radioactivity were present in kidney at all time points (5–60 min after injection). Tissues with the longest retention of [74]As were the lungs, intestinal walls, thyroid and lens. Some of the

[74]As-DMA present in liver and kidney was in the form of complexes, whereas this was not the case in lung or plasma. The authors reported that there was no evidence of *in vivo* demethylation of DMA. Examination of the subcellular distribution of [74]As-DMA-derived radioactivity indicated that it was predominantly (70%–95%) localized in the cytosol.

TMA undergoes more rapid absorption and tissue distribution than does either MMA or DMA in hamsters. Yamauchi et al. (1990) found that tissue levels of TMA peaked 1 h after male Syrian golden hamsters were given a single oral dose of 10 mg As/kg TMA and had returned to control levels by 24 h after dosing. Concentrations were highest in lung, followed by liver, kidney, spleen and brain. Levels of DMA and inorganic arsenic detected in tissues of dosed animals were similar to unexposed controls. Yamauchi et al. (1989b) also reported that clearance of TMAO from both liver and blood was even more rapid than clearance of TMA when a comparable oral dose was given to hamsters.

Vahter et al. (1983) examined [73]As-arsenic tissue distribution in mice and rabbits after intravenous administration of 4 mg As/kg arsenobetaine. Distribution to and clearance from all tissues was rapid, and somewhat faster in mice than in rabbits. Somewhat longer retention in rabbits was attributable to accumulation in muscle, which makes up a larger proportion of their total body mass. Highest tissue concentrations were attained in kidney, liver and pancreas, respectively, in both species; concentrations in testes and epididymis also remained highest at 72 h in both species. In hamsters administered a single oral dose of 36 mg/kg arsenobetaine, tissue concentrations peaked at 1–6 h after administration and declined rapidly thereafter. Highest concentrations were detected in the liver, kidney, lung, spleen, muscle, skin and brain (Yamauchi et al., 1986a).

b) Human studies

Tissue distribution data in humans are derived from limited studies in which human volunteers have ingested [74]As-labelled organoarsenicals. Brown et al. (1990) reported that arsenobetaine is rapidly and widely distributed in soft tissues with no major concentration in any region or organ and that greater than 99% of tracer activity was eliminated from the body within 24 days. Similar studies were unavailable for other organoarsenicals.

6.2.3 *Metabolic transformation*

6.2.3.1 Animal studies

Studies by Yamauchi et al. (1988) demonstrate that MMA undergoes *in vivo* methylation to dimethylated and trimethylated products, but that methylation is not extensive. After a single oral dose of 5, 50 or 250 mg/kg MMA, hamsters excreted respectively 8.4, 1.4 and 0.4% of the dose as DMA and respectively 1.9, trace, and < 0.1% of the dose as TMA in urine. Most of the absorbed MMA was excreted unchanged in urine and this did not differ significantly with dose. There was no evidence that MMA was demethylated in these studies. Similar findings were reported by Hughes & Kenyon (1998) for female $B6C3F_1$ mice administered MMA intravenously. After a single intravenous injection of 0.6 or 60 mg As/kg MMA, respectively 72.5 ± 4.2 and $77.7 \pm 14.1\%$ of the dose was excreted as MMA and respectively 8.1 ± 1.5 and $2.2 \pm 0.7\%$ was excreted as DMA in urine within 24 h. The decrease in DMA excretion with increasing dose that was observed in both hamsters and mice after MMA administration could be due to either dose-dependent saturation or inhibition of MMA methylation (Hughes & Kenyon, 1998).

DMA is methylated to trimethylarsenic compounds to a limited extent in mice, rats and hamsters (Yamauchi & Yamamura, 1984a; Marafante et al., 1987;Yoshida et al., 1997, 1998). Marafante et al. (1987) reported that in mice and hamsters 3.5 ± 0.4 and $6.4 \pm 0.5\%$ respectively of a single oral dose of 40 mg As/kg DMA was eliminated in urine as TMAO within 48 h, TMAO was not detected in the faeces of either species in this study. An unidentified DMA complex was also excreted in both urine (7–11% of the dose) and faeces (4–5% of the dose) in mice and hamsters in this study, with the remainder of the dose excreted as unmetabolized DMA. Hughes & Kenyon (1998) also reported an unidentified and readily oxidizable metabolite in urine of mice administered DMA intravenously. Marafante et al. (1987) speculated that this metabolite might be some type of thiol complex.

Since methylation serves to expedite the excretion of inorganic arsenic, which is more toxic than organoarsenicals, issues such as whether demethylation occurs and if methylation is saturable, inducible, or inhibitable under expected environmental exposure conditions are critical. The fact that radiolabelled inorganic arsenic is not detected in the urine of mice, rats, hamsters and humans after administration of [74]As-DMA indicates that demethylation is insignificant in these species (Vahter et al., 1984; Marafante et al., 1987). However, Yoshida et al. (1997) recently compared the time-course of urinary excretion of DMA and its metabolites after a single oral or intraperitoneal injection of 50 mg/kg DMA to rats. They reported that more arsenite was excreted in urine of rats administered DMA orally than by intraperitoneal injection. The authors interpreted their data as being indicative of *in vivo* demethylation, most likely by intestinal microorganisms. It is worthy of note however, that Hall et al. (1997) found no evidence of demethylation in studies using the caecal microbiota from mice in an *in vitro* anaerobic culture system. Subsequent studies by Yoshida et al. (1998) showed that essentially no inorganic arsenic was excreted in urine of rats exposed to DMA in drinking-water at 100 mg/litre for 7 months.

On the basis of limited studies in hamsters, it appears that neither TMA or TMAO is further methylated or demethylated, but they do undergo *in vivo* redox reactions. Yamauchi et al. (1990) found that ~80% of a 10 mg As/kg oral dose of trimethylarsine (TMA) was oxidized to TMAO and excreted in urine within 120 h of administration. Similarly, when TMAO was administered orally to hamsters at a dose of 10 mg As/kg, only TMAO and no arsenobetaine was eliminated in urine. Interestingly a fraction of the dose (unquantified) was reduced to TMA and excreted in expired air when hamsters were given a single oral or intraperitoneal dose of 50 mg As/kg TMAO (Yamauchi et al., 1989b).

Studies in mice, rats, rabbits and hamsters administered arsenobetaine intravenously or orally indicate that it is not biotransformed or demethylated (Vahter et al., 1983; Yamauchi et al., 1986a). Arsenocholine is also not demethylated, but is metabolized extensively to arsenobetaine. Specifically, Marafante et al. (1984) reported that in mice, rabbits and rats administered 4 mg As/kg arsenocholine intravenously, approximately 40, 50 and 60% of the dose, respectively, was eliminated in urine as arsenobetaine within 48 h. No major difference in urinary excretion or arsenobetaine was noted after oral administration of arsenocholine to rats or mice.

6.2.3.2 Human studies

On the basis of limited data from controlled ingestion studies, it appears that MMA and DMA are metabolized to a similar extent in laboratory animals and humans (see section 6.2.3.1 and Table 15). Buchet et al. (1981a) reported that after a single oral dose of MMA (500 μg As), 87.4% of the total metabolites excreted in urine in 4 days were in the form of MMA and 12.6% were in the form of DMA. In this same study, it was reported that all of the ingested DMA (500 μg As) excreted in the urine was in the form of DMA. However, in a later study, Marafante et al. (1987) reported that 3.5% of a single oral dose of DMA (0.1 mg As/kg) was eliminated in urine as TMAO within 2 days. Metabolic studies in which humans specifically consumed TMA or TMAO alone rather than in seafood were not found.

In common with laboratory animals, humans appear to eliminate arsenobetaine ingested in seafood unchanged in their urine, indicating that arsenobetaine is not metabolized (Tam et al., 1982).

6.2.4 Elimination and excretion

6.2.4.1 Animal studies

Total (urine + faecal) elimination of organoarsenicals is quite rapid in laboratory rodents, with 80% or more of the dose eliminated within 48 h of a single oral or parenteral dose (Table 21). Absorbed MMA and DMA are predominantly eliminated in urine (Table 21). Limited data from studies where multiple dose levels were used (Yamauchi et al., 1988; Hughes & Kenyon, 1998) suggest that urinary elimination is also dose-independent, i.e. the percentage of the dose eliminated in urine does not change with increasing or decreasing dose level.

No studies were identified which directly addressed the issue of biliary elimination of any organoarsenicals. However, given the

Table 21. Cumulative elimination (% of dose) of organoarsenicals in urine and faeces of laboratory animals after oral and parenteral administration

Arsenical	Species	Route	Dose (mg/kg)	Time (h)	Urine	Faeces	Total	Reference
MMA	hamster	oral	5	0–24	38.8	51.6	90.4	Yamauchi et al. (1988)
		oral	50	0–24	28.3	56.0	84.3	
		oral	250	0–24	34.2	46.0	80.2	
MMA	hamster	i.p	50	0–120	82.6	1.0	83.6	Yamauchi et al. (1988)
		oral	50	0–120	36.6	60.9	97.5	
MMA	mouse	i.v	0.6 (As)	0–24	80.6 ± 2.7	3.9 ± 1.4	84.5	Hughes & Kenyon (1998)
		i.v	60 (As)	0–24	79.9 ± 13.6	8.8 ± 2.5	88.7	
DMA	hamster	oral	40	0–120	48.9	36.0	84.9	Yamauchi & Yamamura (1984a)

Table 21 (contd.)

DMA	hamster	oral	40 (As)	0–48	56.3	41.2	97.5	Marafante et al. (1987)
	mouse	oral	40 (As)	0–48	67.7	29.2	96.8	
DMA	mouse	oral	0.4 (As)	0–24	80.2 ± 2.5	15.8 ± 0.6	96.0	Vahter et al.(1984)
	rat	oral	0.4 (As)	0–24	18.2 ± 4.2	2.0 ± 1.0	20.2	
DMA	mouse	i.v.	0.6 (As)	0–24	77.7 ± 8.4	4.4 ± 0.6	82.1	Hughes & Kenyon (1998)
	mouse	i.v	60 (As)	0–24	82.8 ± 5.2	2.3 ± 0.9	85.1	
DMA	rabbit	i.v	0.04 (As)	0–72	93.8 ± 2.5	~2–3	~96	Vahter & Marafante (1983)
TMA	hamster	oral	10 (As)	0–120	79.2 ± 2.7	0.14 ± 0.03	79.4	Yamauchi et al. (1990)
TMAO	hamster	oral	10 (As)	0–120	89.0 ± 9.61	0.56 ± 0.44	89.6	Yamauchi et al. (1989b)

relatively low amounts of MMA and DMA excreted in the faeces (2–9% of the dose) after intravenous administration of these compounds to mice or rabbits (Vahter & Marafante, 1983; Hughes & Kenyon, 1998), it seems unlikely that biliary excretion or other gastric secretory processes contribute significantly to total elimination. Interestingly, however, Hughes & Kenyon (1998) found that the percentage of the dose eliminated in faeces was dose-dependent when either MMA or DMA was administered intravenously to mice (Table 21).

Volatile metabolites of some organoarsenicals are eliminated in expired air after oral administration. After a high oral dose of DMA (1500 mg/kg), mice eliminate dimethylarsine, but not TMA, in expired air (Yamanaka & Okada, 1994). Similarly, in mice orally administered 14 400 mg/kg TMAO, TMA was detected in expired air (Kaise et al., 1989). Hamsters also eliminate TMA in expired air after administration of either TMAO or TMA (Yamauchi et al., 1989b, 1990).

Extensive studies by Vahter et al. (1983) demonstrate that arsenobetaine is rapidly and predominantly eliminated in the urine. After intravenous administration of 4 mg As/kg arsenobetaine 101 ± 5.8%, 94.9 ± 0.8%, and 71.6 ± 0.6% respectively of the dose was eliminated in urine of rats, mice and rabbits with 72 h. The corresponding figures for faecal elimination were 4.5 ± 1.3%, 3.8 ± 1.4%, and 2.3 ± 0.8% of the dose. The pattern of elimination was also very similar in mice administered the same dose of arsenobetaine orally, and the rate of excretion in urine was dose independent in the range of 4–400 mg As/kg arsenobetaine.

Arsenocholine, like arsenobetaine, is predominantly eliminated in the urine of mice, rats and rabbits after intravenous administration (Marafante et al., 1984). However, although the percentage of the dose eliminated in the faeces (2–3%) for the two compounds is quite similar among different animal species, 66% was eliminated in the urine of rabbits compared to 78% in rats and mice within 72 h of administration. Whole-body retention of arsenocholine was consistently significantly greater over a 28-day period in mice dosed intravenously with 4 mg As/kg of either compound. The authors attribute this difference to the fact that arsenocholine can be incorporated into phospholipids whereas arsenobetaine is not (Marafante et al., 1984).

6.2.4.2   Human studies

In common with laboratory animals, humans appear to eliminate orally administered MMA and DMA predominantly in urine. Buchet et al. (1981a) reported that an average of 78.3% and 75.1% of a single oral dose (500 µg As) of MMA and DMA, respectively, was eliminated in urine of human volunteers within a 4-day period. Arsenic ingested in seafood, most probably in the form of arsenobetaine, is predominantly and rapidly eliminated in urine (Table 22). It is worthy of note that the percentage of the dose eliminated in urine after ingestion of arsenic in seafood is quite similar to that seen in laboratory animals dosed orally with arsenobetaine. No studies were identified that specifically addressed the issue of biliary excretion or other routes of elimination for organoarsenicals in humans.

Table 22.   Percentage elimination of As ingested in seafood[a]

Species	As Ingested	No. of subjects	Time (days)	Elimination[b]	Reference
Flounder	5	6	8	77 ± 11 (U)	Freeman et al. (1979)
Flounder	10	15	8	76 ± 8 (U)	Tam et al. (1982)
Plaice	8	8	5	69–85 (U)	Luten et al. (1982)
Cod + labelled arsenobetaine	ND	6	8	92 ± 2 (T)	Brown et al. (1990)

[a]   U = urinary elimination; T = total elimination; ND = not determined
[b]   Elimination figures are percentage mean ± SD or range

6.2.5   Retention and turnover

Vahter et al. (1984) compared the whole-body retention of [74]As-DMA in mice and rats after a single oral dose of 0.4 mg As/kg. In mice, whole-body clearance of DMA was triphasic, with 85% of the dose eliminated with a half-time of 2.5 h, 14% with a half-time of 10 h and the remainder (< 0.5%) with a half-time of 20 days. In rats elimination was biphasic with 45% of the dose having a half-time of ~13 h and the remaining 55% having a half-time of ~50 days. The longer retention of DMA in the rat was attributed to its tendency to accumulate in erythrocytes.

Yamauchi et al. (1990) calculated the biological half-lives after oral administration of organoarsenicals to hamsters from many studies conducted in their laboratory. They reported half-times of 7.4 h for MMA, 5.6 h for DMA, 5.3 h for TMAO, 3.7 h for TMA and 6.1 h for arsenobetaine. No studies that specifically investigated the retention and turnover of organoarsenicals in humans were identified.

6.3 Biomarkers of arsenic exposure

The three most commonly employed biomarkers used to identify or quantify arsenic exposure are total arsenic in hair or nails, blood arsenic, and total or speciated metabolites of arsenic in urine. This section emphasizes the utility and limitations of these biomarkers and provides more limited information on arsenic levels associated with specific environmental exposure concentrations in air and water. Issues related to analytical methods relevant to the use of these biomarkers (e.g. preservation, extraction, storage) are discussed in section 2.4.

6.3.1 Arsenic in hair and nails

Because arsenic accumulates in keratin-rich tissues such as skin, hair and nails as a consequence of its affinity for sulfhydryl groups, arsenic levels in hair and nails may be used as an indicator of past arsenic exposure. Hair and nails have the advantage of being readily and non-invasively sampled, but a major issue of concern is whether external contamination can be removed. Sampling of hair from less readily contaminated sites (e.g. occipital area or nape of neck), and closer to the scalp, can minimize some of these problems. When exposed to water containing high arsenic levels, hair can bind arsenic externally and may not be removed readily by washing procedures. In the studies cited in this section, the issue of possible contamination was apparently adequately addressed in the methodology employed.

Paschal et al. (1989) determined levels of a number of elements in hair (0.5 g occipital new growth hair) of both adults and children without known toxic metal exposure in the USA. The geometric mean levels of arsenic in hair of adults and children did not differ significantly and were 0.035 and 0.032 μg/g, respectively. Wolfsperger et al. (1994) reported that hair of males from both Vienna (Austria) and Rome (Italy) contained significantly more arsenic (μg/g) than the hair of females – 0.12 vs. 0.037 and 0.13 vs. 0.044, respectively. In this same study it was reported that smokers had higher levels of arsenic in hair than non-smokers, although the difference was not statistically significant. Zhuang et al. (1990) reported levels of 0.40 ± 0.22 μg/g in hair of adult male Chinese subjects dying accidentally and with no known history of toxic metal exposure. These authors also reported a significant positive correlation ($r = 0.75$) of hair arsenic with arsenic levels in kidney cortex, but not in lung or liver.

Arsenic levels in both hair and nails are elevated within one to a few weeks after acute poisoning, and return to background levels within a few months (Choucair & Ajox, 1988). Since the rate of hair growth is about 1 cm/month, the segmental distribution of arsenic along the hair shaft has been used to distinguish the between acute and chronic poisoning, as well as to estimate length of time since a poisoning incident (Koons & Peters, 1994).

The arsenic content of fingernails and toenails has also been used as a bioindicators of past arsenic exposure, and fingernail arsenic has been reported to be significantly correlated with hair arsenic content (Lin et al., 1998). Agahian et al. (1990) reported that fingernail arsenic was elevated as a result of occupational arsenic exposure and correlated significantly ($r = 0.89$) with mean arsenic air concentrations.

The use of toenails rather than fingernails has been recommended in some studies because of the larger amount of sample that can generally be obtained (Garland et al., 1993; Karagas et al., 1996). Karagas et al. (1996) reported that toenail arsenic was significantly elevated in individuals using well-water known to be high in arsenic compared to individuals using water from low-arsenic wells with geometric mean ± SE toenail arsenic levels of 0.39 ± 0.12 μg/g and 0.14 ± 0.02 μg/g, respectively. Regression analysis of these data indicated that a 10-fold increase in arsenic concentration in water was associated with a two-fold increase in toenail arsenic levels.

6.3.2 Blood arsenic

Inorganic arsenic is rapidly cleared from blood. It is for this reason that blood arsenic is typically used only as an indicator of very recent or relatively high-level exposure (e.g. in cases of poisoning), or chronic stable exposure (e.g. to drinking-water). The limitation of blood arsenic levels as indicators of low-level exposure or drinking-water is that it is difficult to distinguish the contributions of inorganic arsenic from water and organic arsenic from food (NRC, 1999).

Arsenic concentrations in blood are elevated in individuals with chronic high level exposure to arsenic in drinking-water, but not to the same degree as urinary arsenic. For example, Concha et al. (1998a) reported that in a group of Andean women whose drinking-water contained ~0.65 mg As/litre, median blood arsenic was 0.95 µg/litre and median urinary arsenic concentration was 7.6 µg/litre. In contrast, a similar population whose drinking-water contained ~200 mg As/litre had median blood arsenic levels of 7.6 µg/litre (8-fold higher) and a median urinary arsenic concentration of 303 µg/litre (~40-fold higher).

6.3.3 Arsenic and metabolites in urine

Since arsenic is rapidly metabolized and excreted into the urine, total arsenic, inorganic arsenic and the sum of arsenic metabolites (inorganic arsenic + MMA + DMA) in urine have all been used as biomarkers of recent arsenic exposure. In common with other biomarkers of arsenic exposure, levels of arsenicals in urine may be a consequence of inhalation exposure or ingestion of arsenic from drinking-water, beverages, soil or foodstuffs (NRC, 1999). However, in the case of exposure to arsenic compounds of low solubility, e.g. GaAs, urinary arsenic will reflect the absorbed dose, but not the inhaled amount (Yamauchi et al., 1989a)

In many older studies, total urinary arsenic was used as a biomarker of recent arsenic exposure. However, this is increasingly uncommon because organoarsenicals present in substantial amounts in certain foodstuffs (see sections 5.1 and 5.2) are also excreted in urine. For example, the practically non-toxic compound arsenobetaine is present in mg/kg levels in seafood and excreted mainly unchanged in the urine (Kaise & Fukui, 1992; Le et al., 1993, 1994c). In controlled experiments (e.g. Arbouine & Wilson, 1992; Buchet et al., 1994, 1996), it has been found that consumption of seafood (e.g. marine fishes, crustaceans, bivalves, seaweeds – see section 5.1) by human volunteers is associated with increased total urinary arsenic excretion. Under these conditions, assessment of inorganic arsenic exposure using total urinary arsenic would result in overestimation of inorganic arsenic exposure.

To avoid the potential for overestimation of inorganic arsenic exposure inherent in using total urinary arsenic, most studies now measure speciated metabolites in urine and use either inorganic arsenic or the sum of arsenic metabolites (inorganic arsenic + MMA + DMA) as an index of arsenic exposure. Relatively recently it has been found that adding all arsenic metabolites together can give misleading results unless a careful diet history is taken and/or seafood consumption is prohibited for 2–3 days before urine collection (Buchet et al., 1996). There are two reasons for this. First, some seafood, especially bivalves, contain the arsenic metabolites MMA and DMA, particularly DMA, in fairly high amounts (Velez et al., 1996). Secondly, arsenosugars present in seaweeds and some bivalves are extensively metabolized (either by the body itself or by the gut microbiota) to DMA, which is then excreted in urine (Le et al., 1994c; Ma & Le, 1998). The issue of the extent to which consumption of seafoods and other foods can compromise the estimation of inorganic arsenic exposure by the measurement of arsenic and its metabolites in urine remains an active area of investigation.

7. EFFECTS ON LABORATORY MAMMALS AND *IN VITRO* TEST SYSTEMS

7.1 Inorganic arsenic

7.1.1 *Single exposure*

7.1.1.1 *Acute toxicity data*

a) Oral

The acute oral LD_{50}s of several inorganic arsenicals are listed in Table 23. Trivalent inorganic arsenic is generally more toxic than pentavalent inorganic arsenic. The oral LD_{50} of As_2O_3 (As_2O_3), a trivalent arsenical, administered to three strains of mice, ranges from 26 to 36 mg As/kg body weight (Harrison et al., 1958; Kaise et al., 1985). Because of the low solubility of As_2O_3 in water, Harrison et al. (1958) heated the solution to dissolve the As_2O_3 and there is a question regarding the arsenic valence state because of this treatment. However, Kaise et al. (1985) reported a similar LD_{50} in their study with As_2O_3, but using a younger mouse. They did not indicate whether the solution was heated. The LD_{50} of As_2O_3 administered in food was 145 mg As/kg (Harrison et al., 1958). The symptoms observed from As_2O_3 intoxication include convulsions, retching and haemorrhaging in the intestinal tract. The severity of the symptoms appears to be related to the purity of As_2O_3: pure As_2O_3 has less severe side-effects than the commercial-grade compound, but greater lethality (Harrison et al., 1958).

b) Inhalation

Inhalation is the main route of exposure to gallium arsenide (GaAs), a binary compound used in the semiconductor industry (Sheehy & Jones, 1993). In order to simulate this type of exposure in laboratory animals, Webb et al. (1986) administered GaAs (100 mg/kg) intratracheally to rats and compared its retention and toxicity to that of As_2O_3 (17 mg/kg) and gallium oxide (Ga_2O_3, 65 mg/kg). Ga_2O_3 elicited minimal effects in the lung, despite being retained in this organ (Webb et al., 1986). As would be the case with

Table 23. Acute LD$_{50}$ of inorganic As

Chemical	Species	Age	Sex	Route	LD$_{50}$ (mg As/kg)	LD$_{50}$ (mg/kg)	Reference
As trioxide[a]	Mouse	young	M	oral	26–39	34.1–52.5	Harrison et al. (1958)
As trioxide	Mouse	weanling	M	oral	26	34.5	Kaise et al. (1985)
As trioxide	Rat	adult	M/F	oral	15	20	Harrison et al. (1958)
As trioxide[b]	Rat	adult	M/F	oral	145	188	Harrison et al. (1958)
As trioxide[c]	Rat	adult	M/F	oral	293	385	Done & Peart (1971)
Sodium arsenite[c]	Rat	adult	M/F	oral	24	42	Done & Peart (1971)
Sodium arsenite	Mouse	young	M	i.m.	8	14	Bencko et al. (1978)
Sodium arsenate	Mouse	young	M	i.m.	21	87	Bencko et al. (1978)
Sodium arsenite	Rat	young	–[d]	i.p	4–5[e]	9.7–10.9[e]	Franke & Moxon (1936)
Sodium arsenate	Rat	young	–[d]	i.p	14–18[e]	34–44[e]	Franke & Moxon (1936)
Calcium arsenate	Rat	adult	F	oral	53	298	Gaines (1960)
Lead arsenate	Rat	adult	F	oral	231	1050	Gaines (1960)
Calcium arsenate	Rat	adult	F	dermal	> 400	> 2400	Gaines (1960)
Lead arsenate	Rat	adult	F	dermal	> 500	> 2400	Gaines (1960)

[a] Strains Swiss albino, Dba, C$_3$H, C$_{57}$H$_{46}$, Webster
[b] As in dry feed
[c] As administered in gelatin capsule

[d] Sex not specified
[e] LD$_{75}$

any compound, it is not possible to separate the toxicity of gallium from that of arsenic. Gallium is retained longer than arsenic in the lung of rats. As_2O_3 induced a significantly increased wet and dry lung weight, increased wet lung weight/body weight ratio and an elevation in total pulmonary protein, 4-hydroxyproline content and DNA. These results suggest that arsenic may be acutely fibrogenic in the lung. A significant elevation of lipids, protein and DNA was found in the lungs of GaAs-treated rats. There was also an increase in wet and dry lung weight after GaAs exposure. Decreasing the particle size of GaAs increases the severity of the pulmonary lesions, most probably because of the increased dissolution rate of the smaller particles of GaAs (Webb et al., 1987). Histopathological analysis of lungs of animals from all three exposure groups showed that there was an inflammatory response and pneumocyte hyperplasia, which resulted in thickening of the alveolar walls. The severity of the effect in descending order was GaAs > As_2O_3 >> Ga_2O_3. Although the effect of As_2O_3 was lower than that of GaAs, the dose of As_2O_3 was also lower.

c) Dermal

The acute dermal LD_{50} for the pentavalent arsenicals calcium arsenate and lead arsenate in the rat is > 2400 mg/kg ($\geq$ 400 mg As/kg) (Table 23) (Gaines, 1960).

d) Parenteral

Sodium arsenite is 3–4 times more potent than sodium arsenate in the rat after intraperitoneal administration (Franke & Moxon, 1936) and 2–3 times more potent in the mouse after intramuscular administration with respect to acute lethality (Bencko et al., 1978) (Table 23).

The dose-dependent nephrotoxic effect of sodium arsenate after intravenous administration in the dog was examined by Tsukamoto et al. (1983). Clinical laboratory parameters, renal function and histopathological analysis of kidneys were studied after administration of 0.73, 7.33, or 14.77 mg sodium arsenate/kg. The low dose elicited mild degeneration and vacuolation of the renal tubular epithelium, which was primarily localized in the ascending thick portion of the nephron. No clinical effects or changes in renal function were noted at this dose level. In the medium-dose group the urinalysis showed proteinuria, glucosuria and renal casts in urine sediment, which indicates that the glomerulus and renal tubules were damaged. Vacuolar changes and widespread tubular necrosis were also observed. The high dose of arsenate caused more severe alterations throughout the kidney, and one of three dogs administered this dose died. The clinical and histopathological changes in this organ were indicative of damage to the glomerulus, proximal and distal tubules, and collecting duct. These effects are reversible but may not be completely repaired, because residual scarring and tubular atrophy was evident in one of the two surviving dogs.

Acute arsenic toxicity may be manifested in part by disruption of carbohydrate metabolism. Rats fed a standard diet *ad libitum* survived a subcutaneous dose of As_2O_3 (10 mg As/kg), whereas 80% of animals starved 24 h before arsenic exposure died (Szinicz & Forth, 1988). *In vitro* studies by Szinicz & Forth (1988) showed that As_2O_3 induced a dose-dependent inhibition of gluconeogenesis in isolated hepatocytes and kidney tubules. Arsenic pentoxide (As_2O_5), a pentavalent arsenical, had the same effect, but higher concentrations of it were required. Acetyl coenzyme-A content in the isolated hepatocytes and kidney tubules was also decreased after addition of As_2O_3.

Carbohydrate analysis in guinea-pig liver 1 h after subcutaneous administration of As_2O_3 (10 mg As/kg) showed that there is a significant decrease in fructose-1 6-diphosphate, glycerolaldehyde-3-phosphate and significant increases in phosphoenolpyruvate and pyruvate (Reichl et al., 1988).

Mice administered saline, glucose, or glucose + insulin were more likely to survive a subcutaneous dose of As_2O_3 (12.9 mg/kg) than mice receiving only As_2O_3 (Reichl et al., 1990). All of the mice in the latter group died. Hepatic glucose and glycogen of all animals administered As_2O_3 (and other factors) that died were significantly lower than non-treated mice. Regardless of the treatment, hepatic glucose and glycogen levels were unaffected in surviving mice.

7.1.2 *Short-term exposure*

7.1.2.1 *Oral*

The effect of a 28-day exposure to sodium arsenate (0.025 and 2.5 mg arsenate/litre) in drinking-water of mice was examined by Hughes & Thompson (1996). The mice were exposed to approximately 3 and 300 μg As/kg body weight per day. No treatment effects were observed on food and water consumption or weight gain by the mice. There was a dose-related increase in hepatic vacuolar degeneration. Significantly lower hepatic non-protein sulfhydryl levels, decreased plasma glucose and increased plasma creatinine were observed in the high-dose group. In both the low- and high-dose groups, decreased plasma triglycerides were observed. The elevated plasma creatinine suggested there was an effect by arsenic on the kidney, but the urinalysis showed that the alkaline phosphatase activity in the treated animals was no greater than control. In addition, histopathological analysis of the kidney showed no treatment effect. Thus it is not clear if the kidney was affected by this treatment.

7.1.2.2 *Inhalation*

The solubility as well as the valency of arsenic has an important role in its toxicity. Pershagen et al. (1982) administered As_2O_3 (0.3 mg As), arsenic trisulfide (As_2S_3, 0.5 mg As, trivalent) or calcium arsenate ($Ca(AsO_4)_2$, 0.5 mg As) intratracheally to hamsters once a week for 4 weeks. The particle size mass mean diameters for As_2S_3, As_2O_3, and $Ca(AsO_4)_2$ were 7, 8 and 11 μm, respectively. All of the animals survived the first week after the initial treatment. Between the third and fourth instillation, 80% of the As_2O_3-treated animals died, and 30% of the $Ca(AsO_4)_2$ -treated animals died during the week following the fourth instillation. Two animals in the As_2S_3 group died after the second instillation. Retention of arsenic in the lung, in decreasing order of administered chemical, was $Ca(AsO_4)_2$ $\gg As_2S_3 > As_2O_3$. The higher retention did not appear to be due to differences in particle size distribution of the dusts. The concentration of arsenic was significantly greater in the liver of the As_2O_3-treated animals than in the other groups. The greater toxicity of As_2O_3 may be attributed to its greater solubility, enabling it to distribute throughout the animals and reach target organs at a sufficient concentration to elicit a toxic response.

7.1.2.3   Dermal

There is no recent information on the toxicity of arsenic in laboratory animals from short-term dermal exposure.

7.1.2.4   Parenteral

Total hepatic carbohydrates are significantly decreased at 1 h (94%) and 16 h (76%) after the final repeated subcutaneous administration of As_2O_3 (2.5 mg/kg, twice a day for 5 days) in guinea-pigs (Reichl et al., 1988). The carbohydrate component affected the greatest was glycogen. The carbohydrate depletion by arsenic may be due to inhibition of gluconeogenesis (Szinicz & Forth, 1988) and which would lead to serious toxic effects or even death.

7.1.3   Long-term exposure

7.1.3.1   Oral

The effects of sodium arsenite fed *ad libitum* to dogs were examined by Neiger & Osweiler (1989). Initially, six female dogs/group were fed 0, 1, 2, or 4 mg of sodium arsenite/kg body weight per day. The food intake of the high-dose group was significantly less than the other groups during weeks 1–3. By week 4, the food intake was no longer significantly different between the groups. On day 59, the dosage was doubled to examine feed rejection and compensation. The high-dose group animals were terminated at 119 days, and the remaining animals at 183 days. There was a dose-dependent decrease in feed consumption and body weight of the dogs. Weight loss in pair-fed animals was not different from treated animals, so the loss of body weight in the treated animals was not due sodium arsenite exposure. Two serum enzymes were elevated in dogs examined at study termination, suggesting arsenite-induced hepatotoxicity. Aspartate aminotransferase was elevated in dogs exposed to 4 and 8 mg/kg per day sodium arsenite, and alanine aminotransferase was elevated in dogs exposed to 2, 4 and 8 mg/kg per day sodium arsenite. However, no lesions in the liver were observed after gross or light-microscopic examination.

Minor histological alterations in kidney and liver were observed in rats exposed to sodium arsenate (50 µg As/ml) for 320 days in drinking-water (Carmignani et al., 1983). These alterations were characterized by focal changes in the glomerulus and tubules of the kidney, and swollen hepatocytes localized near the centrilobular vein.

7.1.3.2 Inhalation

There is no recent information on long-term inhalation toxicology studies of inorganic arsenic in laboratory animals.

7.1.3.3 Dermal

There is no recent information on long-term dermal toxicology studies of inorganic arsenic in laboratory animals.

7.1.4 Skin and eye irritation; sensitization

7.1.4.1 Contact sensitivity

Sodium arsenite and sodium arsenate are not allergenic in the guinea-pig maximization test (Wahlberg & Boman, 1986).

7.1.5 Reproductive toxicity, embryotoxicity, and teratogenicity

7.1.5.1 In vivo embryo and fetal toxicity

Inorganic arsenic is toxic to mouse and hamster embryos and fetuses after oral, parenteral or inhalation administration to the dams, with arsenite 3–10-fold more potent than arsenate (Baxley et al., 1981; Willhite, 1981; Hood & Harrison, 1982; Hood & Vedel-Macrander, 1984; Nagymajtenyi et al., 1985; Carpenter, 1987; Domingo et al., 1991; Wlodarczyk et al., 1996). The hamster embryo and fetus is more sensitive to this effect than the mouse. The toxicity is characterized by decreases in fetal weight, crown–rump length, embryo protein content, the number of somites, and growth retardation and lethality. Arsenite-induced lethality is dependent on dose and day of gestation in mice after a single oral dose (Baxley et al., 1981) or in hamsters after an intravenous dose (Willhite, 1981). Arsenite administered orally (20–25 mg/kg) is less effective than parenteral dosing (2.5–5 mg/kg) with respect to mortality (Hood & Harrison, 1982). Nemec et al. (1998) evaluated the developmental toxicity of arsenic acid administered by oral gavage to CD-1 mice and New Zealand white rabbits. Rabbits received doses of 0, 0.19, 0.75 or 3.0 mg/kg per day on gestation days 6–18 and mice received 0, 7.5, 24 or 48 mg/kg per day on gestation days 6–15. Developmental toxicity in the form of increased fetal resorptions and decreased fetal weight were observed only at exposure levels resulting in maternal toxicity (severely decreased weight gain, mortality). On the basis of the no observed adverse effect levels (NOAELs) identified in this study, rabbits (NOAEL = 0.75 mg/kg per day for maternal and developmental toxicity) were more sensitive to the effects of arsenic acid than mice (NOAEL = 7.5 mg/kg per day).

The chelators dimercaprol (BAL) and dimercaptosuccinic acid (DMSA), which are used to treat arsenic poisoning, significantly reduce arsenite-induced prenatal mortality in mice (Hood & Vedel-Macrander, 1984; Domingo et al., 1991). BAL is most effective administered a few hours before arsenite, whereas DMSA is most effective administered within 1 h of arsenite administration.

Ferm & Hanlon (1985) implanted osmotic mini-pumps containing sodium arsenate (150–250 mg/ml) subcutaneously into pregnant hamsters on one of days 4–7 of gestation. The pumps release the solution at a rate of 1 μl/h. The dams were killed on day 13 of gestation and the fetuses examined. Fetal weight decreased and resorption rate increased in the offspring of treated dams, and these effects were dose-dependent and more marked the earlier the arsenate was implanted. There was also a trend of smaller crown–rump length with increased exposure time. The decrease in crown–rump length on day 4 was also dose-dependent.

The developmental effects of GaAs after repeated inhalation exposure were examined in rats by Mast et al. (1990, 1991). Pregnant rats and virgin controls were exposed to 0, 10, 37 or 75 mg/m^3, 6 h/day, during gestation days 4–19. Although the rats exhibited signs of pulmonary toxicity, which included dyspnoea, and grey mottled lungs, there were no effects on maternal body weights. No embryotoxic effects were observed. A concentration-related growth retardation, as shown by decreased fetal body weight, was statistically significant at 37 mg/m^3. The NOAEL for developmental toxicity is 10 mg/m^3. The fetus had an arsenic concentration of

2.2 mg/g at 75 mg/m^3 on gestation day 20, but this level was considerably lower than that detected in maternal blood. Gallium concentration on gestation day 20 was greater in the fetus than in the dam (1.3 vs. 0.5 µg/g at 75 mg/m^3, respectively).

7.1.5.2 In vitro *embryo and fetal toxicity*

Several *in vitro* studies have examined the effect of inorganic arsenic at pre- and post-implantation stages of development. Muller et al. (1986) examined the effect of sodium arsenite on two-cell mouse embryos at pre-implantation stages, which is approximately 30–32 h after conception. Arsenite-induced lethality occurs at a concentration of 100 µmol/litre, but other effects occur at much lower doses. Micronuclei form at 0.7 µmol/litre arsenite, and the ED$_{50}$ for inhibition of blastocyte formation and blastocyte hatching respectively is 0.65 µmol/litre and 0.45 µmol/litre. Cell proliferation is decreased by arsenite in a dose-dependent manner.

After implantation, arsenite and arsenate are toxic (decreases in crown–rump length, number of somites, protein content, head length, yolk sac diameter) and lethal to embryos of mouse (Chaineau et al., 1990; Tabacova et al., 1996) and rat (Mirkes & Cornel, 1992; Mirkes et al., 1994). Tabacova et al. (1996) observed that as gestational age increased when the mouse embryos are isolated and exposed to arsenic, resistance to toxicity or lethality increased. As observed in the *in vivo* studies, arsenite was more potent *in vitro* with respect to embryo and fetal toxicity than arsenate.

7.1.5.3 *Teratogenicity*

a) *In vivo*

There are many older studies of the teratogenic potential of arsenic in animals (IPCS, 1981): many of these were carried out using the intravenous or intraperitoneal routes, which are not considered to be the most appropriate route of arsenic exposure for the fetus. In general the teratogenic effects are seen at levels that produce maternal toxicity.

Inorganic arsenic elicits teratogenic effects in the mouse (Baxley et al., 1981; Morrissey & Mottet, 1983; Nagymajtenyi et al., 1985;

Wlodarczyk et al., 1996), rat (Fisher, 1982) and hamster (Willhite, 1981; Hood & Harrison, 1982; Ferm & Hanlon, 1985; Carpenter, 1987) after oral, parenteral or inhalation administration. The malformations are dependent on the dose of arsenic administered as well as the gestational age. Sodium arsenite is more potent than sodium arsenate in inducing a teratogenic response, and parenteral administration of arsenic is more effective than oral administration. Administration of an acute oral dose of arsenite that is toxic to or near the lethal dose of pregnant mice (40–45 mg/kg) (Baxley et al., 1981) or hamsters (20–25 mg/kg) (Hood & Harrison, 1982) induces a low incidence of teratogenic malformations.

The major teratogenic effect induced by inorganic arsenic in laboratory animals is cephalic axial dysraphic disorder, or neural tube defect. The defect is characterized by exencephaly and encephalocele, which are non-closure and partial closure of the cephalic neural folds, respectively (Carpenter, 1987). Exencephaly and encephalocele occur rarely in laboratory rodents. Other malformations which occur to a minor extent after exposure to arsenic include fused ribs, renal agenesis, micromelia, facial malformations, twisted hindlimb, microphthalmia and anophthalmia.

Administration of inorganic arsenic to dams on days 7–9 of gestation results in neural tube defects in the developing organism. The most sensitive time in mouse embryos to arsenate is when the dams are administered chemical on day 8. Of the fetuses that survived a single dose of sodium arsenate (45 mg/kg) administered intraperitoneally to dams on day 8, 65% or more were exencephalic (Morrissey & Mottet, 1983). After administration of a similar dose of arsenate on day 7 or 9, 3% or fewer of the surviving fetuses were exencephalic.

The neural tube defects seem to result from an apparent arsenic-induced arrest or delay in neural-fold apposition. Takeuchi (1979) examined the changes induced by an embryo-lethal dose of arsenic acid (30 mg/kg arsenic acid) administered to pregnant rats intraperitoneally on day 9 of gestation. At 4 h after exposure, there was some cellular necrosis in the neuroectoderm and mesoderm of the embryos. By 12 h, abnormal mitotic and interphase cells were observed in both tissues, and necrotic cells and debris from these cells were also present. By 24 h, neurulation no longer proceeded, as evidenced by the presence of the V-shaped neural fold that is normally closed by this time.

Willhite (1981) examined the time course of effects on embryos isolated from pregnant hamsters administered sodium arsenate (20 mg/kg) intravenously on day 8 of gestation. There were no differences between embryos from control and arsenate-treated dams at 2 h. Changes became evident at 6 h and more so at 10 h after treatment, with a delay in elevation of the neural fold and closure of the neural tube. There were fewer cells in the cephalic mesoderm of treated embryos than in control embryos of an equivalent gestational age. The author suggested that the damage to the paraxial mesoderm may be associated with the skeletal effects induced by arsenic.

In studies by Morrissey & Mottet (1983), pregnant mice were sacrificed 6–21 h after intraperitoneal administration of sodium arsenate (45 mg/kg) on day 8 of gestation. They observed that the neural folds were widely separated and not positioned for closure in the prospective hindbrain. Necrotic debris was also found primarily in the neuroepithelium of the prospective forebrain and sometimes in the mesenchyme, but it was not clear if this was the main lesion associated with exencephaly.

Results from Lindgren et al. (1984) may explain, in part, the reason for inorganic arsenic-induced neural tube defects. They report that 24 h after intravenous administration of radiolabelled arsenic acid (0.4 µg As/kg) to pregnant mice on day 7 of gestation, the neuroepithelium has the highest concentration of radiolabelled arsenic of all embryonic tissue examined.

Fisher (1982) examined the effect of sodium arsenate (45 mg/kg) on DNA, RNA and protein levels in developing rat embryos after intraperitoneal administration. The rats were killed on day 10, having received arsenate 4 h or 24 h before euthanasia. The embryos were removed and the macromolecule levels were determined then, or at 24 or 42 h after being placed in culture media. *In utero* exposure to arsenate for 4 h did not affect the macromolecule levels. A 24-h *in utero* exposure to arsenate resulted in a significant decrease in DNA, RNA and protein accumulation. The levels were still significantly lower than control at 24 h; by 42 h, the RNA and DNA were still significantly lower, but the protein levels had recovered. After 24 h in culture, and 48 h after exposure to arsenate, morphological changes in the arsenate-exposed embryos included a failure to rotate to a ventroflexed position, failure of closure of the anterior neuropore, no establishment of visceral yolk sac circulation, and no fusion of the allantoic sac in placental formation. The latter effect may reflect problems in formation of the urogenital system.

Nemec et al. (1998) did not observe any teratogenic effects in mice or rabbits orally administered arsenic acid. In this study, mice were administered 0–48 mg/kg per day arsenic acid on gestation days 6–15. Rabbits were administered 0–3 mg/kg per day arsenic acid on gestation days 6–18.

Repeated inhalation exposure of GaAs to pregnant rats (see section 7.1.5.1) on gestation days 4–19 resulted in an increased incidence of skeletal variations (Mast et al., 1990, 1991). This effect became statistically significant at 37 mg/m^3 GaAs. Repeated inhalation exposure of GaAs to pregnant rats (see section 7.1.5.1) on gestation days 4–19 resulted in an increased incidence of skeletal variations.

Treatment of pregnant mice with BAL (subcutaneous, 30 mg/kg) 4 and 8 h before sodium arsenite (intraperitoneal, 12 mg/kg) exposure on day 9 of gestation significantly decreased fetal malformations (Hood & Vedel-Macrander, 1984). Treatment with BAL concurrently with or following arsenite administration decreased fetal malformations, although not significantly.

DMSA (80–320 mg/kg per day) administered for 4 days (0, 24, 48, 72 h) after sodium arsenite administration (intraperitoneal, 12 mg/kg, day 10) decreases the percentage of malformations induced by arsenite in a dose-dependent manner (Domingo et al., 1991). Further experiments showed that the highest dose of DMSA (320 mg/kg) administered once within 1 h after arsenite administration also decreased arsenic-induced malformations. DMSA administered over 1 h after arsenite administration had no effect on the malformations.

b) *In vitro*

Inorganic arsenic is teratogenic to cultured mouse embryos (day 8), with sodium arsenite (1–4 μmol/litre) being approximately

10-fold more effective than sodium arsenate (10–40 μmol/litre) after a 48-h incubation (Chaineau et al., 1990). The most sensitive *in vitro* effect of arsenic is hypoplasia of the prosencephalon. Other effects include failure of neural tube closure and development of limb buds and sensory placode, somite abnormalities and, in the arsenate-exposed embryos, hydropericardium.

Arsenite inhibits chondrogenesis in chick limb bud mesenchymal cells, with complete inhibition at 25 μmol/litre (Lindgren et al., 1984). Arsenate was ineffective at concentrations up to 200 μmol/litre, but when added with arsenite, gave an apparent dose-dependent additive effect.

Sodium arsenite (50 μmol/litre) induces dysmorphology in rat embryos (10 days old) after 2.5 h exposure and followed by a 21.5 h incubation without arsenic. This effect is characterized by hypoplastic prosencepahon, mild swelling of the rhombencephalon, and abnormal somites and flexion of the tail (Mirkes & Cornel, 1992, Mirkes et al., 1994).

Tabacova et al. (1996) examined the teratogenicity of arsenite (1–30 μmol/litre) and arsenate (5–100 μmol/litre) in mouse embryos isolated from pregnant dams on day 9 of gestation. The embryos were incubated with various concentrations of arsenic, for different lengths of time, and at various stages of somite development. Arsenic induced effects such as non-closure of the neural tube, collapsed neural folds, prosencephalic hypoplasia, anophthalmia, pharyngeal arch defects and abnormal somites. The malformation rates were dependent on the dose and oxidation state of arsenic. Arsenite was generally 3–4 times more potent than arsenate in inducing these effects. As the age of the embryos advanced, a higher dose of arsenic was required to elicit the effect. The most sensitive developmental effect to inorganic arsenic was forebrain growth, neural tube closure, eye differentiation, axial rotation (dorso- to ventroflexion) and pharyngeal arch development, which were induced by a 1-h exposure to inorganic arsenic. The authors suggested that, because the *in vitro* malformation pattern is similar to that observed *in vivo* after maternal administration, inorganic arsenic is the toxicant.

7.1.5.4   Gene expression

Wlodarczyk et al. (1996) examined the gene expression of several transcription factors from embryos isolated from pregnant mice administered sodium arsenate (30–45 mg/kg) intraperitoneally – an approximate lethal dose. Expression of several genes is altered by arsenate administered on day 9 of gestation. This day corresponds to the progression of neural tube closure, which is delayed in embryos exposed to inorganic arsenic. In the neuroepithelium of arsenate-exposed embryos, there was significant down-regulation of *Hox 3.1*, and up-regulation of *Pax3*, *Emx-1*, and *creb*. Both *Hox 3.1* and *Pax3* have a role in the regulation of neural cellular adhesion molecules (NCAM), a glycoprotein which affects neural crest cell migration and ultimately neural tube closure (Rutishauser et al., 1988).

7.1.5.5   Induction of heat shock proteins

Arsenic induces the biosynthesis of several proteins in embryos, many of which are also induced by heat shock. The reason for the induction of these so-called heat shock protein (hsp) is not completely understood. It may result in protection of proteins, but it may also be involved in teratogenesis (German, 1984) if the synthesis of proteins critical for development is delayed or reduced.

a)   *In vivo*

Pregnant mice were administered sodium arsenite (0.5 mg/mouse, i.e. for a 30 g mouse, approximately 17 mg/kg body weight) intraperitoneally on day 9–11 of gestation. Two proteins isolated from the embryos were induced and had molecular weights between 45 and 66.2 kDa. Heat shock treatment of pregnant mice induced one embryonic protein with a molecular weight between 45 and 66.2 kDa and a second with a molecular weight between 66.2 and 92.5 kDa (German et al., 1986). Honda et al. (1992) observed in mice administered sodium arsenite (19 mg/kg) intraperitoneally on day 8 and killed 1 day later, that two proteins, hsp70 and hsp105, which are produced constitutively, were increased throughout the embryo. There was a high concentration of these proteins in the neuroepithelial tissue of the embryo after treatment with heat shock or arsenite.

b) In vitro *animal embryos*

Four proteins with molecular weights of 27, 35, 73 and 89 kDa are induced in chick embryo cells by sodium arsenite (50 µmol/litre) or heat shock (Johnston et al., 1980). The mRNA for these proteins is also induced. The induction of the proteins is dose- and time-dependent. For example, the 35 kDa protein is induced at a concentration of 5 µmol/litre sodium arsenite, but the 73 and 89 kDa proteins are minimally induced at this concentration. Only the 27 kDa protein is still induced 24–48 h after treatment.

Mouse embryo cells (gestation day 11) were exposed either to sodium arsenite (50 µmol/litre) for 3 h or heat shock for 10 min. Proteins from cell extracts of these treated cells were analysed by two-dimensional gel electrophoresis (Honda et al., 1992). The synthesis of hsp73 and hsp105 was increased by the exposure to arsenite and heat shock.

In rat embryos (gestation day 10), exposure for 2.5–5 h to an embryotoxic level of sodium arsenite (50 µmol/litre) results in the induction of three hsp (Mirkes & Cornel, 1992). A monoclonal antibody specific for hsp72 recognizes one of the proteins induced by arsenite. Also, mRNA for these heat shock proteins increases in the embryos after exposure to arsenite. Hsp72 is detected 10 h after exposure to arsenite, and maximal levels are observed at 24 h. However, hsp72 was not detected at 48 h, which indicates that this protein does turn over (Mirkes et al., 1994).

An interesting observation of the effects of arsenic on methylation of hsp was reported by Wang & Lazarides (1984). In chick embryo fibroblasts (10–12 days old), arsenite induces the synthesis of hsp70A and 70B and also alters the methylation of lysine and arginine residues respectively in each protein. Arsenite does not affect the amount of intracellular S-adenosyl-L-methionine or the cellular level of methylated protein, with the exception of N^G-monomethyl-arginine, which is increased. The significance of this altered methylation is not known, but the authors suggested that the different methylated forms may have preferential intracellular localization.

c)   In vitro *human fetal tissue*

German et al. (1986) treated human fetal tissue (gestational age 77–84 days) with either sodium arsenite (50 μmol/litre) for 2 h or heat shock for 6 min. The cells were then examined for induction of hsp. Several proteins were induced by both treatments, as well as two with molecular weights < 45 kDa that were induced only by exposure to arsenite.

Honda et al. (1992) treated human chorionic villus cells (gestational age 70–119 days) with sodium arsenite (50 μmol/litre) for 3 h or heat shock for 10 min. In unstressed tissue, hsp70, hsp73, hsp85 and hsp105 were synthesized constitutively. These proteins were induced in the human cells after exposure to sodium arsenite or heat.

7.1.5.6   *Male reproductive toxicity*

The testicular toxicity of GaAs, indium arsenide (InAs) or As_2O_3 administered intratracheally was evaluated in hamsters (Omura et al., 1996a) and rats (Omura et al., 1996b). The animals were dosed twice per week for 6–8 weeks at a level of 7.7 mg/kg for the arsenide compounds and 1.3 mg/kg for As_2O_3. Hamsters are more sensitive than rats to InAs with respect to weight loss, so the dosing of this compound to hamsters was terminated at 6 weeks. The absolute and relative weights of the testes and epididymis of the InAs-treated hamsters were significantly increased, but this effect was probably due to their decreased body weight. As_2O_3 and GaAs had no effect on hamster body weight or weights of the testes or epididymis. In the rat, the weight of testis and epididymis was not affected by any of the compounds. Significant reduction in epididymal sperm count was observed in rats and hamsters after GaAs treatment (Omura et al., 1996a,b) and rats after InAs treatment (Omura et al., 1996b). GaAs also elicited significant spermatid retention in the testes of both species at the post-spermiation stage, with rats being more sensitive than hamsters to this effect. Histopathological examination of the testes showed no tubular changes, but spermatids appeared to be retained in the testes at the post-spermiation stages. A significant proportion of the sperm in the GaAs-treated rats were malformed, with immature and teratic heads or lacking a tail. No effects on the sperm were observed with the other chemicals. Because the serum concentration of gallium was significantly greater than that of arsenic, Omura et al. (1996a) suggested that gallium was involved in the testicular toxicity.

7.1.6 Genotoxicity and related end-points

7.1.6.1 Bacteria

Sodium arsenite is not mutagenic to *Escherichia coli* in tests selecting for tryptophan[+] revertants (Rossman et al., 1980).

GaAs (10–10 000 μg/plate) is not mutagenic in *Salmonella typhimurium* strains TA102 and TA100 (Zeiger et al., 1992).

7.1.6.2 Mammalian cells

Sodium arsenite is not genotoxic to Chinese hamster ovary (CHO) cells (Rossman et al., 1980) or Syrian hamster embryo cells (Lee et al., 1985b) when selecting for ouabain- (ATPase) or thioguanine-resistant (hypoxanthine phosphoribosyl transferase, HPRT) mutants. In the L5178Y mouse lymphoma assay, sodium arsenite is weakly genotoxic at the thymidine kinase locus without metabolic activation (Oberly et al., 1982; Moore et al., 1997a). Sodium arsenate is even a weaker mutagen with (Oberly et al., 1982) and without metabolic activation (Moore et al., 1997a). The type of effects reported by Moore et al. (1997a) were chromosomal aberrations, micronuclei (arsenite only) polyploidy and endo-reduplication.

Sodium arsenate and sodium arsenite induce sister chromatid exchanges and chromosomal aberrations in hamster embryo cells (10^{-7}mol/litre–10^{-4}mol/litre) (Larramendy et al., 1981; Lee et al., 1985b; Kochhar et al., 1996). The aberrations are characterized by chromatid gaps, breaks, and fragmentation, endoreduplication and chromosomal breaks. These clastogenic effects are observed at lower doses of arsenite than arsenate. The difference may be due to greater *in vitro* cellular uptake of arsenite than arsenate (Lerman et al., 1983; Bertolero et al., 1987). GaAs (2.5–10 μg/ml) did not induce micronuclei in Syrian hamster embryo cells (Gibson et al., 1997).

Although it is not a point mutagen, arsenite appears to reduce multilocus deletions (Hei et al., 1998). This type of genotoxic event has most likely not been reported previously because the deletions induced by arsenite were not compatible with cell survival. Hei et al. (1998) treated human hamster hybrid cells, which contain a full complement of hamster chromosomes, plus human chromosome 11, with sodium arsenite (0.5, 1 or 2 µg/ml). Only a small part of chromosome 11 is required for cell viability. They observed a dose-, treatment- and time-dependent increase in mutants arising from deletions of portions of human chromosome 11. Sodium arsenite-dependent deletions of chromosomal marker within chromosome 11 were confirmed by PCR analyses of isolated individual clones.

a) Interaction with other factors

Sodium arsenite appears to act as an antimutagen for spontaneous and ultraviolet (UV) light-induced mutations in some specific strains of *E. coli* (WP44$_s$-NF, WPZuvrA/pkm101) (Rossman et al., 1980; Nunoshiba & Nishioka, 1987). There are at least two potential pathways for this antimutagenic effect (Nunoshiba & Nishioka, 1987). These include inhibition of *umuC* gene expression, which is involved with error-prone repair, and the enhancement of error-free repairs. Arsenite, but not arsenate, is also co-mutagenic with UV light in the excision-proficient *E. coli* strain WP2. The number of mutants/plate is increased 4-fold when bacteria are exposed to arsenite and UV light (Rossman, 1981). Some inorganic arsenic forms (As$_2$O$_3$, sodium arsenite and arsenate) are co-genotoxic with other chemicals and UV light in mammalian cells. There is a synergistic increase in UV-induced chromatid and chromosomal aberrations (Lee et al., 1985a) and mutation at the *HPRT* locus in CHO and V79 cells, after the UV treated cells are exposed to inorganic arsenic (Lee et al., 1985a; Okui & Fujiwara, 1986; Li & Rossman, 1991).

The genotoxic effects at the *HPRT* locus, induced by DNA cross-linking agents such as *cis*-diamminedichloroplatinum (Lee et al., 1986a) and alkylating agents such as methyl methanesulfonate (Lee et al., 1986b) in Chinese hamster cells are enhanced by post-treatment with sodium arsenite. Increased cytotoxicity and inhibition of mitosis and cell proliferation is also observed with methyl methanesulfonate and arsenite post-treatment. Interestingly, pretreatment of cells with arsenite reduces the mutagenic effect of methyl methanesulfonate. The co-genotoxic effect of arsenic with chemicals or UV light does not appear to be associated with any specific DNA lesion (Lee et al., 1986b).

b) Gene amplification

Sodium arsenite and sodium arsenate elicit a genotoxic effect resulting in the growth of mouse 3T6 cells that are resistant to methotrexate (Lee et al., 1988). The number of cells that become resistant is dependent on the dose of arsenic, and arsenite (0.2–6.2 μmol/litre) is more potent than arsenate (1–33 μmol/litre). The resistant cells were found to have amplified copies of the dihydrofolate reductase gene.

c) Transformation

Treating SV40-transformed human keratinocytes with sodium arsenite (6 μmol/litre) resulted in an approximate 3-fold amplification of methotrexate resistance, which is indicative of induction of the dihydrofolate reductase sequences (Rossman & Wolosin, 1992). Amplification of SV40 sequences by arsenite was not observed in these cells. In contrast, several DNA-damaging agents (UV light, mitomycin C, *N*-methyl-*N*-nitro-*N*-nitrosoguanidine, and X-rays) induced amplification of both sequences, but there was no correlation between these chemicals to include both amplification responses.

Sodium arsenate and sodium arsenite induce a dose-dependent transformation of Syrian hamster embryo cells (Lee et al., 1985b) and BALB/3T3 cells (Bertolero et al., 1987). Arsenite is 4–10-fold more potent than arsenate in inducing transformation, partly because of greater cellular uptake of arsenite than arsenate (Bertolero et al., 1987). BALB/3T3 cells transformed by arsenite were tumorigenic in nude mice after their subcutaneous administration (Saffiotti & Bertolero, 1989). The tumours grew rapidly and appeared as fibrosarcomas, but did not metastasize to other sites within the mice. Arsenite may be the transforming agent (Bertolero et al., 1987; Saffiotti & Bertolero, 1989), at least in the BALB/3T3 cells. Arsenate is effectively reduced to arsenite in cell cytosol. In addition, arsenate and arsenite have similar dose response curves for cytotoxicity and transformation when the effects are correlated to the cellular arsenic concentration and not the extracellular dose. Also, arsenite is not methylated by these cells (Bertolero et al., 1987) and arsenobetaine, a methylated arsenical, does not transform BALB/3T3 cells (Sabbioni et al., 1991). However, arsenobetaine is not a mammalian metabolite of inorganic arsenic.

In Syrian hamster embryo cells, neither sodium arsenate nor arsenite is mutagenic at either the ATPase or *HPRT* genetic loci at doses that induced transformation (Lee et al., 1985b).

Studies by Zhao et al. (1997) examined the molecular events in arsenic-induced transformation using the rat liver cell line TRL 1215. Cells are transformed after exposure of 8 weeks or more to arsenite. Fibrosarcomas developed in athymic nude mice at the site of administration of transformed cells and metastasized to the lungs. A permanent genetic change may have occurred in the cells, because they retained their state of transformation and ability to induce tumours in mice, even after removal of arsenite from the culture media.

7.1.6.3 Human cells

Inorganic arsenic induces sister chromatid exchanges, chromosomal aberrations and DNA–protein cross-links in human lymphocytes (Larramendy et al., 1981; Jha et al., 1992; Wiencke & Yager, 1992; Rasmussen & Menzel, 1997) and fibroblasts (Okui & Fujiwara, 1986; Jha et al., 1992; Dong & Luo, 1994). These effects are dependent on dose, and sodium arsenite is more potent than sodium arsenate.

The genotoxic response induced by inorganic arsenic in lymphocytes is variable. Some cultures are more sensitive to arsenite, with a 2–3-fold increase in sister chromatid exchanges over control levels, whereas in other cultures, inorganic arsenic has no effect (Crossen, 1983; Rasmussen & Menzel, 1997).

a) Co-genotoxicity of inorganic arsenic

As observed in mammalian cells, arsenic is co-genotoxic with chemical and electromagnetic radiation in human cells. Wiencke &

Yager (1992) reported that chromosomal aberrations (primarily chromatid deletions and exchanges) induced by the DNA cross-linking agent diepoxybutane are potentiated in lymphocytes after exposure to sodium arsenite. However, arsenite had no effect on diepoxybutane-induced sister chromatid exchanges. Jha et al. (1992) observed a co-genotoxic effect of arsenite with X-rays and UV light in human fibroblast cultures, with an approximate 2-fold increase in chromosomal aberrations in these cells. There was no synergistic effect on sister chromatid exchanges induced by UV light (Jha et al., 1992). Hartwig et al. (1997) showed that As(III) affects NER in UV-irradiated human fibroblasts.

7.1.6.4 In vivo *genotoxicity*

Sodium arsenite and potassium arsenite administered intraperitoneally (2.5–10 mg/kg) to mice induce a linear dose-dependent increase in micronucleated polychromatic erythrocytes (Deknudt et al., 1986; Tinwell et al., 1991). The dominant lethal and sperm abnormality tests were negative with arsenite (Deknudt et al., 1986), which suggests that arsenite does not produce heritable damage. As_2S_3 was ineffective in inducing micronuclei (Tinwell et al., 1991).

Chromosomal aberrations including chromatid gaps and breaks and chromosomal rearrangements are induced in mouse bone marrow cells after oral administration of sodium arsenite (0.1–2.5 mg/kg) (Das et al., 1993; RoyChoudhury et al., 1996). Interestingly, crude garlic extract administered before exposure to arsenite reduces its clastogenic effect (Das et al., 1993; RoyChoudhury et al., 1996). Garlic extract is rich in sulfhydryl groups, which may interact with trivalent arsenic and inhibit its toxic effect.

The hepatic methyl donor status can affect arsenic-induced DNA damage in mice (Tice et al., 1997). The methyl donor status in liver of mice fed a choline-deficient diet is lower than in mice fed a choline-sufficient diet. In choline-deficient mice, urinary excretion of inorganic and organic arsenic is lower after exposure to sodium arsenite (2.5–10 mg/kg). The frequency of micronuclei in bone marrow polychromatic erythrocytes is not affected 24 or 48 h after a single exposure to arsenite (2.5–10 mg/kg) in either choline-sufficient or choline-deficient mice. However, after multiple oral dosing of sodium arsenite (2.5–10 mg/kg, 4 days), a significant increase in frequency of micronucleated erythrocytes at 24 h is observed in both groups of mice. Choline deficiency has no effect on bone marrow toxicity in arsenate-exposed mice. In choline-sufficient mice, there is a significant decrease in DNA migration, determined by the single cell gel assay, in bladder and liver parenchymal cells after one or multiple exposures to arsenite. In choline-deficient mice, there is a significant decrease in DNA migration in skin after one or multiple exposures to arsenite. A decrease in DNA migration can result from DNA cross-linking. Thus, the DNA damage shifts from liver and bladder to skin when the hepatic methyl status of the animal is deficient. Tice et al. (1997) noted that the DNA damage in skin after exposure to arsenite in the choline-deficient mice is consistent with the report of human nutritional deficiency and arsenic-induced skin cancer in Taiwan (Hsueh et al., 1995).

7.1.6.5   *Mechanism of genotoxicity*

Several mechanisms for genotoxicity caused by arsenic have been proposed. These include reactive oxygen species and the inhibition of DNA repair. Because arsenic is not a 'classified' genotoxin in that it does not directly interact with and damage DNA (Rossman et al., 1980), a definitive determination of a mechanism of action has been elusive, and many of the proposed mechanisms are still at the stage of speculation (US EPA, 1992). The exact mechanism leading to such damage is still under debate. The field has been narrowed to two main mechanisms: (1) arsenic-induced oxygen radical damage and (2) arsenic-induced impaired DNA repair processes. Although co-genotoxicity of arsenic with UV and other chemical carcinogens may be more likely to be associated with reduction in repair enzyme activities, direct arsenic toxicity may require both mechanisms to operate.

a)   *Arsenic-induced oxygen radical damage*

Several potential mechanisms for arsenic-induced sister chromatid exchanges and chromosomal aberrations in human cells have been proposed. Nordenson & Beckman (1991) suggested that in lymphocytes, sister chromatid exchanges are mediated by oxygen free radicals. They observed that the addition of superoxide dismutase and catalase to lymphocyte cultures reduced the number of arsenic-induced sister chromatid exchanges. Hei et al. (1998) observed a significant decrease in sodium arsenite-induced genotoxicity (*HPRT* and S1 large deletion mutations) by co-treatment of human–hamster hybrid cells with the radical scavenger dimethyl sulfoxide.

Dong & Luo (1994) suggest that the genotoxic effect of arsenic in human fibroblasts is due to DNA–protein cross-links and not to inhibition of DNA repair. They observed increased unscheduled DNA synthesis in their system, indicating that arsenic induced some type of DNA damage, which was excised, and the resulting gaps were filled with new DNA.

b) Arsenic-induced impaired DNA repair processes

Inhibition of DNA repair may be a mechanism of the co-genotoxic effect of inorganic arsenic with UV light. DNA excision repair of thymine dimers in human fibroblasts is inhibited by inorganic arsenic, with As_2O_3 being more potent than sodium arsenate (Okui & Fujiwara, 1986). Hartwig et al. (1997) showed that As(III) affects nucleotide excision repair in UV-irradiated human fibroblasts. Li & Rossmann (1989) and Jha et al. (1992) proposed that the co-genotoxic effect of arsenic is due to inhibition of DNA ligase. Treatment of cells with arsenite (as low as 10 µmol/litre) inhibits the activity of both nuclear DNA ligase I and II, with the latter enzyme being more sensitive to this effect. At low concentrations of arsenite, the inhibition of DNA ligase II results in misrepair of genotoxic damage to DNA. At higher concentrations of arsenite DNA ligase I is inhibited, which leads to chromosomal gaps, breaks and other types of chromosomal aberrations. However, the observation that higher concentrations of arsenite are required to inhibit DNA ligase *in vitro* than in cells (Li & Rossman, 1989) suggests that arsenite does not affect the enzyme directly, but may affect accessory proteins or cellular control of the ligase (US EPA, 1997). This proposal is supported by a recent study by Hu et al. (1998). They report that for inhibition of the activity of purified human DNA ligase I and III to occur, millimolar concentrations of arsenite or arsenate must be present. The activity of DNA ligase and human DNA polymerase 9, were activated by both arsenicals at doses in the mmol/litre range. On the other hand, total DNA ligase activity in nuclear extracts of human osteosarcoma cells treated for 3 h with 10 µmol/litre sodium arsenite was decreased by 50% compared to nuclear extracts from untreated control cells.

Aneuploidy induced by sodium arsenite in human lymphocytes may be a result of the disruption of microtubule function and spindle formation (Ramirez et al., 1997). Dose-related increases in hyperploid cells were observed after exposure to sodium arsenite (0.001–0.1 µmol/litre). Sodium arsenite inhibits tubulin polymerization and stimulates its depolymerization. The effects on tubulin may affect spindle formation, which results in altered chromosome numbers in the cells.

An enzyme involved with repair of DNA strand breaks, poly-(ADP-ribose) polymerase, is also inhibited by sodium arsenite. Yager & Wiencke (1997) observed a significant decrease in activity of this enzyme in human T-cell lymphoma cells at concentrations > 5 µmol/litre sodium arsenite, with minimal effect on cell viability. The enzyme contains two vicinal dithiol groups, and arsenite may bind one or both groups and inhibit the enzyme. The alteration of DNA by increasing or decreasing its methylation may have a role in the genotoxicity and development of cancer (Counts & Goodman, 1995).

Both hypermethylation and hypomethylation have been observed as a consequence of arsenic exposure. Mass & Wang (1997) examined the effect of arsenic on DNA methylation patterns of the tumour suppressor gene *p53* in the human adenocarcinoma cell line A549. The *p53* gene is primarily involved in cell-cycle regulation, and is responsible for co-ordinating DNA repair activity before mitosis and for the onset of apoptosis in case of extensive DNA damage. Damage of *p53* itself may cause impaired DNA repair and suppression of apoptosis (Levine, 1997). Sodium arsenite (0.08–2 µmol/litre) increases the resistance of DNA in the *p53* promoter region to cleavage by the restriction enzyme *Hpa*II. This enzyme cleaves unmethylated cytosine within the CCGG sequence. Hypermethylation of this region seems to be due to loss of SssI methylase activity rather than to direct methylation by DMA. The increased resistance to cleavage indicates increased methylation of cytosines within the *p53* gene. Methylation of cytosine may alter the ability of transcription factors to bind DNA, which would modify gene expression. Sodium arsenate (3–300 µmol/litre) is less potent in methylating cytosines, and DMA (2–2000 µmol/litre) is ineffective. The hypermethylation induced by arsenite was confirmed by DNA sequencing of the promoter region and using bisulfite to visualize 5-methylcytosine. It appeared that the sequence CpG within the entire genome is methylated, because of diminished ability of *Sss*I methylase to transfer methyl groups from *S*-adenosylmethione (SAM) to DNA. Although the *p53* gene is hypermethylated in the presence of arsenite, it is not known if gene expression is altered.

On the other hand, DNA of sodium-arsenite-transformed rat liver TRL 1215 cells is globally hypomethylated and the effect is both dose and time-dependent (Zhao et al., 1997). DNA hypomethylation might result in aberrant gene expression, and hence transformation of the cells. DNA hypomethylation is not observed in cells treated acutely with arsenite or at doses that do not transform the cells. The levels of the methyl donor SAM were decreased in the transformed cells. Although the levels of *S*-adenosylhomocysteine (SAH), the product formed after the methyl group is transferred from SAM, were unaffected, the SAM/SAH ratio was significantly decreased. In addition, the metallothionein gene, which is controlled by methylation, is hyperexpressed in arsenic-induced transformed cells. DNA methyltransferase activity in the arsenic-transformed cells is decreased by up to 40%, but is not affected by an acute exposure to arsenite. Expression of the DNA methyltransferase gene increased 2-fold in the transformed cells, which is attributed to an attempt by the cells to increase methyltransferase activity. These two reports may appear to be different, but they have a common theme: alteration of DNA methylation by arsenic. Their differences may be due to the fact that Mass & Wang (1997) used a human lung cell line and Zhao et al. (1997) used a rat liver cell line. In addition, Mass & Wang specifically examined the promoter of a specific gene, whereas Zhao et al. examined genomic DNA.

7.1.6.6   *Resistance/hypersensitivity to arsenic cytotoxicity*

Chinese hamster lung V79 cells have been isolated that are resistant or hypersensitive to the cytotoxic effects of arsenite (Wang & Rossman, 1993; Wang et al., 1996b). Exposure of wild-type and resistant cells to a non-toxic dose of arsenite results in a greater inducible resistance to arsenite. The sublines retain their phenotypes for up to 6 months when grown in media without arsenite. Experiments where resistant and sensitive cells are fused indicate that the resistance trait is dominant over the sensitive trait. The resistant cells accumulate less arsenite than wild-type and sensitive cells owing to an energy-dependent efflux of arsenite (Wang et al., 1996b). Analysis of the efflux products shows that arsenite was transported out of the cell. Glutathione (GSH) may have a role in the efflux, because addition of a GSH *S*-transferase inhibitor to the cells reduced arsenite efflux. But no GSH–arsenic complex was detected in cell incubates. However, in other CHO cell lines, there is an association of over-expression of GSH *S*-transferase and greater excretion of arsenic (Wang & Lee, 1993).

Several human cell lines derived from keratinocytes, fibroblasts or tumour cells lack the inducible arsenic-resistance response (Rossman et al., 1997). The human cells were as sensitive to the cytotoxic effects of arsenite as the sensitive Chinese hamster cell line. This lack of inducibility of resistance to arsenic may explain in part the reason why animals are less sensitive to the carcinogenic effects of arsenic than humans.

7.1.7 *Carcinogenicity*

7.1.7.1 *Pulmonary carcinogenicity*

Several studies have examined the pulmonary carcinogenic effect of inorganic arsenic in hamsters (Table 24). The animals were dosed intratracheally, to simulate inhalation of arsenic, once a week for 15 weeks and monitored for tumours throughout their life span. Inorganic arsenic compounds that have been examined by this method include As_2O_3 (Ishinishi et al., 1983; Pershagen et al., 1984; Yamamoto et al., 1987), $Ca(AsO_4)_2$ (Pershagen & Bjorklund, 1985; Yamamoto et al., 1987) and As_2S_3 (Pershagen & Bjorklund, 1985; Yamamoto et al., 1987). Generally, fewer animals survived the 15-week exposure to As_2O_3 (30–50% mortality) than the other two arsenicals, and this effect was dependent on the administered dose. Yamamoto et al. (1987) compared all three arsenicals (total dose, 3.75 mg As) and reported that $Ca(AsO_4)_2$ induced more lung tumours (6 adenomas and 1 adenocarcinoma/25 hamsters) than As_2S_3 (1 adenoma/22 hamsters) and As_2O_3 (1 adenoma/17 hamsters). In the

Table 24. Lung tumours in Syrian golden hamsters after intratracheal instillation

Sex	Dose (mg As)	Survivors after 15 instillations	Number examined	Lung tumours Malignant	Lung tumours Benign	Lung tumour incidence (%)	Reference
As trioxide							
Female	3.75	22/30	20	0	2	10	Ishinishi et al. (1983)
Female	0	22/30	20	0	0	0	Ishinishi et al. (1983)
Male	3.75	18/30	17	1	0	6	Yamamoto et al. (1987)
Male	0	22/22	21	1	0	5	Yamamoto et al. (1987)
Male	3.75	48/67	47	2	3	11	Pershagen et al. (1984)
Male	0	53/68	50	0	2	4	Pershagen et al. (1984)
Female	5.25	10/20	10	0	3	30	Ishinishi et al. (1983)
Female	0	16/20	15	0	0	0	Ishinishi et al. (1983)

Table 24 (contd.)

As trisulfide

Male	3.75	23/30	22	0	1	5	Yamamoto et al. (1987)
Male	0	22/22	21	1	0	5	Yamamoto et al. (1987)
Male	3.75	28/32	28	0	1	4	Pershagen & Bjorklund (1985)
Male	0	26/29	26	0	0	0	Pershagen & Bjorklund (1985)

Calcium arsenate

Male	3.75	27/30	25	1	6	28	Yamamoto et al. (1987)
Male	0	22/22	21	1	0	5	Yamamoto et al. (1987)
Male	3.75	35/41	35	0	4	11	Pershagen & Bjorklund (1985)
Male	0	26/29	26	0	0	0	Pershagen & Bjorklund (1985)

control group of this study, one animal (out of 21) had a lung adenosquamous carcinoma. These results confirmed previous results by Pershagen & Bjorklund (1985), who reported more lung tumours with $Ca(AsO_4)_2$ (4 adenomas/35 hamsters) than with As_2S_3 (1 adenoma/28 hamsters). Pershagen & Bjorklund (1985) concluded that $Ca(AsO_4)_2$ is tumourigenic, but the results for As_2S_3 were inconclusive. The effect of $Ca(AsO_4)_2$ may be due to its longer retention in the lung than the other arsenicals (Pershagen et al., 1982).

As_2O_3 administered intratracheally resulted in 10–30% of the surviving animals with lung adenomas (2 adenomas/20 hamsters, total dose 3.75 mg; 3 adenomas/10 hamsters, total dose, 5.25 mg) and 6% with respiratory tract carcinomas (3 carcinomas/47 hamsters, total dose, 3.6 mg As) (Ishinishi et al., 1983; Pershagen et al., 1984). None of the control animals had pulmonary tumours in either study. Thus, even though arsenic is classified as a human carcinogen, it is difficult to obtain results in animals that will substantiate the effects observed in humans.

Pershagen et al. (1984) examined the interaction in hamsters of As_2O_3 with benzo[*a*]pyrene (B(a)P), a carcinogenic polycyclic aromatic hydrocarbon found in tobacco smoke. The two chemicals were administered intratracheally at dose levels of 3 mg As/kg for As_2O_3 and 6 mg/kg for B(a)P. Papillomas, adenomas and carcinomas were observed in the respiratory tract (larynx, trachea, bronchi and lungs) of all dose groups (As_2O_3, B(a)P, As_2O_3 + B(a)P). There appeared to be a positive interaction when both chemicals were administered, with respect to development of respiratory tract carcinomas. Carcinomas were found in 25 out of 54 animals exposed to both chemicals, 14 out of 40 animals administered B(a)P, and 3 out of 47 animals administered As_2O_3. The As_2O_3 + B(a)P group also had more pulmonary lesions (adenomas, papillomas, adenomatoid lesions) than animals treated with each chemical alone. The combination of As_2O_3 with B(a)P, or perhaps other chemicals found in tobacco smoke, may increase the pulmonary carcinogenic effect of arsenic. Ohyama at al. (1988) examined the carcinogenic effect of GaAs in hamsters after intratracheal administration. The animals received a total dose of 3.75 mg of GaAs, after administration of chemical, once a week for 15 weeks, and 32/33 animals survived the treatment. A total of 30 animals were examined for tumours. No tumours were observed in vehicle-control (30) animals. One malignant hepatic lymphoma was observed in the GaAs-treated animals, but there was no statistically significant increase in the incidence of tumours. The authors concluded that GaAs had no apparent carcinogenicity or tumorigenicity.

7.1.7.2   Skin tumorigenicity

A more recent arsenic carcinogenicity study utilized transgenic animals. Germolec et al. (1997) used the TG.AC female mouse strain (3 months old, 20/group), which contain the v-Ha-*ras* oncogene in their genome. The skin of these mice is genetically initiated, because of the transgene (Leder et al., 1990). Mice were provided with drinking-water containing 200 mg/kg sodium arsenite for 4 weeks, followed by application of the tumour promoter 12-*O*-tetra-decanoylphorbol-13-acetate (TPA, 2.5 µg) on shaved skin, twice a week for 2 weeks. The animals were then monitored for skin papillomas. The incidence of papillomas in arsenic-exposed mice treated with TPA was approximately 4-fold greater by week 6 after promotion, compared to mice that were not exposed to arsenic but had their skin promoted. Papillomas were not found in the skin of control mice or mice treated with arsenic but not promoted. The skin of the arsenic-treated mice showed epidermal thickening and hyperkeratosis.

7.1.7.3   Long-term study in monkeys

A long-term arsenic carcinogenicity study in non-human primates was conducted by Thorgeirsson et al. (1994). In this study cynomolgus monkeys (*Macaca fascicularis*) were dosed orally 5 days per week with sodium arsenate (0.1 mg/kg) for at least 15 years. No malignant tumours developed in the monkeys over the course of the study. This strain of monkeys exhibits 1.5% incidence of spontaneous malignant tumours. Eleven of the monkeys survived the treatment and were killed at the end of the study. No comment was made by the investigators on the causes of death of the other 9 animals: 2 of the monkeys had renal cortical adenomas, 1 had micronodular cirrhosis, 3 had endometriosis, and 3 had hyalinized Langerhans islets of the pancreas. Clinical analysis of one of the latter animals showed that it had signs of diabetes.

7.1.7.4 Long-term study in mice

A preliminary study by Ng et al. (1998c, 1999) examined the effect of sodium arsenate in drinking-water (500 µg As/litre) on 90 female C57BL/6J mice and 140 female metallothionein knock-out transgenic mice. Each group also included 60 control females. The exposure to arsenic lasted up to 26 months (Table 25). The average arsenic intake by the mice was 2–2.5 µg As/day and the daily arsenic dose rate of a 30-g mouse was 0.07–0.08 mg As/kg body weight. Mice that survived 2 years consumed approximately 1.5–1.8 mg of arsenic. Eighty-one percent of the C57BL/6J and 74% of the transgenic mice survived the arsenic treatment, compared to 98% of the control group. One or more tumours were detected in 41% of the C57BL/6J mice and 26% of the transgenic mice. Tumours were found in the gastrointestinal tract, lung, liver, spleen, skin and reproductive system of both strains.

Table 25. Incidence of tumours in control, C57BI/6J and metallothionein knock-out transgenic (MT⁻) mice given sodium arsenate, 500 µg As/litre in drinking-water *ad libitum* for up to 26 months. All figures are percentages

Organ system	C57BL/6J control < 0.1 µg As/litre	C57BL/6J test 500 µg As/litre	MT control < 0.1 µg As/litre	MT test 500 µg As/litre
Lung	0	17.5	0	7.1
Intestinal tract	0	14.4	0	12.9
Liver	0	7.8	0	5.0
Spleen	0	3.3	0	0.7
Reproductive	0	3.3	0	5.0
Skin	0	3.3	0	1.4
Bone	0	2.2	0	0
Eye	0	1.1	0	0

A small percentage of tumours was observed in the skin and eyes of the C57BL/6J mice. Metallothionein does not appear to protect the animals from developing tumours after exposure to arsenic. No macroscopic tumours were observed in the control groups, which is consistent with previous findings (Hoag, 1963, Adkison & Sundberg, 1991). Pathological and histopathological examination confirmed the ovary, lung and skin tumour findings (Ng, 1999). This is the first experimental carcinogenicity study in rodents using a relevant route of exposure and relevant exposure level to have clearly shown a treatment-related increase in tumours. This finding is in contrast to previous studies that have used high levels of exposure and strains more likely to have a high background of tumours. [1]

7.1.8   Other special studies

7.1.8.1   Cardiovascular system

a)   *In vivo*

Baseline cardiovascular parameters such as heart rate, blood pressure, electrocardiogram patterns and cardiovascular responses to several neurohumoral agonists are not altered in rats exposed to sodium arsenate (50 µg As/ml) in drinking-water for 10 or 18 months (Carmignani et al., 1983, 1985). Histological changes in the cardiovascular system were also not evident from this exposure (Carmignani et al., 1983). However, vascular responsiveness to the vasopressor angiotensin I was reduced in these animals, with respect to systolic and diastolic blood pressure (Carmignani et al., 1983). In addition, when adrenaline was administered to anaesthetized rats, at a dose (0.125 µg/kg) which stimulates β-adrenoreceptors, the systolic hypertensive response seen in control animals was reversed in arsenate exposed rats, but the diastolic hypotensive response was increased. At a higher dose of adrenaline (1–2 µg/kg), which stimulates α-adrenoceptors, the blood pressure responses were

---

[1] Dr Nishikawa noted his dissent from the conclusions reached by the group with regard to these studies and considered that it is impossible to evaluate them, partly because of lack of complete histopathological examination and partly because of unusual growth curves. Under the experimental conditions, increased incidence of some tumors, notably of lung, liver, gastrointestinal tract or skin is not possible; Dr Nishikawa also noted that examination of neoplastic lesions in the haematopoietic system, which has been reported previously in this strain of mice (Maekawa et al., 1981), was not made.

unchanged in the arsenate-treated rats. Arsenate did not affect the sensitivity of the baroreflex arc, but hyperactive or hypersensitive sympathetic responses were induced after vagotomy or after administration of the ganglionic blocking drug hexamethonium (Carmignani et al., 1983, 1985). There was also a reduced response to the hypertensive effect of tyramine in the 18-month arsenate-treated rats (Carmignani et al., 1985).

Long-term exposure of sodium arsenite (50 μg As/ml) in drinking-water to rats (18 months) or rabbits (10 months) alters baseline cardiovascular parameters (Carmignani et al., 1985). Arsenite was associated with decreases in stroke volume and cardiac output and an increase in vascular resistance in both species. Aortic blood flow was significantly decreased in the arsenite-exposed rat. Arsenite induces specific changes in the cardiovascular response in each species to neurohumoral and effector agonists. In the rabbit, blood pressure responses to phenylephrine, which stimulates α-adrenoceptors, and bilateral carotid occlusion, were reduced. In the rat, blood pressure response to tyramine was reduced. Carmignani et al. (1985) suggested that the reduced hypertensive responses to various agonists may be due to the increased vascular resistance induced by arsenite.

Significant reduction in vasoreactivity to norepinephrine is observed in rats orally administered As_2O_3 (3 or 15 mg/kg) for 4–28 days (Bekemeier & Hirschelmann, 1989). Vasoreactivity was determined after the last exposure to arsenic by monitoring the perfusion pressure in the femoral artery in the hind leg of the rats. The decrease in reactivity of the blood vessels to adrenaline was dependent on dose and length of exposure to arsenic. However, arsenic did not affect resting perfusion pressure of the artery and there did not appear to be any change in receptor binding of adrenaline. The effect was attributed to impairment of muscle contractility and a decrease in intrinsic sensitivity of the vessel muscle.

b) *In vitro*

Incubation of human umbilical vein endothelial cells with As_2O_3 for < 15 days results in inhibition of cellular hyperplasia (< 50 ng/ml) and glycoprotein synthesis (> 100 ng/ml) (Chen et al.,

1990). Arsenic (0.1–1000 ng/ml) is not directly cytotoxic to these cells after a 5 h incubation and does not affect the expression of two endothelial cell markers, factor VIII related antigen and *Ulex europaeus* agglutinin I binding sites. However, by 72 h, incubations of arsenic at concentrations greater than 100 ng/ml resulted in enlargement and detachment of the cells and a decrease in the two markers. This type of damage may have a role in development of blackfoot disease observed in humans exposed to high levels of arsenic.

As_2O_3 and sodium arsenite induce significant dose-dependent (10–100 µmol/litre) damage to endothelial cells isolated from bovine carotid arteries after a 22 h incubation (Chang et al., 1991). In contrast, the pentavalent arsenicals arsenic acid and p-arsenilic acid had no effect on these cells. Exogenously added GSH, but not glutathione disulfide, decreased the toxicity induced by As_2O_3. The inhibitory effect was dependent on the concentration of GSH and the time it was added. GSH stimulated the cellular biosynthesis of prostacyclin, which is a vasodilator and an inhibitor of platelet aggregation and proliferation of smooth muscle cells. The cyclooxygenase inhibitor aspirin diminished the cytoprotective effect of GSH. The stimulation of prostacyclin by GSH may be involved in protecting the cells from arsenic-induced endothelial cell injury.

Sodium arsenite at concentrations < 5 µmol/litre stimulates [^{3}H]-thymidine incorporation into the DNA of porcine vascular endothelial cells, which is indicative of a mitogenic response (Barchowsky et al., 1996). Higher concentrations of arsenite appear to be cytotoxic to the cells. At mitogenic concentrations of arsenite, oxidants within the cells are increased for 30–60 min after exposure and the thiol levels are increased by 24 h. Within the time frame of the increased oxidants, there is increased nuclear translocation of NFκB, a transcription binding protein. Binding of the proteins to genomic B within the endothelial cells may result in the mitogenic response induced by arsenite.

7.1.8.2 Nervous system

Very few data are available on the effect of arsenic on the nervous system.

As_2O_3 inhibits neurite outgrowth in rat embryo dorsal root ganglion (Windebank, 1986). A 50% inhibition of neurite outgrowth occurs at 9.6 µmol/litre As_2O_3 and 100% inhibition at 15 µmol/litre. The mechanism of this inhibitory effect is not known, but neurite outgrowth is dependent on many process and arsenic may affect any or all of them.

Sodium arsenite is more toxic than sodium arsenate to mouse neuroblastoma cells after a 24 h exposure in culture (Repetto et al., 1994). All parameters examined (cell proliferation, lysosomal function, and membrane integrity) were affected at lower doses of arsenite than arsenate and may be attributable to greater cellular uptake of arsenite (Lerman et al., 1983; Bertolero et al., 1987). There were differences in sensitivity of several parameters to the effects induced by arsenic. Inhibition of several metabolic processes, such as lactate dehydrogenase activity and hexosaminidase activity, occurred at lower concentrations of both arsenicals than the inhibition of cellular proliferation. Because of a minor effect on release of cellular lactate dehydrogenase and the sensitive effect on the metabolic processes, it was suggested that the structural damage elicited by arsenic is less important than inhibition of metabolic processes.

7.1.8.3 Skin

a) Keratinocyte markers

Several distinct markers are expressed by epidermal cells as they differentiate into keratinocytes. Sodium arsenate (in the micromolar range), but not sodium arsenite or dimethylarsinic acid, suppresses expression of one of these markers, involucrin, in cultured non-malignant and malignant human keratinocytes (Kachinskas et al., 1994). Other keratinocyte markers suppressed by arsenate include loricin, filaggrin, spr1 and K10 (Kachinskas et al., 1997). The mechanism of this suppression may involve inhibition of tyrosine phosphatase by arsenate, which induces a dose-dependent increase in phosphorylated proteins (Kachinskas et al., 1994). This inhibition may result in an imbalance between phosphorylation and dephosphorylation reactions, which could then affect signal transduction pathways.

b) Transcription factors

In cells treated with sodium arsenate, DNA binding by the transcription factors AP1 and AP2 is reduced and undetectable, respectively (Kachinskas et al., 1994). Expression of the keratinocyte differentiation marker involucrin is also suppressed by arsenate. The 5'-flanking and promoting region of the involucrin gene contains AP1 and AP2 binding sites, as evidenced by their decreased binding when portions of the gene are deleted or mutated at the two transcriptional sites (Kachinskas et al., 1997). Arsenate appears to alter the transcription factors AP1 and AP2, perhaps through inhibition of tyrosine phosphatase, which ultimately results in suppression of involucrin gene transcription.

Burleson et al. (1996) have reported that in normal human keratinocytes sodium arsenite (4 µmol/litre) stimulates transcription of c-*jun* and c-*fos*. The products of these genes dimerize, forming AP-1 transcription factors. In the arsenite-exposed keratinocytes, there is increased AP-1 DNA binding activity. The difference in effect between the studies of Kachinskas et al. (1994) and Burleson et al. (1996) may be that sodium arsenate was used in the former and sodium arsenite in the latter.

c) Growth factors

The stimulation of growth factors by arsenic may have a role in its carcinogenic effect. Sodium arsenite stimulates interleukin-8 (IL-8) gene expression and thus IL-8 secretion in cultured human keratinocytes after a 24 h incubation (Yen et al., 1996). IL-8 is a peptide that may be an essential autocrine growth factor. Stimulation of IL-8 was not detected at concentrations lower than 28 µmol/litre sodium arsenite, and this dose was not toxic to the cells. There were no obvious signs of toxicity; however, cell viability was not measured in the study.

Other growth factors stimulated in cultured human keratinocytes by sodium arsenite include granulocyte macrophage-colony stimulating factor (GM-CSF), transforming growth factor (TGF)-α, the inflammatory and chemotactic cytokine tumour necrosis factor (TNF)-α (Germolec et al., 1997) and IL-1 (Burleson et al., 1996). Germolec et al. (1997) did not find that arsenite stimulated the cells to secrete IL-8, in contrast to the observations of Yen et al. (1996), but this may be due to the lower arsenite concentrations used. Expression of the proto-oncogene c-*myc*, which is associated with proliferation of keratinocytes (Pietenpol et al., 1990), was also increased in the keratinocytes after exposure to arsenite (Germolec et al., 1997). Proliferation of keratinocytes was increased with low doses of sodium arsenite (1–5 nmol/litre), but higher concentrations decreased cell number. Expression of mRNA transcripts of GM-CSF and TGF-α isolated from the skin of transgenic mice (TG.AC) administered arsenite in their drinking-water and promoted with tetradecanoyl phorbol acetate is increased. Immunostaining of these two growth factors in skin samples from similarly treated transgenic mice is also enhanced. The stimulation of growth factors by arsenic and the sustained cell proliferation that accompanies the stimulation provides a milieu where errors in DNA replication or repair can be amplified.

d)   Organotypic culture of human kertinocytes

Klimecki et al. (1997) used an organotypic culture of human keratinocytes to study arsenic-induced changes in skin. These cultuerythrocytes develop into an "epidermal equivalent", which can proliferate, stratify, cornify and desquamate (Bell et al., 1983). Phenotypic changes resembling early skin lesions in arsenic-exposed humans (Ratnam et al., 1992) develop in these cells after an acute exposure to a pentavalent arsenical mixture consisting of sodium arsenate (0.5 µmol/litre), monomethylarsonic acid (MMA) (0.5 µmol/litre) and dimethylarsinic acid (DMA) (1.5 µmol/litre). These changes include diffuse hyperkeratosis, acanthosis, and nuclei retention in the cornified upper layers of the cells, which is indicative of an irregularity in cell differentiation.

Different cytokeratins are expressed by keratinocytes depending on whether the skin is hyperproliferating or differentiating. After acute exposure to the arsenical mixture there is an enhancement of keratin 16, which is expressed in states of hyperproliferation, and keratins 8 and 18, which are expressed in less differentiated epithelial cells (Klimecki et al., 1997). Keratin 10, which is expressed in differentiated cells, is suppressed by the arsenical mixture, while the marker Ki-76, which is increased in hyperproliferative states, is induced throughout the epidermal layer and not just at the basal cell layer.

The cells show a de-differentiated epidermal phenotype after chronic exposure to the arsenical mixture (Klimecki et al., 1997). A clonal expansion of these cells grew and their structure approximated a rudimentary epithelium. The same effects on the cytokeratins and Ki-67 after acute exposure were observed after chronic exposure to the mixture. Arsenate was the only arsenical of the mixture that could induce a response similar to that of the mixture. This suggests that inorganic arsenic but not organic arsenic induces keratinocyte hyperproliferation and disrupts the process of terminal epidermal differentiation in the epidermis.

7.1.8.4   Immune system

Exposure to sodium arsenate in drinking-water for 10–12 weeks at levels up to 100 mg As/litre increased the survival time and decreased the incidence of progressive tumours in mice administered a tumorigenic cell line (10^5 cells) (Kerkvliet et al., 1980). The tumour cells originated from Moloney sarcoma virus-treated mice. The mechanism of inhibition of tumour progression in the arsenic-exposed mice is not known. Kerkvliet et al. (1980) suggested that arsenic may inhibit cellular transformation by the viruses released from the tumorigenic cells. However, the tumours that did appear grew more rapidly in the arsenic-exposed mice than in control mice. Survival time of the arsenic-exposed mice decreased once the tumours appeared. When a 10-fold greater amount of tumorigenic cells were injected, there was no difference between arsenic-exposed and control mice with respect to survival time and incidence of tumours. Although mature tumour cells had a rapid growth rate in arsenic-treated mice, this did not appear to be a result of arsenic-induced immune suppression. Arsenic had no effect on cell-mediated cytotoxicity in mice injected with the tumorigenic cells. There was a significant dose-related enhancement of tumour cell cytotoxicity by spleen and lymph node cells from mice injected with murine sarcoma virus. Treatment with arsenic did not appear to affect a primary and secondary cytotoxic cell-mediated immune response by the mice, although the high-arsenic group had a significantly lower number of lymphocytes 5 and 9 days after administration of the higher amount of cells.

Sodium arsenite administered to mice in drinking-water (0.5, 2, and 10 mg As /litre) for 3 weeks results in immunosuppression of the humoral response (Blakley et al., 1980). Both primary and secondary immune responses, with respect to the production of immuno-globulin (Ig) M and G, respectively, were suppressed with a maximum effect occurring at 0.5 mg/litre. The immune responses of the arsenic-treated mice were approximately 50% of the control value. There was no correlation between tissue levels of arsenic in liver and kidney and the immunosuppressive effect of arsenite on the humoral immune system.

A single intratracheal administration of GaAs (50–200 mg/kg) affects humoral and cellular immune responses in mice (Sikorski et al., 1989). Dose-dependent decreases are observed in the percentage of lymphocytes in the spleen and peritoneal exudates 14 days after exposure to GaAs. There is also a dose-dependent decrease in the number of splenocyte IgM and IgG antibody-forming cells in mice immunized with the T-dependent antigen sheep erythrocytes (SRBCs). Sodium arsenite (10 mg/kg) administered intratracheally had the same effect on IgM-producing splenocytes, but not IgG-producing cells in immunized mice. The suppressive effect on the antibody forming cells appears to be due to functional alterations in macrophages and T and B lymphocytes (Sikorski et al., 1991b). The immunosuppressive effect is not due to induction of suppressor macrophages (Sikorski et al., 1991b). Further study with the macrophages indicated that GaAs affects the ability of these cells to process and present a particulate antigen like SRBCs (Sikorski et al., 1991a), thus contributing to the observed decrease in antibody production. *In vitro* studies indicated that the arsenic component of GaAs is the predominant species that inhibits the antibody-forming response to SRBCs (Burns et al., 1991). With respect to cell-mediated response, GaAs (50–200 mg/kg) decreased delayed-type hypersensitivity in a dose-dependent manner (Sikorski et al., 1989). In host resistance studies, GaAs has no effect on resistance to *Plasmodium yoelii* or *Streptocococcus pneumoniae*. However, there is a GaAs dose-dependent increase in resistance to the lethal effects of *Listeria monocytogenes*, probably due to the bactericidal activity of arsenic (Burns et al., 1993), but a decrease in resistance to the growth of B16F10 melanoma cells in the lung (Sikorski et al., 1989). Sodium arsenite produces some, but not all of the effects induced by GaAs, which suggests that gallium affects the immune system.

7.1.8.5   Haem biosynthesis and urinary excretion of porphyrins

Acute and subchronic exposure to inorganic arsenic alters haem biosynthesis and urinary excretion of porphyrins in mice and rats. Hepatic δ-aminolaevulinic acid (ALA) synthetase and haem synthetase are significantly decreased in mice and rats after a 6-week exposure to sodium arsenate (20–85 mg As /litre) in drinking-water (Woods & Fowler, 1978). In the mouse, hepatic uroporphyrinogen I synthetase is significantly increased. In both species, urinary excretion of uroporphyrin and coproporphyrin are increased, the former more so, after exposure to arsenate.

There is a dose-dependent increase in urinary uroporphyrin excretion by rats after the first week of exposure to sodium arsenite (5–100 mg As /litre) in drinking-water (Martinez et al., 1983). There was a gradual decrease in the response over time, which was suggestive of an adaptive response.

After an acute subcutaneous exposure of rats to sodium arsenite (12.5–100 µmol/kg) or sodium arsenate (25–200 µmol/kg), hepatic activity of ALA synthetase and haem oxygenase are significantly increased in a dose-dependent manner (Cebrian et al., 1988). Hepatic cytochrome P-450 concentration was also decreased. Arsenite is more potent than arsenate in this response and haem oxygenase activity was increased to a greater extent than ALA synthetase.

After subchronic exposure to sodium arsenite (50 mg/litre) in drinking-water, there is a 1.6-fold increase in ALA synthetase activity 5 days after initiation of treatment (Cebrian et al., 1988). The activity declines to 1.3–1.4-fold activity of control by 20–30 days after initiation of treatment. Haem oxygenase and cytochrome P-450 content are not affected by this treatment, which suggests there is an adaptive response by the rats.

Renal coproporphyrinogen oxidase activity in rats is inhibited in a time- and dose-dependent manner after exposure to sodium arsenate (40–85 mg As /litre) in drinking-water (Woods & Southern, 1989). The decreased activity occurred with an increased concentration of arsenic in the kidneys. No effect on hepatic coproporphyrinogen oxidase was observed, even though hepatic arsenic levels were elevated, but less so than observed in the kidney.

Acute treatment with sodium arsenate (75 mg/kg, intraperitoneal) resulted in a significant increase in urinary coproporphyrin and a significant decrease in coproporphyrinogen oxidase activity in kidney but not liver. *In vitro* incubations with 10 mmol/litre arsenate inhibited hepatic and to a greater extent renal coproporphyrinogen oxidase activity, which suggests this enzyme can be inhibited directly.

Subchronic treatment of mice with sodium arsenite (20 mg/litre) or sodium arsenate (50 mg/litre) in drinking-water results in a time-dependent porphyric response (Garcia-Vargas et al., 1995). Urinary porphyrins were increased by 3 weeks after initial exposure to arsenic, which corresponded to increased activities of hepatic porphobilinogen deaminase and uroporphyrinogen synthetase. Arsenite was more potent in this response than arsenate. Significant decreases in renal uroporphyrinogen decarboxylase and hepatic and renal coproporphyrinogen oxidase were also observed. Goering et al. (1988) reported that an acute intratracheal administration of GaAs (50–200 mg/kg) to rats resulted in a dose-dependent inhibition of δ-aminolaevulinic acid dehydrase activity. This inhibition was concurrent with increased urinary excretion of aminolaevulinic acid. GaAs also inhibited δ-aminolaevulinic acid dehydrase activity in kidney and liver. Gallium nitrate was a more potent *in vitro* inhibitor of δ-aminolaevulinic acid dehydrase activity in blood, liver or kidney post-mitochondrial supernatants than arsenite, suggesting that gallium is the inhibitor of δ-aminolaevulinic acid dehydrase.

The route of exposure (oral vs. intratracheal) of GaAs and sodium arsenate influences the extent of their action on metabolism of porphyrins and haem in various tissues and organs (Kondo & Ichikawa, 1994). For example, intratracheal administration of sodium arsenate or GaAs results in greater suppression of δ-aminolaevulinic acid dehydratase and porphobilinogen deaminase in peripheral erythrocytes than by oral administration of these compounds.

7.1.8.6   Apoptosis

Inorganic arsenic induces apoptosis, or programmed cell death, in CHO cells (Wang et al., 1996a), immature rat thymocytes (Bustamante et al., 1997) and human HL-60 cells (Ochi et al., 1996).

A characteristic marker of apoptosis is internucleosomal DNA cleavage, which was observed in all cell types after exposure to inorganic arsenic. In the HL-60 cells, sodium arsenite (0.05 mmol/litre) produced a greater response than sodium arsenate (0.1 mmol/litre) (8.6 vs. 2.3% apoptosis, respectively). The apoptotic response by both arsenicals was increased when the cells were depleted of GSH by buthione sulfoximine (BSO). However, doses of arsenite greater than 0.01 mmol/litre were very toxic in the GSH-depleted cells, resulting in cell death by necrosis. The mechanism of apoptosis in the thymocytes is unknown, but Wang et al. (1996a) observed that in CHO cells, arsenite induces a cascade of events that leads to apoptosis. This cascade involves the generation of reactive oxygen species, production of hydroxyl radicals via a metal-catalysed Fenton reaction, protein synthesis and activation of protein kinase. Li & Broome (1999) propose that trivalent arsenic induces apoptosis, at least in leukaemia cells, by binding to tubulin. This results in inhibition of tubulin polymerization and eventual formation of microtubules. At doses > 5 µmol/litre, As_2O_3 induces acute necrosis in leukaemia cells. At doses between 0.5 µmol/litre and 5 µmol/litre, As_2O_3 arrests cell proliferation in mitosis. Arsanilic acid, a pentavalent arsenical, does not arrest cell proliferation. Li & Broome suggest that trivalent arsenic binds to cysteine residues (cis-12 and cys-13) of tubulin, which prevents GTP from binding to tubulin. This results in inhibition of microtubule formation, arresting the cells in mitosis, and activation of the genes involved in apoptosis.

7.1.9 *Factors modifying toxicity; toxicity of metabolites*

7.1.9.1 *Interactions with other compounds*

a) Thiols

Scott et al. (1993) and Delnomdedieu et al. (1994a) examined the complexation of arsenic with thiols. GSH non-enzymatically reduces arsenate to arsenite. The reduction requires 2 moles of GSH to 1 mole of arsenate and also results in the formation of oxidized GSH (GSSG). Arsenite then reacts with GSH to form a GSH–arsenite complex. The complex consists of 3 moles GSH per 1 mole arsenite and is termed arsenotriglutathione. The complex is stable over a pH range of 1.5–7.5. Arsenite has a higher affinity for a vicinal thiols, such as that in DMSA, than a dithiol located four carbons apart, such as that in dithiothreitol, or a monothiol such as GSH (Delnomdedieu et al., 1993). Arsenite can also complex with GSH within intact erythrocytes (Delnomdedieu et al., 1994b).

GSH appears to have a role in protecting the kidney from arsenite-induced toxicity. Sodium arsenite (5 mg As/kg) has minimal clinical or histopathological effects after oral administration in the kidney and liver of hamsters, although it decreases the 24-h urine volume by 50% (Hirata et al., 1988, 1990). Depleting GSH levels by administering BSO intraperitoneally 2 h before arsenite administration results in acute renal failure and transient hepatotoxicity. The nephrotoxicity is characterized by oliguria and anuria and increased blood urea nitrogen, plasma creatinine and kidney weight. There is degeneration of epithelial cells in the convoluted tubules and moderate damage to the glomerulus in the kidneys of the BSO + arsenite-treated hamsters. These changes are observed as early 1 h after arsenite administration (and 3 h after BSO), and maximal effects are observed at 24–48 h. The damage to the convoluted tubules is not permanent, because repair of this tissue is observed 48–72 h after arsenite exposure. The hepatotoxicity is characterized by an increase in activities of plasma enzymes (glutamic pyruvic transaminase and glutamic oxaloacetic transaminase), which return to normal levels 48 h after exposure to arsenite. Inorganic arsenic is predominantly excreted in the urine of the BSO + arsenite-treated hamsters, whereas in the arsenite-treated animals, methylated metabolites are primarily excreted. This suggests that GSH may protect the kidney by having a role in the metabolism of inorganic arsenic.

b) Selenium

Selenium is a metalloid found naturally in the environment, and is an essential element. Arsenic and selenium interact such that one can alter the metabolism or toxicity of the other. Selenite and selenate, which are less cytotoxic to cultured fish cells than arsenite and arsenate, antagonize but do not completely eliminate the cytotoxicity of the two arsenicals (Babich et al., 1989). Sodium selenite antagonizes sodium arsenite-induced chromosomal aberrations and sister chromosome exchanges in human lymphocytes (Beckman & Nordenson, 1986). Arsenite enhances the acute toxicity of methylated forms of selenium such as methylselenic acid and trimethylselenonium in the rat (Kraus & Ganther, 1989; Ip & Ganther, 1992), but has no effect on the acute toxicity of inorganic selenium such as sodium selenite (Kraus & Ganther, 1989). Arsenite antagonizes the anti-carcinogenic effect of selenite but enhances the anticarcinogenic effect of trimethylselenonium (Ip & Ganther, 1988, 1992). The mechanism of the synergistic/antagonistic effect of arsenite on selenium is not known, but may be due to inhibition of the methylation of selenium.

c) Cadmium

Another interesting interaction is that between arsenic and cadmium in the rat. Both arsenic and cadmium can be found near copper and zinc smelters (Yanez et al., 1991). The acute toxicity of sodium arsenite administered intraperitoneally followed by cadmium chloride was examined by Diaz-Barriga et al. (1990) and Yanez et al. (1991). Each chemical lowers the LD_{50} of the other, with cadmium affected to a greater extent. When a constant amount of sodium arsenite is administered (10 mg/kg), the cadmium chloride LD_{50} is reduced from 9.5 to 3.2 mg/kg. However, the testicular toxicity induced by cadmium is reduced by the co-exposure to arsenite. Histopathological analysis showed that arsenite alleviated cadmium-induced haemorrhaging in the testes when the compounds were co-administered. With cadmium chloride administered on its own, the GSH level in the testes is decreased by approximately 40%. When arsenite and cadmium are co-administered, the GSH level in the testes is decreased by approximately 25%. Arsenite by itself has no effect on the GSH level in the testes. In the liver, each chemical decreased the GSH level approximately 15%, but after co-administration level decreased by 40%. The significant decrease was attributed to the role of GSH in excretion of the compounds via the bile and the methylation of inorganic arsenic (Diaz-Barriga et al., 1990). In heart tissue, the co-exposure increased the GSH level and lipid peroxidation observed with arsenite on its own, and increased metallothionein levels observed with cadmium chloride on its own (Yanez et al., 1991).

The sequence of administration of sodium arsenite and cadmium chloride to rats has an effect on the outcome of their co-exposure (Hochadel & Waalkes, 1997). Cadmium chloride-induced toxicity (lethality, testicular toxicity and hepatotoxicity) is decreased by a non-toxic dose of arsenite (22.5 μmol/kg) administered subcutaneously 24 h before cadmium exposure. The decreased toxicity may be due to the induction of metallothionein by arsenite. Pretreatment of cadmium chloride (3 μmol/kg) did not affect the toxicity of arsenite (68–90 μmol/kg), even though metallothionein levels were increased 26-fold in the liver. Thus metallothionein apparently has no effect in reducing arsenic-induced toxicity.

7.1.9.2   Biological role of arsenic

Although several studies have attempted to show essentiality of arsenic (Uthus, 1992), a biological role for it has not been demonstrated (NRC, 1999). Arsenic has not been demonstrated to be essential in humans.

Organoarsenical compounds such as arsanilic acid and 3-nitro-4-hydroxyphenylarsonic acid have been administered in poultry rations for growth promotion, improvement of feed conversion, better pigmentation, increased egg production and improved survival (Daghir & Hariri, 1977). However, Proudfoot et al. (1991) reports no improvement of broiler chickens maintained on a diet supplemented with arsanilic acid.

7.1.9.3   Induction of proteins

a)   Metallothionein

Sodium arsenite, and less potently sodium arsenate, induce hepatic metallothionein in mice (Maitani et al., 1987; Kreppel et al., 1993) and rats (Albores et al., 1992) in a dose-dependent manner after oral, intraperitoneal and subcutaneous administration. Both isoforms of metallothionein as well their corresponding mRNA are induced by arsenic (Albores et al., 1992; Kreppel et al., 1993). Arsenite induced metallothionein levels in other organs of the mouse including kidney, heart, lung, spleen, stomach and small intestine. However, the level of induction in these organs is less than the induction of metallothionein in the liver (Kreppel et al., 1993).

Metallothionein contains several sulfhydryl groups, which arsenite might bind. *In vitro* incubations indicated that arsenite can bind to metallothionein, but the binding is pH dependent, with greater binding occurring at pH < 7. At pH 7, zinc can displace arsenite from binding to metallothionein. Arsenite was found not to be associated with hepatic metallothionein after rats were treated with zinc and arsenite (Albores et al., 1992). It appears that inorganic arsenic is an indirect inducer of metallothionein, because it failed to induce metallothionein in mouse hepatocytes (Kreppel et al., 1993). The mechanism of metallothionein induction by arsenic is unclear, but may be due to decreased degradation of metallothionein or increased translation of metallothionein mRNA (Albores et al., 1992). In a recent cancer study (Ng at al., 1999), metallothionein knock-out mice did not show a higher incidence of arsenic-associated tumours than conventional mice. Indicating that metallothionein was not protective in this regard (see section 7.1.7.4).

b) Heat shock or stress-induced proteins

Several proteins induced by heat shock or other stressors are also induced by inorganic arsenic. In the rabbit, intravenous administration of sodium arsenite (0.8 mg/kg) results in induction of a 74-kDa protein in the kidney, liver and heart (Brown & Rush, 1984). Drug-induced hyperthermia also induced a 74-kDa protein in the same organs as well as in the brain. This indicates there are organ-specific differences in the ability of sodium arsenite to induce this protein. The differences may be due to the disposition of arsenic, which generally accumulates in low amounts in the brain (Bertolero et al., 1981).

In murine and human melanoma (Caltabiano et al., 1986) and human skin fibroblast (Keyse & Tyrell, 1989) cell lines, several proteins are induced by sodium arsenite (50 µmol/litre). Interestingly, two of these induced proteins, of 32 kDa (human) and 34 kDa (murine), are not induced by hyperthermia. However, they are induced by stressors such as UV radiation, hydrogen peroxide, heavy metals and thiol-reactive agents (e.g. iodoacetamide). Polyclonal antibodies raised against the human protein cross-react with the murine protein (Caltabiano et al., 1986). This and other evidence suggests that the human and murine genomes that code for these two proteins are closely related. Additional experiments by Keyse & Tyrell (1989) indicated that the 32-kDa human protein is haem oxygenase. It was suggested that this enzyme is induced in response to oxidative stress, and may prevent oxidative damage. Haem oxygenase catalyses the cleavage of haem to form biliverdin, which is metabolized to bilirubin by biliverdin reductase. Bilirubin has antioxidant properties, and can prevent oxidative damage (Stocker et al., 1987). Rats were gavaged with arsenate (20 or 60 mg/kg) or arsenite (1.6 or 24.6 mg/kg), 21 h and 4 h before sacrifice (Brown & Kitchin, 1996; Brown et al., 1997). Rat liver haem oxygenase was induced 6-fold for arsenate and 13-fold for arsenite.

A protein that is induced by arsenite or heat shock in human renal carcinoma cells is P-glycoprotein, which is an expression of the multidrug resistance gene (*MDR1*) (Chin et al., 1990). *MDR1* mRNA levels are also increased when the cells are exposed to arsenite. The P-glycoprotein is an energy-dependent efflux pump that is found in several organs and may transport compounds out of the cells. Cells with increased P-glycoprotein levels are resistant to the antitumour drug vinblastine.

c) Metabolic enzymes

Enzymes involved in the metabolism of xenobiotics are affected by arsenic, and this effect appears to be isozyme selective. In the guinea-pig, sodium arsenite (75 μmol/kg) administered subcutaneously inhibits cytochrome P-450 1A1 activity in liver, lung and kidney and cytochrome P-450 2B4 in lung (Falkner et al., 1993b). The activity of both enzymes is decreased 12 h and 24 h after administration of arsenite, and returns to control levels in the lung by 72 h, but the hepatic and renal 1A1 activity is still decreased. The inhibition of these enzymes is consistent with the increased haem oxygenase activity observed in all the tissues examined as well as the decreased P-450 content in kidney and liver. On the other hand, activity of cytochrome P-450 4B1 in the lung is increased at 72 h, but is not affected in the liver. Also, induction of cytochrome P-450 1A1 by β-naphthoflavone in the lung is potentiated by arsenite, but the induction of this enzyme in liver and kidney is antagonized by arsenite.

In rat lung, cytochrome P-450 1A1 is selectively increased after acute subcutaneous exposure to sodium arsenite (75 μmol/kg) (Albores et al., 1995). In contrast, the enzymatic activity of P-450

1A1 is inhibited by arsenite in kidney and liver. Arsenite does not inhibit pulmonary cytochrome P-450 2B4, but does inhibit it in the liver. Incubating rat lung microsomes with arsenite at concentrations up to 1 mmol/litre does not affect cytochrome P-450 1A1 activity, which suggests that the isozyme-selective modulation of cytochrome P-450 by arsenite is by an indirect mechanism.

Non-haem enzymes induced by acute subcutaneous exposure to sodium arsenite (75 μmol/kg) include GSH *S*-transferase in guinea-pig lung (Falkner et al., 1993b) and rat kidney (Falkner et al., 1993a) and NAD(P)H:quinone acceptor reductase (QOR) in guinea-pig liver, lung and kidney (Falkner et al., 1993b) and rat liver and kidney (Falkner et al., 1993a). The induction of QOR may be a result of the disposition of arsenic and not a tissue-selective induction. The reason for induction of both enzymes is that the gene for the transferase and QOR share the same response elements.

GSH reductase, either purified from yeast or in rabbit erythrocyte lysate, is inhibited by the arsenic complex arseno-triglutathione (Styblo & Thomas, 1995). The complex, which forms by the non-enzymatic reaction of arsenite with GSH (Scott et al., 1993; Delnomdedieu et al., 1994a) is a mixed-type inhibitor (K_i = 0.34 mmol/litre) of the reductase and is 10-fold more potent than either arsenite or GSH. The result of the inhibition of GSH reductase may be an altered cellular oxidation state, because of less GSSG is reduced to GSH. The inhibition of the reductase may lead to serious cytotoxic effects.

7.1.10 *Potential mechanisms of toxicity – mode of action*

7.1.10.1 *Toxicity of trivalent inorganic arsenic*

Trivalent inorganic arsenicals, such as arsenite, readily react with sulfhydryl groups such as GSH and cysteine (Scott et al., 1993; Delnomdedieu et al., 1994a). The complex between arsenic and vicinal sulfhydryl reagent is particularly strong. The activity of enzymes or receptors is due in part to the functional groups on amino acids such as the sulfhydryl group on cysteine or coenzymes such as lipoic acid, which has vicinal thiol groups. Thus, if arsenite binds to a critical thiol or dithiol, the enzyme may be inhibited (Aposhian, 1989). Arsenite inhibits pyruvate dehydrogenase (Peters, 1955;

Szinicz & Forth, 1988), a lipoic-acid-dependent enzyme involved in gluconeogenesis. The acute toxicity of inorganic arsenic may result in part from inhibition of gluconeogenesis and ultimately depletion of carbohydrates from the organism (Reichl et al., 1988; Szinicz & Forth, 1988). However, binding of arsenite to protein at non-essential sites may be a detoxication mechanism (Aposhian, 1989). Arsenite inhibits the binding of steroids to the glucocorticoid receptor, but not other steroid receptors (Lopez et al., 1990; Simons et al., 1990). The glucocorticoid receptor has vicinal thiols that are involved with steroid binding (Simons et al., 1990).

7.1.10.2   *Toxicity of pentavalent inorganic arsenic*

A mechanism of toxicity of pentavalent inorganic arsenic, such as arsenate, is its reduction to a trivalent form, such as arsenite. The reduction of arsenate to arsenite occurs *in vivo* (see Chapter 6). Arsenite is more toxic than arsenate, as evidenced by the lower amount of it needed to elicit a toxic response.

Another potential mechanism is the replacement of phosphate with arsenate. Kenney & Kaplan (1988) have reported that in the human erythrocyte, arsenate can replace phosphate in the sodium pump and the anion exchange transport system. In KB oral epidermoid carcinoma cells, arsenate accumulates at a greater rate when the cells are grown in phosphate-free media (Huang & Lee, 1996). Arsenate uptake by these cells is inhibited by phosphate in a dose-dependent manner. Arsenate can form esters with glucose and gluconate (Lagunas, 1980; Gresser, 1981), forming glucose-6-arsenate and 6-arsenogluconate, respectively. These compounds resemble glucose-6-phosphate and 6-phosphogluconate. Glucose–6-phosphate and glucose-6-arsenate have similar K_m and V_{max} values as substrates for glucose-6-phosphate dehydrogenase and each can inhibit hexokinase.

Arsenate uncouples *in vitro* oxidative phosphorylation because it has a similar structure to phosphate. Two mechanisms for this effect, termed arsenolysis, have been proposed. During glycolysis, arsenate can substitute for phosphate to form 1-arsenato-3-phospho-D-glycer-ate, instead of 1,3-biphospho-D-glycerate, from D-glyceraldehyde-3-phosphate. The arsenic anhydride is unstable and hydrolyses to arsenate and 3-phosphoglycerate. Normally adenosine-5'-

triphosphate (ATP) is generated in this reaction, but with arsenate present instead of phosphate, ATP is not formed (Crane & Lipmann, 1953; Aposhian, 1989). Adenosine-5'-diphosphate-arsenate is synthesized by submitochondrial particles from adenosine-5'-diphosphate (ADP) and arsenate in the presence of succinate (Gresser, 1981). ADP-arsenate hydrolyses more easily than ATP. The formation and hydrolysis of ADP-arsenate results in arsenolysis. The depletion of ATP in rabbit erythrocytes exposed *in vitro* to arsenate has been reported (Delnomdedieu et al., 1994b).

7.1.10.3   Carcinogenicity

Because trivalent inorganic arsenic has greater reactivity and toxicity than pentavalent inorganic arsenic, it is generally believed that trivalent form is the carcinogen.

Arsenic is not a point mutagen but does induce chromosomal abnormalities including changes in structure and number of chromosomes, endoreduplication and sister chromatid exchanges (section 7.1.6). DNA repair is inhibited by arsenic, and this inhibition can result in a co-mutagenic effect with X-rays, UV radiation and several chemicals (section 7.1.6.5). However, concentrations of arsenite that are required to inhibit DNA ligase activity *in vitro* are higher than that needed to inhibit repair within cells. This suggests that arsenite does not directly inhibit DNA ligase, but affects repair processes controlled by the cell (Li & Rossmann, 1989; US EPA, 1997; Hu et al., 1998). Hypermethylation of DNA, particularly the promoter region, can result in inactivation of tumour suppressor genes or genes involved in DNA repair (US EPA, 1997).

Rossman & Wang (1999) isolated two cDNAs from arsenite-resistant Chinese hamster V79 cells. One of these cDNAs is almost homologous with the rat tumour suppressor gene *fau.* This tumour suppressor gene contains a ubiquitin-like region fused to ribosomal protein (Michiels et al., 1993). Klemperer & Pickart (1989) have shown that arsenite inhibits the ubiquitin-dependent proteolytic pathway. Rossman & Wang (1999) suggest that the gene product, or a component within the ubiquitin system, is targeted by arsenic, resulting in alterations that may result in genotoxicity and carcinogenicity.

Mass & Wang (1997) showed that arsenite increased the methylation of the tumour suppressor gene *p53* (section 7.1.6.8). In contrast, arsenite induced hypermethylation of DNA, as observed by Mass & Wang (1997). Zhao et al. (1997) have shown that exposure of arsenite to rat liver cells results in global DNA hypomethylation. The rat liver cells are transformed by the exposure to arsenite, which may have resulted from aberrant gene expression (section 7.1.6.8). With respect to oxidative stress, arsenite induces metallothionein (section 7.1.9.3a) and hsp (section 7.1.9.3b). Catalase and superoxide dismutase reduce arsenite-induced micronuclei in CHO cells (Wang & Huang, 1994) and sister chromatid exchanges in human lymphocytes (Nordenson & Beckman, 1991). Antioxidants such as vitamin E, methylamine and benzyl alcohol reduce the killing of human fibroblasts by arsenite (Lee & Ho, 1994).

Increased cell proliferation has been observed with dimethyl-arsinic acid (Murai et al., 1993; Wanibuchi et al., 1996). Induction of ornithine decarboxylase, another indicator of cellular proliferation, was observed in rat liver after administration of arsenite (Brown & Kitchin, 1996). The co-carcinogenic effect of inorganic arsenic was proposed from the observed co-genotoxic effect of arsenite and the inhibition of DNA repair. Dimethylarsinic acid is also a tumour promoter in several organs. The panel concluded that each of these modes of action could operate, that more than one may act at the same or different dose levels, and there is little evidence of favouring one over any other mode of action.

A US expert panel has recently evaluated various theories including chromosomal abnormalities, DNA repair, DNA methylation, oxidative stress, cell proliferation and co-carcinogenesis (ERG, 1997).

7.2 Organic arsenic compounds

7.2.1 Single exposure

7.2.1.1 Acute toxicity data

a) Oral

The acute oral LD_{50}s of MMA (pentavalent), DMA (pentavalent) and other organic arsenicals in mice are listed in Tables 26–28. The

Table 26. Acute LD$_{50}$ of MMA in rats, mice and rabbits

Species	Compound	Sex	Age	Route	LD$_{50}$ (mg/kg)	Reference
Rat	MMA	M	adult	oral	1101	Gaines & Linder (1986)
Rat	MMA	F	adult	oral	961	Gaines & Linder (1986)
Rat	MMA	F	weanling	oral	> 2200	Gaines & Linder (1986)
Mouse	MMA	M	weanling	oral	1800	Kaise et al. (1989)
Rabbit	MSMA[a]	M	adult	oral	102	Jaghabir et al. (1988)
Rat	Ansar 170 HC[b]	M	adult	oral	1105	Gaines & Linder (1986)
Rat	Ansar 170 HC	F	adult	oral	1059	Gaines & Linder (1986)
Rat	Ansar 170 HC	F	weanling	oral	1746	Gaines & Linder (1986)
Rat	Ansar 8100[c]	M	adult	oral	928	Gaines & Linder (1986)
Rat	Ansar 8100	F	adult	oral	821	Gaines & Linder (1986)
Rat	Ansar 8100	F	weanling	oral	1698	Gaines & Linder (1986)
Rat	Ansar 8100	M	adult	i.p.	600	Stevens et al. (1979)
Rat	Ansar 8100	F	adult	i.p.	561	Stevens et al. (1979)
Mouse	Ansar 8100	M	adult	i.p.	600	Stevens et al. (1979)
Mouse	Ansar 8100	F	adult	i.p.	681	Stevens et al. (1979)

[a] MSMA, monosodium salt of MMA
[b] Ansar 170 HC, 58.4% MSMA
[c] Ansar 8100, 80.1% DSMA (disodium salt of MMA)

values range from a low of 580 mg/kg for tetramethylarsonium chloride, an arsenical found in marine organisms (Shiomi et al., 1987, 1988b), to a high of > 10 000 mg/kg for arsenobetaine (Kaise et al., 1985) and TMAO (Kaise et al., 1989). No deaths were observed in mice administered arsenobetaine (Kaise et al., 1985). Common signs of toxicity of the organic arsenicals in mice include depression of motility and respiration, irritability, ataxia and convulsions. Death appears to be due to respiratory depression (Kaise et al., 1985, 1989). For DMA and TMAO, a period of increased spontaneous motility preceded the death of the mice (Kaise et al., 1989). More than half of the mice that received a lethal dose of MMA, DMA or TMAO had diarrhoea, which may have contributed to their deaths (Kaise et al., 1989).

Table 27. Acute LD_{50} of DMA in mice and rats

Species	Sex	Age	Route	LD_{50} (mg/kg)	Reference
Rat	M	adult	oral	1315	Gaines & Linder (1986)
Rat	F	adult	oral	644	Gaines & Linder (1986)
Rat	M	weanling	oral	1433	Gaines & Linder (1986)
Mouse	M	weanling	oral	1800	Kaise et al. (1989)
Rat	M	adult	i.p.	720	Stevens et al. (1979)
Rat	F	adult	i.p.	520	Stevens et al. (1979)
Rat	F	adult	i.v.	470	Stevens et al. (1979)
Mouse	M	adult	i.p.	520	Stevens et al. (1979)
Mouse	F	adult	i.p.	600	Stevens et al. (1979)

The acute oral LD_{50}s of MMA and DMA in rats, based on sex and age, are listed in Tables 26 and 27. Adult females are more sensitive to DMA than adult and weanling males (Gaines & Linder, 1986). Female weanling rats are less sensitive to MMA than adult rats (Gaines & Linder, 1986).

Table 28. Acute LD$_{50}$ of various organoarsenicals in mice and rats

Chemical	Species	Sex	Age	Route	LD$_{50}$ (mg/kg)	Reference
Trimethyl arsine oxide	mouse	M	weanling	oral	10 600	Kaise et al. (1989)
Arsenobetaine	mouse	M	weanling	oral	>10 000	Kaise et al. (1985)
Tetramethylarsonium chloride	mouse	M	weanling	oral	580	Shiomi et al. (1988b)
Tetramethylarsonium chloride	mouse	M	weanling	i.p.	114	Shiomi et al. (1988b)
Tetramethylarsonium chloride	mouse	M	weanling	i.v.	53	Shiomi et al. (1988b)
Tetramethylarsonium iodide	mouse	M	weanling	oral	890	Shiomi et al. (1988b)
Tetramethylarsonium iodide	mouse	M	weanling	i.p.	175	Shiomi et al. (1988b)
Tetramethylarsonium iodide	mouse	M	weanling	i.v.	82	Shiomi et al. (1988b)
Copper acetoarsenite	rat	F	adult	dermal	>2400	Gaines (1960)

Jaghabir et al. (1988) determined the acute oral LD_{50} of MMA (monosodium salt) in the rabbit to be 102 mg/kg. General weakness and decreased activity, appetite and urine volume were observed in all of the dosed animals. The urine was discoloured in all treated animals and several had diarrhoea. The cause of death in these animals was not determined.

b)   Inhalation

The effect of a 2 h inhalation exposure to technical-grade DMA and MMA in rats and mice was investigated by Stevens et al. (1979). The LC_{50} values are listed in Table 29. Stevens et al. (1979) were unable to determine an LC_{50} for DMA in male rats, because none died at a concentration of 6.9 mg/litre, although some rats (2/10) died at a concentration of 4.1 mg/litre. Female rats are more sensitive to an inhalation exposure to DMA than males, as evidenced by their lower LC_{50} of 3.9 mg/litre. The LC_{50} of DMA in male and female mice is > 6.4 mg/litre, which was the highest concentration tested. Only 1 out of 20 mice died from this exposure. Effects during the exposure to DMA included respiratory distress, rhinorrhoea and porphyrin-like encrustation of the eyes of both species. After exposure to DMA, some of the animals developed diarrhoea and had decreased weight gain. Observations of the female rats that died from the exposure to DMA included impacted caecum, blood in the intestine, bright red lungs, and dark spots in the lung. There was no consistent pattern of gross pathological lesions in these animals.

Table 29. LC_{50} of MMA and DMA in adult mice and rats[a]

Chemical	Species	Sex	LC_{50} (mg/litre)
MMA[b]	rat	M/F	3.7
MMA[b]	mouse	M/F	3.1
DMA[c]	rat	M	> 6.9
DMA[c]	rat	F	> 3.9
DMA[c]	mouse	M/F	> 6.4

[a]   Stevens et al. (1979)
[b]   Ansar 8100, 80.1% DSMA (disodium salt of MMA)
[c]   Phytar ®138, 65.5% DMA

No deaths were observed in either species after exposure to MMA (disodium salt) up to a concentration of 6.9 mg/litre. Respiratory distress was observed during the exposure, but respiration returned to normal afterwards. MMA (disodium salt) is a greater respiratory irritant (RD_{50} of 1.54 mg/litre) in mice, measured as decreased respiratory rate, than purified DMA (RD_{50} of 3.15 mg/litre).

c) Dermal

The acute dermal LD_{50} of the arsenical pesticide copper acetoarsenite (Paris green) is > 2400 mg/kg (Gaines, 1960).

MMA was tested for dermal irritation in male rabbits and classified as a mild irritant (Jaghabir et al., 1988).

d) Parenteral

The LD_{50}s of various organic arsenicals in mice and rats after intraperitoneal or intravenous administration are listed in Tables 26–28. In mice, the LD_{50}s range from a low of 53 mg/kg for tetramethylarsonium chloride after intravenous administration (Shiomi et al., 1988b) to a high of 681 mg/kg for technical-grade MMA (disodium salt) after intraperitoneal administration (Stevens et al., 1979). In the rat, the LD_{50}s range from a low of 470 mg/kg of DMA in females after intravenous administration to a high of 720 mg/kg of DMA in males after intraperitoneal administration (Stevens et al., 1979). The LD_{50}s for the organic arsenicals are lower after parenteral administration than after oral administration.

7.2.2 Short-term exposure

7.2.2.1 Oral

Jaghabir et al. (1989) dosed rabbits orally with MMA (5–20 mg/kg, monosodium salt) for 40 days and several organs were examined for histopathological alterations. In all of the treated rabbits, there were gross signs of change in the digestive tract (hyperaemia, overdistension), liver (hyperaemia) and kidney (swollen). Histopathological changes in the liver and kidney were mild to moderate, dose-related and appeared reversible. The changes in liver included cellular degeneration and periportal inflammation. In the kidney, the changes included tubular nephrosis, interstitial nephritis and vascular hyperaemia.

Murai et al. (1993) examined the effect of administering DMA to female and male rats (57–113 mg/kg) orally 5 times a week for 4 weeks. In both sexes they observed a dose-related decrease in body weight and survival rate. Female rats are more sensitive than males to the exposure to DMA, as shown by their higher mortality and renal lesions, which emerged sooner than in the males. The renal lesions included proximal tubular degeneration, papillary and cortical necrosis, and hyperplasia of the epithelium covering the renal papillae. Another lesion observed in the surviving rats was ulcerative colitis in the caecum. The animals of both sexes that died had significant necrosis of their proximal tubules, which led to renal failure and death in the animals.

7.2.3 Long-term exposure

7.2.3.1 Oral

Roxarsone (4-hydroxy-3-nitrophenylarsonic acid) is used as a growth promoter and to treat diseases in swine and poultry. Rats and mice of both sexes were administered roxarsone (50–800 mg/kg) in their diets for up to 13 weeks to evaluate the toxic effects of this arsenical (Abdo et al., 1989). There was a significant decrease in body weight gain in mice fed 800 mg/kg and rats fed 200–800 mg/kg roxarsone. However, the mice and rats in the 800 mg/kg group consumed the greatest amount of food, so feed aversion was not a cause of the weight loss. Both sexes were affected equally. Mortalities occurred in both sexes at 800 mg/kg in rats and 400–800 mg/kg in mice. Clinical signs of toxicity at the 800 mg/kg dose level included trembling in both species, ataxia in rats, and pale skin colour in mice. Roxarsone did not induce any haematological or clinical chemistry effects in either species. The only pathological lesion induced by roxarsone was in the rat kidney, which was characterized by tubular necrosis, haemorrhage and mineralization of the outer medulla. Overall, it appears that rats are more susceptible to roxarsone toxicity than mice.

Prukop & Savage (1986) administered commercial-grade MMA (monosodium salt) orally to male and female mice. The dose levels were 11.9 and 119 mg/kg and the animals were dosed 3 times a week for 10 weeks. There was no effect on survival of the animals or terminal haematological parameters from exposure to MMA. The mice in the high-dose group had significantly lower weight gain than control and the low-dose group. This effect was attributed to the lower number of litters produced in this group. Additional studies showed that fewer litters were produced by untreated females mated with treated males (119 mg/kg) than control animals by an unknown mechanism.

7.2.3.2 Inhalation

There is no recent information on the toxic effects that may result from long-term inhalation of organic arsenic in laboratory animals.

7.2.3.3 Dermal

There is no recent information on the toxic effects that may result from long-term dermal exposure to organic arsenic in laboratory animals.

7.2.4 Skin and eye irritation; sensitization

There is no recent information on irritation or sensitization from exposure to organic arsenic in laboratory animals.

7.2.5 Reproductive toxicity, embryotoxicity, and teratogenicity

The methylated arsenicals MMA and DMA have been examined for toxic effects in embryos and fetuses, because they are *in vivo* metabolites of inorganic arsenic, which is a reproductive toxicant in laboratory animals.

7.2.5.1 In vivo *embryo and fetal toxicity*

A single intravenous administration of MMA (disodium salt) or DMA (sodium salt) on day 8 of gestation at dose levels of 20–100 mg/kg elicited a low resorption rate ($\leq$10%) in pregnant hamsters (Willhite, 1981). At higher doses of DMA (sodium salt, 900–1000 mg/kg) administered intraperitoneally to pregnant hamsters on one of days 8–12 of gestation higher resorption rates occur, ranging from 30–100% of the litters (Hood et al., 1982). The disodium salt of MMA (500 mg/kg) is less toxic after intraperitoneal administration than DMA, with 6–21% of the litters resorbed. Fetal growth was retarded after administration of MMA on days 9, 10 or 12.

DMA administered orally to pregnant mice (200–600 mg/kg per day) and rats (7.5–60 mg/kg per day) on days 7–16 of gestation resulted in significant fetal mortality in mice at 600 mg/kg per day and rats at 50–60 mg/kg per day (Rogers et al., 1981). A significant decrease in fetal weight gain was observed in mice at 400–600 mg/kg and rats at 40–60 mg/kg.

7.2.5.2 Teratogenicity

MMA (disodium salt, 20–100 mg/kg) and DMA (sodium salt, 20–100 mg/kg) induced a low percentage of malformations (≤6%) after intravenous administration on day 8 of gestation in pregnant hamsters (Willhite, 1981). The effects were characterized by fused ribs, renal agenesis or encephalocele, with the latter anomaly observed only with DMA. Neither MMA nor DMA caused maternal toxicity after intravenous administration in these animals.

DMA (sodium salt, 900–100 mg/kg) induced more (3–100% of fetuses) gross (cleft palate and lip, micromelia, syndactyly, exencephaly, talipes) and skeletal (fused ribs) fetal malformations than MMA (disodium salt, 500 mg/kg) (< 4% of fetuses) after intraperitoneal administration on one of days 8–12 of gestation (Hood et al., 1982). Up to 50% and 37% respectively of the pregnant hamsters died after intraperitoneal administration of DMA and MMA (Hood et al., 1982).

The effect of continuous oral exposure of DMA during days 7–16 of gestation in pregnant mice (200–600 mg/kg per day) and rats (7.5–60 mg/kg per day) was examined by Rogers et al. (1981). Cleft palate in the mice was the major teratogenic response of DMA, and was observed at the two highest doses. There was also a significantly decreased incidence of supernumerary ribs. In the 400 mg/kg per day group, four mouse fetuses had irregular palatine rugae. In the rat, the average number of sternal and caudal ossifications was decreased at the two highest doses and the percentage of irregular palatine rugae increased significantly with dose. A dose-related decrease in maternal weight gain and increase in lethality occurred at the highest dose for the mice (59%) and the two highest doses for the rat (14–67%).

7.2.6 Genotoxicity and related end-points

7.2.6.1 Bacteria

DMA (10 mmol/litre) is mutagenic in *E. coli* B tester strains after at least a 3 h incubation, which suggested to Yamanaka et al. (1989c) that a DMA metabolite was involved in this effect. Dimethylarsine and trimethylarsine were detected in the gas phase of the cell suspensions containing DMA. Only dimethylarsine was mutagenic when the two organoarsines were incubated with the bacteria. The mutagenic effect of dimethylarsine required oxygen and was reduced in a nitrogen atmosphere, which suggested that a reactive species formed from the reaction of dimethylarsine and oxygen, perhaps a peroxyl radical, is the mutagen.

7.2.6.2 Mammalian cells

Genotoxic effects induced by the organic arsenicals include excess tetraploids (DMA, TMAO), and mitotic arrest (MMA, DMA and TMAO) in Chinese hamster lung (V79) cells (Endo et al., 1992; Eguchi et al., 1997). Arsenocholine and arsenobetaine are not genotoxic in V79 cells (Eguchi et al., 1997). DMA was the only compound of several inorganic and organic arsenicals that could inhibit human lymphocyte phytohaemagglutinin-M stimulated mitogenesis (Endo et al., 1992). In mouse lymphoma cells (L5178Y/TK$^{+/-}$), incubation of organic arsenicals for 4 h induced cytotoxicity and clastogenicity. MMA was more potent than DMA, but less so than the inorganic arsenicals (Moore et al., 1997a).

7.2.6.3 Human cells

DNA single-strand scissions and DNA–protein cross-links occur in human alveolar type II (L-132) cells after incubation with DMA (10 mmol/litre) (Tezuka et al., 1993; Yamanaka et al., 1993, 1995, 1997; Rin et al., 1995). It has been proposed that the genotoxic effects are due to a DMA peroxyl radical, which is formed by the reaction of dimethylarsine, a metabolite of DMA, with oxygen (Yamanaka et al., 1989b, 1990).

Tezuka et al. (1993) proposed that DMA modifies DNA before the appearance of the single-strand scissions, by forming a DMA–DNA adduct. This hypothesis is based on the observation of decreased DNA synthesis soon after exposure to DMA, and a shorter nascent DNA chain length of DMA-exposed cells. The proposed adduct renders the DNA susceptible to single-strand scissions induced by UV light (Tezuka et al., 1993) or superoxide radicals (Rin et al., 1995). Apurinic/apyriminidic sites may also form where the initial damage occurred (Yamanaka et al., 1995). The damaged DNA may undergo β-elimination to form DNA single-strand scissions or a Schiff-base reaction to form DNA–protein cross-links.

Chromatid breaks were observed in human umbilical cord fibroblasts after 24 h exposure to organic arsenicals. The rank order of potency of these arsenicals, which is less than that of inorganic arsenic, was DMA > MMA > TMAO. Minor clastogenic effects were induced by arsenobetaine, arsenocholine, the arsenosugar 2′,3′-dihydroxypropyl-5-deoxy-5-dimethylarsinoyl-β-D-riboside, and tetramethylarsonium iodide (Oya-Ohta et al., 1996).

7.2.6.4 In vivo *genotoxicity*

After a single oral administration of DMA (1.5 g/kg), DNA single-strand scissions and DNA–protein cross-links are specifically induced in mouse and rat lung (Yamanaka et al., 1989a,b, 1991, 1993). The damage is not observed until 12 h after exposure and is repaired by 24 h. Brown et al. (1997) also observed increased DNA single-strand scissions in rat lung after administration of two doses of DMA (387 mg/kg) at 21 and 4 h before sacrifice. Single-strand scissions are not observed in mouse and rat liver (Yamanaka et al., 1989a; Brown et al., 1997) or mouse kidney (Yamanaka et al., 1989a) after exposure to DMA. One drawback with several of these studies is the high dose (1.5 g/kg) of DMA administered. It is questionable whether levels of DMA this high would reach the lung after exposure to inorganic arsenic, the precursor to DMA. However, Brown et al. (1997a) did observe a significant increase in DNA single-strand scissions in rat lung after two doses of 387 mg/kg of DMA were administered.

7.2.6.5 *Apoptosis*

DMA (1–5 mmol/litre) induces apoptosis in human renal carcinoma HL–60 cells (Ochi et al., 1996). MMA and TMAO are ineffective in inducing this effect in these cells. The apoptotic response (> 10%) induced by DMA is greater than that observed with sodium arsenite, but greater doses of DMA are required. GSH may have a role in the DMA-induced apoptotic response. Depletion of GSH by BSO in the cells diminished DMA-induced apoptosis. This is in contrast to the increased apoptotic response to inorganic arsenic in GSH-depleted cells.

7.2.7 **Carcinogenicity**

7.2.7.1 *Bladder*

Wei et al. (1999) administered DMA (0, 12.5, 50, and 200 mg/litre) in drinking-water for 2 years to male F344/DuCrj rats. Limited experimental details were given in this study. No bladder tumours were observed in the control group. The incidence of bladder tumours in the exposed groups was dose-dependent (12.5 mg/litre: 0/33 rats; 50 mg/litre: 8/31 rats; 200 mg/litre: 12/31). Treatment did not affect urinary pH, and no bladder calculi were observed in these animals. The kinetics of DMA in rats is different from that in other animals and humans.

7.2.7.2 *Promotion*

In a two-stage carcinogenesis study, Yamamoto et al. (1995) first gave rats five carcinogens (diethylnitrosamine, *N*-methyl-*N*-nitrosourea, 1,2-dimethylhydrazine, *N*-butyl-*N*-(4-hydroxybutyl) nitrosamine, *N*-*bis*(2-hydroxypropyl)nitrosamine). This treatment was followed by exposing the rats to DMA in their drinking-water (50–400 mg/litre) for 24 weeks. DMA promoted tumours only in initiated animals and these occurred in urinary bladder, kidney, liver and thyroid gland. No lung tumours were observed in animals treated with the five carcinogens and promoted with DMA. In initiated animals DMA also significantly increased the number of pre-neoplastic lesions in liver (foci positive for GSH *S*-transferase placental form) and kidney (atypical tubules). At the two highest doses of DMA, the number of hepatic foci could not be determined because of the great number of tumours present in the liver.

Yamanaka et al. (1996) administered the initiator 4-nitro-quinoline 1-oxide (4-NQO) subcutaneously (10 mg/kg) to ddY mice, followed by 25-week exposure to DMA (200 and 400 mg/litre) in drinking-water. The percentage of mice with pulmonary tumour nodules was greater and the number of tumours per mouse was significantly increased in the 4-NQO + DMA group than for 4-NQO on its own, or 4-NQO plus a different promoter (glycerol).

Wanibuchi et al. (1996) examined the promoting effect of DMA in rat urinary bladder after initiation with *N*-butyl-*N*-(4-hydroxy-butyl)nitrosamine. In one experiment, male rats were exposed to the initiator in drinking-water for 4 weeks, followed by various levels of DMA (0, 2, 10, 25, 50 100 mg/litre) in drinking-water for 32 weeks. A DMA dose-dependent increase in preneoplastic lesions, papillomas and carcinomas in the bladder was observed in the initiated and promoted animals. Microscopic examination of the urinary bladder of mice treated with DMA by itself for 8 weeks showed increased alteration of surfaces of the epithelial cells and labelling of cells with 5-bromo-2'-deoxyuridine. There was a lower labelling index in the higher DMA dose group, which was suggested to result from the higher toxicity of the DMA in this group.

DMA administered to male rats in drinking-water (0, 25, 50 and 100 mg/litre) significantly increased ornithine decarboxylase activity, a biomarker of cell proliferation, and formation of 8-hydroxydeoxyguanosine, a marker of oxygen-radical-mediated DNA damage, in line with dose-dependent increase of pre-neoplastic liver cell foci after initiation with diethylnitrosamine in a medium-term bioassay (Wanibuchi et al., 1997).

7.2.8 Factors modifying toxicity; toxicity of metabolites

7.2.8.1 Interaction with thiols

The pentavalent organic arsenicals MMA and DMA are reduced by thiols such as GSH to the trivalent form which can then complex with the thiols (Cullen et al., 1984a; Scott et al., 1993; Delnomdedieu et al., 1994a). During the reduction, oxidized thiol such as GSSG is formed. One mole of trivalent MMA will complex with two moles of GSH. One mole of trivalent DMA will complex with one mole of GSH. MMA and DMA have been found to be bound to proteins *in vivo* (Styblo et al., 1995). Because of the reactivity of trivalent arsenicals with sulfhydryl groups, the possibility exists that MMA and DMA were reduced to trivalent forms *in vivo* and then reacted with sulfhydryl groups on the proteins.

7.2.8.2 Inhibition of GSH reductase

The trivalent organic arsenicals methylarsonous diiodide and dimethylarsinous iodide, as well as their corresponding thiol complexes (cysteinyl and glutathione), are potent inhibitors of GSH reductase (Styblo et al., 1997). The mode of inhibition of these arsenicals is competitive, and their K_is range from 0.009 to 0.834 mmol/litre. The trivalent organoarsenicals and complexes are at least 10-fold more potent inhibitors than are arsenite and arsenate. The pentavalent organic arsenicals MMA and DMA did not inhibit GSH reductase at concentrations < 20 mmol/litre. The methylation of inorganic arsenic may yield metabolites that alter the cellular oxidation status by potently inhibiting the reduction of glutathione disulfide. The alteration of the oxidation status of the cell by these arsenicals may lead to more serious cytotoxic effects.

7.2.8.3 Induction of proteins

a) Metallothionein

MMA and DMA induce metallothionein in the liver of mice after intraperitoneal (Maitani et al., 1987) and subcutaneous (Kreppel et al., 1993) administration. MMA can induce a greater of amount of metallothionein than arsenite, arsenate and DMA (Maitani et al., 1987; Kreppel et al., 1993). However, compared to arsenite, greater concentrations of MMA (50 times more) and DMA (120 times more) are required to induce a response (Kreppel et al., 1993). Both organic arsenicals were ineffective in inducing metallothionein *in vitro* in mouse hepatocyte primary cultures (Kreppel et al., 1993).

b) Heat shock proteins

DMA (10 mmol/litre, 6 h) induces several hsp in cultured human alveolar cells (L-132) (Kato et al., 1997). The induction by DMA is time- and dose-dependent. Hsp72 is the primary hsp induced by

DMA, heat shock (42 °C, 3 h) and arsenite (50 μmol/litre). Other proteins induced by these stressors include hsp45, hsp73, hsp90 and hsp110. There is a difference in the distribution of hsp72 induced by DMA, heat shock and arsenite. With heat shock, hsp72 accumulated primarily in the nucleolus of the cells. In the case of DMA, accumulation of hsp72 is in the nucleolus and nucleoplasm. Hsp72 is primarily distributed in the cytoplasm after exposure to arsenite. Hsp72 induced by DMA was also more tightly bound to nuclear macromolecules than hsp72 induced by heat shock. The authors suggested the specific nuclear accumulation of hsp72 induced by DMA may be a response to the DNA damage elicited by DMA. Rats were gavaged with MMA (226 or 679 mg/kg) or DMA (126 or 387 mg/kg) 21 and 4 h before sacrifice. Unlike the results with inorganic arsenic (section 7.1.9.3b), hepatic haem oxygenase is not induced by the organic arsenicals.

7.2.9 Potential mechanisms of toxicity: mode of action

7.2.9.1 Acute toxicity

Trivalent organic arsenicals react with sulfhydryl groups, as observed with trivalent inorganic arsenicals. *In vitro* binding of trivalent MMA and DMA to protein occurs to a greater extent than with the pentavalent organic forms (Styblo & Thomas, 1997). MMA and DMA have been found to be bound to protein of rat liver cytosol incubated with arsenite (Styblo et al., 1995) and in liver and kidney of mice administered arsenite (Styblo et al., 1996). These compounds would be in the trivalent oxidation state when bound to protein.

Methylated trivalent arsenicals are potent inhibitors of GSH reductase (Styblo et al., 1997). The activity of these chemicals is greater than inorganic trivalent arsenic and the pentavalent organic arsenicals. GSH reductase contains five cysteine residues in each dimeric unit (Collinson & Dawes, 1995), which may provide a binding site for trivalent arsenic to inactivate the enzyme.

Pentavalent organic arsenicals are reduced *in vitro* by thiols to trivalent organic arsenicals which then bind other thiols (Cullen et al., 1984a; Delnomdedieu et al., 1994a). The reduction of organic pentavalent arsenicals to their trivalent forms, as observed with inorganic pentavalent arsenicals, is a potential mechanism of action of the pentavalent organic arsenicals.

7.2.9.2   Carcinogenicity

One drawback with most carcinogenicity studies is the high dose levels of DMA used, ranging from 10 to 100 mg/kg. DMA is the primary metabolite of inorganic arsenic excreted by most mammals. However, it is unlikely that exposure to the levels used in the promotion studies would occur *in vivo* after exposure to inorganic arsenic. Also, because such high doses of DMA were used, the possibility exists that small amounts of inorganic arsenic were present.

DMA has been suggested to be an initiatior, on the basis of the DNA damage it induced in rat lung (Brown et al., 1997). Wei et al. (1999) have recently reported that DMA is a rat bladder carcinogen after a 2 year drinking-water exposure. Most studies have focused on the tumour-promoting activity of DMA. DMA promotes tumour development in several different organs (Yamamoto et al., 1995; Wanibuchi et al., 1996; Yamanaka et al., 1996). Yamanaka et al. (1996) has suggested that DMA may be a tumour progressor. The nitrosamine-initiated tumours were primarily benign, but after exposure to DMA the tumours progressed to adenocarcinomas.

One potential mechanism of tumour promotion by DMA is increased cell proliferation, as observed in the bladder (Wanibuchi et al., 1996), kidney (Murai et al., 1993) and liver (Wanibuchi et al., 1997) of uninitiated rats. A second mechanism could be DMA-induced oxidative stress. A DMA peroxyl radical has been detected *in vitro* (Yamanaka et al., 1990) and DMA induces oxidative damage in the lung of mice (Yamanaka et al., 1991), in the liver of rats (Wanibuchi et al., 1997) and heat shock (section 7.2.8.3b) and other stress-related proteins such as metallothionein (section 7.2.8.3b). Trivalent organic arsenicals inhibit GSH reductase, which might result in a decreased ability of cells to protect against oxidants.

8.   EFFECTS ON HUMANS

Arsenic has long been known because of its acute and long-term toxicity. The first indications for the latter came mainly from its medicinal uses for different purposes. Arsenic has effects on widely different organ systems in the body. It has produced serious effects in humans after both oral and inhalation exposure, it has many end-points, and exposure is widespread all over the world. A peculiarity of arsenic carcinogenicity is that the information mainly comes from experience with exposed humans: it has been unusually difficult to find any animal models.

The health effects of arsenic have been reviewed by many national and international organizations (IARC, 1973, 1980, 1987; IPCS, 1981; ATSDR, 1993, 2000; NRC, 1999).

8.1   Short-term effects

Ingestion of large doses of arsenic may lead to acute symptoms within 30–60 min, but the effects may be delayed when the arsenic is taken with food. Acute gastrointestinal syndrome is the most common presentation of acute arsenic poisoning. This syndrome starts with a metallic or garlic-like taste associated with dry mouth, burning lips and dysphagia. Violent vomiting may ensue and may eventually lead to haematemesis. Gastrointestinal symptoms, which are caused by paralysis of the capillary control in the intestinal tract, may lead to a decrease in blood volume, lowered blood pressure and electrolyte imbalance. Thus, after the initial gastrointestinal problems, multi-organ failure may occur, including renal failure, respiratory failure, failure of vital cardiovascular and brain functions, and death. Survivors of the acute toxicity often develop bone marrow suppression (anaemia and leukopenia), haemolysis, hepatomegaly, melanosis and polyneuropathy resulting from damage to the peripheral nervous system. Polyneuropathy is usually more severe in the sensory nerves, but may also affect the motor neurones (IPCS, 1981; ATSDR, 2000).

Fatal arsenic poisonings have been described after oral exposure to estimated doses of 2 g (Levin-Scherz et al., 1987), 8 g

(Benramdane et al., 1999) and 21 g (Civantos et al., 1995), and cases with non-fatal outcome (usually after treatment and often with permanent neurological sequelae) have been reported after oral doses of 1–4 g (Fincher & Koerker, 1987; Fesmire et al., 1988; Moore et al., 1994) up to 8–16 g arsenic (Mathieu et al., 1992; Bartolome et al., 1999). Serious, non-fatal intoxications in infants have been observed after doses of 0.7 mg of arsensic trioxide (As_2O_3) (0.05 mg/kg) (Cullen et al., 1995), 9–14 mg (Watson et al., 1981) and 2400 mg (4 mg/kg) (Brayer et al., 1997). Incidents of continuous or repeated oral exposure to arsenic over a short period of time have been described. When they drank water containing 108 mg As/litre for 1 week 2 out of 9 exposed persons died, 4 developed encephalopathy and 8 gastrointestinal symptoms (Armstrong et al., 1984). No deaths, but symptoms mainly from the gastrointestinal tract and skin, were observed among 220 patients studied among 447 who had been exposed to arsenic in soy sauce at a level of 100 mg/litre for 2–3 weeks; the estimated daily dose of arsenic was 3 mg (Mizuta et al., 1956). In a mass poisoning in Japan, where 12 000 infants were fed with milk powder inadvertently contaminated with arsenic at a level of 15–24 mg/kg, leading to an estimated daily dose of 1.3–3.6 mg for a period of varying duration, 130 of the infants died (Hamamoto, 1955).

8.2 Long-term effects: historical introduction

A case of lung cancer associated with exposure to arsenical dust was brought to the notice of the British Factory Department, and some further cases were detected in the early 1940s (Hill & Faning, 1948). These reports were followed by an investigation of the matter, and a remarkably elevated relative cancer mortality rate from lung and skin cancer was observed in a sheep-dip factory manufacturing sodium arsenite (Hill & Faning, 1948). Several further case series also reported unexpectedly high lung cancer mortality in different occupational exposure situations (Osburn, 1957, 1969; Roth, 1958; Galy et al., 1963a,b; Pinto & Bennet, 1963; Latarjet et al., 1964; Lee & Fraumeni, 1969).

Chronic skin effects of arsenic, including pigmentation changes, hyperkeratosis and skin cancer, from medicinal use but also from drinking-water, were reported as early as the 19th century (for references, see Hutchinson, 1887; Geyer, 1898; Dubreuilh, 1910). A large number of case series on arsenical skin cancer after exposure via drinking-water were published from Argentina, Chile, Mexico and Taiwan in the early 1900s (for references, see Zaldivar, 1974).

An endemic peripheral vascular disease (PVD), known as wu chiao ping or blackfoot disease (BFD), leading to progressive gangrene of the legs, has been known in Taiwan since the 1920s. It has increased in prevalence since the 1950s, and has been the subject of intense investigation since the late 1950s (Wu et al., 1961; Chen & Wu, 1962).

8.3   Levels of arsenic in drinking-water in epidemiological studies

Extensive information concerning health effects of ingestion of inorganic arsenic in drinking-water comes from a series of studies performed in Taiwan. In the late 1960s, exposure to arsenic from drinking-water was suggested to be the cause of BFD (Ch'i & Blackwell, 1968). Since the 1910s artesian wells which contain high concentrations of arsenic have been used as a source of drinking-water in the area. In 1956 reservoir water was introduced to replace artesian wells as the source of drinking-water. A contemporary account (Tseng et al., 1968) reported that by early 1966 most of the villages had drinking-water with a low arsenic concentration. Another assessment (Chen & Wang, 1990), however, based on official health statistics, presents a view that the public water supply system served only 50% of the total Taiwanese population in 1974–1976, and because the water supply system primarily served metropolitan precincts, its coverage in urban and rural townships was as low as 30% in 1975. According to data from the Taiwan Water Supply Corporation (Tsai et al., 1998), the coverage of tap-water supply was 44% in Peimen, 41% in Hsuechia, 17% in Putai and 0% in Ichu in 1957. These figures increased respectively to 85%, 79%, 55% and 25% in 1967; to 97%, 88%, 60% and 61% in 1977, and to 95%, 94%, 71% and 85% in 1981.

An early study by Chen et al. (1962), using the mercury bromide method, reported on the basis of 34 samples that the median well-water arsenic concentration in four BFD-endemic villages was 780 µg As/litre (range 350–1100). In another early report (Kuo, 1968) the mean well-water arsenic concentration in 11 villages in the endemic area was reported to be 520 µg/litre (range 342–896 µg/litre). Similar concentrations of arsenic (mean 590, range 240–960 µg/litre) were reported by Blackwell et al. (1961) in 13 deep well-water samples. In a later, more extensive report, based on 126 analyses from 29 villages in the BFD-endemic area, the average arsenic concentration was 500 µg/litre, village averages varying between 54 and 831 µg/litre; approximately 50% were between 400 and 700 µg/litre (Kuo, 1968). In a survey in 1964–1966 of the arsenic concentration in artesian wells in the BFD-endemic area, a total of 114 wells was studied; the arsenic concentration was between 10 and 1820 µg/litre, and more than 50% of the wells had a concentration between 300 and 700 µg/litre (Tseng et al., 1968). Within a single village, the variation between individual wells was quite marked: in Tung-Kuo the range was 10–700 µg/litre, and in Kuan-Ho 200–900 µg/litre (Kuo, 1968).

On the basis of a survey by Lo (1975), Chiang et al. (1988) reported that in three BFD-endemic villages, Peimen, Hsuechia and Putai, the arsenic content exceeded 50 µg/litre in respectively 81%, 27%, and 58% of the wells; concentrations in excess of 350 µg/litre were found in 62%, 7%, and 8% of the wells respectively.

From national surveys performed in 1974–1976 (Lo, 1975; Lo et al., 1977), Chen et al. (1985) extracted the arsenic well-water concentration for the villages in the BFD-endemic area, and concluded that in 29.1% of the wells the arsenic concentration exceeded 50 µg/litre and in 5.2% it exceeded 350 µg/litre. The highest reported value for the BFD-endemic area was stated to be 2500 µg/litre. For the rest of Taiwan, 5% of wells had an arsenic concentration of 50 µg/litre or more, and 0.3% had 350 µg/litre or more. For Taiwan as a whole, the figures were 18.7% and 2.7% (Lo, 1975).

The views on the temporal consistency of the arsenic concentration in the wells differ. Tseng et al. (1968) report that the arsenic concentration varied with time: one well had a concentration of 528 µg/litre in June 1962, 530 µg/litre in June 1963 and 1192 µg/litre in February 1964. The variation of the arsenic concentration in six measurements from one well (interval between measurements not indicated) was from 544 to 976 µg/litre (Kuo, 1968). On the other hand, Chen & Wang (1990) report that the three consecutive surveys – in the 1960s, in 1971–1973, and in 1974–1976 – gave very consistent results for the same wells.

The accuracy and sensitivity of the methods employed for the analysis of arsenic in the studies described above is not clear. (This is true of most drinking-water studies throughout the world at that time.) The analytical series in Taiwan in the 1960s (Kuo, 1968; Tseng et al., 1968) were performed using the Natelson method, which has later been estimated to yield an imprecision (standard deviation) of 10% at concentrations of approximately 40 µg/litre or higher (Greshonig & Irgolic, 1997). In the first, limited series (Chen et al., 1962), and in the more extensive surveys done in the 1970s (Lo, 1975; Lo et al., 1977) the standard mercuric bromide staining method was used, which was later estimated to have an imprecision of < 20% for concentrations of ≥200 µg/litre, and to be quite unreliable for concentrations ≤100 µg/litre (Greshonig & Irgolic, 1997).

Historical records of arsenic concentrations in drinking-water were available for 1950–1992 in Region II of Chile (Rivara et al., 1997). The annual 'province-weighted average' water arsenic levels were approximately 200 µg/litre in the years 1950–1957, 650 µg/litre for 1958–1970, 200 µg/litre for 1971, 540 µg/litre for 1972–1977, 100 µg/litre for 1978–1987 and 50 µg/litre thereafter. There was a marked variation between the different locations within the region: for the period 1958–1970, when the exposure was highest, the average was 860 µg/litre for Antofagasta, but ≤250 µg/litre for all other measurement areas. There is no information on the number of measurements actually performed, or on the methods used.

The assessment of exposure in studies in Argentina (Hopenhayn-Rich et al., 1996c, 1998) drew on official records of arsenic concentrations, which were based on measurements in the 1930s, two scientific sampling studies, and one local water survey in the 1970s. In the 1930s survey, 42/61 and 49/57 measurements of arsenic in drinking-water were above the detection limit (40 µg/litre) in the two counties in the high-exposure group. The highest measured concentration was 533 µg/litre and the average drinking-water concentration of the measurements above 40 µg/litre in the two "high-exposure" counties was 178 µg/litre; the authors note, however, that this should not be considered to be representative of the population exposure (Hopenhayn-Rich et al., 1996d).

8.4 Vascular diseases

Exposure to arsenic has been associated with several different vascular effects, in both large and small vessels. Most of the early work on arsenic and vascular disease related to effects in small vessels, whereas later research has been primarily directed at effects in larger vessels, such as the coronary and cerebral arteries. Some work has also investigated links between arsenic exposure and vascular disease risk factors, such as hypertension, diabetes and hyperlipidaemia (Table 30).

8.4.1 Peripheral vascular disease

A series of Taiwanese studies has found that exposure to drinking-water arsenic is associated with the development of BFD (Chen et al., 1988b; Wu et al., 1989). This condition is characterized by an insidious onset of coldness and numbness in one or both feet, progressing on to ulceration, black discolouration and dry gangrene. There are two main pathological types; thromboangiitis obliterans and arteriosclerosis obliterans. In a case–control study of 241 cases of BFD, a significant exposure–response relationship with increasing duration of residence in the area of arsenic contaminated artesian well-water was seen (Chen et al., 1988b). However, other risk factors were also thought to play a role in the development of BFD: the risk of BFD was inversely related to the frequency of eggs, meat, and vegetables in the diet, and directly related to the frequency of consumption of sweet potatoes. The odds ratios for the lowest egg, vegetable, and meat consumption, and highest sweet potato consumption, were 7.2, 1.8, 4.0, and 3.3, respectively. All four parameters reflect undernourishment, and may indicate that this is a contributing factor in the pathogenesis of BFD. In another part of this study, a cohort of 789 BFD patients followed for 15 years had a significant increase in mortality from PVDs as compared both with the general Taiwanese population and with residents of the BFD-endemic area. However, no adjustment for potential confounders, such as smoking, was undertaken.

Table 30. Effects of As on vascular system

Study design	Study population	Source and level of As exposure	Health effects, metric of exposure and measure of association	Comments	Reference
Case-referent	241 BFD patients and 759 age–sex–residence matched controls	well-water As concentrations ≤ 1140 µg/litre, with progress-ive decrease since 1956	exp. time (yr) PVD (OR) <1 1.0 1–29 3.0 >30 3.4 $p < 0.001$ for trend	OR adjusted for nutritional factors, family history of BFD, education, and evidence of skin lesions	Chen et al. (1988b)
Cohort	789 BFD patients	well-water As concentration ≤ 1140 µg/litre, with progressive decrease since 1956	end-point $SMR_{national}$ SMR_{local} PVD 1243*** 351*** CVD 209*** 160** CVA. 118 NS 107 NS *** $p < 0.001$; ** $p < 0.01$	no adjustment for potential confounders	Chen et al. (1988b)

Table 30 (contd.)

			age adjusted mortality rates per 100 000 As exposure			
Ecological	mortality and population data for 1973–1986 in 42 villages in Taiwan	well-water As concentration ≤1140 µg/litre, with progressive decrease since 1956	< 0.30	0.30–0.59	≥ 0.60 mg/kg	no increase in cerebrovascular accidents in either males or females at any exposure dose used published Taiwan data from 1964 to 1966; the Natelson method was used (Tseng et al., 1968; Kuo, 1968).
			all vascular diseases:			Wu et al. (1989)
			males 364	421	573	
			females 278	371	386	
			PVDs:			
			males 23	58	60	
			females 18	48	35	
			cardiovascular diseases:			
			males 126	154	260	
			females 1	153	145	
Cross-sectional	382 men and 516 women residing in villages in BFD-endemic area	well-water As concentration ≤1140 µg/litre, with progressive decrease since 1956	hypertension cumulative exposure (mg · litre^{-1} year)		OR	exposure determined from residential history and village median well-water As concentration, based on the analysis of Kuo (1968; 126 samples from 29 villages, Natelson method) ORs adjusted for age, sex, disease status of diabetes, proteinuria, body mass index, fasting serum triglyceride levels
			0		1.0	Chen et al. (1995)
			0.1–6.3		0.8 (0.2–3.2)	
			6.4–10.8		2.3 (0.8–6.8)	
			10.9–14.7		3.4 (1.2–9.2)	
			14.8–18.5		3.8 (1.4–10.3)	
			> 18.5		2.9 (1.1–7.3)	
			unknown		1.5 (0.6–4.2)	

Table 30 (contd.)

Study design	Study population	Source and level of As exposure	Health effects, metric of exposure and measure of association		Comments	Reference
Cross-sectional	582 residents of BFD-endemic area	drinking-water As range 1–1097 µg/litre, 50% between 300 and 700 µg/litre	exposure category (mg/litre year) 0 1–19 > 20	PVD OR (CI 95%) 1 3.1 (0.9–10.4) 4.8 (1.4–16.7)	142 water samples from 114 well analysed for As used ratio of ankle and brachial systolic arterial pressure as indicator of PVD. Measurement by Doppler ultrasound. Those with ABI of > 1.20 excluded because of possible misclassification of PVD. adjusted for age, sex, body mass index, cigarette smoking, diabetes mellitus, hypertension, plasma lipids	Tseng et al. (1996)

Table 30 (contd.)

Ecological	residents of 60 villages in As endemic area in Taiwan 1 355 915 person years	well-water As concentration ≤1140 µg/litre, with progress-ive decrease since 1956	cumulative mortality from IHD (1973–1986) from birth to age 79 exposure category (As mg/litre) — cumulative mortality % < 0.10 — 3.4 0.10–0.34 — 3.5 0.35–0.59 — 4.7 ≥ 0.60 — 6.6	exposure determined from village median well-water As con-centration, based on the analysis of Kuo (1968; 126 samples from 29 villages)	Chen et al. (1996)
Cohort	263 BFD patients and 2293 referents from the 60 villages above	same as above	Exposure category mg/litre · year — Relative risk of IHD (CI) 0 — 1.00 < 10 — 2.2 (0.46–10.2) 10.0–19.9 — 3.3 (0.83–13.4) 20 + — 4.9 (1.4–17.7)	exposure determined from village median well-water As con-centration, based on the analysis of Kuo (1968; 126 samples from 29 villages). small number of deaths. Cox propor-tional hazard model adjusted for age, sex, smoking, body mass index, serum choles-terol, serum triglycer-ide level, hypertension, diabetes mellitus BFD. relative risk of BFD patients vs. non-BFD, 2.48 (1.1.4–5.4)	Chen et al. (1996)

Table 30 (contd.)

Study design	Study population	Source and level of As exposure	Health effects, metric of exposure and measure of association				Comments	Reference
Case-referent	74 cases of ISHD and 193 referents from the population of the Chen et al. (1995) study	well-water As concentration ≤ 1140 µg/litre, with progressive decrease since 1956	Duration of drinking As-containing water				exposure determined from village median well-water As concentration, based on the analysis of Kuo (1968; 126 samples from 29 villages).	Hsueh et al. (1998)
			Year	IHD	OR	(CI)		
			> 13	1.0				
			13–29	2.6	(1.0–6.4)			
			≥ 30	2.9	(1.0–8.3)			
Ecological	4 townships in BFD-endemic area, mortality in 1971–1994 compared to local and national rates	well-water As concentration ≤140 µg/litre, with progressive decrease since 1956	mortality compared to local rates				OR age- and sex adjusted; no significant association with cumulative As exposure	Tsai et al. (1999)
				SMR	CI			
			hypertension	73	62–85			
			IHD	175	159–192			
			CVD	114	108–121		national statistics were used to calculate expected deaths. 99% of causes of deaths based on diagnosis of a physician. Overlaps with earlier studies in the BFD-endemic area.	
			vasc. dis.	356	291–430			

Table 30 (contd.)

Study type	Description	Exposure	Exposure category (µg/litre)	CVD OR (CV)	Cerebral infarction OR (CV)	Comments	Reference
Cross-sectional	8102 males and females from the Lanyang Basin on the north-east coast of Taiwan	As in drinking-water	< 0.1 0.1–50 50–299.9 ≥ 300	1.0 2.5 (1.5–4.5) 2.8 (1.6–5.0) 3.6 (1.8–7.1)	3.4 (1.6–7.3) 4.5 (2.0–9.9) 6.9 (2.9–16.4)	OR adjusted for age, sex, smoking, alcohol intake, hypertension and diabetes. exposure category determined by median As concentration of well-water.	Chiou et al. (1997a)
Ecological	mortality study from 30 US counties, 1968–1984	As in drinking-water	Diseases of arteries, arterioles and capillaries Exposure category (µg/litre) 5–10 10–20 > 20	SMRs (CI) Males 110 (110–120) 110 (100–110) 160 (150–180)	Females 110 (110–120) 110 (100–120) 190 (170–210)	no effects were observed for all circulatory diseases, IHD or cerebral vascular disease. expected numbers of deaths generated using US mortality rates. As concentrations were from public water supply records	Engel & Smith (1994)

Table 30 (contd.)

Study design	Study population	Source and level of As exposure	Health effects, metric of exposure and measure of association			Comments	Reference
				Males SMR (CI)	Females SMR (CI)		
Cohort	4058 members of The Church of Jesus Christ of Latter Day Saints in Millard County, Utah	range of exposure 3.5–620 µg/litre; median exposures range from 14 to 166 µg/litre depending on location	cerebrovasc. dis.	79 (62–99)	87 (71–106)	for 2073 cohort members, "most" had at least 20 yr history of exposure in their respective towns. The balance of the cohort (n = 1985) were included if they had spent any length of time in the As-affected community. existing and historic As concentrations used. death rates for the state of Utah for the years, 1960 to, 1992 were used to generate the expected deaths. no indication of exposure–response relationship for any of the vascular health effects.	Lewis et al. (1999)
			all heart dis.	80 (73–88)	81 (72–91)		
			IHD.	76 (67–85)	64 (53–76)		
			dis. art. capill. arteriosclerosis	93 (61–135)	86 (52–132)		
			aortic aneurysm	124 (69–204)	118 (68–188)		
			hyp. heart dis.	76 (35–144)	48 (6–173)		
			other heart dis.	220 (136–336)	173 (111–258)		
				94 (71–122)	143 (111–180)		

246

Table 30 (contd.)

| Cross-sectional | 1595 people from 4 villages in Bangladesh: 1481 exposed to As and 114 non-exposed controls | As in drinking-water. For 39, 36, 18, and 7%, the exposure was <0.5, 0.5–1, and >1 mg/litre, and unknown, respectively. | Exposure category (mg/litre · year)
0
<5
5–10
>10 | PR* for hypertension (CI)
0.8 (0.3–1.7)
1.5 (0.7–2.9)
2.2 (1.1–4.4)
3.0 (1.5–5.8) | exposure for the highest exposure group likely to be overestimated because of introduction of low-As water into one community, which was not considered in the analysis

Used existing As water measurements (measured by flow-injection hydride generation AAS). hypertension defined as >140 mmHg systolic BP together with >90 mmHg diastolic BP study limited to the 1595 individuals out of 1794 eligible, who were at home at the time of the interview. 114 persons were considered unexposed and were used as the reference group. *PR, Mantel–Haenszel prevalence ratio adjusted for age, sex and BMI | Rahman et al. (1999a) |

Table 30 (contd.)

Study design	Study population	Source and level of As exposure	Health effects, metric of exposure and measure of association	Comments	Reference
Cohort	478 patients treated with Fowler's solution for 2 weeks–12 years in 1946–1960 and followed until, 1990	Cumulative dose <500 mg, 500–999 mg, 1000–1999 mg; ≥2000 mg	Mortality from vascular diseases 	SMR	CI CVD 91 74–110 IHD 85 60–110 Cerebrovasc. disease 72 40–110	SMRs for the whole group. No dose–response relationship observed, but the numbers were small	Cuzick et al. (1992)

Occupational exposure

Study design	Study population	Source and level of As exposure	Health effects, metric of exposure and measure of association			Comments	Reference
Cohort	2802 men who worked in the smelter for ≥1 yr during 1940–1964, vital status followed 1941–1986	ambient air in a smelter	Cum. exp. ($mg/m^3 \cdot yr$)	IHD cases	SMR	Exposure assessed from industrial hygiene data (available from 1938) and extrapolation from urinary As concentrations	Enterline et al. (1995)
			<0.75	55	108		
			0.75	67	103		
			2.0	74	107		
			4.0	87	122		
			8.0	91	128		
			20	46	132		
			≥45	8	90		

Table 30 (contd.)

Cohort	2802 men who worked in the smelter for ≥1 yr during 1940–1964, vital status followed 1940–1976 (same cohort as in Enterline et al. (1995), but a shorter follow-up time)	ambient air in a smelter	Cum. exp. (mg/m³ · yr) / IHD RR / CI: <0.75 — 1.0 0.75–1.999 — 0.9 — 0.64–1.3 2.0–3.999 — 1.1 — 0.78–1.6 4.0–7.999 — 1.4 — 0.98–2.0 8.0–19.999 — 1.7 — 1.2–2.5 >20 — 1.5 — 0.95–2.5	20-year lag and work status included in the model. No effects found for cerebrovascular disease.	Hertz-Picciotto et al. (2000)
Cohort	8104 white males employed for ≥1 year before 1957, vital status followed 1938–1987	ambient air in a smelter	Arteriosclerosis and coronary heart disease: SMR 105 (CI 99–110); Cerebrovascular disease: SMR 103 (CI 93–115)		Lubin et al. (2000)
Cohort	3916 men who worked ≥3 mo in the smelter in 1928–1967. Vital status followed until 1981	Ambient air in a smelter. Categories for cumulative exposure <0.25, 0.25–15, 15–100 and ≥100 mg³ yr.	IHD SMR 107 (CI 97–117); Cerebrovascular disease SMR 106 (CI 88–126)	in an earlier report (Axelson et al., 1978), a two-fold increase in mortality from cardiovascular disease	Järup et al. (1989)

Table 30 (contd.)

Study design	Study population	Source and level of As exposure	Health effects, metric of exposure and measure of association	Comments	Reference
Cohort	839 copper smelter workers	ambient air in a smelter	7 deaths from heart diseases vs. 14.9 expected		Tokudome & Kuratsune (1976)
Cohort	1974 gold miners	airborne exposure to As, radon, silica	IHD SMR 103 (173 expected cases)		Armstrong et al. (1979)
Cohort	1330 gold mine and refinery workers	airborne exposure to As, radon and silica	SMR for "diseases of the circulatory system" 54 (CI 39–73)		Simonato et al. (1994)
Cohort	611 workers at a pesticide plant	inhalation exposure to As	SMR for all circulatory diseases 80 (CI 65–98)		Sobel et al. (1988)
Cross-sectional	32 As-exposed workers and 26 non-exposed referents	average urinary As 35.9 for the exposed and 14.5 µmol/mol creatinine for the referents	average systolic BP 128 among the exposed and 120 among the referents, $p = 0.023$	exposed group included taxidermists garden fence makers, weekend cottage constructors, wood impregnators, electric pole impregnators, new house constructors	Jensen & Hansen (1998)

An increasing risk of PVD was also found in an ecological study of 42 villages in the BFD-endemic area in south-western Taiwan (Wu et al., 1989). This study found that age-adjusted mortality rates for PVD increased in an exposure–response relationship with increasing median concentrations of drinking-water arsenic from artesian wells at < 0.30, 0.30–0.59, and ≥0.60 mg/litre for males and females. As this was an ecological study, no individual measures of arsenic exposure were available. In addition, the rates were not adjusted for potential confounders, such as cigarette smoking.

A further Taiwanese study attempted to investigate the association between long-term arsenic exposure and PVD morbidity, rather than mortality, using Doppler ultrasound to measure the ankle–brachial index (blood pressure ratio between ankle and brachium, ABI) (Tseng et al., 1996). A cross-sectional study was undertaken, recruiting participants in a previous cohort study. Of the 941 subjects in the original cohort, 582 (62%) took part in the cross-sectional study, so a possible selection bias may have been operating. The study had several advantages over previous Taiwanese studies of BFD, including the use of an objective and more sensitive measure of PVD (i.e. ABI) rather than the physical examination used in previous studies, individual measures of arsenic exposure and the ability to adjust for potential confounders. The study found that the risk of PVD increased with increasing cumulative exposure to arsenic, with a statistically significant increase for the high subgroup (≥20 (mg/litre) year). This association persisted when different cut-off points for ABI were used to diagnose PVD.

No association was seen between cumulative arsenic exposure for any of the serum lipids among the 533 individuals studied for these end-points (Tseng et al., 1997). In a case–referent study among 45 healthy residents of the BFD area and 51 referents, it was observed that the perfusion of the big toe, as measured by laser Doppler flowmetry, was weaker among the arsenic-exposed (Tseng et al., 1995).

Swedish copper-smelter workers exposed to arsenic ($n = 47$), with a mean average exposure of 23 years, had a higher prevalence of Raynaud's phenomenon, indicated by a vasospastic tendency in their fingers after localized cooling, as compared with 48 controls (Lagerkvist et al., 1986). The vasospastic tendency did not disappear during the summer vacation, and thus appeared to be related to long-term rather than short-term exposure to arsenic. However, the vasospastic tendency appeared to diminish over the course of several years, after the exposure to arsenic was reduced (Lagerkvist et al., 1988).

8.4.2 Cardio- and cerebrovascular disease

A cohort of 789 BFD patients, followed for 15 years, had a significant increase in mortality from cardiovascular diseases but not cerebrovascular disease (CVD), as compared both with the general Taiwanese population and with residents of the BFD-endemic area (Table 30). However, no adjustment for potential confounders, such as smoking, was undertaken (Chen et al., 1988b).

The finding of increasing risk of cardiovascular disease mortality was also found in an ecological study of 42 villages in the BFD-endemic area in south-western Taiwan (Wu et al., 1989). This study found that age-adjusted mortality rates for males and females for all vascular diseases combined increased in a exposure–response relationship with increasing median concentrations of drinking-water arsenic from artesian wells at < 0.30, 0.30–0.59, and ≥0.60 mg/litre. The age-adjusted mortality rates for all vascular diseases and cardiovascular diseases were significantly increased along this exposure gradient. Although the rates increased across exposure groups for CVD, there was no significant exposure–response relationship for either males or females. As this was an ecological study, no individual measures of arsenic exposure were available.

Chen et al. (1996) assessed the relationship between ischaemic heart disease (IHD) mortality and long-term arsenic exposure, using two different study designs. The first was an ecological study, which examined the mortality rates of IHD in 60 villages located in a BFD-endemic area in Taiwan. They found a monotonic biological gradient relationship between arsenic exposure in artesian well-water and IHD mortality rates in these villages. The second part of this study was a cohort study of 263 BFD patients and 2293 non-BFD patients recruited from three of the villages with the highest BFD prevalence in Taiwan. This cohort was followed up for an average period of 5 years and an exposure–response relationship between cumulative arsenic intake and mortality from IHD was found. The relative risks were 2.2, 3.3 and 4.9 respectively for those with cumulative exposures of 0.1–9.9 mg/litre, 10–19.9 mg/litre, and ≥20 mg/litre compared to those without exposure, after adjustment for age, sex, cigarette smoking, BMI, serum levels of cholesterol and triglycerides, hypertension and diabetes (Chen et al., 1996). The exposure to arsenic of 74 cases of IHD (as diagnosed from ECG and a standardized questionnaire), and of 193 referents without IHD, was compared (Hsueh et al., 1998). There was a borderline significant increase of IHD with increasing duration of use of arsenic-containing drinking-water, and a non-significant association with cumulative arsenic exposure.

In the most recent ecological study in Taiwan (Tsai et al., 1999), mortality from different causes during 1971–1994 was studied in the area investigated in the first study (Chen et al., 1985), and compared to local rates in both the Chiayi-Tainan county and the whole of Taiwan. The total number of deaths and person-years for the study group was 20 067 and 2 913 382. Age- and sex-specific mortality rates were calculated for each disease for the years 1971–1994. There was an excess mortality from IHD with a standardized mortality ratio (SMR) of 175 (CI 159–192), and a very small but significant excess in the mortality from CVD (SMR 114, CI 108–121; local rates).

A study by Chiou et al. (1997a) attempted to elucidate the exposure–response relationship between CVD and ingested arsenic via drinking-water in the north-east coast of Taiwan, an area with elevated drinking-water arsenic concentration, but different from the BFD-endemic area on the south-western coast. The population in this cross-sectional study consisted of 8102 men and women from 3901 households. The CVD status was assessed through initial home interviews and validated by review of medical records; 139 CVD patients were found, including 95 with cerebral infarction. Individual exposure information was obtained by measuring the arsenic concentration in the well-water for each household. Exposure categories were 0, 0.1–50.0, 50.1–299.9, and ≥ 300 μg/litre. This study concluded that a exposure–response relationship exists between the arsenic concentration in well-water and the prevalence of CVD after adjustment for age, sex, hypertension, diabetes mellitus, cigarette smoking and alcohol consumption. This relationship was even more prominent when only the cerebral infarction subgroup was analysed.

An ecological mortality study by Engel & Smith (1994) was carried out in 30 counties in the USA with weighted mean concentrations > 5 µg As/litre in drinking-water. This study compared mortality due to several vascular diseases (arteriosclerosis, aortic aneurysm, congenital vascular anomalies, IHD and CVD) in these counties with the expected numbers of deaths generated by US mortality rates. The study found excess mortality rates for males and females for diseases of the arteries, arterioles and capillaries, especially for the highest exposure subgroup (> 20 µg/litre). When this group of diseases was divided into its three main subgroups, the most consistent elevations for the highest exposure group were found for arteriosclerosis mortality, less consistent elevations for mortality from aortic aneurysm, and no elevations for mortality from all other diseases of the arteries, arterioles and capillaries. No elevation in SMRs for either sex in any exposure group was found for all circulatory diseases, IHD or CVD.

A cohort study on the relationship between drinking-water arsenic and different causes of mortality was conducted among members of the Church of Jesus Christ of Latter-day Saints (Mormons) in 7 communities in Utah (Lewis et al., 1999). The total number of cohort members was 4058; there were altogether 2203 decedents. By the time of the closing date of the follow-up (1996), 70% of the cohort members had attained the age of 60, and for 67% of the decedents, the time in the cohort was ≥ 40 years. Three hundred individuals (7.4%) were lost to follow-up, and were considered at risk until the last known residence date. Exposure to arsenic was determined from analyses of arsenic in drinking-water, performed by the state health laboratory between 1976 and 1997; the number of the samples for the 7 communities was altogether 151, of which 60 were from the year 1997. The cumulative exposure to arsenic for each individuals was computed on the basis of the residence history from the church records, and the median arsenic concentration of the locality. "Most" of the 2073 members of the cohort had at least 20 years of exposure in their respective town. The balance of the cohort (*n* = 1985) were included if they had spent "any length of time" in the arsenic affected community. The median drinking-water arsenic concentrations were between 14 and

166 µg/litre for the different localities, and the maximal recorded concentration was 620 µg/litre. For the community of Hinckley, which provided 29.4% of the cohort participants, and which had the highest median drinking-water concentration, a new water source, low in arsenic, was brought into use in 1981, but only the analytical data before this date were used in the calculations. It is therefore likely that the arsenic exposure represents an overestimation. The observed numbers of deaths from different causes were compared to data for the state of Utah. The expected numbers from the years 1950–1954 were used for those who died (number not given) before 1950, and the expected number from 1990–1992 for those who died after 1992. In men, the overall mortality (SMR 91, CI 86–96) and the mortality from non-malignant respiratory disease were lower than expected (SMR 68, CI 54–85). A similar tendency was observed in women, but was not significant. The study found a deficit in the mortality from CVD, all heart disease and IHD, but a significant excess of deaths from hypertensive heart disease among men and women, and all other heart disease (apart from IHD and hypertensive heart disease) among women. The increases of hypertensive heart diseases showed no exposure–response relationship. The low smoking rates among the church members may have explained the low SMRs for those vascular causes of death related to cigarette smoking (Villanueva & Kogevinas, 1999).

Cuzick et al. (1992) studied the causes of death during 1945–1992 among 478 patients treated with Fowler's solution during the period 1945–1965. Nineteen patients had emigrated and 31 were lost to follow-up; how they were considered in the analysis is not indicated. A total of 188 patients had died before their 85th birthday, and were included in the analysis. Expected values were based on age-, sex-, and calendar year-adjusted rates for England and Wales. The total arsenic dose was calculated from the original treatment records; no data on smoking was available. This study found no association between arsenic exposure and mortality from all circulatory diseases, IHD or CVD. The total exposure of the members of this cohort was lower than that of the major drinking-water cohorts.

255

The relationship between arsenic exposure and vascular diseases has also been studied in some of the occupational cohorts[1]. These studies are described more fully in section 8.6. In the Tacoma smelter cohort (Enterline et al., 1995) there was a significant excess of IHD, with a weak exposure–response relationship. In a further analysis of this cohort, where attempts were made to adjust for the healthy worker survivor effect, this association was strengthened with a clear exposure–response relationship (Hertz-Picciotto et al., 2000). No significant increase for mortality from CVD was found. In an earlier report on the cohort (Enterline & Marsh, 1982), no excess mortality from heart disease was observed. No significant increase in the mortality from arteriosclerosis and IHD or from CVD was observed among the members of the Montana smelter cohort (Lubin et al., 2000). When the analysis was repeated with an attempt to adjust for the healthy worker survivor effect, there were no changes to the initial findings (Lubin & Fraumeni, 2000). In the first report on the Rönnskär cohort, an arsenic-exposure-related 2-fold increase in the mortality from cardiovascular disease was observed (Axelson et al., 1978). However, in the most recent update, no relationship between exposure to arsenic and IHD or CVD was observed (Järup et al., 1989). In the Japanese smelter cohort (Tokudome & Kuratsune, 1976), there was a deficit of the mortality from heart diseases (7 observed and 14.90 expected cases). Mortality from IHD in a cohort of Australian gold-miners was not different from that expected (Armstrong et al., 1979). In the French gold-miner cohort (Simonato et al., 1994) the mortality from the diseases of the circulatory system was significantly lower than expected. The mortality from the diseases of the circulatory system was also significantly lower than expected in the US pesticide production worker cohort (Sobel et al., 1988).

---

[1]   It should be noted, however, that most studies have used SMRs as risk estimates for the exposure response relationships. SMRs are indirectly standardized rate ratios and are thus not directly comparable, unless the risks are homogenous over age strata, or the age structure is similar in the subgroups compared. This is of particular concern for cumulative exposure estimates where there is an inherent heterogeneity in age over exposure subgroups. However, when there are large differences in risk between exposure categories, this theoretical objection is probably less important.

8.4.3 *Hypertension*

A cross-sectional study was performed by Chen et al. (1995) to examine the association between long-term exposure to inorganic arsenic and the prevalence of hypertension (Table 30). Hypertension was defined as a systolic blood pressure > 160 mmHg or diastolic blood pressure > 95 mmHg, or a reported history of hypertension regularly treated with antihypertensive drugs. Researchers studied a total of 382 men and 516 women residing in villages in the BFD-endemic area in Taiwan, representing 83% of those invited to take part. The age-adjusted prevalence of hypertension was 17.3% (95% CI 13.1–21.5) for men and 18.0% (95% CI 14.1–21.9) for women. The long-term arsenic exposure was calculated from the history of artesian well-water consumption obtained through standardized interviews based on a structured questionnaire and the arsenic concentrations in well-water measured in the 1960s (Kuo, 1968; Natelson method). In this study, residents in the BFD-endemic area had a significantly increased age- and sex-adjusted prevalence of hypertension compared with residents in non-endemic areas. Prevalence odds ratios (POR) for hypertension appeared to follow a exposure–response relationship, as they increased significantly with cumulative arsenic exposures. Odds ratios for the three highest categories remained statistically significant after adjustment for age, sex, diabetes mellitus, proteinuria, body mass index (BMI) and serum triglyceride level (Chen et al., 1995).

There was a statistically significant deficit in the mortality from hypertension in the update of the ecological study in Taiwan (Tsai et al., 1999), with an SMR of 73 (CI 62–83). It should be noted that there was, however, an excess mortality from IHD, and that the number of deaths from hypertension itself was unusually high (239 deaths, whereas there were only 283 deaths from IHD).

A study by Rahman et al. (1999a) in Bangladesh compared the prevalence of hypertension (assessed by blood pressure measurements) among residents with arsenic exposure and those without. A total of 1481 subjects exposed to arsenic-contaminated drinking-water and 114 unexposed subjects were analysed for their time-weighted mean arsenic levels and divided into categories: 0 mg/litre (control) (no detection limit given), < 0.5 mg/litre, 0.5–1.0 mg/litre and > 1.0 mg/litre, and alternatively as cumulative exposures of 0,

< 1.0, 1.0–5.0, 5.0–10.0, and > 10.0 (mg/litre) · year. These exposure categories were assessed with respect to their prevalence of hypertension (a systolic blood pressure of ≥140 mmHg in combination with a diastolic blood pressure of ≥90 mg Hg). It was found that the prevalence ratios, adjusted for age, sex, and BMI, were 1.2, 2.2, 2.5, and 0.8, 1.5, 2.2, and 3.0 in relation to arsenic exposure in mg/litre and (mg/litre) · year respectively. The exposure–response relationships were significant ($p < 0.001$) for both series of risk estimates.

In a study among a group of 40 Danish workers exposed to arsenic (average urinary arsenic level three times that of the referents), the systolic blood pressure was found to be 8 mmHg higher than that among referents ($p = 0.023$) (Jensen & Hansen, 1998).

8.5 Diabetes mellitus

Lai et al. (1994) assessed the relationship between ingested inorganic arsenic and prevalence of diabetes mellitus in a cross-sectional study (Table 30). The authors examined 891 adults residing in the BFD-endemic area in Taiwan. Diabetic status was determined through oral glucose tolerance test or a history of diabetes regularly treated with sulfonylurea agents or insulin. The rate of diabetes among the 891 study subjects was twice that of the rates previously reported for residents in Taipei and the entire Taiwan population, after adjustment for age and sex. The authors also estimated the cumulative exposure to arsenic from a detailed history of residential addresses and duration of artesian well-water obtained through standardized questionnaires and personal interviews. Prevalence of diabetes, after adjusting for age, sex, BMI and physical activity level increased with increasing arsenic exposure with odds ratios of 6.6 and 10.1 for the two cumulative exposure groups respectively (0.1–15 and > 15 (mg/litre) · year).

There was an excess mortality from diabetes among the arsenic exposed population in the most recent ecological study in Taiwan (Tsai et al., 1999; for study description, see section 8.7 on cancer), with an SMR of 135 (CI 116–155).

In a study in Bangladesh, people with skin keratosis in six districts with arsenic-contaminated drinking-water were identified as a study group. This group was then divided into three drinking-water arsenic concentration strata, on the basis of mean arsenic level in drinking-water over the lifetime of the subject. A non-exposed group was identified in a door-to-door survey in Dhaka (Rahman et al., 1998). This study showed elevated risks for diabetes for those exposed to arsenic in their drinking-water (prevalence ratio = 5.9 after controlling for age, BMI, and sex) as compared with the unexposed. There was also a strong exposure–response relationship among the three exposure subgroups (Table 31).[2]

In another cross-sectional study in the same villages (Rahman et al., 1999b; for study description see section 8.4.3, Rahman et al., 1999a), The prevalence ratios of glucosuria, adjusted for age and sex, were 0.4, 0.9, 1.2 and 1.7 for individuals without skin lesions, and 0.8, 1.7, 2.1, and 2.9 for those with skin lesions, in the cumulative arsenic exposure categories of > 1, 1–5, 5–10, and > 10 (mg/litre) · year, respectively. The exposure–response relationships were significant ($p < 0.001$) for both series groups.

In the Utah mortality study (Lewis et al., 1999; see section 8.4.2 above) no significant excess number of deaths from diabetes mellitus was found in men (SMR = 79) or women (SMR = 123). However, in the USA diabetes is a condition with a low case fatality rate, so an association with diabetes mellitus may not be observed in a mortality study.

In order to investigate the role of occupational arsenic exposure in the pathogenesis of diabetes mellitus, Rahman & Axelson (1995) conducted a small (12 exposed cases) case–referent study in the Rönnskär cohort (Axelson et al., 1978). An elevated risk of diabetes mellitus associated with arsenic exposure was observed: OR 2.0, 4.2, and 7.0 for the exposure categories $<< 0.5$, < 0.5 and > 0.5 mg/m^3; confidence intervals for all included unity, and the trend was of borderline significance ($p = 0.03$).

---

[2]   After the Task Group meeting, the secretariat became aware of a further study reporting an association between exposure to arsenic in drinking water and diabetes mellitus (Tseng et al., 2000a,b)

Table 31. Diabetes mellitus among As-exposed populations

Study design	Study population	Source and level of As exposure	Health effects, metric of exposure and measure of association		Comments	Reference
Cross-sectional	891 adult residents in BFD-endemic area in Taiwan	drinking-water <1.14 mg/litre, decreasing with progressive use of reservoir water starting in 1956	diabetes mellitus exposure category (mg/litre · yr) 0 0.1–15.0 >15 significant at $p \leq 0.05$ level	OR (95% CI) 1.0 6.6 (0.9, 51.0) 10.0 (1.3, 77.9) *	no exposure index for 19% of subjects– excluded from analysis. water analyses by the Natelson method from Kuo (1968) used diabetes mellitus status established by subject receiving regular insulin or sulfonylurea treatment or glucose tolerance test. ORs adjusted for age, sex, BMI and physical activity level.	Lai et al. (1994)
Ecological	4 townships in BFD-endemic area in Taiwan, mortality in 1971–1994, compared to local and national rates	drinking-water <1.14 mg/litre, decreasing with progressive use of reservoir water starting in 1956	diabetes mellitus SMR for females and males combined 135 (116–155) compared to local rates 114 (98–131) compared to national rates		national statistics used to calculate expected deaths. 99% of causes of deaths based on physician diagnosis	Tsai et al. (1999)

260

Table 31 (contd.)

Cohort		Health measure SMR (95% CI)	
4058 members of the Church of Jesus Christ of Latter Day Saints in Millard County, Utah	range of exposure 3.5–620 µg/litre; median exposures 14–166 µg/litre depending on location	Diabetes mellitus: Males = 79 (48, 122) Females = 123 (86, 171) Significant at $p \leq 0.05$ level	existing and historic As concentrations used. death rates for Utah 1960–1992 were used to generate the expected deaths. no indication of exposure–response relationship for any of the vascular health effects. exposure for the highest exposure group likely to be overestimated because of introduction of low-As water into one community, which was not considered in the analysis	Lewis et al. (1999)

Table 31 (contd.)

Study design	Study population	Source and level of As exposure	Health effects, metric of exposure and measure of association		Comments	Reference
Cross-sectional	163 keratotic subjects over 30 yr of age from Bangladesh districts. Comparison group was 854 subjects from Dhaka.		Diabetes mellitus Mantel–Haenszel weighted prevalence ratio		only keratotic subjects were recruited as exposed comparison group is described as unexposed people, recruited in a door-to-door survey in Dhaka diabetes mellitus status established by glucose tolerance test. water testing at one point in time only; analytical procedure not given	Rahman et al. (1998)
			Exp. category (TWA mg/litre)	Prev. ratio (CI)		
			< 0.5	2.6 (1.2, 5.7)		
			0.5–1.0	3.9 (1.8, 8.2)		
			> 1.0	8.8 (2.7, 28.4)		

Table 31 (contd.)

Study design	Study population	Source and level of As exposure	Health effects, metric of exposure and measure of association		Comments	Reference
Cross-sectional	163 keratotic subjects over 30 yr of age from Bangladesh districts. Comparison group was 854 subjects from Dhaka.		Diabetes mellitus Mantel–Haenszel weighted prevalence ratio		only keratotic subjects were recruited as exposed	Rahman et al. (1998)
			Exp. category (TWA mg/litre)	Prev. ratio (CI)	comparison group is described as unexposed people, recruited in a door-to-door survey in Dhaka	
			< 0.5	2.6 (1.2, 5.7)	diabetes mellitus status established by glucose tolerance test.	
			0.5–1.0	3.9 (1.8, 8.2)	water testing at one point in time only; analy- tical procedure not given	
			> 1.0	8.8 (2.7, 28.4)		

Table 31 (contd.)

Study design	Study population	Source and level of As exposure	Health effects, metric of exposure and measure of association	Comments	Reference
Case–referent	240 cases of diabetes and 2216 referents living in an area of glass industry	exposure to airborne As	OR for those occupationally exposed 1.4 (CI 0.9–2.1)		Rahman et al. (1996)
Cross-sectional	32 As-exposed workers and 26 non-exposed referents	average urinary As 35.9 for the exposed and 14.5 µmol/mol creatinine for the referents	average glycosylated Hb 5.7% among the exposed, and 4.4 % among the referents, $p < 0.001$	The exposed group included taxidermists, garden fence makers, week-end cottage constructors, wood impregnators, electric pole impregnators, new house constructors	Jensen & Hansen (1998)

Rahman et al. (1996) also conducted a case–referent analysis in the glass industry area in Sweden on 240 individuals who had diabetes as the underlying or contributing cause of death on the death certificate, and 2216 controls who died of other causes during 1950–1982. They found a slightly elevated risk of dying from diabetes among glasswork employees considered to be exposed to arsenic on the basis of their occupational histories (OR 1.4, 95% CI 0.9–2.2).

In a study among a group of 40 Danish workers exposed to arsenic (average urinary arsenic level 22.3 µmol/mol creatinine; twice that of the referents), the blood concentration of glycosylated haemoglobin (used as a marker of long-term blood glucose level) was 25% higher ($p < 0.001$) than that among referents, and showed a significant trend with increasing urinary arsenic concentrations (Jensen & Hansen, 1998).

8.6 Neurotoxicity

Polyneuropathy is often among the sequelae of an acute oral arsenic poisoning (Heyman et al., 1956), and was reported as early as the 18th century (for references see Geyer, 1898). Sensory nerve (median, ulnar and sural nerves) conduction is often affected more (absent or low action potentials) than the motor nerves (primarily low amplitude in action potential, slowing or prolonged nerve conduction velocity) (Murphy et al., 1981; Oh, 1991). The conduction velocity may decrease during several weeks after short-term exposure, but conduction velocity changes were observed 3 days after a large dose (Ramirez-Campos et al., 1998). If the patient survives, the electrophysiological changes show a slow recovery. Histological examination of the nerves involved typically reveals wallerian degeneration (Murphy et al., 1981; Goebel et al., 1990; Oh, 1991). More rarely, arsenic intoxication may also lead into prolonged toxic encephalopathy (Freeman & Couch, 1956; Fincher & Koerker, 1987). Cases with neuropsychological and neurophysiological damage after occupational exposure to arsenic have also been described (Becket et al., 1986; Bolla-Wilson & Bleecker, 1987; Morton & Caron, 1989).

Few epidemiological studies have investigated whether a lower level long-term exposure to arsenic may also lead to neurotoxicity.

Hindmarsh et al. (1977) assessed the effect of drinking-water with high arsenic concentration on electromyographic abnormalities. Out of 110 persons exposed to elevated arsenic concentrations in drinking-water, 32 were studied using electromyography (EMG), and compared to 12 non-exposed referents. There was a positive relationship between EMG abnormalities and well and hair arsenic concentrations. Among those using water with > 1 mg As/litre, the frequency of EMG abnormalities was 50%.

In a cross-sectional study on 211 people in Fairbanks, Alaska, "brief clinical investigation" of peripheral nervous system function (not specified) did not reveal neuropathy related to estimated daily arsenic dose from drinking-water; the latter was estimated from well-water arsenic concentration, and reported use of well-water and bottled water. The estimated average arsenic exposure in the highest exposure category was 0.3 mg/day (Harrington et al., 1978).

Workers at a copper-smelting plant exposed to As_2O_3 were examined for peripheral neuropathy (Feldman et al., 1979). A total of 70 factory workers and 41 non-arsenic workers were evaluated. Among the exposed workers there was an association between exposure to arsenic (quantitated in urine, hair and nails) and a higher number of peripheral neuropathological disorders (sensory and motor neuropathy) and electrophysiological abnormalities (reduced nerve conduction velocity and amplitude measurements). Of the arsenic-exposed workers 30% had sensory and 13% motor neuropathy, compared to 12% of the non-exposed group with sensory and none with motor neuropathy.

Power station workers were exposed to fuel coal with a high content of arsenic (Buchancova et al., 1998). The author describes a variety of clinical symptoms potentially associated with arsenic: sensory and motor polyneuropathy, pseudoneurasthenic syndrome, toxic encephalopathy and nasal septum perforation. However, workers were also exposed to manganese and lead, both known neurotoxic chemicals.

A girl who was accidentally exposed to copper acetoarsenite (Paris green) used as a pesticide had severe clinical signs of arsenic poisoning including Mees' bands in fingers and toenails, encephalopathy, epileptic seizures and demyelinating polyneuropathy with a severe motor deficit (Brouwer et al., 1992). Although her family was also exposed, as indicated by elevated arsenic in their urine, they remained asymptomatic. Further analysis indicated that she was deficient in 5,10-methylene-tetrahydrofolate reductase (MTHFR), which is involved in the conversion of 5,10-methylene tetrahydrofolate to 5-methyl tetrahydrofolate and is important for myelin biosynthesis. MTFH deficiency may have led to decreased synthesis of S-adenosylmethione (SAM), which is a methyl donor for arsenic methylation, and thus to increased toxicity of arsenic.

8.7 Cancer

8.7.1 Exposure via inhalation

Investigation into elevated cancer risk amongst copper-smelter workers was initiated during the early 1960s. The emphasis in the studies on inhalation exposure to arsenic and cancer has been in respiratory cancer, mainly lung cancer (Table 32).

8.7.1.1 Lung cancer

a) Non-ferrous smelters

There are three occupational cohorts in which exposure assessments allow evaluation of the relationship between exposure to arsenic and lung cancer, namely those of the copper smelters in Tacoma, Washington (USA), Anaconda, Montana (USA), and Rönnskär (Sweden). These studies are described below in more detail, and other studies on the relationship between arsenic exposure and cancer are presented in a more condensed form.

Results from the Tacoma copper smelter have been published in a series of papers (Pinto & Bennett, 1963; Pinto et al., 1977, 1978; Enterline & Marsh, 1980, 1982; Enterline et al., 1987a, 1995). In the most recent update (Enterline et al., 1995), the vital status of 2802 men who worked at the smelter for a year or more during the period 1940–1964 was followed for the period 1941–1986; exposure assessment was extended to 1984. The vital status was determined for 98.5% of the cohort, and of the 1583 known deaths, death certificates were obtained for 96.6%. The expected numbers of deaths for various diseases were calculated from age- and time-

Table 32. Cancer risk in studies on occupationally exposed populations

Study design	Study population	Source and level of As exposure; other exposures	Metric of exposure, measure of association Lung / respiratory cancer			Cancer at other sites; comments	Reference
			Cum. exp. cases	Mean cum.exp.	SMR		
Cohort	2802 Tacoma Smelter workers who worked ≥1 yr in 1940–1964, followed for vital status 1941–1986	cumulative exposure in mg/m^3 · yr from < 0.75 to > 45	<0.75–0.41 0.75–1.31 2.0–2.93 4.0–5.71 8.0–12.3 20–28.3 45–59.0 ** = p<0.01;	154 176** 210** 212** 252** 284** 316* * = p < 0.05	22 30 36 36 39 20 5	large intestine SMR = 162 (p < 0.01); rectum SMR = 176 NS; kidney SMR = 164 NS; liver SMR 21 (1 case) exposure assessment based on urinary As measurements from the equation air As = 0.0064 × (urinary As)$^{1.942}$	Enterline et al. (1995)

Table 32 (contd.)

Cohort		Cum. exp. decile	RR	CI		Lubin et al. (2000)
8014 white smelter workers who worked ≥1 year in the Anaconda smelter before 1957; follow-up for vital status, 1938–1989	estimated exposure 0.29, 0.58 and 11.3 mg/m^{-3} in areas of light, medium, and heavy exposure; maximum follow-up 52 years	1	1.00		cancer of the digestive organs and peritoneum SMR = 94 (CI 83–107); stomach 116 (CI 91–149); kidney 57 (33–101); skin 53 (26–106); liver 82 (50–133); bladder 128 (93–176) Cumulative exposure calculated as 0.29 × L + 0.58 × M + 11.3 × H, where L, M, and H are years worked in areas where exposure was considered to be light (or unknown), medium, or heavy, respectively. An earlier study estimated these exposures to have been 0.38, 7.03, and 61.99, respectively (Lee-Feldstein, 1989) (Poisson regression analysis with an internal reference group)	
		2	1.0	0.6–1.8		
		3	1.0	0.6–1.1		
		4	2.1	1.2–3.9		
		5	2.6	1.4–4.6		
		6	2.4	1.3–4.3		
		7	1.7	1.0–3.2		
		8	3.4	1.9–6.1		
		9	2.7	1.5–5.0		
		10	4.0	2.2–7.1		

Table 32 (contd.)

Study design	Study population	Source and level of As exposure; other exposures	Metric of exposure, measure of association Lung / respiratory cancer			Cancer at other sites; comments	Reference
			Cum. exp. (mg/m · yr)	SMR	95%CI		
Cohort	3916 smelter workers who worked ≥ 3 mo in the Rönnskär smelter between 1928 and 1967 and were followed for vital status 1947–1981	airborne exposure in a smelter	< 0.25	271	148; 454	digestive organs SMR = 117 (130 cases); urogenital organs SMR = 109 (124 cases). No statistical analysis reported. measurement data on exposure generally available since 1951; for earlier times, estimates were based on production figures, and for the period 1945–1951 on a few measurements	Järup et al. (1989); Sandström et al. (1989)
			0.25–< 1	360	192; 615		
			1–< 5	238	139; 382		
			5–< 15	338	189; 558		
			15–< 50	461	309; 662		
			50–< 100	728	267; 1585		
			100 +	1137	588; 1986		

Table 32 (contd).

Cohort	839 copper Smelter workers	lung cancer SMR 1189**	stomach cancer SMR 68 (10 cases); large intestine excl. rectum SMR 508 (3 cases)	Tokudome & Kuratsune (1976)
Cohort	1974 gold-miners; 25 551 person-years	airborne As, radon, diesel exhaust	respiratory cancer SMR 140 **	stomach SMR 40 (4 cases); colorectal SMR 80 (9 cases); bladder SMR 60 (2 cases)	Armstrong et al. (1979)
Cohort	2228 metal refinery workers in 8 refineries	airborne As, approx. 70 $\mu g/m^3$ in the smelter with highest exposure	lung cancer SMR 211; excess limited to 1 refinery out of 8 studied.	no other cancer sites reported estimated exposure to sulfur dioxide not related to lung cancer mortality	Enterline et al. (1987b)

271

Table 32 (contd.)

Study design	Study population	Source and level of As exposure; other exposures	Metric of exposure, measure of association Lung / respiratory cancer	Cancer at other sites; comments	Reference
Cohort	5408 gold-miners	As, radon, diesel exhaust	lung cancer SMR 140, 95% CI 122–159 for workers who had not mined uranium or nickel, and had started work at a gold-mine before 1946	no other sites reported	Kusiak et al. (1991, 1993)
Cohort	1330 men who had worked ≥3 mo in gold-mine and refinery after 1954, followed for vital status 1972–1987	As, radon, silica	lung cancer SMR 213 for miners	stomach cancer SMR 115 (3 cases), kidney cancer SMR 0 (0.79 expected); bladder cancer SMR 74 (1 case)	Simonato et al. (1994)
Cohort	611 pesticide manufacturers	As and other pesticides	lung cancer SMR 225 (CI 156–312)	digest. syst. SMR 106 (58–117); bladder SMR 72 (1–403); kidney SMR 0 (0–231)	Sobel et al. (1988)

specific rates for white males in the state of Washington (all studied workers were males and "nearly all" were white). No significant differences were observed in expected numbers calculated on the basis of county-specific rates.

Exposure to arsenic was estimated from departmental measurements of arsenic in air from the annual company reports, available since 1938 (the factory began operation in 1913), and from measurements of urinary arsenic since 1948. Before 1971, the air arsenic concentrations came from "spot" samples and "tape" samples (apparently surface sampling), thereafter from personal air sampling. An empirical relationship between air and urinary arsenic was developed, based on 28 pairs of arithmetic mean arsenic concentrations in air in 11 departments and geometric mean concentrations of arsenic in the urine of workers in that department:

$$\text{Air arsenic} = 0.0064 \times (\text{urine arsenic})^{1.942}$$

Using this equation, urinary arsenic concentrations were transformed into air data for departments for which no air data were available. For exposure before the year 1938, data from that year were used. For each worker, cumulative exposure in $(\mu g/m^3) \cdot$ year was then calculated, on the basis of individual history of work in different departments (Enterline et al., 1987a).

An increase in lung cancer risk related to cumulative arsenic exposure was observed, which reached an SMR of 316 in the highest exposure category (Table 32). When the SMR is plotted against cumulative arsenic exposure on an arithmetic exposure scale (Fig. 4), relatively larger increments in respiratory cancer risk are observed at low exposure levels, i.e. the exposure–response curve is concave downward. This had already been found in the previous report from the same cohort, where the follow-up time was 10 years shorter (Enterline et al., 1987a). The lung cancer SMR was 188 in the group with < 20 years after the first exposure, and 217 among those with > 20 years since first exposure, indicating a rather short latency period. However, when lung cancer SMR was plotted against measured urinary arsenic concentrations, a linear relationship was observed (Enterline et al., 1987a).

An elevated risk of lung cancer among workers in the Anaconda copper smelter in Montana was originally reported by Lee & Fraumeni (1969). Updates and further cohort and nested case–referent analyses were published later (Lubin et al., 1981; Welch et al., 1982; Brown & Chu, 1983a,b; Lee-Feldstein, 1983, 1986, 1989; Lubin et al., 2000).

The study population of the latest cohort update (Lubin et al., 2000) consisted of 8014 white males, who were employed for ≥12 months before 1957. Their vital status was followed from 1 January 1938 to 31 December 1987; a total of 4930 (63%) were deceased, including 446 from respiratory cancer. The vital status at the end of the follow-up period was not known for 1175 workers (15%), and they were assumed to be alive at the end of the study period (except the 81 workers born before 1900, who were assumed to have died). Industrial hygiene data (702 measurements), collected between 1943 and 1958, were used to categorize each work site to an exposure category on a scale 1–10, and work areas were then grouped as representing "light", "medium" or "heavy" exposure. Based in addition on estimates of workers' daily exposure time, time-weighted average (TWA) exposures for each category were created, and were considered to be 0.29, 0.58 and 11.3 mg/m^3 arsenic for the "light", "medium", and "heavy" exposure category (Lubin et al., 2000). It should be noted that in earlier reports on this cohort the TWA exposure estimates used were different, notably for the "heavy" exposure category (0.38, 7.03, and 61.99 mg/m^3, respectively). For each worker, the cumulative exposure was estimated from the time of working in different work areas. The authors note that industrial hygiene measurements were actually available for less than half of the 29 working areas; no data were collected before 1943, and the measurements were often performed when an industrial hygiene control measure was instituted or after a process change occurred, and most often in areas where arsenic was thought to be a hazard. The locations for sampling were not randomly selected.

Altogether 446 deaths from respiratory cancer (SMR 155; CI 141–170) were observed. A trend of increasing risk with increasing estimated exposure was seen (Table 32); the risk increased linearly with time of employment in each exposure category.

The elevated lung cancer incidence among workers of the Rönnskär smelter in northern Sweden was originally reported in a population-based case–referent study in St Örjan parish in 1978 (Axelson et al., 1978). Since then, studies using both cohort and case–referent approaches have been published (Wall, 1980; Pershagen et al., 1981, 1987; Järup et al., 1989; Sandström et al., 1989; Järup & Pershagen, 1991; Sandström & Wall, 1993). The cohort consisted of 3916 male smelter workers, who had worked for at least 3 months at the smelter between 1928 and 1967. The vital status of all but 15 (0.4%) of them was verified. Mortality of different causes, as defined on death certificates, was compared to local rates. Reference rates were not available for the period before 1951, but the contribution of deaths during this period (89 out of a total of 1275, i.e. 7%) was minor. Air concentrations of arsenic were estimated by the factory industrial hygienists. The first measurements were carried out in 1945, and from 1951 exposure data were more generally available; production figures were used to extrapolate exposures before 1951. Each work site was characterized by an exposure level during three consecutive time periods, and the workers' cumulative exposure was assessed on the basis of their working history in these different work sites.

The SMRs were very similar whether they were calculated with no latency, 10 years minimum latency or 10 years minimum latency with exposure lagged 5 years. A dose-dependent increase in the mortality from lung cancer was observed (Table 32), and a statistically significantly increased risk was observed even in the lowest exposure category, $< 0.25 \ (\text{mg/m}^3) \cdot$ year. A sensitivity analysis showed that the SMRs were fairly robust, particularly among the workers with low and medium exposure (Järup, 1992). Even when the exposures before 1940 were reduced dramatically (assuming there was a large overestimation of the early exposures), these SMRs changed only marginally. As expected, the SMRs in the highest exposure group increased as the early exposures were reduced. An overestimation of the early exposures would thus tend to decrease the strength of the exposure–response association. However, in a nested case–referent study on the interaction between smoking and arsenic exposure as cancer-causing agents (Järup & Pershagen, 1991), little increased risk of lung cancer due to arsenic exposure was observed among smokers or non-smokers in exposure categories $< 15 \ (\text{mg/m}^3) \cdot$ year. Little difference was observed in the SMRs for workers hired before 1940, in 1940–1949, or after 1949, when the estimated level of exposure was similar, meaning that a longer follow-up did not increase the apparent risk. In most subcohorts, and in the total cohort, the mortality increased with increasing average intensity of exposure, but no clear-cut trend was observed for the duration of exposure. Exposure to sulfur dioxide was also assessed. The lung cancer risk was elevated in all groups exposed to sulfur dioxide, but there was no exposure–response with the estimated cumulative sulfur dioxide exposure.

In a cancer incidence study (Sandström et al., 1989), partly overlapping with the mortality study, the cancer risk of the smelter workers over a moving 5-year period was observed to decrease steadily from 1976–1979 to 1980–1984. Further follow-up of an expanded Rönnskär cohort (n = 6 334) by Sandström & Wall (1992) showed a decreasing trend in lung cancer incidence and mortality, but there was still an elevated lung cancer incidence among the workers when compared with Swedish men.

A very high excess of lung cancer (SMR 2500; 10 observed and 0.40 expected cases in the heavy exposure category), which was related to duration and level of exposure, was observed in the copper smelter of a Japanese metal refinery (Tokudome & Kuratsune, 1976); the study was prompted by an earlier case–referent study that demonstrated an excess lung cancer rate among copper-smelter workers (Kuratsune et al., 1974). There was an approximately 3-fold increase in the relative death rate from lung cancer among employees of a copper smelter in Utah, in comparison to workers of the same company not employed in the smelter (mainly mine and concentrator workers), and also in comparison to Utah state figures (Rencher et al., 1977). The risk was related to all estimated exposure parameters (cumulative exposure to arsenic, sulfuric acid, lead and copper), and was similar for smokers and non-smokers. This refinery was a part of a cohort study in eight copper smelters (Enterline et al., 1987b), the SMR for respiratory cancer < 20 years since first exposure was 170 (11 deaths), and ≥ 20 years 108 (39 deaths) (reported in Enterline et al., 1995). In this study, the only smelter with an appreciable exposure to arsenic was the Utah one, and this was the only one with a statistically significant excess in lung cancer.

b)   Pesticide manufacture and application

Ott et al. (1974) conducted a proportionate mortality study of decedents who had worked at a factory producing arsenical pesticides, mainly lead arsenate, calcium arsenate, copper aceto-arsenite and magnesium arsenate. The cause of death of 173 workers who had worked at least 1 day in jobs with presumed arsenic exposure was compared to that of 1809 decedents (age- and calendar-year-adjusted) from the same factory, with no exposure to arsenic or asbestos. The exposure of the workers was analysed from a job exposure matrix covering the working history. The proportionate mortality ratio (PMR) for lung cancer increased with estimated exposure, from a PMR of 200 at an exposure level of 1–1.9 $(mg/m^3) \cdot$ month to a PMR of 700 at the highest cumulative exposure group $\geq 96 \, (mg/m^3) \cdot$ month. Ott et al. (1974) also conducted a cohort study at the pesticide plant. The cohort was expanded and updated through December 1982 (Sobel et al., 1988) to include 611 workers altogether; the mortality was compared to age- and calendar-time standardized data on US white males. A significant excess of lung cancer mortality was observed (35 observed vs. 15.6 expected cases; SMR 225, 95% CI 156–312). The small number of deaths made analyses by duration and latency difficult; analysis by exposure level or cumulative exposure was not reported.

In a cohort study of pesticide manufacturing workers in Baltimore, the vital status of 1050 men and 343 women was followed from 1946 through 1977 (Mabuchi et al., 1979, 1980). The vital status was determined for 86.9% of men and 66.8% of women; the non-traced subjects were counted as being alive at the time of ending the follow-up. Cause-specific mortality was compared to that of Baltimore city whites, age- and calendar time adjusted, and 23 lung cancer deaths were identified, which represents an excess lung cancer mortality (SMR 168 based on Baltimore City whites, or 265 based on US whites; $p < 0.05$ for both). There was an exposure–response with presumed cumulative exposure (no relevant measurement data on exposure were available), the SMR reaching 2750 in the highest exposure category (3 lung cancer deaths). No exposure–response was observed with presumed cumulative exposure to non-arsenical pesticides.

In an autopsy series of 163 winegrowers from the Moselle area (Lüchtrath, 1983), 130 cases of cancer in internal organs were observed. Of these, 108 were lung cancers. In an age- and sex-adjusted control group of 163 people, there were 23 malignant tumours, out of which 14 were lung tumours. Exposure to arsenic was considered to be by inhalation of arsenic-containing insecticide, but to a much larger extent, by drinking arsenic-contaminated "Haustrunk" (a wine substitute made from already pressed grapes), which was estimated to lead to a daily intake of about 3–30 mg arsenic.

In 1938 a cohort of 1231 people living in the Wenatchee area in Washington, where lead arsenate was extensively used in orchards, was identified to study the health effects of this exposure. The mortality experience of this cohort was reported by Nelson et al. (1973), Wicklund et al. (1988) and Tollestrup et al. (1995). No difference in lung cancer mortality was observed between orchardists exposed to arsenical insecticides and consumers who were not significantly exposed to arsenicals (hazard ratio 0.59, 95% CI 0.19–1.85) (Tollestrup et al., 1995). It is likely that the overall exposure to arsenic for orchardists was low. A case–control study included all white male orchardists (n = 155) who died in Washington state between 1968 and 1980 from respiratory cancer, using orchardists who died of other causes as controls (n = 155) (Wicklund et al., 1988). Lead arsenate exposure did not differ between cases and controls, and smoking habits were similar.

c) Miners and other

In a cohort study on tin-miners in the UK (Hodgson & Jones, 1990), 13 workers had worked in arsenic calcining. Three of them had died of cancer of the trachea, bronchus, lung or pleura (0.55 expected, SMR 550, $p < 0.05$), and two of stomach cancer (0.2 expected, SMR 890, $p < 0.05$). A very high lung cancer mortality has been demonstrated among tin-mine workers exposed to arsenic and radon in Yunnan, China (Taylor et al., 1989; Qiao et al., 1997). The lung cancer risk increased with estimated cumulative exposure to arsenic (Qiao et al., 1997). A 2-fold excess (SMR 213; 95% CI 148–296) in lung cancer mortality was observed among workers in a gold-mine and refinery in France, mainly among workers with a history of exposure to arsenic, diesel exhaust, radon and silica. There was little change in the relative risk with length of employment, and the risk was similar among refinery workers and miners (Simonato et al., 1994). An exposure-related increase in the lung cancer mortality was also observed among gold-miners in Ontario, exposed to arsenic and radon daughters (Kusiak et al., 1991,

1993). Similarly, lung cancer mortality among Australian gold-miners was higher than that expected from the experience of all Western Australian men (SMR 140, 59 observed and 40.8 expected cases, $p < 0.01$). The gold-miners were exposed to arsenic, radon daughters and silica, and apparently smoked more than the referent population (Armstrong et al., 1979).

Female hat-makers, probably exposed to arsenic while making felt hats, had an elevated risk of lung cancer (6 cases but no controls were hat-makers) in a case–referent study (376 cases with 892 controls) on occupational risk factors of lung cancer in Italy (Buiatti et al., 1985).

A cohort mortality study of workers in a Russian fertilizer plant, including 2039 men and 2957 women, showed an excess mortality from all cancers combined (SMR 143) and lung cancer (SMR 186) for the male production workers (Bulbulyan et al., 1996). Excess mortality from all cancers and stomach cancer was found for the workers with the highest average exposure to arsenic, and excess lung cancer mortality was attributed to exposure to arsenic.

d) Interactions of arsenic exposure and tobacco smoking

Hertz-Picciotto et al. (1992) assessed the joint effect of smoking and arsenic exposure on the basis of published case–control and cohort studies on arsenic-exposed populations. There were six studies on two overlapping smelter populations, where a direct evaluation of the interaction could be assessed (Rencher et al., 1977; Pershagen et al., 1981; Enterline, 1983; Pershagen, 1985; Enterline et al., 1987b; Järup & Pershagen, 1991). The excess relative risk was assessed by:

$$ERR = \frac{R_{AB} - R_{aB} - R_{Ab} + R_{ab}}{R_{AB}},$$

where R is the ratio of cases to referents (case–referent studies) or the absolute risk (cohort studies), and the subscripts AB/ab denote the two exposures, present (upper case), or absent (lower case). In all the studies, the ERR exceed the simple additive effect by 30–54%, indicating a synergism between smoking and arsenic exposure.

e)   Lung cancer in the vicinity of arsenic-emitting industries

Mortality rates for lung cancer for white men and women in 1950–1960 were significantly higher in US counties with copper, lead, or zinc smelting and refining industries (Blot & Fraumeni, 1975), and a 2-fold mortality of lung cancer was observed among people with residence near a zinc smelter, and in areas with high topsoil concentrations of arsenic, cadmium, copper, lead and manganese (Brown et al., 1984). A slightly higher mortality of lung cancer was observed among male residents of Rouyn-Noranda, a community with a copper smelter, than among male residents of a referent community (SMR 150) or Quebec (Canada) as a whole (SMR 120); no such difference was observed among women (only 7 exposed cases) (Cordier et al., 1983). Although the lung cancer mortality between 1935 and 1969 in women living in three geographically defined areas in the vicinity of an arsenic-emitting smelter was not different from that expected from nationwide expected figures, there was a positive trend following predicted exposure levels (Frost et al., 1987). No difference was observed between the frequency of lung cancer and that of other cancers in the vicinity of non-ferrous smelters (a lead-zinc smelter and 10 copper smelters) in the USA (Greaves et al., 1981).

The lung cancer mortality among people living in the vicinity of a copper smelter in Rönnskär (Sweden) was studied in a cohort and a case–referent study (Pershagen et al., 1977; Pershagen, 1985). In the cohort study, a significantly higher mortality from lung cancer was observed among men living close to the smelter than among men in a reference area (Pershagen et al., 1977). The difference disappeared, however, when men working in the smelter were excluded. In the case–referent study, the odds ratio (OR) for residence in the exposed area was 2.0 (95% CI 1.2–3.4), and it was not explained by occupation in the smelter, or by differences in smoking habits (Pershagen, 1985). Lung cancer mortality was higher in men living in the vicinity of a factory producing arsenical pesticides in Baltimore (Maryland, USA) (Matanoski et al., 1981). No association between the distance from a smelter and lung cancer risk was observed in a case–referent study where 575 lung cancer cases were compared with 1490 breast and prostate cases collected from 1944 to 1973 in El Paso, Texas (USA), where a smelter had been operating since 1887 (Rom et al., 1982).

In a study of lung cancer mortality in 6 Arizona (USA) copper smelter towns, using 185 lung cancer cases and 2 matched controls per case from decedent residents during 1979–1990, information on lifetime residential, occupational, and smoking history was obtained (Marsh et al., 1997, 1998). Historical environmental exposures to smelter emissions were linked with residential histories to derive individual profiles of residential exposure. Occupational histories were characterized by potential exposure to smelter emissions, asbestos and ionizing radiation. No statistically significant associations were observed between lung cancer risk and residential exposure to smelter emissions, when adjustment for potential confounding factors (gender, Hispanic ethnicity, and smoking) were made. The authors concluded that the study provided little evidence of a positive association between lung cancer mortality and residential exposure to smelter emissions.

A Chinese case–control study including 1249 lung cancer patients and 1345 population-based controls showed 3-fold elevated risks among smelter workers (Xu et al., 1989, 1991). Soil levels of arsenic rose with increasing proximity to the Shenyang copper smelter, and, after controlling for smoking and work experience in the smelter, elevated risks of lung cancer were found among men, but not women, living within 1 km of its central stacks.

It has been noted that epidemiological studies designed to detect lung cancer risk and other health effects in communities surrounding arsenic-producing copper smelters usually have insufficient statistical power to detect the small increases in risk that may occur (Hughes et al., 1988). Most such studies have little power to detect relative risks under 2.0. The authors argue that these studies may be a good and economical first investigation but, because of the lack of statistical power, null findings do not rule out the possibility of excess risks that may be significant from a public health viewpoint.

f)   Exposure–response relationships

Sufficient information on the levels of exposure to ensure reliable assessment of the exposure–response relationships can be found only in the three copper smelter cohorts: Tacoma, Anaconda and Rönnskär. In all, there was an increase in lung cancer risk with increasing exposure (Table 32, Fig. 4). The risk seems to increase

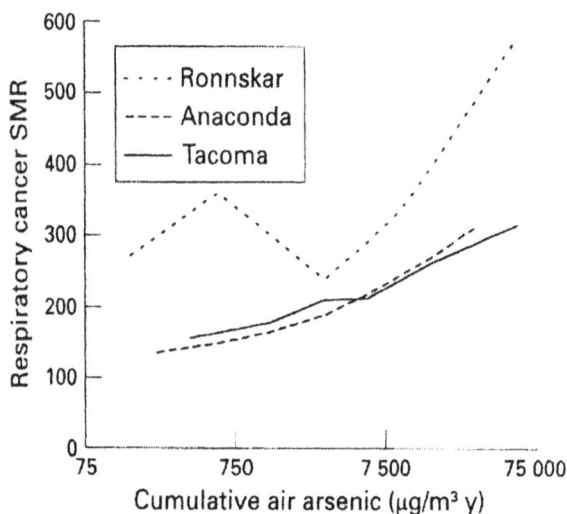

Fig. 4. Respiratory cancer risk in the three copper smelter cohorts
(Enterline et al., 1995)

more rapidly with dose at low cumulative dose levels than at higher exposures, and the general form of the exposure–response is rather similar in the three studies (Fig 4; note logarithmic scale of the x-axis). The shape of the exposure–response curve has been further analysed and discussed by Hertz-Picciotto & Smith (1993), who note that all of the studies with quantitative data are consistent with a supralinear exposure–response relationship. Neither toxicokinetic mechanisms nor confounding from age, smoking, or other workplace carcinogens that differ by exposure level appears likely to explain this curvilinearity. The authors argue that a plausible explanation may be synergism (with smoking) which varies in magnitude according to the level of arsenic exposure. Another possible explanation may be a long-term survivorship in higher-exposure jobs among the healthier, less susceptible individuals (Hertz-Picciotto & Smith, 1993; Arrighi & Hertz-Picciotto, 1996). It is also plausible that exposure estimate errors are more prominent at higher exposure levels as a result of past industrial hygiene sampling or worker protection practices (Hertz-Picciotto & Smith, 1993), which is consistent with the findings of the sensitivity analysis of the Rönnskär data (Järup, 1992), and the update of the Anaconda cohort (Lubin et al., 2000).

Analyses performed on a subcohort at the Tacoma smelter suggested that there was strong evidence of confounding by year of initial hire, with a non-linear exposure–response evident only among workers hired before 1940 (Viren & Silvers, 1999). Among workers hired after 1940, analyses showed that a linear dose–response provided a clearly superior fit.

Re-analysis of the Anaconda cohort (Lubin et al., 2000) is also in favour of a linear exposure-risk relationship, and ascribes the apparent non-linearity to overestimation of the high exposures (mainly because of use of protective devices).

32. All three main studies demonstrate a statistically significant excess risk of lung cancer at exposure levels of approximately ≥ 0.75 $(mg/m^3) \cdot$ year.

8.7.1.2 *Cancer at other sites* (Table 32)

34. Autopsy series (Roth, 1955, 1957a,b) on wine growers (exposure from wine consumption and arsenical pesticide application) have linked arsenic exposure to hepatic angiosarcoma. Among 168 people diagnosed with hepatic angiosarcoma in the USA, occupational exposure to arsenic was found in 4 cases and exposure to Fowler's solution in 6 (Falk et al., 1981a,b); among 43 cases diagnosed in the state of New York between 1958 and 1979, 6 had a possible history of occupational exposure to arsenic (Vianna et al., 1981).

Although statistically not significant, arsenic exposure was associated with an increased risk of kidney cancer in the Tacoma cohort (SMR 164, 11 observed and 6.73 expected cases) (Enterline et al., 1995) (Table 32). The kidney cancer was not in excess in the, 1938–1987 follow-up of the Anaconda cohort (SMR 57, CI 33–101) (Lubin et al., 2000). No cases of kidney cancer (1.48 expected) were observed in a French gold-miner cohort (Simonato et al., 1994) or in the US pesticide-producer cohort (1.8 expected) (Sobel et al., 1988). No significant relationship was observed between arsenic exposure and incidence of cancer in 1958–1982 of the large "urogenital organs" category in the Swedish cohort (Sandström et al., 1989).

Hill & Faning (1948) found a remarkably elevated relative skin cancer mortality among workers in a sheep-dip factory manu-facturing sodium arsenite. Among 47 autopsies of winegrowers with arsenic intoxication (exposure from wine consumption and arsenical pesticide application), 13 of the subjects had a total of 40 skin cancers (Roth, 1958). A case series on skin cancer among arsenic-exposed wine growers has been published (Thiers et al., 1967). No excess skin cancer was observed in the Anaconda cohort (SMR 53, CI 26–106) (Lubin et al, 2000).

Wong et al. (1992) compared skin cancer incidence in two counties in Montana (USA) considered potentially exposed to arsenic and two counties in the same state considered not exposed, during the period 1 January 1980 to 30 June 1986. Arsenic exposure in the counties potentially exposed to arsenic would have occurred through inhalation or ingestion of arsenic-contaminated soil and or dust resulting from the presence of an open-pit copper mine in Butte, and a copper smelter at Anaconda. Skin cancer cases were identified from surgical and biopsy specimen reports from pathologists' and hospital records and dermatologists' office records. Age-, race- and sex-specific population data for the four counties were obtained from the US Census Bureau (1980 census). The age-adjusted annual skin cancer rates were found to be higher in the control counties than in the two exposed counties. No evidence is provided by the authors as to how much arsenic exposure may have been experienced by the residents of the counties where there was "potential exposure", considerably limiting the interpretation of the study's results.

Significant relationship was observed between arsenic exposure and incidence of cancer in 1958–1982 in the large "digestive organs" category in the Swedish smelter cohort (Sandström et al., 1989). No increase in the risk of cancer of digestive organs and peritoneum was observed in the Anaconda cohort (SMR 94, CI 83–107) (Lubin et al, 2000).

No excess of stomach cancer was observed in the Japanese smelter cohort (SMR 68, 10 observed and 14.71 expected cases

(Tokudome & Kuratsune, 1976). The SMR for stomach cancer was 116 (CI 91–149) in the Anaconda smelter cohort (Lubin et al., 2000).

There was an increase in the cancer of the large intestine (SMR 162, 38 observed and 23.48 expected cases, $p < 0.01$) in the Tacoma smelter cohort (Enterline et al., 1995).

There was similarly a moderate excess of rectal cancer (SMR 176, 15 observed and 8.52 expected cases, NS) in the Tacoma cohort (Enterline et al., 1995), and in the French gold-miner cohort (SMR 280, 95% CI 113–577) (Simonato et al., 1994).

Although statistically non-significant, arsenic exposure was associated with an increased risk of cancer of the buccal cavity and pharynx in the Tacoma refinery cohort (SMR 169, 12 observed and 7.12 expected cases) (Enterline et al., 1995), but no such excess was observed in the Anaconda smelter cohort in 1964–1977 (SMR 97, CI 66–143) (Lubin et al., 2000) or in the US pesticide producer cohort (0 observed, 1.5 expected) (Sobel et al., 1988).

There was an association of arsenic exposure with cancer of the bone in the Tacoma smelter cohort (SMR 456, 5 observed and 1.10 expected cases; two of the cases apparently were not primary bone tumours) (Enterline et al., 1995).

8.7.2 Exposure via drinking-water (Table 33)

The first ecological study by Chen et al. (1985) investigated the cancer mortality in 84 communities in four townships (Peimen, Hsuechia, Putai, and Ichu) on south-western coast of Taiwan in, 1968–82, and compared the age-adjusted rates to figures of all Taiwan. During the study period, the population of Taiwan increased by ~35%, whereas that in the BFD-endemic area decreased by ~15%, mainly because of emigration. As it is mandatory to register each death using a standardized death certificate, the authors state that the statistics are "very complete". They also report that 85% of all cancer deaths except liver cancer deaths are confirmed by histological or cytological analysis.

The mortality (SMR) from lung, liver, kidney, bladder and skin cancers was substantially elevated in the BFD-endemic area; there was a moderate excess in the mortality from colon cancer, and no excess in the mortality from nasopharyngeal, oesophageal, stomach, small-intestinal or rectal cancer, or leukaemia (Chen et al., 1985).

Chen et al. (1986) also carried out a case–referent study on cancer in this population. Cases were people who died of bladder, lung or liver cancer, confirmed diagnostically either by biopsy or by other tests. Controls were selected from the same geographical areas as the cases, frequency-matched on age and sex. Structured questionnaires included history of artesian water use, socio-demographic variables, dietary and lifestyle habits, as well as medical history. Proxy interviews of relatives were used for deceased cases. The total number of bladder, lung and liver cancer cases were 70, 77 and 65, respectively (response rates of 93%, 90% and 93%), with the same controls used for all the cases ($n = 368$, 92% response rate).

The ORs of liver, lung and bladder cancer cases showed an increasing risk with increasing duration of exposure (see Table 33). The trend remained significant for the cancers of lung and bladder after controlling for age, sex, cigarette smoking, tea drinking, vegetarian habit, vegetable consumption frequency and consumption of fermented beans.

Chen et al. (1988b) followed the vital status of a cohort of 871 BFD patients from 1968 to 1984; 84 (9.6%) were lost to follow-up, for 8 the cause of death was ill-defined and for 6 there was an undefined cancer on the death certificate. The cause-specific mortality was compared to that of the whole of Taiwan, and to that of the BFD-endemic area. The SMR of cancer of the lung, liver, kidney, bladder and skin was markedly elevated among BFD patients in comparison to the Taiwanese population, and somewhat less in comparison to the population in the endemic area without BFD. An excess (SMR 381, 4 cases, $p < 0.05$) of colon cancer was observed in comparison to the national figures; which was not significant in comparison to referents from the BFD-endemic area. The SMR from stomach cancer was not significantly elevated (Chen et al., 1988b).

A further ecological study in south-western Taiwan (Wu et al., 1989) investigated the mortality in 1973–1986 from vascular diseases and cancer of residents in 42 villages in Peimen, Hsuechia,

Putai, Yensui and Hsiyang townships, where data on well-water arsenic concentrations had been measured in the early 1960s. This study covers 27 of the 84 townships studied in the first study, with the addition of two other townships, Yensui and Hsiyang. The overlap of this study with the first study cannot be fully assessed. The villages were classified into three categories on the basis of the median well-water arsenic concentrations: < 300 µg/litre, 300–590 µg/litre, and > 600 µg/litre, as determined in the 1964–1966 survey. Death certificates were used to ascertain cause of death, and person-years for the same time period were calculated on the basis of demographic reports. A further exposure–response analysis from these data is presented in a further study (Chen et al., 1992), where the lowest exposure category is divided into two (< 100 and 100–300 µg/litre). Although the follow-up time and the overall person-years of follow-up for these two analyses were identical, the numbers of deaths were larger for all sites in the later analysis: 304 vs. 268 for lung cancer, 202 vs. 174 for liver cancer, 202 vs. 181 for bladder cancer and 64 vs. 59 for kidney cancer. Furthermore, a preliminary report of the same study (Chen et al., 1988a) gives still another (lowest) risk estimate for all cancer sites (based on 1031 cancer deaths, compared to 1152 in the study by Wu et al., 1989).

The age-adjusted mortality rates from lung, liver, kidney and bladder cancer, as well as those from skin cancer, showed an association with the village median well-water arsenic concentration in both men and women. Where the number of cases allowed a meaningful interpretation, the monotonic exposure–response relationship continued at the lower exposure levels studied (< 100 and 100–300 µg/litre) in the later report (Chen et al., 1992). No significant association was observed for cancers of the nasopharynx, oesophagus, stomach, colon or uterine cervix, or leukaemia (Wu et al., 1989).

Chen & Wang (1990) performed a further ecological study on drinking-water arsenic concentration and mortality from malignant neoplasms in 1972–1983, where the unit of study was all 314 precincts and townships in Taiwan where the well arsenic concentration had been analysed in 1974–1976. Nearly all cancer deaths among the arsenic exposed in this study seem to be included in the study by Chen et al. (1985). The average precinct or township well-water arsenic concentration was taken as the indicator of

Table 33. Studies on cancer after exposure via oral route (drinking-water, unless otherwise stated). For cancer of skin, see Table 34

Study design	Study population	Source and level of As exposure	Health effects, metric of exposure and measure of association		Comments	Reference
Ecologi-cal	BFD-endemic area of Taiwan; mortality 1968–1982	drinking-water up to 1.14 mg/litre, decreasing with bringing into use of reservoir water starting in 1956	SMR (CI), males bladder 1100 (933–1267) kidney 772 (537–1007) liver 170 (151–189) colon 160 (117–203) lung: 320 (286–354)	females 2009 (1702–2316) 1119 (938–1400) 229 (192–266) 168 (126–210) 413 (360–466)	small intestine, oeso-phagus, rectum, stomach, nasopha-rynx, leukaemia, thyroid were not significantly ele-vated in males or females; population of Taiwan as the reference	Chen et al. (1985)
Case–referent	69 bladder, 76 lung, 65 liver cancer decedents in Taiwan in 1980–1982. 65 live controls matched by age and sex	<40 years of use of arte-sian water in BFD-endemic area up to 1.14 mg As per litre	OR for years of use of As-contaminated water: site 1–20 21–40 >40 none 1.0 1.3 1.7 bladder 1.0 1.1 1.5 lung 1.0 0.9 1.1 liver 1.0	4.1 (p <0.01) 3.0 (p <0.01) 2.0 (p <0.1)	deceased cancer cases; ORs adjusted for age, sex, ciga-rette smoking, tea drinking, vegetarian habit, vegetable con-sumption frequency and fermented bean consumption frequency, when the factor was significant at p < 0.1; referents from the same area	Chen et al. (1986)

Table 33 (contd.)

			SMR	national ref. rate		local ref. rate			
Cohort	cohort of 789 BFD patients (15 years and 7278 person years of follow-up)	drinking-water concentrations 350–1140 µg/litre	bladder	3880	(p< 0.001)	255	(p < 0.01)	10.6% lost to follow-up	Chen et al. (1988b)
			kidney	1953	(NS)	160	(NS)		
			prostate	1729	(p< 0.001)	268	(NS)		
			lung	1049	(p< 0.001)	284	(p < 0.01)		
			liver	466	(p< 0.001)	248	(p < 0.01)		
			colon	381	(p < 0.05)	230	(NS)		
			oesophagus	305	(NS)	222	(NS)		
			stomach	194	(NS)	202	(NS)		

			age-adjusted mortality rates per 10^5 in males by well As concentration (µg/litre)					
				<300	300–599	≥600	p	
Ecological	used data from the 1964–1966 survey of 155 wells in 42 villages and used village medians in the analysis	mortality and population data from 1973–1986 in 42 villages in Taiwan	*males:*					observed numbers of deaths smaller than in the Chen et al. (1992) study, although the person-years are identical. no significant association for leukaemia or cancer of naso-pharynx, oesophagus, stomach, colon or uterine cervix
			bladder	22.6	61.0	92.7	<0.001	Wu et al. (1989)
			kidney	8.4	18.9	25.3	<0.05	
			lung	49.2	100.7	104.8	<0.001	
			liver	47.8	67.6	86.7	<0.05	
			prostate	1.0	9.0	9.2	<0.05	
			females:					
			bladder	25.6	57.0	111.3	<0.001	
			kidney	3.4	19.4	58.0	<0.001	
			lung	36.7	60.8	122.2	<0.001	
			liver	21.4	24.2	31.8	NS	

Table 33 (contd.)

Study design	Study population	Source and level of As exposure	Health effects, metric of exposure and measure of association	Comments	Reference
Ecological	mortality from malignant neoplasms in 1972–1983 in 314 precincts and townships in Taiwan	74% of precincts had <5% wells with ≥50 µg/litre As, 15% has 5–14% and 12% had ≥15% such wells. Village Mean used in analysis	statistically significant association between As level in well-water and mortality from the cancer of the lung, liver, kidney, bladder, skin, prostate and nasopharynx after adjustment for indices of urbanization and industrialization	nearly all cancer deaths among the As-exposed included in the Chen et al. (1985) study no numerical risk estimates given.	Chen & Wang (1990)
Ecological	incident bladder cancer cases 1981–1985 identified from tumour registry in 4 BFD-endemic and 2 neighbouring counties vs. whole Taiwan	As-contaminated water in the BFD-endemic area in Taiwan	average annual age-adjusted incidence of bladder cancer per 100 000: 23.5 in the 4 counties, 4.45 in the neighbouring counties, and 2.29 in the whole of Taiwan	tumour registry not validated	Chiang et al. (1993)

Table 33 (contd.)

			cum. expos. mg/litre-yr	SMR (CI)		
Cohort	263 BFD patients and 2293 residents in Taiwan follow-up of 7 years.	cumulative As exposure for drinking-water from village median well As concentration as determined in the 1964–1966 survey	0 0.1–19.9 20+	bladder cancer 100 160 (44–560) 360* (110–1220)	lung cancer: 100 274 (69–1100) 401 (100–1612)	adjusted for age, sex, smoking, BFD; deaths not over-lapping with older studies in Taiwan; cases of BFD and referents largely from different villages	Chiou et al. (1995)
Ecologi-cal	243 Taiwanese townships—approxi-mately 11.4 million resi-dents. Incident cases of urothelial and kidney cancer, 1980–1987	As measured in over 80 000 wells from 1974–1976, in 78% of townships average As content was non-detect-able, in 91% <50 and in 99.5% <640 µg/litre	estimated rate difference per 10^5 for 1% increase in the proportion of wells in the highest exposure category (640 µg/litre): transitional cancer/bladder transitional cell/kidney transitional cell/ureter all urethral cancer	males 0.57 0.03 0.11 0.056	females 0.33 0.14 0.10 0.027	mercuric bromide method used to analyse As; smoking not included in the models as not good predictor for any cancer in this study; tumour registry not validated	Guo et al. (1997)

Table 33 (contd.)

Study design	Study population	Source and level of As exposure	Health effects, metric of exposure and measure of association	Comments	Reference
Ecological	4 townships in BFD-endemic area in Taiwan, mortality in 1971–1994, compared to local and national	drinking-water rates up to 1.14 mg/litre, decreasing with bringing into use of reservoir water starting in 1956	cancer SMRs for males and females combined, compared to local rates SMR (CI) / SMR (CI) all malignant 219 (211–228) oesophagus 167 (130–212) lung 310 (288–334) stomach 136 (117–146) bone 246 (177–334) sml intestine 210 (120–354) prostate 252 (186–334) colon 149 (120–183) bladder 892 (796–996) liver 183 (169–198) kidney 676 (546–827) nasal 300 (214–409) lymphoma 163 (123–211) laryngeal 178 (120–255) leukaemia 134 (104–170)	age- and sex-specific mortality rates based on population data from Ministry of Interior, deaths from computer database on deaths; 99% of causes of death based on physician diagnosis; all cancers confirmed by pathological examination; overlaps with earlier Taiwanese studies	Tsai et al. (1999)
Ecological	mortality and population data from, 1973–1986 in 42 villages in Taiwan	drinking-water concentrations 350–1140 µg/litre	SMRs men and women well-water As (µg/l) bladder cancer lung cancer <50 1002 156 50–100 415 143 100–200 1047 243 200–300 766 308 300–400 744 197 400–500 2968 365 500–600 1490 332 600+ 3270 514	used data from the 1964–1966 survey of 155 wells in 42 villages and used village medians in the analysis	Morales et al. (2000)

Table 33 (contd.)

			Results	Comments	Reference
Ecological	residents in Cordoba vs. rest of Argentina	in the high-exposure group, in two selected towns, 42/61 and 49/57 measurements ≥ 40 µg/litre; highest measured concentration 533 µg/litre	SMRs (95% CI by exposure group) *males* low intermediate high bladder 80 (66–096) 128 (105–153) 214 (178–253) lung 92 (85–098) 154 (144–164) 177 (163–190) kidney 87 (66–110) 133 (102–168) 157 (117–205) *females* bladder 122 (86–167) 139 (93–199) 182 (119–264) lung 124 (106–142) 134 (112–158) 216 (183–252) kidney 100 (71–137) 136 (94–189) 181 (119–264)	no smoking data, but no difference in COPD, used as surrogate, between exposure groups; cancer of liver, stomach or skin not significantly related to As exposure	Hopenhayn-Rich et al. (1996c, 1998)
Ecological	Region 2 (high exposure) Northern Chile compared to Region 8 (low exposure); mortality 1952–1990	drinking-water As concentration in 1950–1992 ND up to 860 µg/litre in different locations in Region 2; average concentration <200 before 1958, >500 in 1959–1977, <100 thereafter	cancer mortality rate ratio CI all cancer 1.2 1.17–1.21 lung 5.6 5.3–6.3 bladder 6.7 5.9–7.7 kidney 2.7 2.4–3.1 larynx 3.2 2.7–4.0 liver 1.1 1.0–1.2	air levels of As were measured in some locations and concentrations up to 2.7 µg/m³ were observed at Chuquichamata, a copper smelter area in Region 2	Rivara et al. (1997)

Table 33 (contd.)

Study design	Study population	Source and level of As exposure	Health effects, metric of exposure and measure of association			Comments	Reference
Ecological	Region 2 Chile (1989–1993) compared to the rest of Chile.	drinking-water avg. 43–568 µg/litre, 1950–1994; exposure decreased over time, from 569 µg/litre (1955–1969) to 43 µg/litre (1990–1994)	SMR (CI) bladder cancer lung cancer kidney cancer liver cancer	males 600 (480–740) 380 (350–410) 160 (110–210) 110 (80–150)	females 820 (630–1050) 301 (270–370) 270 (190–380) 110 (80–150)	routinely collected As concentration measurements; population partially overlaps that of Rivara (1997)	Smith et al. (1998)
Case-control	three regions in northern Chile. 151 lung cancer cases in 1994–1996, histologically confirmed, 2 referents per case	drinking-water levels measured in 1950–1996 by water companies	Mean lifetime exposure mg/litre 0–0.01 0.01–0.029 0.03–0.049 0.05–0.199 0.2–0.40	OR (CI) 1 1.7 (0.5–5.1) 3.9 (1.2–13.4) 5.5 (2.2–13.5) 9.0 (3.6–22)		some gaps in exposure measures in some years; patients with skin lesions had higher risk of lung cancer; adjusted for age, sex and smoking status, occupational history	Ferreccio et al. (1998, 2000)

Table 33 (contd.)

Cohort	residents of Niigata, Japan (n = 467).	drinking-water conta-minated with As from a factory in 1955–1959; water analy-sed for As in 1959	SMR for ≥ 1 mg/litre compared to 0 mg/litre all causes of death 174 110–274 all cancer 482 209–1114 lung cancer 1972 434–895 000 mortality from "urinary" cancer significantly elevated, SMR 627 (CI 171–1839)	97.2% of residents in 1959 followed for vital status, 1959–1992; RR controlled for smoking and age; smoking was corre-lated with gender	Tsuda et al. (1989, 1995)
Case–control	117 newly diagnosed histologically confirmed cases of bladder cancer in Utah (USA) and 266 population referents	drinking-water; cumu-lative dose categories: <19 mg; 19–<33 mg; 33–<53 mg; ≥53 mg	OR for bladder cancer, adjusted for sex, age, smoking, exposure to chlorinated water, history of bladder infection, high risk occupation, education level, urbanization, in different cumulative exposure groups: 1.00; 1.6 (0.8–3.2); 1.0 (0.4–2.0); 1.4 (0.7–2.9)	among ever-smokers with As exposure 10–19 years earlier, an association between OR of quartiles of total proportion or As-containing drinking-water of total daily fluid intake.	Bates et al. (1995)

295

Table 33 (contd.)

Study design	Study population	Source and level of As exposure	Health effects, metric of exposure and measure of association	Comments	Reference
Ecological	residents in areas of Belgium with various exposures to As	exposure from air (0.3 µg/m^3 annual mean); and water (20–50 µg As per litre). Daily geometric mean U-As 35 µg in the most exposed group (smelter area), 7–12 µg/d in the less exposed	cancer of lung, kidney, bladder and leukaemia studied; increased RR, 1.3 (1.14–1.43) observed for lung cancer in males in smelter area compared to a lesser-exposed group	directly standardized rate ratios (SRRs) used; authors explained the increased lung cancer risk by occupational exposure; other "As-linked" diagnosis was analysed showing no elevated risks	Buchet & Lison (1998)

Table 33 (contd.)

Ecological	cancer incidence in 22 areas in Victoria (Australia) in 1982–1991. Population size in 1986 was 152 246	soil/water As elevated in some parts, medians for low water As areas 1–2 µg/litre, and 13–1077 for high water-As areas (median of medians, 80 µg/litre)	for cancers of narsenical cavity, lung, bladder, stomach, colon, rectum, Hodgkin's lymphoma, non-Hodgkin's lymphoma, multiple myeloma, acute and chronic lymphatic leukaemia, and acute myeloid leukaemia SIR were <120, and the confidence interval included unity; for prostate cancer SIR was 114 (CI 105–123, melanoma 136 (124–148), breast 110 (103–148), and for chronic myeloid leukaemia 154 (113–210; for liver cancer, SIR was 53 (CI 34–82)	no information on coverage or frequency of water As sampling; postal codes that were used for calculating expected cases represent large geographic areas and may lead to random misclassification; rainwater reservoirs at least at present are an important alternative source of drinking-water, again leading to exposure misclassification	Hinwood et al. (1999)

Table 33 (contd.)

Study design	Study population	Source and level of As exposure	Health effects, metric of exposure and measure of association		Comments	Reference
			males	females		
Cohort	Mormons in Millard County, Utah (USA)	median range: 14–166 µg/litre	SMR (CI):		historic As concentrations in drinking-water used; death rates for the state of Utah for 1960–1992 were used to generate the expected deaths. Decreased SMR for lung and bladder cancer and all cancers may be due to lower prevalence of smoking among cohort members than in the reference population of Utah. Exposure for the highest exposure group likely to be overestimated because of introduction of low-As water into one community, which was not considered in the analysis	Lewis et al. (1999)
			all causes 91 (86–96)	96 (92–104)		
			non-malign respiratory 68 (54–85)	93 (70–120)		
			all cancer 82 (70–95)	73 (61–87)		
			large intestine 50 (28–99)	74 (40–124)		
			biliary tract and liver 85 (18–248)	142 (57–293)		
			respiratory system 57 (38–82)	44 (16–95)		
			prostate 145 (107–191)			
			kidney 175 (80–332)	160 (44–411)		
			bladder and other urinary organs 42 (8–122)	81 (10–293)		

Table 33 (contd.)

Case–referent	61 bladder and 49 kidney cancer cases and 275 referents not serviced by municipal drinking-water supply, Finland	5% of reference group had As in drinking-water >5 µg/litre and 1% (11/275) had consumed >10 µg/litre. As in drinking-water <0.05 to maximum 64 µg/litre detection limit 0.05 µg/litre	age-, sex- and smoking-adjusted risk ratios for bladder cancer when exposure 3–9 years before diagnosis		no association between cumulative As exposure and bladder cancer. no association between well-water As and kidney cancer	Kurttio et al. (1999)
			As in water (µg/litre)	RR	CI	
			<0.1	1.0		
			0.1–0.5	1.5	0.8–3.1	
			≥ 0.5	2.4	1.1–5.4	

Cohort	478 patients treated with Fowler's solution for 2 weeks– 12 years in 1946–1960 and followed until 1990	exposure from treatment with Fowler's solution; cumulative dose <500 mg, 500–999 mg, 1000–1999 mg; ≥2000 mg	mortality from cancer in the entire cohort		no dose response observed	Cuzick et al. (1992)
				SMR	CI	
			all cancer	95	17–130	
			bladder	307	101–730	
			liver	123	40–470	
			haematopoietic system	38	1–200	
			digestive organs	119	70–190	
			stomach	99	30–170	
			respiratory system	100	50–170	
			skin	244	8–1400	

arsenic exposure (continuous variable), and the information concerning death numbers and causes of death, as well as midyear population by age, sex, calendar year and precinct, were obtained from the Taiwan provincial department of health. The relationship between arsenic exposure, and mortality from cancer at 21 sites was analysed using multiple linear regression, adjusting for urbanization and industrialization of the precinct or township.

Multivariate analysis, adjusting for indices of urbanization and industrialization, revealed a statistically significant association between arsenic level in well-water and mortality from cancer of the lung, liver, kidney, bladder, skin, prostate and nasopharynx. Mortality from liver cancer was three times higher for men than for women (Chen & Wang, 1990).

In a cancer registry study in Taiwan, incident cases of bladder cancer in 1981–1985 were identified from the National Cancer Registration Centre, and the population was estimated from yearly registration in four counties endemic for BFD (Putai, Peimen, Hsuehchia, Yihjwu) The average annual age-adjusted incidence of bladder cancer in the four counties was 23.5×10^{-5}; in the whole of Taiwan it was 2.29×10^{-5} (Chiang et al., 1993). In two neighbouring counties (Jiangjiun and Yanshoei), also considered to be endemic for BFD, the annual incidence of bladder cancer was 4.45×10^{-5}. The computerized tumour registry began in 1982, and the registry was not validated.

Chiou et al. (1995) studied cancer incidence during a 7-year period in the BFD-endemic area in south-western Taiwan among 263 people with BFD and 2293 healthy controls from the area. Their cancer incidence was monitored from annual health examinations, home visits, household registration, national death certification and the national cancer registry. This study apparently represents a population independent of the earlier Taiwanese studies, since deaths from 1986 to 1993 were analysed. It should be noted that during the follow-up 7 of the healthy controls were diagnosed with BFD, and that 257/263 of the cases, but only 753/2293 referents, were identified in the earlier study (Chen et al., 1988b) from the townships of Peimen, Hsuechia, Putai, or Ichu, the rest coming from three further villages and identified later, in 1988. The prevalence of BFD was 5.57, 3.87, 2.02, and 0.64/1000 for the different townships in the first population, but 9.6–13.6/1000 in the latter. History of residence, use of artesian well-water for drinking, tobacco-smoking habits and other possible confounding factors were analysed from question-naires. The personal arsenic exposure was assessed from residence history, use of artesian well-water, and median village well-water arsenic concentrations defined in the survey in the 1960s (see above). Age, status of BFD (diseased or not), the concentration of arsenic in drinking-water, the duration of drinking artesian well-water, and estimated cumulative arsenic exposure of drinking artesian well-water (measured in (mg/litre) · year), were statistically significantly related to the risk of lung and bladder cancer risk in a multivariate analysis. In addition, tobacco smoking was related to risk of lung cancer, but not bladder cancer.

In an ecological study on bladder cancer incidence in 243 townships with (previous) high drinking-water arsenic concentration (Guo et al., 1997), the number of wells with a specified arsenic concentration (from the survey in 1974–1976 using the standard mercury bromide method, see above) was studied as a determinant of bladder cancer incidence in 1980–1987. A small part of the original Chen et al. (1985) study population, and an unknown large part of the population in the Wu et al. (1989) study, are included in this study. The percentage of wells in the highest arsenic-exposure category (of six categories, > 640 µg/litre) was significantly related to the incidence of transitional cell carcinoma of the bladder, ureter, urethra and kidney, and of adenocarcinoma of the bladder, but not to squamous cell carcinoma of the bladder, or of renal cell carcinoma, or of nephroblastoma. The tumour registry was not validated.

Age-adjusted mortality rate ratios for cancer of lung, liver, bladder and skin combined were studied in 3-year time periods in four townships in the BFD-endemic area in Taiwan from 1971 to 1994, when there apparently was a time-dependent decrease in the drinking-water concentration of arsenic. A gradual decrease in the risk was observed in men over 40 years of age, but for men less than 40 years of age little change was observed; for women both over and under 40 years, the risks were lowest during the last two 3-year periods studied (Tsai et al., 1998).

In the update of the ecological study (Chen et al., 1985) in Taiwan (Tsai et al., 1999), statistically significantly elevated mortalities were observed from cancer of the lung, larynx, oesophagus, stomach, small intestine, colon, rectum, liver, nose, larynx, lung, bone, prostate, bladder, kidney and skin, as well as from lymphoma and leukaemia. Mortality from nasopharyngeal, buccal or pharyngeal cancer was not elevated.

The arsenic exposure–response relationships for lung, bladder, and liver cancer were recently modelled using data from the BFD-endemic area (Morales et al. 2000). Exposure groupings were based on individual well-water arsenic concentrations for each village. Depending on the model and whether or not a comparison population was used in the analysis, risk estimates varied widely (many by at least an order of magnitude). Independent of the modelling, the authors reported excess lung and bladder cancer risks at exposure concentrations < 50 µg/litre (see Table 33).

The relationship between drinking-water arsenic and mortality from cancers at several sites in 26 rural counties in Córdoba (Argentina) in 1986–1991 was the subject of an ecological study (Hopenhayn-Rich et al., 1996c, 1998). Mortality rates in this area were compared with expected figures based on 1991 data for the whole of Argentina. The counties were divided into three drinking-water arsenic strata on the basis of a limited number of measurements and the relative number of reports on arsenical skin diseases. In two counties in the high-exposure group, 42/61 and 49/57 measurements were above the detection limit (40 µg/litre); the highest measured concentration was 533 µg/litre and the average drinking-water concentration of the measurements above 40 µg/litre in the two "high-exposure" counties was 178 µg/litre; the authors note, however, that this should not be considered to be representative of the population exposure (Hopenhayn-Rich et al., 1996d). To control for the potential confounding effect of smoking, mortality from chronic obstructive pulmonary disease was also assessed. The mortality from lung, kidney and bladder cancer was lowest in the counties with presumed lowest drinking-water arsenic concentration, intermediate in the medium exposure counties, and high in the high-exposure counties. The relative risk of chronic obstructive pulmonary disease was below unity for most groups and if anything, inversely related to arsenic exposure. The mortality from stomach, liver or skin cancer showed no clear-cut relationship with presumed arsenic exposure.

In an ecological study on arsenic exposure and cancer in Chile, mortality in Region II during 1950–1992 was compared to the more southerly Region VIII (Rivara et al., 1997). Region II includes copper mining and refining centres, and in one location annual average air arsenic concentrations were high (up to 2.7 $\mu g/m^3$); for the other two for which measurement data were available, they were approximately 0.2 and 0.02 $\mu g/m^3$ (Rivara et al., 1997). The annual province-weighted water arsenic levels were approximately 200 $\mu g/litre$ in the years 1950–1957, 650 $\mu g/litre$ for 1958–1970, 200 $\mu g/litre$ for 1971, 540 $\mu g/litre$ for 1972–1977, 100 $\mu g/litre$ for 1978–1987 and 50 $\mu g/litre$ thereafter. No information was available on other cancer risk factors, and the population in the arsenic-contaminated Region II was considerably younger than that in the reference region, mainly because of the extensive migration of the older population – which apparently would tend to bias the risk estimates toward lower values. The all-cancer mortality was slightly elevated in Region II, because of an excess in cancers of the lung, larynx, bladder, kidney and skin; no other cancer site showed an excess.

In another study in Chile (Smith et al., 1998), mortality from bladder, lung, kidney, skin cancer in 1989–1993 in Region II (where drinking-water arsenic levels had been high), were compared to age-adjusted mortality rates from the rest of Chile, using mortality data from 1991 and census data from 1992 for the age distribution. Smoking habits were available from a national survey in the two largest cities in the region. Apparently, all cancer cases in this study were also included in the earlier study (Rivara et al., 1997). The SMR for lung cancer in Region II with elevated arsenic drinking-water concentration was 380 (CI 350–410) in men, and 310 (CI 270–370) in women. Mortality from chronic obstructive pulmonary disease was not elevated; neither did the limited information on smoking indicate smoking to be a confounder. The SMR for bladder cancer was 600 (CI 480–740) in men, and 820 (CI 630–950) in women, that for skin cancer 770 (CI 470–1190) in men, and 320 (CI 130–660) in women, compared to the figures for the whole of Chile.

A cohort study was carried out among the residents in an arsenic-polluted area in Japan (Tsuda et al., 1987, 1989, 1990, 1995), where the pollution had arisen from a factory producing arsenic trisulfide, with waste waters contaminating the groundwater. A list of 467 residents living in the vicinity of the factory was made by the government in 1959. The latest update (Tsuda et al., 1995) identified 454 of these original residents, and followed their vital status (100% follow-up) from 1959 to 1992. The analysis of the cause-specific mortality was performed on those 443 identified as having used water from wells analysed for arsenic content in 1959. The mortality experience was compared to that of the Niigata prefecture, which has a population of approximately 2.5 million. The decedents were divided in three strata on the basis of the drinking-water concentration of arsenic in 1959: < 50 µg/litre, 50–990 µg/litre and > 1 mg/litre. The highest measured arsenic concentrations were < 3 mg/litre, and although the polluting industry had been in production for more than 45 years, it was considered likely that the exposure had been very low before 1954 and practically stopped in 1959.On the basis of information about the protection technology, it was considered that exposure to arsenic by inhalation had been very low. The number of deaths from all causes was 105 (100.5 expected) and the number from all cancer was 34 (SMR 148, CI 106–207). A significantly elevated mortality from lung cancer was observed in the highest exposure group (SMR 1569 (CI 738–3102) The total number of lung cancer cases was 9, of which 8 were in the highest exposure category. Three urinary tract cancers were observed in the highest exposure category, which represents an excess (SMR 3118, CI 862–9175), on the basis of expected numbers for kidney and bladder cancer combined for the whole Japan. A strong association was observed between arsenic-induced lesions identified in 1959 and subsequent mortality from lung cancer.

As a part of the US National Bladder Cancer Study, a case–referent study (Bates et al., 1995) was performed in communities in Utah, where measurements of drinking-water arsenic were available for the years 1978–1979. Patients diagnosed with bladder cancer during a period of 1 year around 1978 were included in the study. Two referents per case were frequency-matched by age, sex, and geographic area. For each case, two referents were identified. Two indices of arsenic exposure were used: (1) total of arsenic intake (in micrograms), and (2) the proportion of the arsenic-containing drinking-water to the total daily fluid intake multiplied by the total (µg As/litre) · year. In non-smokers, no relationship was observed between bladder cancer and either arsenic measure. Among ever smokers with exposure for 10–19 years before 1978, the linear trend test for the four quartiles of the second index of exposure was statistically significant ($p < 0.05$). The drinking-water arsenic concentrations were relatively low: 0.5–160 µg/litre, with an average of 5.0 µg/litre, 92% of the measurements < 10 µg/litre and 98.9% < 50 µg/litre.

In a hospital-based case–referent study in northern Chile (Regions I–III), drinking-water arsenic exposure was compared between 151 lung cancer cases and 419 referents (167 with cancer and 242 with other diseases) (Ferreccio et al., 1998, 2000). Drinking-water arsenic concentration was assessed from the records of municipal water companies covering the years 1950–1994, and information on residence, health and employment history from a questionnaire. The OR for lung cancer, adjusted for age, sex and smoking (ever/never), was related to drinking-water arsenic levels in the five exposure strata, and reached statistical significance in the highest exposure stratum. (see Table 33).

Mortality from cancer and other diseases was studied in an ecological study in northern Belgium, where the population is exposed to arsenic (among other elements) because of a conglomeration of non-ferrous metal smelters (Buchet & Lison, 1998). The study covered deaths between 1981 and 1991 identified from the national statistics. In the area with highest exposure, the drinking-water concentration of arsenic was stated to be between 20 and 50 µg/litre (no details provided). In the area with intermediate exposure this figure was < 20 µg/litre, and in the least exposed area, < 5 µg/litre; local mortality rates (for the Maaseik and Turnhout districts) were used as the reference. No relationship with arsenic exposure, and mortality from cancer of the lung, kidney or bladder was observed.

In an ecological study in Victoria (Australia), Hinwood et al. (1999) investigated the cancer incidence in 1982–1991 in 22 different areas where measurements were available on arsenic concentrations in soil and water. Population information was derived from the 1986 census, and areas were assigned water arsenic concentrations on the basis of their postal codes. Median water arsenic concentration in the high-exposure areas varied between 13 and 1077 µg/litre. Standardized incidence rates (SIRs) were below 120, and the CI included unity for most cancers, including lung and bladder. The SIRs for prostate and breast cancer, as well as for melanoma and chronic myeloid leukaemia, were slightly elevated. For liver cancer there was a statistically significant deficit. The authors note that the crude exposure assessment was likely to lead to random misclassification, and thus under-estimation of the risk.

In the cohort study among members of the Church of Jesus Christ of Latter-day Saints (Mormons) in Utah (Lewis et al., 1999; see section 8.4.2) there was a lower than expected mortality from respiratory cancer in both men and women. For bladder cancer there was also a deficit, but it was not statistically significant. As the smoking habits of Mormons are different from those of the general population (also indicated by the lower-than-expected risks of chronic pulmonary diseases, see above), the data on lung and bladder cancer in this study are difficult to interpret. The mortality from kidney cancer was slightly (statistically not significantly) elevated. No excess in cancer of the stomach or large intestine was observed either. The mortality from prostate cancer was higher than expected but was lowest in the low-exposure group.

In a case–referent study in Finland (Kurttio et al., 1999), the relationship between drinking-water and the incidence of kidney and bladder cancer in 1981–1995 was studied in a cohort of people who had used drilled well-water as drinking-water in 1967–1980. The arsenic exposure history was reconstructed from questionnaire data on residence and analysis of arsenic in the well in 1996, with the assumption that the arsenic content had remained stable over the years. The final study population consisted of 61 cases of bladder cancer and 49 of kidney cancer, and an age- and sex-matched reference group of 275 people. The arsenic levels in the water were low (median 0.14 µg/litre, range < 0.05–64 µg/litre); 95th percentile for cases of bladder and kidney cancer and the reference cohort was respectively 3.0, 1.8, and 4.5 µg/litre. Cases of bladder cancer tended to have a higher arsenic exposure during years 3–9 before diagnosis. This reached statistical significance in the high-dose group (well-water arsenic ≥ 0.5 µg/litre), when the well-water arsenic was used as the indicator of arsenic exposure (but not, if the cumulative arsenic dose was used in the calculation). A weaker relationship was observed between bladder cancer incidence and arsenic exposure when a latency period of ≥ 10 years was applied, and no association was observed between cumulative arsenic exposure and bladder cancer. No relationship was observed between arsenic exposure and kidney cancer.

In the study on the health effects of Fowler's solution as a medical treatment (Cuzick et al., 1992; for study description and limitations, see section 8.4.2), the mortality of bladder cancer was elevated, but only weakly related to the total arsenic dose administered. Mortality from respiratory or stomach cancer was not elevated.

8.7.3 Dermal effects, including skin cancer (Table 34)

Assessment of the association between arsenic exposure and skin cancer risk is hampered by the low case fatality rate of non-melanoma skin cancer: mortality studies are likely to markedly underestimate the incidence of the disease.

Skin cancer often arises from a keratotic change, the developed forms of which are classified as Bowen's disease; keratosis in turn may be preceded by disturbances in the skin pigmentation (hyper- and/or hypopigmentation). Studies on all these changes are included in the section below (Table 34).

Several case reports and series have suggested that arsenic from medicinal use, drinking-water and occupational exposure may be related to skin diseases, including cancer (see section 8.2; for further references, see Neubauer, 1947; Hill & Faning, 1948; Sommers & McManus, 1953; Sanderson, 1963; Minkowitz, 1964; Zaldivar, 1974). An early study also suggested a dose–response relationship among patients treated with arsenicals (Fierz, 1965)

In a survey of the health status of 40 421 habitants of 37 villages in the BFD-endemic area of Taiwan the prevalence of skin cancer was 10.6×10^{-3}, and showed a relationship with the village well-water arsenic concentration (2.6, 10.1 and 21.4×10^{-3} in villages with well arsenic concentration < 300, 300–600 and > 600 µg/litre, respectively. Prevalence of hyperpigmentation was approximately 18%, and that of keratosis and BFD 7% and 9%, respectively (Tseng et al., 1968; Tseng, 1977).

Table 34. Effects of As exposure on the skin

Study design	Study population	Source and level of As exposure	Health effects, metric of exposure and measure of association	Comments	Reference
Cross-sectional	40 421 males and females in 37 villages in As high exposure area in south-west Taiwan and 7500 in low exposure area	142 samples from 114 wells analysed for As: variation 1–1097 µg/litre 50% 300–700 µg/litre	Prevalence (10^{-3}) of hyperpigmentation 183.5, of keratosis 71.0 in high exposure area, 0 for both in low exposure area (1–17 µg/litre) As conc. (µg/litre) — Prevalence of skin cancer (10^{-3}) <300 — 4 (M), 1.3 (F) 300–600 — 14.4 (M), 6.3 (F) >600 — 31.0 (M), 12.1 (F) unknown — 16.3(M), 4.7 (F)	reference cited for As analysis; exposure–response effect was seen across age and gender	Tseng et al. (1968); Tseng (1977)
Ecological	As-exposed areas of Taiwan	drinking-water up to 1.14 mg/litre, decreasing with bringing into use of reservoir water from 1956	SMR values 1973–1986 mortality in As-exposed area of SW Taiwan: skin: 534 (379–689) (M) 652 (469–835) (F)	population of Taiwan as the reference	Chen et al. (1985)

Table 34 (contd.)

Cohort	cohort of 789 BFD patients (7278 person-years of observation)	drinking-water concentrations 350–1140 µg/litre	Taiwan reference pop. SMR 2846 ($p < 0.01$) local reference pop. SMR 451 ($p < 0.05$)	10.6% lost on follow-up	Chen et al. (1988b)
Ecological	mortality from malignant neoplasms in 1972–1983 in 314 precincts and townships in Taiwan	74% of precincts had <5% wells with ≥ 50 µg/litre As, 15% has 5–14% and 12% had ≥15% such wells; village mean used in analysis.	statistically significant association between As level in well-water and mortality from skin cancer after adjustment for indices of urbanization and industrialization		Chen & Wang (1990)
Ecological	mortality and population data 1973–1986 in 42 villages in Taiwan	used published Taiwanese data 1964–1966 and village medians in the analysis	age-adjusted mortality rates per 10^5 by well As concentration (µg/litre): see sub-table below	observed numbers of deaths smaller than in the study by Chen et al. (1992), although the person-years are identical	Wu et al. (1989)

age-adjusted mortality rates per 10^5 by well As concentration (µg/litre)

	300	300–599	≥600	p
males:	2.03	14.01	32.41	<0.001
females:	1.73	14.75	18.66	<0.05

Table 34 (contd.)

Study design	Study population	Source and level of As exposure	Health effects, metric of exposure and measure of association			Comments	Reference
Cross-sec-tional	1571 residents >30 years of age from high As exposure areas of Taiwan	median As in well-water 0.70–0.93 mg per litre in early 1960s	cum. expos. (mg/litre-yr) ≤4 5–24 ≥25	Prevalence 1.0 6.7 13.8	OR (1.1–59) (1.1–77)	drinking-water As concentration estimates based on a 1960s study (Kuo, 1968), using the Natelson method. 68.8% participation rate, for ~25% cumulative exposure history not known. Exposure—response between duration of consumption of sweet potato and prevalence of skin cancer. ORs adjusted for age and sex, duration of consumption of sweet potato, working in rice fields and hepatitis B–surface antigen	Hsueh et al. (1995)

310

Table 34 (contd.)

Ecological	243 Taiwanese townships – approximately 11.4 million residents; incident cases of urothelial and kidney cancer 1980–1987	As measured in over 80 000 wells, 1974–1976; in 78% of townships average As content was non-detectable, in 91%, <50 and in 99.5% <640 µg/litre	no relationship between skin cancer incidence and the mean township well-water As concentration; positive association between skin cancer and percentage of wells in the highest concentration category (>640 µg/litre); negative association between skin cancer and percentage of wells in the lowest concentration category	Used data from 1970s survey on As in well-water, using mercuric bromide method to analyse As. Smoking not included in the models as not good predictor for any cancer in this study. Potential bias from source of case ascertainment, i.e. tumour registry not validated	Guo et al. (1998)
Ecological	4 townships in BFD-endemic area in Taiwan, mortality in, 1971–1994, compared to local and national rates	drinking-water up to 1.14 mg/litre, decreasing with bringing into use of reservoir water from 1956	skin cancer SMR for females and males combined, compared to local rates: SMR 597 CI 483 374–615 (local rates) 462–760 (national rates)	age- and sex-specific mortality rates based on population data from Ministry of Interior, deaths from computer database on deaths. 99% of causes of death based on physician confirmed by pathological examination. Overlaps with earlier Taiwanese studies	Tsai et al. (1999)

311

Table 34 (contd.)

Study design	Study population	Source and level of As exposure	Health effects, metric of exposure and measure of association	Comments	Reference
Ecological	Cordoba residents vs. rest of Argentina	in the high-exposure group, in two selected towns, 42/61 and 49/57 measurements ≥40 µg/litre; highest measured concentration 533 µg/litre	exposure SMR (CI) males, females low 204 (138–289), 85 (42–151) medium 149 (83–245), 82 (32–168) high 149 (71–273), 278 (161–444)		Hopenhayn-Rich et al. (1998)
Ecological	Region II (higher As exposure) in northern Chile compared to Region VIII (low exposure)	drinking-water As concentration varied during 1950–1992, ranging from ND to 860 µg/litre through the time period in different locations in Region II.	mortality rate ratio (CI Region II vs. Region VIII) 4.3 (2.3–5.1)	air levels of As measured in some locations and were considerably elevated at Chuquicamata copper smelter in Region II.	Rivara et al. (1997)

Table 34 (contd.)

Ecological	Chile: Region II compared to the rest of Chile, 1989–1993	drinking-water avg. 43–568 µg/litre (1950–1994) exposure decreased over time: 569 µg/litre (1955–69) to 43 µg/litre (1990–94)	males females	SMR (CI) 770 (470–1190) 320 (130–660)	measurements taken by water company. Population partially overlaps that of Rivara (1997)	Smith et al. (1998)
Cross-sectional	one-third of households in two towns in North Mexico, one with As contaminated drinking-water, the other without	average water As 400 (SD 114) µg/litre for the exposed, based on 20 samples in 1975–1978. For the referents, mean (SD) 5 (7) µg/litre		prevalence of hypopigmentation, hyperpigmentation, palmoplantar keratosis, papular keratosis and cancer 17.6, 12.2, 11.2, 5.1 and 1.4% among the exposed, and 2.2, 1.9, 0.3, 0.0, 0.0 % among the referents	prevalence rates not age-standardized, but among the referents, the proportion of >60 year-olds greater than among the exposed	Cebrian et al. (1983)

Table 34 (contd.)

Study design	Study population	Source and level of As exposure	Health effects, metric of exposure and measure of association			Comments	Reference
Cross-sectional	7683 inhabitants in 25 villages in West-Bengal in 1995–1996; exposure to As probably started in the late 1960s	for 45%, drinking-water As was <50 µg/litre, for 69% <200 µg/litre, for 88% < 500 µg/litre and for 99.8% <800 µg/litre	prevalence of keratosis and hyperpigmentation			keratosis but not hyperpigmentation more prevalent among individuals with body weight in the lowest quintile. drinking-water source of each recruited household analysed for As using hydride generation AAS	Mazumder et al. (1998)
			As-conc (µg/litre)	keratosis (M) (F)	hyperpigmentation males females		
			<50	0.2 0.3	0 0.4		
			50–99	1.5 0.8	0.4 3.2		
			100–149	1.6 5.7	1.2 11.0		
			150–199	4.7 5.1	2.3 7.8		
			200–349	4.9 6.5	2.0 13.1		
			350–499	9.0 9.5	2.7 15.7		
			500–799	8.9 5.3	3.1 13.8		
			≥800	10.7 11.5	8.3 22.7		
Cross-sectional	1481 subjects in 4 villages in Bangladesh	well-water As concentration at the time of the study was 10–2040 µg/litre	well-water As concentration-dependent increase in the prevalence of skin lesions (hyper- or hypopigmentation, or keratosis)			data on individual skin lesion types not given. As concentration from previous studies, analyses by hydride generation AAS. Number or representativity of analyses not discussed	Tondel et al. (1999)
			skin lesion prevalence (%)				
			As-conc (µg/litre)	males	females		
			<150	18.6	17.9		
			151–350	21.9	20.5		
			351–550	32.9	32.1		
			551–1000	36.8	34.0		
			≥1000	37.0	24.9		

The SMR from skin cancer in the BFD-endemic area was 534 (CI 379–689) in men and 652 (CI 469–835) in women in the first ecological study in south-western Taiwan (Chen et al., 1985). The skin cancer SMR among BFD patients in this population during a 15-year follow-up was 2846 in comparison to the Taiwanese population, and 451 in comparison to the population in the endemic area without BFD (Chen et al., 1988b). Multivariate analysis, adjusting for indices of urbanization and industrialization, revealed a statistically significant association between arsenic level in well-water and mortality from skin cancer (Chen & Wang, 1990). In the second ecological study in this area (Wu et al., 1989), age-adjusted mortality rates from skin cancer for men were 2.03×10^{-5}, 14.01×10^{-5} and 32.41×10^{-5} in villages with a median well arsenic concentration of < 300, 300–600 and > 600 µg/litre, respectively, and there was a similar exposure–response relationship for women (for study descriptions, see section 8.4).

In 1988–1989 a survey was carried out to investigate the relationship of arsenic exposure to skin cancer in three villages in the BFD-endemic area of Taiwan (Hsueh et al., 1995), in which 1571 habitants of the three villages, who lived in one of the villages no less than 5 days a week (out of a total population of 2258), were interviewed for their drinking-water consumption and other personal history. Of these, 1081 (68.8%) participated in a physical examination, and were included in the analysis. Altogether 66 cases of skin cancer were diagnosed, and the age- and sex-adjusted prevalence OR of skin cancer was related to all parameters of arsenic exposure, i.e. village well-water mean arsenic concentration (from analyses in the early 1960s), duration of living in the BFD-endemic area, duration of drinking artesian well-water and cumulative arsenic exposure. For well-water arsenic concentration, the OR was 3.5 (CI 0.7–17.0) and 5.0 (1.1–23.8) for concentrations 0–700 and > 700 µg/litre, respectively.

Skin cancer incidence in 1980–1987 was studied in the 243 townships in Taiwan, where the arsenic concentrations in some 83 000 wells had been investigated in the 1974–1976 survey (Guo et al., 1998). A total of 1547 skin cancer cases were identified, and demographic data on the study population was obtained from the Department of Internal Affairs. Sex-specific age-adjusted standardized incidence rates were calculated for each township. No relationship was observed between skin cancer incidence and the mean township arsenic concentration in the well-water. However, when a multiple-variable analysis was applied, using the percentage of wells within a township with a specified range or arsenic concentrations as the parameter describing exposure, a positive association was observed at the highest arsenic exposure category (> 640 µg/litre) for both men and women.

In the most recent ecological study in the area (Tsai et al., 1999), an elevated mortality from skin cancer (SMR 483, CI 374–615, and 597 in comparison to the local, and national figures for men and women combined), was also observed (for study description, see section 8.4.2).

In the ecological study in Argentina (Hopenhayn-Rich et al., 1998; for study description, see section 8.7.2), the mortality from skin cancer showed a negative association with arsenic exposure in men and a positive association in women.

In the mortality study in Chile in 1989–1993 (Smith et al., 1998), the SMR for skin cancer in Region II with elevated arsenic drinking-water concentration was 770 (CI 470–1190) in men, and 320 (130–660) in women, compared to the figures for the whole of Chile. The excess skin cancer mortality in Region II was similar in the comparison with Region VIII (Rivara et al., 1997), in which cases from 1950–1992 were studied (RR 4.3, CI 2.3–5.1 for both sexes combined) (For study descriptions, see section 8.7.2.)

Cebrian et al. (1983) conducted a cross-sectional study of skin lesions in two towns in Mexico. The average arsenic concentration of water samples in the exposed town was 0.411 mg/litre; the average concentration in the control town was 0.005 mg/litre. The subjects examined were selected for examination by systematic sampling of the populations in the two towns: 296 individuals from the exposed population of 998 and 318 individuals from the control population of 1488 were physically examined. In the exposed town, there were 52 cases (17.6%) of hypopigmentation, 36 cases of hyperpigmentation (12.2%), 33 cases (11.2%) of palmoplantar keratosis, 15 cases (5.1%) of papular keratosis, and 4 cases (1.4%) of ulcerative zones (skin cancer). In the control town, there were 7 cases (2.2%) of hypopigmentation, 6 cases (1.9%) of hyperpigmentation, 1 case (0.3%) of palmoplantar keratosis, no cases of papular keratosis, and no cases of ulcerative zones (skin cancer). The prevalence of all the skin lesions was significantly elevated in the exposed town ($p < 0.001$, with the exception of ulcerative zones which had $p = 0.04$) and was generally found to increase with age. Non-specific symptoms (e.g. nausea, epigastric pain, colic abdominal pain, diarrhoea, headache, and oedema) were found to be more prevalent in the exposed town than the control town and more common among those with skin lesions.

The relationship between skin keratosis and hyperpigmentation and consumption of arsenic-contaminated drinking-water was investigated in a cross-sectional study of a population in West Bengal (India) (Mazumder et al., 1998). There were 7683 participants from areas of both high and low arsenic-exposure. Each participant was questioned about drinking-water sources, water intake, diet, medical symptoms, height, weight, and other variables. Participants were medically examined, with a careful inspection for arsenic skin lesions. Water samples were collected from each tube-well used by the households of the participants in the study. The age-adjusted prevalence of keratosis rose from zero in the lowest exposure level (< 50 µg As/litre) to 8.3 per 100 for females drinking-water containing > 800 µg As/litre. For males, the age-adjusted prevalence of keratosis increased from 0.2 per 100 in the lowest exposure category to 10.7 per 100 for males in the highest exposure level (> 800 µg As/litre). For females, the age-adjusted prevalence of hyperpigmentation rose from 0.3 per 100 in the lowest exposure category to 11.5 per 100 in the highest exposure category; for males the age-adjusted prevalence rose from 0.4 per 100 in the lowest exposure category to 22.7 per 100 in the highest exposure category. Comparison by dose per body weight found that men had roughly 2–3 times the prevalence of both keratosis and hyperpigmentation compared to women apparently ingesting the same dose of arsenic from drinking-water. Subjects below 80% of their body weight for their age and sex had a 1.6-fold (CI 1.0–2.4) increase in the prevalence of keratoses, suggesting that malnutrition may play a role in increasing susceptibility. No such difference was observed for hyperpigmentation. Twelve subjects with keratosis drank water containing < 100 µg As/litre; 29 with hyperpigmentation drank water containing < 100 µg As/litre.

Tondel et al. (1999) examined 1481 subjects ≥30 years of age in four villages in Bangladesh. All were determined to have had a history of arsenic exposure through arsenic-contaminated drinking-water. Arsenic concentrations in the drinking-water ranged from 10 to 2040 µg As/litre. Of the 1481 people examined, 430 were found to have skin lesions (pigmentation changes or keratosis). The age-adjusted prevalence rate of skin lesions was found to increase from 18.6 per 100 in the lowest exposure category (≤150 µg As/litre) to 37.0 per 100 in the highest exposure category (> 1000 µg As/litre) for males and from 17.9 per 100 in the lowest exposure category to 24.9 per 100 in the highest exposure category for females. The trend was statistically significant for both males and females. When the exposure was considered by dose (µg/litre · kg), there was also an increase in the age-adjusted prevalence rate of skin lesions for both males and females across dose groups, the trend being statistically significant.

A clinical study of 11 families from a village in northern Chile, supplied by water containing up to 800 µg As/litre, found skin changes in 6 of 44 subjects despite good nutritional status (Smith et al., 2000). Arsenic exposure in this village is reported to have been present for thousands of years, suggesting that there has been no adaptation to arsenic exposure by the population.

Arsenite can induce an irritative contact dermatitis after occupational exposure (Goncalo et al., 1980), but dermal sensitization to inorganic arsenic appears to be a rare occurrence. Barbaud et al. (1995) reported on the contact hypersensitivity of arsenic in a crystal factory employee. A patch test was done with various compounds that he came in contact with at work, and arsenate was the only chemical that tested positive.

8.8 Reproductive toxicity (Table 35)

A series of reproductive outcomes have been examined among female employees and women living close to the Rönnskär copper smelter in Sweden. Nordstrom et al. (1978a) compared birth weights of all offspring of women employed at the Rönnskär smelter during a 2-year period (1975–1976), categorized in three main groups according to their work location: factory, laboratory and administration. Information was also collected for infants of women who lived in four areas close to the smelter but at increasing distances. Births from the University Hospital in Umeå, a city distant from the smelter, were used as external controls. The average birth weight of infants of Rönnskär employees and of women living closer to the smelter were significantly lower than those from the two more distant areas and from Umeå. Among employees, those working in the laboratory had larger babies than those in the factory or in administration. In general, the effect was mainly observed in higher pregnancy orders (second born or later). Contrary to what is generally observed, there was a decrease in the average birth weight of offspring of higher parity among employees.

Nordstrom et al. (1978b) reviewed data from hospital files on over 4427 pregnancies of women born in or after 1930 who lived in four areas of increasing distances from the smelter. A control group of 4544 pregnancies was used from a hospital in Umeå, a non-exposed town. There was a clear dose–response relationship between the occurrence of spontaneous abortions and residential proximity to the smelter. In particular, women in the closest town (< 10 km from the smelter) had the highest rates (11% vs. 7.6%, $p < 0.005$). In the most exposed area, 4 of 20 women with abortions had had 2 abortions and no normal pregnancies. No women with such reproductive history were found in the other three areas.

In a further study (Nordstrom et al., 1979a), the previous analysis (Nordstrom et al., 1978a,b) was expanded to cover 662 births among women employed in or living near the smelter and in the unexposed town of Umeå in 1930–1959. Personal questionnaires were used to assess exposure and confounding factors. The average birth weight of babies born to employees was significantly lower than those born in Umeå (p < 0.05); in addition, birth weight was lower if the mother worked in the highest exposure categories (e.g. smelting and cleaning operations) rather than in lower exposure jobs. Most differences were found for the birth weight of third or later-born infants, which the authors proposed to be caused by the cumulative exposure with age. The differences were not found to be confounded by the gestational ages. Spontaneous abortions were highest when the mother was employed during pregnancy (14%) or before pregnancy and living near the smelter. Within employment categories, significantly higher rates of abortion were observed in high exposure jobs (28% vs. 14%, NS). The abortion rate was even

Table 35. Reproductive toxicity of As

Study design	Study population	Source and level of As exposure	Health effects, metric of exposure and measure of association		Comments	Reference
Ecological	offspring born to women employed at the Ronnskar smelter in Sweden 1975–1976, and women in two areas near the smelter (A+B), and more distant (C+D) born after 1930; Control group at Umea, Sweden, for the years 1955, 1965 and, 1975	occupational for smelter employees, environmental for nearby residents	average birth weight g: employees 3391 3395 3412 3495 3470 Umea (control) 3460	p <0.05 <0.001 <0.01 NS NS	other exposures in the smelter such as lead and copper likely; no control for these or other potential confounding factors (maternal age, lifestyle, medical)	Nordstrom et al. (1978a)

320

Table 35 (contd.)

		distance from smelter	area	spontaneous abortions (%)	total pregnancies		Reference
Ecological	all pregnant women born after 1930 in areas around smelter	distance from smelter: areas A and B <10 km, C and D 10–15 km	A B C D	11 9.2 8.2 7.0	1358 791 969 1118	low percentage of abortions in all groups; no exposure data for As; no adjustment for potential cofounders, particularly socio-economic status; possible co-exposure	Nordstrom et al. (1978b)
Cohort	662 women employed at Ronnskar smelter, or living close to smelter, born 1930–1959	employment status during pregnancy (3 work locations), and residential distance from smelter	average birth weight (g) employed high exp. job home <10 km home >10 km control (Umea) abortion: employed during pregnancy father and mother employed high exposure jobs	all 3366 3087 3406 3411 3460 14%; 19% 28%	>2nd born 3213 3061 3397 3435 3568	data on individual smoking habits and other variables not given; other exposures in the smelter and surrounding area likely	Nordstrom et al. (1979a)

321

Table 35 (contd.)

Study design	Study population	Source and level of As exposure	Health effects, metric of exposure and measure of association		Comments	Reference
Cohort	offspring of Female employees at Ronnskar smelter in Sweden ($n = 1291$)	emission at smelter contained As	rate of malformations: mother employed mother not employed subgroup of mothers (born after 1930): emp. before pregnancy emp. during pregnancy	5.8% 2.2% ($p < 0.05$) 3.2% 5.1% ($p < 0.025$)	congenital malformations identified from question-naires and medical rec-ords; potential exposures to other metals likely lead, cadmium); no control for other potential confounding factors (maternal age, life-style, medical)	Nordstrom et al. (1979b)
Case–referent	live newborns ($n = 270$), diagnosed with severe congenital heat disease in Massachu-setts (USA), 1980–1983. Controls selected ran-domly from all Massachusetts births ($n = 665$)	drinking-water. As measures obtained from public water supplies; limit of detection 0.8 μg/litre; highest level 22 μg/litre.	POR comparing those with measures above to below detection levels all congenital heart disease coarctation of the aorta patent ductus arteriosus conotruncal defect ventricular septal defect	1.0 (0.6–1.6) 3.4 (1.3–8.9) 1.2 (0.6–2.6); 0.9 (0.5–1.7) 1.3 (0.6–2.8)	As levels in water quite low; controlled for other chemicals in the water, parental education, source of water (surface vs. ground); no controls for other potential confounders	Zierler et al. (1988)

322

Table 35 (contd.)

Case–referent	births at 2 Massachusetts hospitals, 1976–1978 cases: spontaneous abortions by 27 weeks gestation ($n = 158$); controls from same hospital ($n = 690$)	drinking-water contaminants analysed: As, pH, alkalinity, hardness, silica, chloride, ammonia, nitrate, nitrite, a number of other metals	spontaneous abortion As (μg/litre) undetected: 0.8–1.3 1.4–1.9	OR(CI) 1 1.1 (0.6–1.8) 1.5 (0.4–4.7)	outcomes restricted to spontaneous abortions after the fact at hospital (method may under estimate numbers); OR adjusted for maternal age, educational level, history of spontaneous abortion, and other measured exposures measured, i.e. not organic contaminants As measurements from Massachusetts Department of Environmental Quality Engineering Interval from sample analysis to date of conception was reported to range from 5 d to 3.5 yr (median was 1.6 yr for cases and 2.2 for controls)	Aschengrau et al. (1989)

Table 35 (contd.)

Study design	Study population	Source and level of As exposure	Health effects, metric of exposure and measure of association	Comments	Reference
Ecologi- cal	2 areas in southern Hungary; study period 1980–1987	drinking-water, high exposure defined as up to 0.1 mg/litre	spontaneous abortions: rate ratio = 1.4 ($p < 0.05$) stillbirths: rate ratio = 2.8 ($p < 0.05$)	no information on case ascertainment procedures or their competencies, or concomitant exposures or population characteristics. Without giving figures, it is stated that no difference was observed in the frequency of cancer on PVD between the two areas	Borzsonyi et al. (1992)
Ecologi- cal	rates of tox- aemia in Srednogorie (town close to smelter) and Bulgaria	living near copper smelter	rate of toxaemia/1000 births: Srednogorie 8.0 Bulgaria 2.5	no control for other factors associated with toxemia	Tabacova et al. (1994a)

Table 35 (contd.)

Cross-sectional	Bulgaria births (n = 34) in area around smelter; non-smelter control area (n = 15)	As in sedimented dust measurement 0.047–0.37 mg/m²	mean birth weight: smelter area 3012 g; non-smelter 3193 g; placenta As concentration smelter area 0.023 mg/kg; non-smelter area 0.007 mg/kg; $p < 0.001$	other metals present (lead and cadmium) were measured and no significant difference found in placental concentrations between smelter and non-smelter areas; differences in birth weights were observed between smoking and non-smoking group — Tabacova et al. (1994b)
Ecological	residents of US counties with mean As water levels >5 µg/litre (n = 30 counties), 1968–1984	drinking-water range: 5.–91.5 µg/litre Exposure groups: low: 5–10 µg/litre. med: 10–20 µg/litre. High: >20 µg/litre	SMRs for congenital anomalies exp/group — male — female heart low 120 (110–130) 100 (90–110) medium 90 (110–130) 100 (90–110) high 90 (60–120) 130 (100–180) circulatory system low 100 (80–150) 110 (90–110) medium 110 (80–150) 100 (70–140) high 130 (70–240) 200 (110–340)	no controls for other exposures; drinking-water measures provided by water companies; SMRs based on comparison to US population — Engel & Smith (1994)

Table 35 (contd.)

Study design	Study population	Source and level of As exposure	Health effects, metric of exposure and measure of association	Comments	Reference
Case–referent	cases: stillbirths delivered in Texas hospital (n = 119), 1983–1993. controls: randomly selected from same hospital, frequency matched on year of delivery (n = 267)	air levels surrounding As pesticide plant, estimated from an atmospheric dispersion model	POR for stillbirths low 0.7 (0.4–1.3) med 1.2 (0.6–2.3) high 4.0 (1.2–13.7) Hispanics in high exposure: 7.8 (1.6–38.6)	modelling based on actual measures, but no further measures to validate the model; small numbers in sub-groups by ethnicity; exposure levels based on address at time of delivery; 0, 0–10, 10–100, >100 ng/m^3	Ihrig et al. (1998)
Ecological	3 Chilean cities: Santiago, Antofagasta, Valparaiso; Births 1950–1996	drinking-water; Antofagasta had high exposure 1958–1970 (800 µg/litre in city's water supply)	trends in late fetal, neonatal and postneonatal mortality across 4-yr intervals; rate differences indicate greatest contrast during high As period, suggesting As role in increased infant mortality	As data from previous routine measurements; other potential confounders not available.	Hopenhayn-Rich et al. (1999, 2000)

higher if the father also worked at the smelter, not just the mother (19% vs. 14%; statistical significance not provided). Individual smoking data was not given, although no differences were found between the smoking rates of the different groups.

Nordstrom et al. (1979b), in an investigation of the occurrence of congenital malformations, found rates of 5.8% among infants born to female employees who worked at the smelter during pregnancy, compared to 2.2% among female employees who did not work ($p < 0.005$) and 3% among residents in the larger region. Multiple malformations were four times more common among employees working during pregnancy than among residents in the larger region.

Most of these studies by Nordstrom and co-workers had an ecological design with little or no information on other factors. Although arsenic exposures in and around the Rönnskär smelter were high, confounding from lead or copper could not be excluded. In addition, no adjustments were made for the effects of other potential confounding risk factors, such as maternal age, which is known to have a strong relationship to spontaneous abortion and congenital anomalies.

Zierler et al. (1988) compared 270 cases of infants born with congenital heart disease and 665 controls from Massachusetts (USA). The POR, adjusted for all measured contaminants, source of water, and maternal education, for any congenital heart disease in relation to any arsenic exposure above the detection limit of 0.8 µg/litre, was not elevated. However, for a specific malformation, coarctation of the aorta, there was a significant POR of 3.4 (1.3–8.9). The exposure was quite low, the 90th percentile level being 1 µg/litre.

On the basis of information from a previous case–control study of spontaneous abortions in Boston, Aschengrau et al. (1989) examined 286 women who experienced spontaneous abortions and 1391 controls in relation to the content of their water supplies. An adjusted odds ratio of 1.5 was found for the group with the highest arsenic concentrations. However, this exposure group had low levels of arsenic in water (1.4–1.9 µg/litre), close to or lower than laboratory analytical detection limits, and the possibility of chance or unaccounted confounders could not be discounted.

A study in an area of south-east Hungary (Borzsonyi et al., 1992) with exposure to arsenic from drinking-water examined the rates of spontaneous abortions and stillbirths for the period 1980–1987. Two populations were compared: one with levels of arsenic in drinking-water > 100 µg/litre (n = 25 648 people) and one control area with low arsenic levels (n = 20 836) (no information on analytical method, or timing or frequency of sampling was available). Both outcomes were significantly higher in the exposed groups, with a 1.4-fold increase in spontaneous abortions (p = 0.007) and a 2.8-fold increase in stillbirths (p = 0.028). Although both populations were reported to be similar in several characteristics, such as smoking, lifestyle, occupation and socio-economic status, no information was provided, and other important factors such as smoking and maternal age were not accounted for. Furthermore, no mention was made of other potential environmental exposures; in populations of roughly similar size, the number of live births during the study period, 1980–1987 was 5218 in the high- and 2112 in the low-exposure area; it was stated that no significant differences were observed in the cancer frequency; the frequency of spontaneous abortions was unusually low, 7 and 5%, respectively.

A retrospective analysis (Tabacova et al., 1994a) compared the 5-year incidence rate of toxaemia of pregnancy in Srednogorie, a small town in Bulgaria located 2 km from a copper smelter, to the Bulgarian national rates. Data was derived from local and central morbidity registers. The incidence of toxaemia was more than three times greater around the smelter (8.0 vs. 2.5 per 1000 births). However, this report did not present any information regarding other factors or exposures that could be associated with the rate differences. The study also followed 71 pregnancies in Strednogorie, including measurement of urinary arsenic. The levels reported do not appear to reflect high environmental arsenic exposures.

Another study in Bulgaria (Tabacova et al., 1994b) included 34 maternal–infant pairs from the smelter area and 15 from a non-smelter area free from industrial exposures. Information regarding lifestyle characteristics, occupation, residence and medical history was ascertained by personal interviews. Samples of maternal and cord blood and placenta were obtained for analysis of arsenic, cadmium and lead concentrations. Infants born in the proximity of the smelter had lower birth weight (3012 vs. 3193 g). Although the study was small, these differences were also observed when divided into smoking and parity sub-groups. Placental concentrations of arsenic were three times higher ($p < 0.001$) in the smelter area, but cadmium levels were also elevated although the difference was not statistically significant. There was no difference in average lead levels in the two exposure areas.

A study conducted in the USA (Engel & Smith, 1994) investigated mortality from vascular diseases in the 30 counties with the highest average levels of arsenic in drinking-water for the period 1968–1984. The levels ranged from 5.4 µg/litre in Pierce County, Washington to 92 µg/litre in Churchill County, Nevada. When counties were grouped in three arsenic exposure categories, defined as 5–10, 10–20 and > 20 µg/litre, there appeared to be an increase in mortality from congenital anomalies of the heart only for females in the highest exposure group (SMR = 130, 90% CI (100–180), and for both sexes for congenital anomalies of the circulatory system (female SMR = 200, CI (110–340); male SMR = 130, CI (70–240). Slight increases in congenital anomalies of the heart and other anomalies of the circulatory system were found for two counties in the highest exposure group (29 and 46 µg As/litre in water, respectively), but none were found for Churchill County, which has the highest arsenic levels (92.5 µg/litre).

A hospital case–control study in the USA investigated the occurrence of stillbirths in relation to residential proximity to an arsenical pesticide production plant in Texas (Ihrig et al., 1998). Exposure was categorized in three groups according to arsenic air levels. An increasing, but not significant, trend in the risk of stillbirths was observed. The number of stillbirths was significantly elevated for the high-exposure group. When stratified by ethnicity, however, the findings remained significant for Hispanics only. Other exposures from the chemical plant were possible and were not measured in the study.

An ecological study examined infant mortality rates in three Chilean cities over a 46-year period (1950–1996) (Hopenhayn-Rich et al., 1999, 2000) (Fig. 4). Antofagasta, in northern Chile, experienced very high arsenic levels in drinking-water for a period of 12 years. In 1958 a new water source which contained arsenic concentrations around 800 µg/litre was introduced as the main

Neonatal Mortality Rates

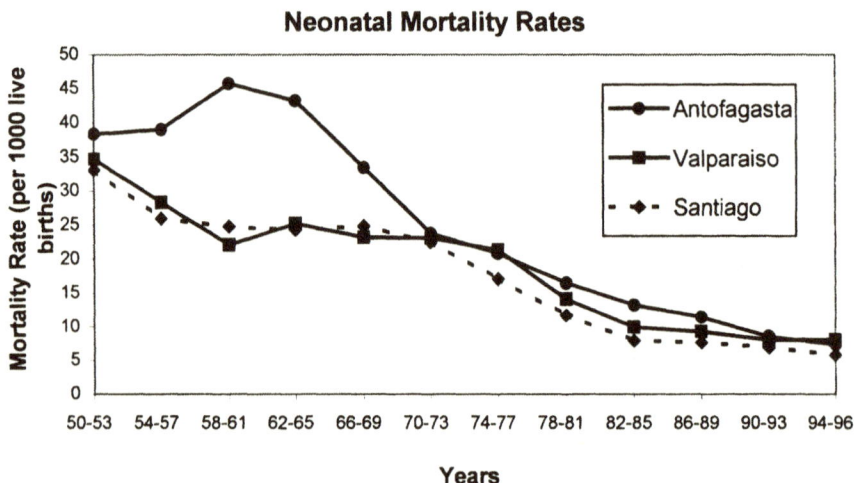

Fig. 5. Neonatal mortality rates in different areas in Chile
(Hopenhayn-Rich et al., 1999)

supplier of public water. In 1970, because of the overt signs of arsenicism observed in several studies, an arsenic removal plant was installed, and levels decreased initially to around 100 µg/litre, and then gradually over time to around 40 µg/litre. The changes in late fetal, neonatal and post-neonatal mortality rates over time in Antofagasta were compared to those in Valparaiso, another Chilean city with similar demographic characteristics but low in arsenic, and the capital, Santiago, with similarly low concentrations arsenic in the drinking-water. A close temporal relationship was observed between the high arsenic period and a rise in mortality rates in Antofagasta, whereas the other two cities had a steady decline in infant mortality. Fig. 5 shows the rates for neonatal mortality. Although data on other contaminants or factors related to infant mortality were not presented, the temporal relationship strongly suggests a role for arsenic exposure.

8.9 Genotoxicity and related end-points (Tables 36 and 45)

Genotoxicity studies in relation to arsenic exposure have included exposed and unexposed individuals from several populations, and have based their analysis on various tissues, including blood, buccal and bladder cells as well as sections from tumour biopsies. In several cases there appears to be at least a partial overlap of study subjects. For example, Ostrosky-Wegman et al. (1991), Gonsebatt et al. (1994, 1997) present results from different assays on samples from individuals from the Lagunera region in Mexico. Kuo et al. (1997), Hsu et al. (1997) and others focus their tumour analyses on samples from patients from the high-arsenic area of south-western Taiwan with different types of skin cancer.

The studies can also be categorized into main groups according to the focus of their investigation, such as *p53* mutations in tumour samples; sister chromatid exchange (SCE), chromosome aberrations (CA) and replication index (RI) in cultured lymphocytes; or micronuclei (MN) in exfoliated bladder and buccal cells as possible target tissues from direct exposure to arsenic from drinking-water.

Warner et al. (1994) compared the MN frequency in exfoliated buccal and bladder cells of individuals living in Nevada (USA) who relied on individual, private wells for their water supply. The frequency of MN in bladder cells was higher among the exposed than the unexposed subjects. This difference was more marked in men than in women. The lack of an effect in women was attributed to the fact that their urine contains other types of exfoliated cells in addition to bladder cells, thus diluting the effect towards the null. There was a significant positive correlation between urinary arsenic concentrations and the frequency of MN. No effect was observed in buccal cells. Moore et al. (1996) used fluorescent *in situ* hybridization (FISH) with a centromeric probe to examine MN formation in the exfoliated bladder cells in the same population from Nevada. They observed a 65% increase in frequency of MN containing acentric fragments, and a 37% increase in MN containing whole chromosomes, which is suggestive of both clastogenic and weak aneuploidogenic effects of arsenic. The frequencies of these two anomalies were associated with inorganic arsenic and its methylated metabolites excreted in urine. However, they were not significantly different from the control group.

331

Table 36. Genotoxicity of As in exposed humans

Study design	Study population, end-points measured	Source and level of As exposure	Health effects, metric of exposure and measure of association	Comments	Reference
Cross-sectional	18 exposed and 18 referents in Nevada MN in bladder and buccal cells	high exposure: well-water As concentration >500 (average 1312) µg/litre; referents, average 16 µg/litre	bladder cells MN/100 cells (SE) males — exposed 5.00 (1.50), referents 2.14 (0.46), Freq. ratio (CI) 2.34 (1.27, 4.29) females — 1.82 (0.53), 1.28 (0.31), 1.43 (0.76, 2.65) both acentric and whole chromosomes increased. no effect on MN in buccal cells	referents' age- and smoking-matched	Warner et al. (1994)
Cross-sectional	same as Warner et al. (1994) absence/presence of centromeres by FISH in bladder cells	high exposure: well-water As concentration >500 (average 1312) µg/litre; referents, average 16 µg/litre	MN+ (%) — expos. 0.190, ref (p) 0.102 (0.08); MN−(%) — exp. 0.167, ref (p) 0.081 (0.07) males females 0.078, 0.072 (0.31), 0.057, 0.041 (0.48) MN+/−, aberrations with/without centromere		Moore et al. (1996)

Table 36 (contd.)

		U-As (µg/litre)	MN+*	CI	MN-*	CI		
Cross-sectional	70 Chilean males with high water As exposure and 50 referents bladder cell MN using FISH for centromeres	high exposure: 600 µg/litre As, referents 15 µg/litre. average U-As levels 616 and 66 µg/litre, respectively					exposure stratified by quintiles	Moore et al. (1997b)
		<54	1.0		1.0			
		54–137	2.3	1.0–5.0	4.7	1.7–13.2		
		137–415	2.0	0.9–4.5	7.5	2.8–20.3		
		415–729	3.1	1.4–6.6	5.2	1.9–14.6		
		>729	0.9	0.4–2.2	1.0	0.3–3.3		
		*prevalence ratio						
Intervention	34 men from the exposed group in the previous study (Moore et al., 1997b) bladder cell MN	water with 45 µg/litre As supplied to participants for 8 weeks. U-As decreased from 742 to 225 µg/litre	MN frequency decreased from 2.6 to 1.8/1000 (prevalence ratio 0.7, $p < 0.05$); for those whose U-As was <700 µg/litre at the beginning of the intervention, the decrease was from 3.5 to 1.5 (prevalence ratio 0.4, $p = 0.002$).					Moore et al. (1997c)

Table 36 (contd.)

Study design	Study population, end-points measured	Source and level of As exposure	Health effects, metric of exposure and measure of association			Comments	Reference
Cross-sec-tional	13 exposed and 15 less exposed habitants in North Mexico CA, SCE, HPRT mutations in lymphocytes	average drinking-water As concentration for the exposed 390 µg/litre, 19–60 µg/litre for the referents	CA% (SD) high exp. 2.55 (1.73) low exp. 3.00 (2.82) all non-significant; *HPRT mutations	SCE(SD) 9.10 (2.7) 8.80 (1.6)	Vf (SD)* 2.42 (2.26) 5.03 (2.99)	complex chromosomal aberrations (dicentrics, rings, translocations) increased among the more heavily exposed (0.73% vs. 0.16%)	Ostrosky-Wegman et al. (1991)
Cross-sec-tional	33 exposed and 30 referents in Lagunera region, Mexico labelling index, mitotic index, replication index in lymphocytes	water As in average 412 µg/litre for the exposed and 37 µg/litre for the referents	labelling index controls 3.37 (SE 0.61), exposed without skin lesions: 3.95 (0.56), exposed with skin lesions 2.42 (0.49; $p < 0.05$) mitotic index at 72 h for controls 3.78 (SE 0.34), for exposed 6.34 (0.45; $p < 0.01$); no difference at 48 or 72 h replication index lower in exposed females at 48, 60 and 72 h; no difference among males				Gonsebatt et al. (1994)

334

Table 36 (contd.)

			CA(SE)	MN$_{buccal}$ (SE)	MN$_{bladder}$ (SE)		
Cross-sectional	35 exposed volunteers and 35 referents in Lagunera region, Mexico CA in lymphocytes, MN in buccal and bladder cells	water As in average 410 µg/litre for the exposed and 30 µg/litre for the referents	Ref. 2.96 (0.54) Exposed 7.12 (1.00)* Skin les. 7.38 (1.46) * $p < 0.05$	0.56 (0.13) 2.21 (0.47)* 3.28 (0.96)*	0.48 (0.10) 2.22 (0.99)* 4.64 (2.59)*	statistical comparisons included other variables, such as smoking, age and gender	Gonsebatt et al. (1997)
Cross-sectional	282 non-smoking exposed inhabitants and 155 referents from neighbouring province in Argentina lymphocyte SCE	water As for the exposed province ≥130 µg/litre; for the reference area, ≤20 µg/litre duration of exposure ≥20 years	lymphocyte SCE/cell among exposed 10.46 (SD 1.02) and among referents 7.49 (SD 0.97, $p < 0.001$) correlation between urinary As and SCE: R^2 0.64 for females and 0.33 for males			exposed considerably younger than referents (mean ages, 38.9 vs. 56.7 years) age or sex not considered in the analysis	Lerda (1994)

Table 36 (contd.)

Study design	Study population, end-points measured	Source and level of As exposure	Health effects, metric of exposure and measure of association		Comments	Reference
			MN/1000 (SE)	SCE/cell (SD)		
Cross-sectional	12 exposed women and 10 exposed children, and 10 referent women and 12 referent children in Argentina. MN and SCE in lymphocytes; FISH for aberration type	drinking-water As 0.2–0.5 mg/litre for the exposed; U-As median 260 µg/litre for exposed women and 310 for exposed children, and 8 and 13 µg/litre for the non-exposed	exp. children: 35 (46) exp. women: 41 (4.9) ref. children: 5.6 (1.6) ref. women: 8.5 (3.4) no differences in chromosomal translocations; aneuploidy more frequent (0.21 vs. 0%) among the exposed	4.4 (1.1) 5.7 (1.3) 4.6 (1.2) 5.5 (1.3)	MN frequency unusually low among the referents. As metabolite pattern different from that reported earlier for white populations	Dulout et al. (1996)

Table 36 (contd.)

			CA incl gaps (SD);p	CA excl gaps (SD);p	
Cross-sectional	32 current and 10 ex-users of arsenic-containing well water plus 8 referents in Finland lymphocyte CA	median well-water arsenic concentration 410 µg/litre for the exposed (all >1 µg/litre) and <1 µg/litre for the referents	current users 6.9 (3.4) ex-users 4.2 (1.9) referents 8.6 (3.6); 0.02 cum. dose mg/lifetime* ≤1.894 6.0 (2.9) ≥1.894 8.6 (4.3); 0.02 As in urine µg/litre* <206 6.1 (3.0) >206 8.9 (4.4); 0.02 *cut-off point 75th percentile. P-values from analysis of variance	3.5 (2.5) 1.9 (1.3) 3.6 (1.7); 0.1 2.8 (4.3) 4.5 (3.0); 0.02 2.8 (1.8) 4.8 (3.0); 0.008	Mäki-Paakkanen et al. (1998)
Cross-sectional	26 individuals well-water with Bowen's disease with known drinking-water arsenic exposure and 22 non-exposed BfD patients from BFD endemic area. p53 over-expression and proliferation in the tumour	arsenic concentration ≤1140 µg/litre, with progressive decrease since 1956	>10% immunohistochemical staining for p53 protein in tumours from 11/26 exposed and 2/22 non-exposed (p = 0.01) no difference in cell proliferative activity		Kuo et al. (1997)

Table 36 (contd.)

Study design	Study population, end-points measured	Source and level of As exposure	Health effects, metric of exposure and measure of association			Comments	Reference
Cross-sec-tional	15 cases of Bowen disease and 34 referents from the BFD endemic area	well-water arsenic con-centration ≤ 1140 µg/litre, with progres-sive decrease since 1956	patients referents	SCE/cell (SE); p 8.42 (51) 6.94 (0.37) ; <0.05	HFC % (SE); p 17.89 (2.83) 8.59 (1.66) < 0.05	referents matched for age, sex and residence	Hsu et al. (1997)
Cross-sec-tional	22 patients with cancer, 10 with BFD, 8 with cancer and BFD, 26 healthy indi-viduals from the BFD endemic area, and 23 healthy non-exposed referents	well-water arsenic concentration ≤1140 µg/litre, with progres-sive decrease since 1956	SCE frequencies not different among different groups. mitomycin-induced SCE frequencies higher among individuals from the BFD area than among referents not exposed to arsenic				Liou et al. (1996)

Table 36 (contd.)

			cases	referents	p		
Pros- pective	686 resi- dents of the Taiwan BFD- endemic area, of whom 31 developed cancer during a 4-year- follow up period SCE and CA in lympho- cytes	well-water As concentration ≤1140 µg/litre, with progres- sive decrease since 1956	SCE/cell (SD) 6.73 (1.53) CA (chromosome-type) 2.6 (1.7) CA (chromatid-type) 3.3 (1.8) CA tota 6.1 (2.4)	6.22 (1.11) 0.9 (1.0) 3.4 (2.0) 4.4 (2.6)	0.36 0.01 NS 0.018	for 9 of the 31 CA could not be analysed; final ana- lysis done on 22 cases and 22 referents	Liou et al. (1999)
Cross- sec- tional	13 cases of urothelial cancer (age 37–74 years) from BFD- endemic area in Taiwan	well-water As concentration ≤1140 µg/litre, with progres- sive decrease since 1956	8/13 cases had a mutation in exons 5–8 of the p53 gene; 9/10 point mutations were transitions			authors conclude that the mutation pattern observed is not different from those observed in transitional cell tumours in patients without As exposure	Shibata et al. (1994)

Table 36 (contd.)

Study design	Study population, end-points measured	Source and level of As exposure	Health effects, metric of exposure and measure of association	Comments	Reference
Cross-sec-tional	26 skin biop-sies from 16 Bowen's disease patients from the BFD-en-demic area	well-water As concentration ≤1140 µg/litre, with progres-sive decrease since 1956	no mutations observed in exons 5–8 in p53, or in codons 12, 13 or 61 H-, K- or N-ras oncogenes	no information on personal exposure level or duration of the study subjects, nor any personal characteristics such as age, sex or smoking	Hsieh et al. (1994)
	ras and p53 mutations				
Cross-sec-tional	23 patients with Bowen's disease, 7 with barsen-sical cell carcinoma, and 9 with squamous cell carci-noma from the BFD-en-demic area	well-water As concentration ≤1140 µg/litre, with progres-sive decrease since 1956	9/23 (39%) of Bowen's disease patients, 23/7 (29%) of BCC cases, and 5/9 (56%) of SCC cases had mutations in the p53 gene	authors concluded that the p53 gene mutation rates, sites and types in As-related skin cancer are signifi-cantly different from those in UV-induced skin cancer (from earlier studies)	Hsu et al. (1999)

In a larger study, differences in the frequency of bladder cell MN were investigated in volunteers from two northern Chilean communities with contrasting arsenic levels in drinking-water (Moore et al., 1997b). Bladder cell MN frequencies were higher in exposed men, and the difference increased when the analyses excluded individuals from the low exposure area with higher than background urinary arsenic levels (> 50 µg As/litre) (ratio = 2.0, 95% CI 1.3–3.1, $p < 0.001$). Smoking status did not affect the results. The prevalence of centromere positive cells was twice as high for the high-exposure group as for the modified low-exposure group ($p = 0.02$). Similar findings were reported for centromere-negative cells (1.2/1000) vs. 0.2/1000, $p = 0.001$). The increase in the centromere-negative cells in the high-exposure group indicates an increase in the proportion of chromosomal breakage vs. chromosomal lagging. The prevalence of MN increased from with exposure in exposure quintiles 1–4 (p for trend < 0.001). In the highest quintile, MN frequency returned to baseline, suggesting that MN formation may be inhibited at high doses owing to cytotoxicity or cytostasis.

An intervention study was conducted as an extension of the work described above (Moore et al., 1997b). A selected group of families from the high-exposed town were provided with water lower in arsenic (45 µg/litre) for 2 months (Moore et al., 1997c). Overall, MN frequency among smokers (but not among non-smokers) decreased, but for men originally in the highest quintile of exposure there was an increase in the MN frequency, which was attributed to a decrease in cytotoxicity.

Ostrosky-Wegman et al. (1991) compared various markers of effect in peripheral lymphocytes from individuals with high exposure and low exposure to arsenic in drinking-water in the Lagunera region of Mexico. There was no significant difference between the two groups in the rate of chromosomal aberrations, or SCEs. However, the average frequency of complex chromosomal aberrations (Cas; chromosome exchanges) was greater in the high-exposure group. There was also a slowdown in lymphocyte proliferation in the high-exposure group, indicating an alteration in the immune response. The HGPRT locus assay indicated that arsenic did not induce gene mutations.

The effects of chronic arsenic exposure on lymphocyte proliferation were further investigated in a larger group from the same Lagunera region of Mexico (Gonsebatt et al., 1994). Blood and urine samples were obtained from exposed and unexposed adult residents from towns with different water arsenic levels. Exposure was assessed by questionnaires, and by urinary and water arsenic levels. Lymphocyte proliferation was performed at different culture times using labelling indexes (LI), mitotic indexes (MI) and replication indexes (RI). The results showed slower cell kinetics among exposed individuals overall, but variations in sub-groups were observed: mean LIs were higher among those with skin lesions, RIs were lower in exposed individuals, but this effect was significant only in females. No effects were found when correlating LIs, MIs or RIs with respect to age, duration of residence or urinary arsenic concentrations.

In a subsequent study in the same area, Gonsebatt et al. (1997) examined the rates and types of CAs in lymphocytes, and the MN frequency in exfoliated buccal and bladder cells, in volunteers from the same towns (there is significant overlap of study subjects with Gonsebatt et al., 1994). In lymphocytes, the CA per cell, and the percentage of cells with CA, were higher in the exposed group than in the controls. The most frequent types of CA observed were chromatid deletions. The mean frequency of MN/1000 cells was also higher in the exposed group than in the control group, for both buccal cells and urothelial cells. Exposed individuals bearing skin lesions had higher frequencies of MN, but not CAs, than those without lesions.

A study in Argentina examined the frequency of SCEs in relation to water and urinary arsenic levels of arsenic-exposed and unexposed persons (Lerda, 1994). Among the exposed, hyper-keratosis, melanosis, and basal carcinomas were observed. Water samples were obtained from each subject's home and analysed for arsenic content. The mean SCE frequency was higher in the exposed than in the unexposed group. It should be noted that the control group was on average younger than the exposed group. The frequency of SCEs was not correlated with age, sex, or exposure to pesticides, lead, or other metals.

The clastogenic and aneugenic potential for arsenic exposure, not contaminated with lead, cadmium, industrial pollution, or pesticides, was investigated in cultured lymphocytes from women and children from populations exposed to high arsenic levels from drinking-water in north-western Argentina (Dulout et al., 1996). The frequency of MN in lymphocytes from the exposed group was substantially higher than from the control group, both in children and in women. No differences were found for SCEs or for cell cycle progression analysis in relation to exposure level. The analysis of CAs indicated no differences in chromosomal translocations, but the frequency of numerical aberrations (aneuploidy–trisomy) was higher in the exposed group as supported by FISH analysis. The authors note that this population displays some unique characteristics which may be associated with distinct polymorphisms, since the frequency of MN in the low-exposure groups were about half of those previously reported for whites. In addition, patterns of urinary arsenic metabolites appear to be different than those of other studied populations.

In a study conducted in Finland, individuals exposed to arsenic from well-water were compared to unexposed controls, with respect to urinary arsenic and frequency of CAs (Mäki-Paakkanen et al., 1998) Exposed subjects were classified as current users and ex-users of arsenic-laden water; the latter group had stopped using their well-water 2–4 months before the study began and were drinking from a low-arsenic source. The non-exposed group had the highest CA frequency, followed by current users and ex-users. However, when dichotomized into two urinary arsenic concentration or cumulative lifetime exposure groups, CAs were highest in the high-exposure group. When urinary arsenic was entered as a continuous variable in a regression model (adjusting for gender, age, smoking and cell culture batch), the correlation was stronger for CA excluding gaps (all subjects $r = 0.23$, $p = 0.02$; current users $r = 0.30$, $p = 0.06$).

A comparative investigation assessed the activity of the *p53* gene in formalin-fixed biopsy samples from 48 individuals diagnosed with Bowen's disease with ($n = 26$) or without ($n = 22$) a history of exposure to arsenic (Kuo et al., 1997). The rates of *p53* positivity were higher in the arsenic-exposed group. In a sub-group of 5 cases which had multiple lesions examined, the *p53* status remained 100% consistent within individuals, adding strength to the association. The *p53* overexpression did not seem linked to the proliferative activity.

343

Patients with Bowen's disease in the BFD-endemic area in Taiwan were found to have a higher frequency of SCEs and HFCs and a lower RI (Hsu et al., 1997). *In vitro* treatment with sodium arsenite increased SCEs, HFCs and RI, but only the decrease in RI was statistically more pronounced in arsenic-induced Bowen's disease patients than controls.

Liou et al. (1996) also examined the rates of SCEs and mitomycin C-induced SCEs in patients with cancer only (17 skin, 5 other types), BFD only, with both BFD and cancer (6 with skin cancer), and healthy controls from the BFD area. In addition, healthy non-arsenic-exposed workers were used as external controls. The baseline SCEs did not differ among the 5 groups studied. Although smokers consistently had higher SCEs than non-smokers across all groups, no differences were observed between groups when stratified by smoking status. After *in vitro* treatment with mitomycin, the increase in SCEs was the same in lymphocytes from the 4 BFD-area groups (among which there were no differences in lifetime exposure to arsenic in well-water), but they were overall higher than in the external controls.

A nested case–control study was performed in the BFD-endemic area of Taiwan (Liou et al., 1999). A cohort of 686 residents was assembled, and after 4 years, 31 people had developed cancer. Of these, blood samples obtained at the beginning of the cohort study were successfully processed for 22 cases. Controls were selected from among members of the cohort who had not developed cancer, matched on sex, age, village of residence and smoking. No differences were found in overall frequencies of SCEs. The frequency of total CAs was significantly higher among the cases. When categorized by specific type of CA, all chromosome-type CAs were significant except exchanges, whereas none of the chromatid-type CAs were significant. The limitation of this investigation is the apparent uniformity of arsenic exposure among cases and controls, making it hard to link the CA frequency with arsenic, rather than to other factors, or genetic susceptibility.

An investigation focused on *p53* analysis was performed on 13 urothelial tumours from residents of the BFC area in Taiwan (Shibata et al., 1994); 11 cases were transitional cell carcinomas and 2 were squamous cell carcinomas. DNA was extracted from archival tissue and analysed for mutations in the *p53* gene by SSCP, and if mutations were identified, these were subsequently sequenced. Eight cases (62%) showed mutations and 9 of the 10 point mutations observed were transitions. The type of mutations was not different from the *p53* mutational spectra of other transitional cell carcinomas. However, 2 of the mutations were at a mutational hot spot for colon cancer, but not previously associated with transitional cell carcinomas except in cases associated with inflammatory agents (e.g. phenacetin, schistosomiasis); and 3 of the tumours contained double mutations, a relatively rare mutagenic event in human cancers.

Hsieh et al. (1994) examined 26 skin biopsies from 16 patients from the BFD-endemic area of Taiwan with arsenic-related Bowen's disease, looking for p53 mutations by SSCP on exons 5–8, and for mutations of H-, K-, and N-*ras* genes. No mutations were detected in any of the samples analysed. These results are contrary to what is found in UV-related skin tumours, where mutations in all the genes studied have been detected.

To understand the role of the *p53* gene in the process of carcinogenesis of arsenic-induced skin cancers, samples of tumour specimens and normal skin were collected from 23 patients with Bowen's disease, 7 with basal cell carcinomas and 9 with squamous cell carcinoma (Hsu et al., 1999). All the tumours were from patients from the BFD-endemic area in Taiwan, and were collected from areas of the body not usually exposed to sunlight. Direct sequencing of the *p53* gene on exons 2–11 showed six types of *p53* mutations, with 38% of the type G:C A:T transition, as well as the finding of the same hot spot in 3 cases (codon 175). The frequency of *p53* mutations differed by tumour type, although the groups were quite small: 39% of Bowen's disease cases, 29% of basal cell carcinoma cases, and 56% of squamous cell carcinoma cases. Overall, the *p53* gene mutation rates, sites and types in arsenic-related skin cancer were significantly different from those in UV-induced skin cancer.

9. EFFECTS ON OTHER ORGANISMS IN THE ENVIRONMENT

9.1 Laboratory experiments

The toxic effects of arsenicals are significantly modified by numerous biological and abiotic factors. The toxicity of arsenic in the environment is affected by temperature, pH, Eh, organic content, phosphate concentration, adsorption to solid matrices, the presence of other substances and toxicants, duration of exposure and the arsenic species present. In general, inorganic arsenicals are more toxic than organoarsenicals to biota, and trivalent species are more toxic than pentavalent species for both inorganic and organic arsenic compounds (NAS, 1977; NRCC, 1978; Eisler, 1988).

9.1.1 Microorganisms

9.1.1.1 Water

Aquatic microorganisms show a wide range of sensitivities to arsenic species, with arsenite generally being more toxic than arsenate. Resistance to arsenic species has been reported in microorganisms. The toxicity of arsenate is decreased by increasing the phosphate concentration, whereas the inhibitory effect of arsenite is independent of phosphate.

Bringmann & Kühn (1977, 1978) exposed the bacterium *P. putida* for 16 h and the cyanobacteria *Microcystis aeroginosa* for 8 days to arsenate. The toxic thresholds for the inhibition of cell multiplication were found to be 8.1 and 7.3 mg As(V)/litre respectively.

Growth under phosphate-limiting conditions (20 µmol/litre) of the blue-green alga *Synechococcus leopoliensis* was unaffected by arsenate < 15 mg As(V)/litre (200 µmol/litre) (Budd & Craig, 1981). Arsenite concentrations > 3.75 mg As(III)/litre (50 µmol/litre) inhibited the growth of *S. leopoliensis* but the inhibition was transitory with growth resuming after a lag period the length of which was related to the arsenite exposure. The growth of algae which had been previously exposed to arsenite (0.75 mg As(III)/litre (10 μmol/litre) for 12 h) was unaffected at 15 mg As(III)/litre (200 μmol/litre) (Budd et al., 1986).

Arsenate at ≥30 mg As(V)/litre completely inhibited growth of a marine cyanobacterium (*Phormidium* sp.) when incubated in the absence of phosphate (Takahashi et al., 1990). However, at a phosphate concentration of 50 μmol/litre no effect on growth was observed at 150 mg As(V)/litre. Thiel (1988) reports that arsenate was a poor non-competitive inhibitor of phosphate transport in the cyanobacterium *Anabaena variabilis* with a K_i of 82.5 mg As(V)/litre (1.1 mmol/litre). In cells starved of phosphate for 3 days, arsenate was almost completely non-competitive with a K_i of 5.6 mg As(V)/litre (75 μmol/litre). Preincubation of phosphate-starved cells with arsenate caused subsequent inhibition of phosphate transport, suggesting that intracellular arsenate inhibited phosphate transport.

Toxicity of arsenic to microalgae is summarized in Table 37. $EC_{50}s$, based on growth, range from 48 μg As(V)/litre to 202 mg As(V)/litre (14 days). Blum (1966) found that arsenate competitively inhibits phosphate uptake by green algae *Euglena gracilis*. The authors report that arsenate has almost as high an affinity for the transport system as does phosphate, although arsenate is accumulated at a much lower rate than phosphate.

Hörnström (1990) examined the toxicity of arsenate to a variety of algal species in 72–96-h tests. The most tolerant group was the Chrysophyceae, with four of the five species tested having a no-observed-effect concentration (NOEC) of 500 μg As(V)/litre. Chlorophyceae including *Chlamydomonas* sp. and *Scenedesmus denticulatus*, Bacillariophyceae and Cryptophyceae show a lowest observed effect concentration (LOEC) of 50 μg/litre. The most sensitive alga was a Chrysophyte *Stichogloea doederleinii* with a LOEC of 5 μg As(V)/litre.

Bringmann & Kuhn (1977) exposed the green alga *Scenedesmus quadricauda* to arsenate for 8 days. They found the toxic threshold for the inhibition of cell multiplication to be 3.5 mg As(V)/litre. Maeda et al. (1985) found that the growth of microalgae (*Chlorella vulgaris*) isolated from an arsenic-polluted environment was unaffected at arsenate concentrations of 2000 mg As(V)/litre.

Table 37. Toxicity of As to microalgae[a]

Organism	Stat/flow	Tempera-ture (°C)	pH	Salt	Duration	EC50 (mg As/litre)	Reference
Freshwater							
Green algae							
Scenedesmus quadricauda	stat	25	8.0	arsenate	12 d	61 n (59.3–70.1)	Fargasova (1994a)
Scenedesmus obliquus	stat	24	7.0	arsenate	14 d	0.048	Vocke et al. (1980)
Ankistrodesmus falcatus	stat	24	7.0	arsenate	14 d	0.26	Vocke et al. (1980)
Selenastrum capricornutum	stat			arsenite	96 h	31.2	US EPA (1985)
	stat			arsenate	96 h	0.69	US EPA (1985)
	stat	24	7.0	arsenate	14 d	30.8	Vocke et al. (1980)
Chlamydomonas reinhardii	stat	20	7.0	arsenate	14 d	202	Jurewicz & Buikema (1980)
Marine							
Diatom							
Nitzschia closterium	stat	21	8.1	arsenate	72 h	>2	Florence et al. (1994)
	stat	21	8.1	arsenite	72 h	0.007	Florence et al. (1994)

[a] stat = static conditions (water unchanged for duration of test; EC$_{50}$s based on growth); n = based on nominal concentrations

Michnowicz & Weaks (1984) studied the effects of pH on the toxicity of arsenate to *Selenastrum capricornutum* in 14-day tests. Growth was significantly enhanced at pH 6, 8, 10 and 12 compared with pH 4, with optimum growth at pH 10. Addition of arsenate (0.2 mg/litre) inhibited growth over the range of pH values tested; growth of cultures at pH 8 was significantly higher than at pH 4. No significant difference in arsenate-induced growth inhibition was observed at 7 and 14 days, indicating that no arsenic toxicity occurred after 7 days.

In 14-day growth inhibition tests with *Chlamydomonas reinhardii* stimulation of the biomass occurred at arsenate concentrations of < 151 mg As(V)/litre with maximum stimulation (73% above controls) at 100 mg As(V)/litre (Jurewicz & Buikema, 1980).

Conway (1978) found no significant effect on growth or micronutrient utilization of the freshwater diatom *Asterionella formosa* exposed to 160 μg As(V)/litre (as arsenate) for < 23 days.

Growth and survival of the marine microalgae *Tetraselmis chui* and *Hymenomonas carterae* were not affected during a 6-day exposure to concentrations as high as 1 mg As/litre of arsenite or arsenate (Bottino et al., 1978).

Growth of *Dunaliella* sp. was inhibited during exposure for 3 days to arsenate concentrations of ≥100 μg As(V)/litre, but during continued exposure growth rate recovers and is normal after 12 days at concentrations < 2 mg As(V)/litre (Yamaoka & Takimura, 1986).

Hollibaugh et al. (1980) studied the toxicity of arsenic to *Thalassiosira aestevalis* in 4-day tests. Arsenate had no effect on growth at 75 μg As(V)/litre (1000 nmol/litre) in a high-phosphate medium. Growth was reported to be repressed by both arsenate and arsenite at > 22.5 μg As/litre (300 nmol/litre) in a low-phosphate medium; however, no statistical analysis was carried out on these results.

Sanders (1979b) exposed the diatom *Skeletonema costatum* to arsenate, arsenite and DMA for 6–8 days. Growth was significantly inhibited at arsenate and arsenite concentrations of 12.5 μg As/litre

(167 nmol/litre). DMA had no effect on growth at 9.8 µg As/litre (130 nmol/litre). Arsenate and arsenite additions of ≥5 µg As/litre (67 nmol/litre) caused significant inhibition of ^{14}C uptake by *Skeletonema* during both log and stationary phases; however, DMA had no significant effect on carbon uptake at 25.5 µg As/litre (340 nmol/litre). Additions of phosphate (20 µmol/litre) to the media eliminated the arsenate inhibition of carbon uptake.

Knauer et al. (1999) investigated the toxicity of arsenic species to natural phytoplankton assemblages from contaminated lakes within the Aberjona watershed (USA) using short-term photosynthesis bioassays. The toxicity of the arsenic species generally decreased in the order arsenate > arsenite > DMA. Toxicity of arsenate to phytoplankton collected from an unpolluted lake was highest (EC_{50} = 3 × 10^{-7} mol/litre), whereas algae from polluted lakes were more tolerant (EC_{50} = 3 × 10^{-6} mol/litre). The sensitivities of the different algal communities to arsenite were similar (EC_{50} = 5 × 10^{-5} mol/litre). Long-term studies with cultures of natural phytoplankton communities exposed to low levels of arsenate (1–15 µg As(V)/litre) showed that pentavalent arsenic differentially inhibits certain plants, causing a change in species composition, succession and predator–prey relationships (Sanders & Vermersch, 1982; Sanders & Cibik, 1985, 1988; Sanders, 1986). Arsenate inputs caused declines in large centric diatoms and replacement by a smaller diatom (*Th. pseudonana*) and small flagellates. Sanders & Riedel (1987) suggest two mechanisms by which algal species may be less sensitive to arsenate and, therefore, dominate exposed communities. Resistant species may have a higher affinity for phosphate and thus a lower uptake rate of arsenate, or some species may be able to transform it intracellularly into a less toxic form (Sanders & Riedel, 1987).

Blanck & Wängberg (1988b) established 1 h IC_{20}s for arsenate, based on inhibition of photosynthesis, for both natural marine periphyton communities and laboratory-established communities under low-phosphate conditions. Although there were substantial differences between the communities in terms of structure and biomass, IC_{20}s were similar at 30 µg As(V)/litre (0.4 µmol/litre) and 45 µg As(V)/litre (0.6 µmol/litre) respectively. In longer-term studies (3 weeks) similar values for IC_{20}s (15–60 µg As(V)/litre; 0.2–0.8 µmol/litre) were established for changes in species composition, and reductions in carbon, nitrogen and chlorophyll *a* content. Blanck & Wängberg (1988a) found that marine periphyton communities previously exposed to arsenate concentrations from 7.5 to 22.5 μg As(V)/litre (0.1–0.3 μmol/litre) showed increased resistance to arsenate. The authors concluded that arsenate exerts a selection pressure on the community, leading to the replacement of sensitive species with tolerant ones which causes the overall arsenate tolerance of the community to increase. Pre-exposure to arsenate at 750 μg As(V)/litre (10 μmol/litre) increased arsenate tolerance of the community by a factor of 16 000 (Blanck & Wängberg, 1991).

Wängberg & Blanck (1990) found that increasing the phosphate concentration from 0.1 to 0.8 μmol/litre decreased the toxicity of arsenate to marine periphyton communities in 72 h tests by > 3000-fold EC$_{20}$ values based on carbon dioxide fixation increased from 22.5 μg: As(V)/litre to > 75 mg/litre (0.3 μmol/litre to > 1 mmol/litre). Similar changes in toxicity were observed during natural upwelling episodes during which both phosphate and nitrate concentrations increased.

9.1.1.2 Soil

Soil microorganisms show wide variation in resistance to arsenic species.

The toxicity of arsenate (As(V)), as measured by retardation or inhibition of growth in 8-week tests, showed wide variation among different species of fungi. Toxicity was consistently reduced by the addition of phosphate with both arsenate-sensitive and arsenate-tolerant strains. *Poria monticola*, an arsenate-sensitive fungus, was completely inhibited by 187.5 mg As(V)/kg (0.0025 mol/litre) but was progressively less inhibited as the phosphate concentration increased and some growth occurred at 3000 mg As(V)/kg (0.04 mol/litre) when 0.16 mol/litre potassium phosphate was added. An arsenate-tolerant fungus, *Cladosporium herbarum*, showed 36% reduction in growth at 6000 mg As(V)/kg (0.08 mol/litre) but when 0.01 mol/litre phosphate was added there was no effect on growth at arsenate concentrations of 48 g As(V)/kg (0.64 mol/litre). Addition of phosphate also reduced the toxicity of arsenite (As(III)) but not DMA (Da Costa, 1972). The counteracting effect of phosphate on arsenate toxicity was found to occur with all of the fungi tested and with the bacteria *Bacillus subtilis* and *P. aeruginosa*.

Sharples et al. (1999) found $EC_{50}s$ for arsenate, based on growth inhibition, for the endomycorrhizal fungus *Hymenoscyphus ericae* and the ectomycorrhizal fungus *Hebeloma crustuliniforme* to be 99.6 mg As(V)/litre (1.33 mol/m^3) and 24.7 mg As(V)/litre (0.33 mol/m^3) respectively. The presence of phosphate (0.01–1.0 mol/m^3) in the media ameliorated the toxic effects of arsenate.

9.1.1.3 Bacterial resistance to arsenic

Burton (1987) collected heterotrophic bacteria from a variety of contaminated sites and found that < 0.21% were resistant to arsenite at 750 mg As(III)/litre (10 mmol/litre). Resistance to arsenite was much lower than that reported for selenite (54%) at the same sites.

Huysmans & Frankenberger (1990) isolated arsenic-resistant bacteria from contaminated agricultural drainage water. Plasmid-mediated arsenic-resistant bacteria have now been widely found in various sources. Several of the systems have been cloned and sequenced; closely related arsenic resistance systems are found on plasmids and the chromosome of *Escherichia coli* and plasmids of *Staphylococcus* (Rosenstein et al., 1992; Ji & Silver, 1995). The mechanisms of bacterial resistance are reviewed and discussed in detail by Ji & Silver (1995). Genes on bacterial plasmids have been identified that encode specific resistance systems for arsenic. The chromosomally encoded arsenical resistance (*ars*) operon was found in all strains of *E. coli* but not in *Salmonella typhimurium*, *P. aeruginosa* or *B. subtilis* (Carlin et al., 1995). Bröer et al. (1993) demonstrated energy-dependent accelerated arsenite efflux from *Staphylococcus aureus* cells with the cloned resistance determinant. In Gram-negative bacteria, the efflux pump consists of a complex formed by an ATPase (ArsA) associated with a membrane anion channel (ArsB). Arsenate is converted to arsenite by a soluble reductase (ArsC). Proteins ArsA and ArsB, but not the ATPase, are also found in Gram-positive bacteria (Cervantes, 1995). Other than plasmid arsenic resistance determinants, some bacteria also have the ability to oxidize arsenite to arsenate enzymatically (see section 4.2.1).

Suzuki et al. (1997) isolated the acidophilic bacterium *Acidiphilium multivorum* from acid mine drainage in Japan. Bacteria were found to be resistant to arsenite at concentrations < 1125 mg As(III)/litre (15 mmol/litre). Baldi et al. (1995) isolated arsenic-resistant bacteria (750 mg/litre sodium arsenite) from mosses growing near geothermal plants (south-west of Siena, Italy).

Growth of *P. putida* was not impaired by arsenate concentrations of 1000 mg As(V)/litre (Maeda et al., 1990b). Huysmans & Frankenberger (1990) isolated bacteria from agricultural drain water and exposed the organisms to arsenic compounds for 48 h. Arsenate, MMA and DMA had no effect on colony-forming units at concentrations < 1000 mg/litre. However, arsenite concentrations > 1 mg/litre inhibited the population. Similar results were found with sediment, with no effect at an arsenate concentration of 1000 mg As/kg and increases in colony-forming units at MMA or DMA concentrations > 25 mg/kg, but arsenite concentrations > 100 mg/kg caused a decline in colony-forming units. In further tests, the arsenic-resistant bacteria showed a high tolerance to a variety of metals and antibiotics.

9.1.2 Aquatic organisms

9.1.2.1 Macroalgae

Growth was significantly reduced at 212 µg As(III)/litre, and at 300 µg/litre all plants died. Phosphate concentrations of < 9.1 µmol/litre had no effect on arsenite toxicity. Concentrations of 10 mg/litre arsenate did not cause mortality, but sexual reproduction did not occur at this concentration. The toxicity of arsenate increased as the phosphate concentration decreased (Thursby & Steele, 1984).

9.1.2.2 Aquatic plants

Jenner & Janssen-Mommen (1993) studied the effect of arsenite and arsenate on the growth of the duckweed *Lemna minor*. They reported 14-day EC_{50} values, based on growth inhibition, of 0.63 mg As(III)/litre and 22.2 mg As(V)/litre, and NOECs of < 0.75 mg As(III)/litre and < 4 mg As(V)/litre. Sarkar & Jana (1986) found the floating aquatic heterosporous fern *Azolla pinnata* to be resistant to arsenic in 28-day tests. No significant effect on growth, photosynthesis, chlorophyll and protein content or tissue permeability were observed at arsenate concentrations of 1 mg As(V)/litre.

9.1.2.3 Invertebrates

Acute toxicity of inorganic arsenic to freshwater and marine invertebrates is summarized in Tables 38 and 39 respectively. In general, arsenite appears to be more toxic than arsenate in acute tests. The 48-h LC/EC_{50} values range from 0.68 to 73.5 mg/litre for trivalent arsenic and from 3.6 to 49.6 mg/litre for pentavalent arsenic. However, the lowest acute value is for the nauplius stage of the marine copepod *Tigriopus brevicornis*, with a 96-h LC_{50} of 10.9 µg As(V)/litre. Acute toxicity of organic arsenicals to aquatic invertebrates is summarized in Table 40. The 48-h LC_{50} values for MMA range from 17.4 mg As/litre to 2361 mg As/litre.

Schaefer & Pipes (1973) found that the acute toxicity of arsenate to rotifers (*Philodina roseola*) increased with increasing temperature (5 to 35 °C). For example, the 96-h LC_{50} was 18 mg As(V)/litre at 5 °C and 6.6 mg As(V)/litre at 35 °C. Bryant et al. (1985) exposed three estuarine invertebrates (*Corophium volutator*, *Macoma balthica* and *Tubifex costatus*) to pentavalent arsenic for < 384 h. Median survival times decreased as temperature (5, 10 and 15 °C) and concentration of arsenic (1 to 128 mg As(V)/litre) increased but salinity changes (5 to 35 g/litre) had no significant effect. The presence of sediment consistently reduced the acute toxicity (48 h) of trivalent arsenic to *D. magna* in repeated tests. The toxicity had been reduced by a factor of 5 within 6 days and by a factor of 16 within 49 days (Burton et al., 1987). Golding et al. (1997) found that the freshwater snail *Potamopyrgus antipodarum* was insensitive to arsenic in avoidance tests. The 48-h EC_{50}s for arsenite and arsenate, based on immobilization, were 34.6 mg As(III)/litre and 325 mg As(V)/litre for snails from a contaminated site (0.3 mg As/litre), and 19.3 mg As(III)/litre and 194 mg As(V)/litre for a control site. The authors report that previously exposed snails appear to be more sensitive to the presence of arsenate and arsenite; in avoidance experiments snails from a contaminated site showed significant avoidance at 15 mg As(III)/litre whereas control snails responded at 28 mg As(III)/litre.

Spehar et al. (1980) found no significant effect on survival or young production of *D. magna* exposed to either arsenite, arsenate, MMA (disodium salt) or DMA (sodium salt) at 1 mg/litre in 14-day tests. In a 7-day test on *Ceriodaphnia dubia* the maxiumum allowable toxicant concentration (MATC) for arsenite was found to be 1.1 mg As(III)/litre, with production of young being the most sensitive parameter (Spehar & Fiandt, 1986). In 15-day tests there was no effect on the survival of the estuarine copepod *Eurytemora affinis* at arsenate concentrations of 50 µg/litre; however, concentrations of ≥100 µg/litre caused a significant increase in the mortality of juveniles. Adult copepod survival was significantly reduced at arsenate concentrations of 1 mg/litre (Sanders, 1986).

On the basis of survival and reproductive impairment of *D. magna*, 21-day EC_{50}s were found to be 2.9 and 1.4 mg As(V)/litre (as arsenate) respectively at a water hardness of 45 mg $CaCO_3$/litre (Biesinger & Christensen, 1972). However, Naddy et al. (1995) found no significant effect of arsenate on fecundity of *C. dubia* at 1.42 mg As(V)/litre when tested at 120 mg $CaCO_3$/litre. Enserink et al. (1991) report a 21-day LC_{50} for arsenic pentoxide of 5.8 mg As(V)/litre (semi-static test) for D. magna with an EC_{50} of 3.2 mg As(V)/litre on the basis of population effects (survival, body growth, rate of population increase and maximum yield) under flow-through conditions. Lima et al. (1984) found that daphnid survival, production of young and mean total length of adults were significantly reduced at ≥1320 µg As(III)/litre (as arsenite) in 28-day tests. The NOEC was between 633 and 1320 µg/litre for these parameters.

Ettajani et al. (1996) found no effect of arsenic-spiked sediments (20.5 mg As(V)/kg) or dissolved (10 µg As(V)/litre) arsenate on survival of oysters (*Crassostrea gigas*). Structural alterations of mitochondria and nuclei in arsenic-exposed oysters were observed by electron microscopy.

Naqvi & Flagge (1990) exposed crayfish to MMA at a concentration of 46 mg As/litre for 168 days. There was no significant effect on the number of eggs laid, but arsenic significantly reduced the number of eggs that hatched. No significant effect was observed on growth or moulting frequency of newly hatched crayfish exposed to 6.9 mg As/litre (as MMA).

Cowell (1965) found that arsenite concentrations of 4 mg As(III)/litre caused significant reductions in populations of zooplankton (rotifers, copepods and cladocerans) in experimental

Table 38. Toxicity of inorganic As to freshwater invertebrates

Organism	Size/ age	Stat/ flow	Temperature (°C)	Hardness (mg/litre)	pH	As species	Duration (h)	LC_{50} (mg As/litre)	Reference
Snail *Aplexa hypnorum*	adult	stat	25	49.5	7.4–7.7	As_2O_3	96	18.6 m	Holcombe et al. (1983)
Tubificid worm *Tubifex tubifex*	20 mm	stat	20	NS	7.4	arsenate	96	127.4 (108.8–134.3) n	Fargasova (1994a)
Amphipod *Gammarus pseudolimnaeus*	NS	flow	18.5	46.3–49.9	7.2–8.1	arsenite	96	0.87 m	Lima et al. (1984)
Water fleas	NS	stat	18	45	7.7	arsenate	48	7.4 n	Biesinger & Christensen (1972)

Table 38 (contd).

Species									Reference
Daphnia magna	NS	stat	15.6	46.3–49.9	7.2–8.1	arsenite	48	1.5 (1.2–1.9) ! m	Lima et al. (1984)
	NS	stat	15.6	46.3–49.9	7.2–8.1	arsenite	48	4.6 (3.7–5.8) !! m	
	6–24 h	stat	20	NS	7.3	arsenate	48	44.7 (35.2–50.9) n	Fargasova (1994a)
	NS	stat	NS	NS	NS	arsenite	48	2.1 & 6.6	Burton et al. (1987)
	1st instar	stat	15	44	7.4	arsenite	48	1.7 (1.3–2.4) @	Mayer & Ellersieck (1986)
Daphnia pulex	NS	stat	NS	NS	NS	arsenate	48	3.6 (3.3–3.9) m	Jurewicz & Buikema (1980)
	>24 h	stat	17	120	6.8	arsenate	48	49.6 (48.7–50.5) @ n	Passino & Novak (1984)
Ceriodaphnia dubia	<24 h	stat	25	100	8.2	arsenite	48	1.5 (1.2–1.7)	Spehar & Fiandt (1986)

357

Table 38 (contd.)

Organism	Size/age	Stat/flow	Temperature (°C)	Hardness (mg/litre)	pH	As species	Duration (h)	LC_{50} (mg As/litre)	Reference
Simocephalus serrulatus	1st instar	stat	16	44	7.4	arsenite	48	0.8 (0.6–1.1) @	Mayer & Ellersieck (1986)
Cladoceran *Bosmina longirostris*	<24 h	stat	17	120	6.8	arsenate	96	0.85 (0.7–1.0) @ n	Passino & Novak (1984)
Midges *Chironomus tentans*	3rd instar	stat	14	25	6.3		48	0.68 @ n	Khangarot & Ray (1989)
Tanytarsus dissimilis	3rd/4th instar	stat	24	47	7.2–7.7	As_2O_3	48	73.5 (72–75) m	Holcombe et al. (1983)
Stonefly *Pteronarcys californica*	1st year class	stat	15	44	7.4	arsenite	96	21.9 (17.3–27.7)	Mayer & Ellersieck (1986)

stat = static conditions (water unchanged for duration of test); hardness expressed as mg $CaCO_3$/litre; @ = EC_{50}s based on immobilization; ! = unfed; !! = fed; n = based on nominal concentrations; m = based on measured concentrations

Table 39. Toxicity of inorganic As to marine invertebrates

Organism	Size/age	Stat/flow	Temper-ature (°C)	Salinity (g/litre)	pH	As species	Duration (h)	LC_{50} (mg As/litre)	Reference
Eastern oyster									
Crassostrea virginica	juvenile	flow	15	22	NS	As_2O_3	96	> 0.75 # n	Mayer (1987)
	juvenile	flow	13	31	NS	arsenate	96	> 0.4 # n	Mayer (1987)
Pacific oyster									
Crassostrea gigas	embryo	stat	20	34	8.1	As_2O_3	48	0.33 ## n	Martin et al. (1981)
Mussel									
Mytilus edulis	embryo	stat	17	34	8.1	As_2O_3	48	> 3.0 ## n	Martin et al. (1981)
Bay scallop									
Argopecten irradians	juvenile	stat$	20	25	NS	arsenite	48	4.4 n	Nelson et al. (1976)
	juvenile	stat$	20	25	NS	arsenite	96	3.5 (2.1–5.8) n	
Polychaete worm									
Neanthes arenaceodentata	NS	flow	NS	NS	NS	arsenite	NS	10.1 m	US EPA (1985)

Table 39 (contd.)

Organism	Size/ age	Stat/ flow	Temperature (°C)	Salinity (g/litre)	pH	As species	Duration (h)	LC_{50} (mg As/litre)	Reference
Dungeness crab									
Cancer magister	zoeae	stat	15	34	8.1	As_2O_3	96	0.23 n	Martin et al. (1981)
Intertidal crab									
Scylla serrata	60–70 mm*	stat$	26.5–29.5	NS	7.0–7.2	As_2O_3	48	23 (18–29.4) n	Krishnaja et al. (1987)
	60–70 mm*	stat$	26.5–29.5	NS	7.0–7.2	As_2O_3	96	17 (13.4–21.4) n	
Amphipods									
Elasmopus bampo	8–12 mm	stat	19	NS	NS	As_2O_3	96	2.8 (1.8–4.3) n	Reish (1993)
Corophium insidiosum	8–12 mm	stat	19	NS	NS	As_2O_3	96	1.1 (0.8–1.6) n	Reish (1993)
Copepods									
Acartia clausi	NS	stat	NS	NS	NS	arsenite	96	0.51 n	US EPA (1985)
	nauplius	stat$	20	35	7.7–8.1	arsenate	96	0.011 (0.009–0.013) n	Forget et al. (1998)
Tigriopus brevicornis	copepodid	stat$	20	35	7.7–8.1	arsenate	96	0.02 (0.018–0.022) n	Forget et al. (1998)
	ovigerous female	stat$	20	35	7.7–8.1	arsenate	96	0.028 (0.025–0.03) n	Forget et al. (1998)

Table 39 (contd.)

Harpacticoid copepod									
Nitocra spinipes	adult	stat	20–22	3	NS	arsenite	96	3.5 (2.8–4.3) n	Bengtsson & Bergström (1987)
	adult	stat	20–22	7	NS	arsenate	96	3.0 (2.1–4.2) n	
Mysid shrimp									
Mysidopsis bahia	NS	flow	NS	NS	NS	arsenite	96	1.7 m	US EPA (1985)
	NS	flow	NS	NS	NS	arsenate	96	2.3 m	US EPA (1985)
Pink shrimp									
Penaeus duorarum	juvenile	stat	19	24	NS	As_2O_3	48	>30 @ n	Mayer (1987)
	juvenile	stat	19	24	NS	arsenate	48	>15 @ n	Mayer (1987)
White shrimp									
Penaeus setiferus	juvenile	stat	22	25	8.3–8.7	As trisulfide	96	24.8 (19.1–35.2) n	Curtis et al. (1979)

stat = static conditions (water unchanged for duration of test); stat$ = static renewal conditions (water replaced on a regular basis); flow = flow-through conditions (As concentration continuously maintained);NS = not stated; * = carapace breadth; # = EC_{50}s based on inhibition of shell deposition; ## = EC_{50}s based on abnormal development; @ = EC_{50}s based on immobilization; n = based on nominal concentrations; m = based on measured concentrations

Table 40. Toxicity of organic As to aquatic invertebrates

Organism	Size/ age	Stat/ flow	Temper- ature (°C)	Hardness (mg/litre)	pH	Arsenical	Duration (h)	LC$_{50}$ (mg As/litre)	Reference
Scud									
Gammarus fasciatus mature		stat	15	44	7.4	MMA (34.8%)	96	>16	Mayer & Ellersieck (1986
Cladoceran									
Alonella sp.	NS	stat	20	15	7.8	MMA	48	18.2	Naqvi et al. (1985)
Calanoid									
Diaptomus sp.	NS	stat	20	15	7.8	MMA	48	17.4	Naqvi et al. (1985)
Cyclopoid									
Eucyclops sp.	NS	stat	20	15	7.8	MMA	48	44.6	Naqvi et al. (1985)
Ostracod									
Cypria sp.	NS	stat	20	15	7.8	MMA	48	45.5	Naqvi et al. (1985)
Crayfish									
Procambarus sp.	NS	stat	NS	NS	NS	MMA	48	2361 n	Anderson et al. (1975)
Procambarus sp.	NS	stat	NS	NS	NS	MMA	96	509 n	Anderson et al. (1975)
Procambarus clarkii juvenile		stat	21–27	NS	5.8–7.8	MMA	96	46.7	Naqvi et al. (1987)
adult		stat	21–27	NS	5.8–7.8	MMA	96	472	Naqvi et al. (1987)

stat = static conditions (water unchanged for duration of test); NS = not stated; hardness expressed as mg CaCO$_3$/litre; MMA = mono-methylarsonic acid (administered as the monosodium salt); EC$_{50}$s based on growth; n = based on nominal concentrations

362

ponds. Sanders (1986) studied the effect of arsenate (15 µg/litre) on zooplankton survival in natural phytoplankton assemblages over 24 days. After 10 days the arsenate-treated tanks contained phytoplankton assemblages dominated by the centric diatom *Th. pseudonana*, whereas control tanks contained a variety of algal species, although flagellates became increasingly more important. By the end of the experiment zooplankton density in control tanks was 2–3 times higher than in arsenate-treated tanks. No significant effect on survival of a natural assemblage of copepods was observed when organisms were exposed to arsenate at < 10 mg As(V)/litre for 2 weeks; however, arsenite caused significant mortality at both 4 and 10 mg As(III)/litre (Borgmann et al., 1980).

9.1.2.4   Vertebrates

Acute toxicity of arsenic to freshwater and marine fish is summarized in Tables 41 and 42 respectively. The 96-h LC_{50}s for freshwater fish range from 10.8 to 91 mg/litre for trivalent arsenic and from 4.8 to > 360 mg/litre for pentavalent arsenic. In marine fish 96-h LC_{50}s range from 12.7 to 28.5 mg As(III)/litre and from 21.4 to 157 mg As(V)/litre. Acute toxicity of organic arsenicals to fish is summarized in Table 43. LC_{50}s for the organic arsenical MMA range from 1.9 to 1412 mg As/litre.

McGeachy & Dixon (1989) found no effect of temperature on arsenite toxicity to rainbow trout (*O. mykiss*) with 144-day LC_{50}s of 17.7 and 20.7 mg/litre at 5 and 15 °C respectively. However, arsenate toxicity was increased from an LC_{50} of 114.1 mg/litre at 5 °C to 58 mg/litre at 15 °C.

Most of the data on the effects of arsenic on fish are based on acute toxicity tests which measure fish mortality over 96 h. Some studies have also examined sub-lethal effects such as growth, avoidance behaviour and fertilization/hatching.

Nichols et al. (1984) exposed coho salmon (*O. kisutch*) fry to As_2O_3 (30, 100 or 300 µg As(III)/litre) for 6 months in freshwater; smolting fry were transferred to seawater (salinity 28 g/litre) for a further 6 months. Survival and growth were unaffected by arsenic exposure. Migration of trout released after the arsenic exposure was significantly reduced at the highest concentration.

Table 41. Toxicity of inorganic As to freshwater fish

Organism	Size/ age	Stat/ flow	Temper- ature (°C)	Hardness (mg/litre)	pH	As species	Duration (h)	LC$_{50}$ (mg As/litre)	Reference
Chinook salmon									
Oncorhynchus tshawytscha	0.5 g	stat	12	211	7.0–8.3	As$_2$O$_3$	96	25.1 (19.3–32.7) n	Hamilton & Buhl
	0.5 g	stat	12	211	3.4–7.2	As pentoxide	96	90 (70–116) n	Buhl
	1 g	stat	12	211	6.9–7.4	As pentoxide	96	167 (120–233) * n	(1990)
Coho salmon									
Oncorhynchus kisutch	alevin	stat	12	41	7.1–8.0	arsenite	96	49.4 (42.1–57.9) n	Buhl &
	juvenile	stat	12	41	7.1–8.0	arsenite	96	18.5 (14.5–23.7) n	Hamilton (1991)
	alevin	stat	12	41	7.1–8.0	arsenate	96	306 (216–433) n	Buhl &
	0.41 g	stat	12	41	7.1–8.0	arsenate	96	43.6 (32.5–58.5) n	Hamilton
	0.47 g	stat	12	41	7.1–8.0	As pentoxide	96	58.5 (49.4–69.3) n	(1990)
Rainbow trout									
Oncorhynchus mykiss	2.6 g	stat	12	44	7.4	arsenite	96	13.3 (7.8–22.5)	Mayer & Ellersieck (1986)
	alevin	stat	12	41	7.1–8.0	arsenite	96	91 (69.9–119) n	Buhl &
	juvenile	stat	12	41	7.1–8.0	arsenite	96	16 (12.7–20.1) n	Hamilton (1991)
	alevin	stat	12	41	7.1–8.0	arsenate	96	>360 n	Buhl &
	0.6 g	stat	12	41	7.1–8.0	arsenate	96	67.5 (56.1–81.2) n	Hamilton (1990)

Table 41 (contd.)

Brook trout *Salvelinus fontinalis*	adult	flow	15	152	7.8	arsenite	96	15 m	Cardwell et al. (1976)
Bluegill *Lepomis macrochirus*	0.8 g	stat	24	50	7.3–8.0	arsenite	96	15.2 n	Inglis & Davis (1972)
	0.8 g	stat	24	210	7.5–8.0	arsenite	96	16.2 n	
	0.8 g	stat	24	370	7.7–8.0	arsenite	96	15.4 n	
	1 g	stat	24	44	7.4	arsenite	96	17.3 (12.3–24.4)	Mayer & Ellersieck (1986)
	juvenile	flow	25	147	7.8	arsenite	96	41.6 m	Cardwell et al. (1976)
Fathead minnow *Pimephales promelas*	juvenile	stat	22	25	8.3–8.7	As trisulfide	96	82.4 (65.4–105.9) n	Curtis et al. (1979)
	juvenile	flow	25	149	7.8	arsenite	96	15.6 m	Cardwell et al. (1976)
	NS	flow	23–25.8	46.3–49.9		arsenite	96	14.1 (12.5–15.9) m	Lima et al. (1984)
	0.15 g	flow	22–28	43.9	6.0–8.1	arsenite	96	12.6 (9.9–15.9)	Spehar & Fiandt (1986)
						arsenate	96	25.6	US EPA (1985)

Table 41 (contd.)

Organism	Size/age	Stat/flow	Temperature (°C)	Hardness (mg/litre)	pH	As species	Duration (h)	LC$_{50}$ (mg As/litre)	Reference
Goldfish *Carassius auratus*	4–8 cm	stat	23	NS	6.0–6.9	arsenate	48	32 (24.6–41.6) n	Weir & Hine (1970)
	juvenile	flow	25	148	7.6	arsenite	96	26 m	Cardwell et al. (1976)
						arsenate	96	34	US EPA (1985)
Golden shiner *Notemigonus crysoleucas*	NS	flow	NS	72.2	7.5	arsenite	96	12.5 (10.0–14.6) m	Hartwell et al. (1989)
Striped bass *Morone saxatilis*	63 d	stat	20	40	8.1	As pentoxide	96	40.5 (32–51.2) n	Palawski et al. (1985)
	63 d	stat	20	285	7.9	As pentoxide	96	30.5 (21.3–43.7) n	
Colorado squawfish *Ptychocheilus lucius*	larvae	stat	25	144	8.1	arsenate	96	105 (74–164) n	Hamilton & Buhl (1997)
Razorback sucker *Xyrauchen texanus*	larvae	stat	25	144	8.1	arsenate	96	17.8 (13.7–21.3) n	Hamilton & Buhl (1997)

Table 41 (contd.)

Flagfish									
Jordanella floridae	fry	flow	25	NS	NS	arsenite	96	28 m	Cardwell et al. (1976)
	juvenile	flow	23–25.8	46.3–49.9	NS	arsenite	96	14.4 (12.7–16.3) m	Lima et al. (1984)
Arctic grayling									
Thymallus arcticus	alevin	stat	12	41	7.1–8.0	arsenite	96	27.7 (23.5–32.7) n	Buhl & Hamilton (1991)
	juvenile	stat	12	41	7.1–8.0	arsenite	96	13.7 (11.6–16.1) n	Buhl & Hamilton (1990)
	fry	stat	12	41.3	7.1–8.0	As pentoxide	96	5.0 (3.5–7.2) n	Buhl & Hamilton (1990)
	0.2 g	stat	12	41.3	7.1–8.0	As pentoxide	96	4.8 (3.8–5.9) n	Buhl & Hamilton (1990)
	0.34 g	stat	12	41.3	7.1–8.0	As pentoxide	96	5.5 (4.1–7.5) n	Buhl & Hamilton (1990)
	alevin	stat	12	41.3	7.1–8.0	arsenate	96	102 (75–141) n	Buhl & Hamilton (1990)
	alevin	stat	12	41.3	7.1–8.0	arsenate	96	197 (145–267) n	Buhl & Hamilton (1990)
	0.85 g	stat	12	41.3	7.1–8.0	arsenate	96	47.7 (35–65) n	Buhl & Hamilton (1990)
	0.97 g	stat	12	41.3	7.1–8.0	arsenate	96	32.5 (25.3–41.6) n	Buhl & Hamilton (1990)
	1.85 g	stat	12	41.3	7.1–8.0	arsenate	96	30.9 (21.9–43.6) n	Buhl & Hamilton (1990)

Table 41 (contd.)

Organism	Size/age	Stat/flow	Temperature (°C)	Hardness (mg/litre)	pH	As species	Duration (h)	LC$_{50}$ (mg As/litre)	Reference
Channel catfish *Ictalurus punctatus*	juvenile	flow	25	140	8.0	arsenite	96	18 m	Cardwell et al. (1976)
Murrel *Channa punctatus*	fingerling	stat	24	124	7.2	As$_2$O$_3$	96	10.8 (9.9–12.7) n	Shukla et al. (1987)
Mosquito fish *Gambusia affinis*	NS	stat	20	NS	NS	arsenate	96	49 (44–54)	Jurewicz & Buikema (1980)
Deepwater ciscoe (chub) *Coregonus* sp.	fry	stat	7	NS	NS	As$_2$O$_3$	96	17 (13–22) n	Passino & Kramer (1980)

Stat = static conditions (water unchanged for duration of test); flow = flow-through conditions (As concentration continuously maintained); NS = not stated; hardness expressed as mg CaCO$_3$/litre; * = pH buffered; n = based on nominal concentrations; m = based on measured concentrations

Table 42. Toxicity of inorganic As to marine fish

Organism	Size/ age	Stat/ flow	Temper- ature (°C)	Salinity (g/litre)	pH	As species	Duration (h)	LC_{50} (mg As/litre)	Reference
Chinook salmon									
Oncorhynchus tshawytscha	1.99 g	stat	12	brackish	6.7–8.4	As_2O_3	96	21.4 (18.1–25.3) n	Hamilton & Buhl (1990)
	1.99 g	stat	12	brackish	2.7–6.0	As pentoxide	96	66.5 (55.4–79.8) n	Buhl (1990)
	1.99 g	stat	12	brackish	6.9–7.4	As pentoxide	96	157 (119–208) * n	Hamilton & Buhl (1990)
Sheepshead minnow									
Cyprinodon variegatus	NS	flow	NS	NS	NS	Arsenite	NS	12.7 m	US EPA (1985)
Atlantic silverside									
Menidia menidia	NS	stat	NS	NS	NS	arsenite	NS	16 n	US EPA (1985)
Fourspine stickleback									
Apeltes quadracus	NS	stat	NS	NS	NS	arsenite	NS	15 n	US EPA (1985)
Longnose killifish									
Fundulus similis	juvenile	stat	16	31	NS	As_2O_3	48	>30 n	Mayer (1987)
	juvenile	stat	16	31	NS	arsenate	48	>15 n	Mayer (1987)

Table 42 (contd.)

Organism	Size/age	Stat/flow	Temperature (°C)	Salinity (g/litre)	pH	As species	Duration (h)	LC$_{50}$ (mg As/litre)	Reference
Striped bass *Morone saxatilis*	1.8 g	stat	20	22	NS	arsenate	96	10.3 (6.4–13.5)	Dwyer et al. (1992)
Therapon jarbua	0.2–0.7 g	stat	NS	36	NS	arsenite	96	3.38	Krishna-kumari et al. (1983)
Dab *Limanda limanda*	16.9 g	flow	12	34.6	7.7	arsenite	96	28.5 (22.7–36.0) m	Taylor et al. (1985)
Grey mullet *Chelon labrosus*	0.87 g	flow	12	34.6	7.7	arsenite	96	27.3 (23.4–30.2) m	Taylor et al. (1985)

stat = static conditions (water unchanged for duration of test); flow = flow-through conditions (As concentration continuously maintained); NS = not stated; salinity measured in mg/litre; * = pH buffered; n = based on nominal concentrations; m = based on measured concentrations

Table 43. Toxicity of MMA to freshwater fish

Organism	Size/age	Stat/flow	Temperature (°C)	Hardness (mg/litre)	pH	Arsenical	Duration (h)	LC_{50} (mg As/litre)	Reference
Rainbow trout									
Oncorhynchus mykiss	0.6 g	stat	12	44	7.4	MMA (51.2%)	96	18.5 (14.2–23.9)	Mayer & Ellersieck (1986)
Bluegill									
Lepomis macrochirus	0.9 g	stat	18	44	7.1	MMA (34.8%)	96	1.9 n	Mayer & Ellersieck (1986)
	1.0 g	stat	17	44	7.1	MMA (37.7%)	96	8.6 (4.4–16.7)	
	1.0 g	stat	22	44	7.4	MMA (51.2%)	96	> 23.7	
Fathead minnow									
Pimephales promelas	0.9 g	stat	18	44	7.1	MMA (34.8%)	96	2.1 (0.8–5.8) n	Mayer & Ellersieck (1986)
Goldfish									
Carassius auratus	0.9 g	stat	18	44	7.1	MMA (34.8%)	96	5 (3.9–6.3) n	Mayer & Ellersieck (1986)

371

Table 43 (contd.)

Organism	Size/ age	Stat/ flow	Temper- ature (°C)	Hardness (mg/litre)	pH	Arsenical	Duration (h)	LC_{50} (mg As/litre)	Reference
Channel catfish *Ictalurus lacustris*	NS	stat	NS	NS	NS	MMA	96	1412 n	Anderson et al. (1975)
	2.1 g	stat	17	44	7.1	MMA (37.7%)	96	4.7 (3.5–6.3)	Mayer & Ellersieck (1986)
Black bass *Micropterus dolmieu*	NS	stat	NS	NS	NS	MMA	96	417 n	Anderson et al. (1975)

stat = static conditions (water unchanged for duration of test); NS = not stated; hardness expressed as mg $CaCO_3$/litre; MMA = monomethylarsonic acid; n = based on nominal concentrations

Lima et al. (1984) exposed fathead minnow (*Pimephales promelas*) and flagfish (*Jordanella floridae*) to arsenite in 29-day and 31-day tests respectively. Growth was significantly reduced at concentrations of 4300 and 4120 µg As(III)/litre for the two species respectively. The NOEC, based on growth, was between 2130 and 4300 µg As(III)/litre for *P. promelas* and between 2130 and 4120 µg As(III)/litre for *J. floridae*. Spehar et al. (1980) found no significant effect on survival of rainbow trout exposed to either arsenite (961 µg As(III)/litre), arsenate (973 µg As(V)/litre), MMA (970 µg As/litre) or DMA (846 µg As/litre) in 28-day tests. Growth of fingerling freshwater murrel (*Channa punctatus*) was significantly reduced by As_2O_3 at 7 mg As(III)/litre during a 31-day test (Shukla et al., 1987). In 32-day tests on *P. promelas*, growth was found to be the most sensitive parameter, the MATC being 3.3 mg As(III)/litre (Spehar & Fiandt, 1986).

Cockell & Hilton (1985) found no adverse effect on rainbow trout fed diets containing 10–90 mg As(V)/kg arsenate for 16 weeks or 120–1600 mg As/kg DMA or arsanilic acid for 8 weeks. In 28-day tests a significant reduction in growth was observed at all dietary concentrations of 137–1477 mg As(III) or As(V)/kg (as As_2O_3 and sodium arsenate). However, no significant adverse effects could be found at concentrations of < 1497 mg As/kg for the organic arsenicals DMA and arsanilic acid (Cockell & Hilton, 1988). Cockell et al. (1991) found the MATC for arsenate to be between 13 and 33 mg As(V)/kg diet (0.281 to 0.525 mg As/kg body weight) in 12-week and 24-week tests. The most sensitive indicator of chronic dietary toxicity was inflammation of the gallbladder wall. Oladimeji et al. (1984) reported that rainbow trout fed on a diet containing arsenite at 20 or 30 mg As(III)/kg (equivalent to 0.2, 0.4 and 0.6 mg/kg fish wet weight per day) for 8 weeks showed significantly reduced growth. No effect on growth was noted at 10 mg As(III)/kg.

Weir & Hine (1970) found a significant impairment of avoidance behaviour at 100 µg As(V)/litre (as arsenate) but no effect at 50 µg/litre on goldfish (*Carassius auratus*). The avoidance threshold for golden shiner (*Notemigonus crysoleucas*) was 28 µg As(III)/litre (as arsenite) in flow-through tests (Hartwell et al., 1989).

Birge et al. (1978) exposed rainbow trout (*O. mykiss*), largemouth bass (*Micropterus salmoides*) and marbled salamander

(*Ambystoma opacum*) to arsenite from fertilization to 4 days after hatching. Treatment periods were 8 days for bass and salamander, and 28 days for trout. LC_{50}s were 0.54, 42.1 and 4.45 mg As(III)/litre for trout, bass and salamander respectively. Similar embryo–larval stage tests carried out by Birge (1978) revealed 7-day LC_{50} values of 0.49 and 0.04 mg As(III)/litre for goldfish (*C. auratus*) and narrow-mouthed toad (*Gastrophryne carolinensis*) respectively. Khangarot et al. (1985) found 48-day and 96-h LC_{50}s for *Rana hexadactyla* tadpoles to be 0.27 and 0.25 mg As(III)/litre respectively in static tests with As_2O_3.

9.1.3 Terrestrial organisms

9.1.3.1 Plants

The phytotoxic actions of inorganic and organic arsenicals are modified by the environment in which the plant is growing. Dominant factors include iron, aluminium, calcium and phosphate content of the soil, and pH. Levels of soil arsenic reported to be toxic to plants span a broad range. In general, arsenate is less toxic to plants than arsenite. The primary mechanism of arsenite toxicity is considered to be due to its binding with protein sulfhydryl groups (Peoples, 1975). Arsenite penetrates the plant cuticle to a greater degree than arsenate (NAS, 1977). One of the first indications of plant injury by sodium arsenite is wilting caused by loss of turgor, whereas stress due to sodium arsenate involves chlorosis but not rapid loss of turgor (NAS, 1977). Arsenate is known to uncouple phosphorylation by replacing phosphate; the coupled oxidative phosphorylation of adenosine diphosphate is blocked. Organoarsenicals such as DMA enter plants mostly by absorption of sprays; uptake from the soil contributes only a minor fraction. The phytotoxicity of organoarsenical herbicides is characterized by chlorosis, cessation of growth, gradual browning, dehydration, and death (NAS, 1977). There is a much better correlation between plant growth and available arsenic than between plant growth and total arsenic (Woolson et al., 1971; Woolson, 1973; Walsh & Keeney, 1975).

Davis et al. (1978) grew barley in sand treated with a nutrient solution to which arsenate had been added. The upper critical level (minimum concentration in actively growing tissues of a plant at which yield is reduced) in the leaves and shoots of barley plants was 20 mg As(V)/kg (dry weight). Fargasová (1994b) grew mustard seeds (*Sinapis alba*) in nutrient solution and calculated a 72-h LC_{50}, based on germination, and an EC_{50}, based on root growth inhibition, at 30.2 mg As(V)/litre and 5.5 mg As(V)/litre respectively. Neumann et al. (1998) performed tests with alfalfa seedlings inoculated with *Rhizobium meliloti* for 14 days. EC_{50}s, based on a reduction in nodulation, were 2.6, 5.5 and 20.1 µmol/litre As for arsenate, arsenite and arsenic pentoxide respectively. Inhibition of nodulation was two orders of magnitude more sensitive than the inhibition of root and shoot growth. Vaughan & Greensdale (1998) found that seedling emergence of *Lactuca sativa* was more sensitive to arsenite (EC_{50} 6.15 mg/kg, 120 h exposure) than arsenate (EC_{50} 26.1 mg/kg). Similarly, root elongation of *L. sativa* was inhibited by lower levels of arsenite (EC_{50} 0.6 mg/litre, 120 h exposure) than arsenate (EC_{50} 2.3 mg/litre).

Carbonell-Barrachina et al. (1998) grew the perennial marsh plants *Spartina patens* and *S. alterniflora* in nutrient solution at arsenate, arsenite, DMA and MMA concentrations of 0.2, 0.8 and 2 mg/litre for 30 days. Plant growth (total dry biomass production) was significantly reduced at all three exposure concentrations when the source of arsenic was DMA. However, arsenate at the lower two concentrations, and arsenite at the lowest concentration, caused a significant increase in growth and this was accompanied by an increase in plant phosphate content. The highest arsenic concentration caused a significant reduction in growth when plants were exposed to either arsenate or arsenite. MMA caused a significant increase in growth of *S. patens* at both 0.2 and 0.8 mg As/litre; however, all exposures of MMA to *S. alterniflora* caused significant decreases in growth.

Sheppard (1992) reviewed phytotoxic levels of arsenic in soil and found that the source of arsenic had the largest effect on mean toxic levels in soils with inorganic and waste forms of arsenic significantly less toxic than organic sources. Waste forms of arsenic include solid mineral forms where much of the arsenic would be occluded. For inorganic sources, arsenic is 5-fold more toxic in sands (toxicity threshold 40 mg/kg) than in clay soils (toxicity threshold 200 mg/kg), on the basis of geometric means.

Woolson (1973) found that growth was reduced by 50% at arsenate levels ranging from 0.7 to 87 mg As(V)/kg (in edible tissues) for six different vegetable varieties. The authors report that available arsenic accounts for 64–83% of the variation in plant growth. Woolson et al. (1971) grew corn seedlings for 4 weeks in Lakeland loamy soil contaminated with arsenic compounds. Growth was reduced by 50% at a sodium arsenate concentration of 42 mg As(V)/kg. Iron, aluminium and calcium arsenate compounds were between 13% and 62% less toxic.

Jiang & Singh (1994) found that application rates of arsenite and arsenate at $\geq$50 mg As(III) or As(V)/kg caused significant reductions in the yield of ryegrass (*Lolium perenne*) and barley (*Hordeum vulgare*). However, application of an NPK fertilizer containing < 3000 mg As(III)/kg did not affect yield, although increases in arsenic residues were observed. The yield reduction due to arsenic application was greater in sand than in loam soil. The higher arsenic retention capacity of the latter soil was due to its higher content of iron and aluminium oxides.

Anastasia & Kender (1973) grew lowbush blueberry (*Vaccinium angustifolium*) in a loamy sand containing As_2O_3. Growth was significantly inhibited at $\leq$69.5 mg As(III)/kg soil; significant reductions in growth were observed at foliar concentrations of 6.7 mg As/kg.

Deuel & Swoboda (1972b) determined toxicity levels for As_2O_3 to cotton (*G. hirsutum*) and soybean (*Glycine max*) in a 6-week greenhouse experiment using fine sandy loam and clay soils. Significant reductions in cotton yield were observed at 56 and 280 kg As(III)/ha for the two soil types respectively. Soybeans were found to be more sensitive, with significant reductions in yield at 28 and 168 kg As(III)/ha respectively. The application rates at which significant yield reductions were observed were related to water-soluble arsenic levels of 8 and 28 mg As(III)/litre for cotton and 3 and 12 mg As(III)/litre for soybeans. The authors found that the critical plant tissue concentrations with regard to yield were $\geq$4.4 mg As/kg for cotton and $\geq$1 mg/kg for soybeans. Weaver et al. (1984) found that 90 mg As(III)/kg (as As_2O_3) essentially prevented growth of Bermuda grass (*Cynodon dactylon*) on silt loam and fine sand, and significantly reduced growth on clay soil. No significant effect on growth was observed at 45 mg As(III)/kg for the clay and at 10 mg/kg for the other two soil types.

Jacobs et al. (1970b) found significant reductions in the yield of peas (*Pisum sativum*), snap beans (*Phaseolus vulgaris*) and sweet corn (*Zea mays*) at sodium arsenite application rates ≥80 kg As(III)/ha in field studies on sandy soil. Woolson & Isensee (1981) report significant reductions in the yield of soybean (*G. max*) and radish (*Raphanus sativus*) at 89.6 kg As(III)/ha (10 times the recommended application rate) each year during a 7-year trial. DMA (112 kg/ha) and MMA (56 kg/ha) caused significant reductions in yield during some years but not consistently throughout the experiment.

Schweizer (1967) observed a significant reduction in growth (height) of cotton (*G. hirsutum*) after 4 weeks at ≥80 mg DSMA/kg in a silt loam. Significant adverse effects on the fresh weight and height of rice and soybean, and on the fresh weight alone of corn, cotton and oats, were noted at 50 mg DSMA/kg. Baker et al. (1969) found no effect of MSMA and DSMA on cotton when applied to weeds as a directed spray at 0.4–0.6 kg/ha; however, when applied topically to the cotton plants both arsenic compounds caused significant reductions in yield. Arle & Hamilton (1971) applied MSMA and DSMA topically to cotton plants. Single applications of DSMA at rates of 2.2–9 kg/ha had no significant effect on cotton seed yield; however, MSMA significantly reduced yield at ≥ 6.7 kg/ha. Two or more applications of 2.2 kg MSMA/ha also caused significant reductions in yield. Significant adverse effects were observed on average cotton boll components with multiple applications of DSMA, and to a greater extent MSMA. Keeley & Thullen (1971) found that MSMA (with or without 0.5% surfactant) was significantly more toxic to cotton at 13 °C than at 20 °C or 31 °C when applied to the foliage at a rate of 3.36 kg/ha. Similarly, DSMA was significantly more toxic at 13 °C than at higher temperatures but only with the addition of 0.5% surfactant.

Since arsenic has been found to persist in soil at toxic levels after application of arsenate insecticides, much work has been focused on the alleviation of arsenic toxicity by a variety of means. Laboratory studies have shown that the addition of sufficient phosphate can depress uptake of arsenate by plants. Woolson et al. (1973) found that when sufficient phosphate was added to a sandy loam soil (P/As ratio 7) improved yields were observed. However, at very high arsenic concentrations (1000 mg/kg) phosphate did not overcome arsenate toxicity. Woolson et al. (1973) pointed out that in soils in which added phosphate desorbs arsenate, leaching of the soil after addition of phosphate may be a viable approach to removing arsenic from the root zone. Steevans et al. (1972) attempted to alleviate arsenic toxicity on a sandy soil by application of 4 t/ha of ferric sulfate or aluminium sulfate. The iron treatment had a slight beneficial effect but the aluminium treatment actually depressed yields further.

Tolerance of plants to arsenic has been identified. Macnair & Cumbes (1987) found that the growth of the grass *Holcus lanatus* (non-tolerant strain) was significantly reduced at a nutrient solution arsenate concentration of 5 mg As(V)/litre (0.067 mmol/litre) in a 7-day test. However, no effect on growth was observed in a tolerant strain at 10 mg As(V)/litre (0.133 mmol/litre). The addition of phosphate (0.1 mmol/litre) at the lowest arsenate concentration (5 mg/litre, 0.067 mmol/litre) reduced arsenic toxicity in non-tolerant strains significantly more than for tolerant plants. Similarly, Meharg & Macnair (1991b) found that the addition of phosphate at 0.5 mol/m^3 protected plants against arsenic toxicity in non-tolerant strains. Meharg & Macnair (1990) showed that arsenate tolerance in *H. lanatus* was due to an altered phosphate and arsenate uptake system, where the high-affinity uptake system is suppressed or absent in tolerant plants. Similar mechanisms of tolerance have been identified in the grasses *Deschampsia cespitosa* and *Agrostis capillaris* (Meharg & Macnair, 1991a).

9.1.3.2 Invertebrates

Goldstein & Babich (1989) report arsenite and arsenate 7-day LC$_{50}$ values of 40.5 mg As(III)/kg (0.54 mmol/litre) and 59.3 mg As(V)/kg (0.79 mmol/litre) for the fruit fly *Drosophila melanogaster*. The 48-h LC$_{50}$s for As(III) and As(V) in drinking-water were 80 and 110 mg/litre respectively for the house fly *Musca domestica*. In the cabbage looper moth (*Trichoplusia ni*) 48-hLC$_{50}$s were 320 mg As(III)/kg diet and 794 mg As(V)/kg diet (Zaman & Pardini, 1995). Robertson & McLean (1985) found the dietary LC$_{50}$ for As$_2$O$_3$ to sixth instar western spruce budworm (*Choristoneura occidentalis*) to be 1932 mg As(III)/kg.

Meharg et al. (1998) maintained earthworms (*Lumbricus terrestris*) in sandy loam soil dosed with arsenate concentrations ranging from 15 to 500 mg/kg (dry weight). The natural soil arsenic level was 1.2 mg/kg. A plot of the concentration causing 50% mortality (LC_{50}) with time showed a steep decline between 2 and 8 days with the LC_{50} ranging from 400 to 100 mg/kg. In a second experiment toxicity was found to increase with depth of soil, with the 4-day LC_{50} ranging from 300 mg/kg at a depth of 0–70 mm to < 100 mg/kg at 500–700 mm.

Vaughan & Greenslade (1998) investigated the effect of arsenite and arsenate on survival and reproduction in three species of Collembola (springtails). The EC_{50} for inhibition of reproduction over a 28-day exposure of *Folsomia candida* by arsenite was 3 mg/kg, compared with 119 mg/kg for arsenate. This species was the most sensitive Collembola to arsenite, followed by *Proisotoma minuta* (EC_{50} 4.4 mg/kg) and *Sinella communis* (EC_{50} 9.9 mg/kg).

Watson et al. (1976) exposed fourth-instar nymphs of meadow katydids (*Conocephalus fasciatus*) to drinking-water containing arsenical compounds. They reported 7-day LD_{50}s for the organic arsenicals cacodylic acid and Phytar 560 of 12.1 and 1.3 mg As/litre respectively; 14-day values were 2.6 and 0.4 mg As/litre respectively. The inorganic As_2O_3 and arsenic pentoxide showed 7- and 14-day LD_{50}s of 3.1 and 1.2 mg As(III)/litre, and 4.5 and 1.5 mg As(V)/litre respectively.

Bertholf & Pilson (1941) found the median lethal dose for arsenate to honey bees to be 0.6 µg As(V)/bee. The authors found a decrease in toxicity with an increase in particle size. At a fine particle size (3 µm) the median lethal dose was 0.7 µg As/bee whereas a coarse particle size (28 µm) resulted in a median lethal dose of 1.3 µg/bee.

9.1.3.3   Vertebrates

Toxicity of arsenic to birds is summarized in Table 44. Signs of acute arsenite poisoning in birds include ataxia, asthenia, slowness, jerkiness, falling, hyporeactivity, fluffed feathers, ptosis, huddled position, loss of righting reflex, immobility and tetanic seizures (Hudson et al., 1984).

Table 44. Toxicity of As to birds

Species	Age	Arsenical	Parameter	Concentration (mg As/kg)	Reference
Mallard *Anas platyrhynchos*	3 mo.	Arsenite	LD[a]$_{50}$	323 (149–699)	Hudson et al. (1984)
	3–4 mo.	Silvisar 510*	LD$_{50}$	> 2400	Hudson et al. (1984)
California quail *Callipepla californica*	9–12 mo.	Arsenite	LD$_{50}$	47.6 (34.3–66)	Hudson et al. (1984)
Pheasant *Phasianus colchicus*	3–4 mo.	Arsenite	LD$_{50}$	386 (221–671)	Hudson et al. (1984)
Chukar *Alectoris chukar*	4 mo.	Silvisar 510*	LD$_{50}$	≥ 2000	Hudson et al. (1984)
Japanese quail *Coturnix coturnix japonica*	14 days	MMA	5-day LC[b]$_{50}$	> 5000	Hill & Camardese (1986)

[a] LD$_{50}$ expressed as mg/kg body weight (single oral dose)
[b] LC$_{50}$ expressed as mg/kg diet
Silvisar 510=54.3%, total As = 27.14% (DMA 46%, triethanolamine DMA 8.3%); MMA monomethylarsonic acid

Holcman & Stibilj (1997) fed Rhode Island Red hens on a diet containing up to 30 mg As(III)/kg (as As_2O_3) for 19 days. No significant effects were found on feed consumption, number of eggs per hen, body weight or average egg weight.

Haegele & Tucker (1974) dosed female mallards (*Anas platyrhynchos*) with a single oral dose of 100 mg sodium arsenite/kg body weight. Eggshell thickness was reduced within 3 days but had recovered to normal 5 days after dosing. The authors suggest that this temporary reduction in eggshell thickness could be caused by a decrease in food consumption; however, food consumption was not monitored during the test.

Camardese et al. (1990) fed mallard ducklings (*A. platyrhynchos*) on a diet containing 30, 100 or 300 mg As(V)/kg (as sodium arsenate) for 10 weeks. Arsenic accumulated significantly in brain and liver of ducklings fed 100 or 300 mg/kg but did not result in histopathological lesions. In a similar experiment Whitworth et al. (1991) found that the highest exposure concentration caused a significant increase in resting time and a significant decrease in time spent in alert behaviours. Ducklings on 300 mg As(V)/kg spent significantly more time under the heat lamp. Arsenate had no effect on feeding behaviour. No significant effect on survival or growth was observed in mallard ducklings fed on a diet containing 200 mg As(V)/kg (as sodium arsenate) for 4 weeks. However, ducklings maintained on a protein-restricted diet at the same arsenate dose showed significant reductions in survival and growth, and an increased incidence of histopathological lesions (Hoffman et al., 1992). Stanley et al. (1994) found that 400 mg As(V)/kg (as sodium arsenate) significantly reduced mallard duckling growth but did not affect survival rate.

Hudson et al. (1984) found the LD_{50} for the mule deer (*Odocoileus hemionus*) to be > 320 mg/kg for the organic arsenical Silvisar 510 (Phytar 560).

Savabieasfahani et al. (1998) exposed cotton rats (*Sigmodon hispidus*) to 5 or 10 mg/litre sodium arsenite in drinking-water for 6 weeks. Food intake was significantly depressed at both dose levels. However, there was no significant dose-related effect of arsenite on organ weights or immune function.

9.2 Field observations

9.2.1 Microorganisms

Arsenic resistance is often found in microorganisms inhabiting arsenic-contaminated environments (see section 9.1.1.1).

9.2.2 Aquatic organisms

Crearley (1973) found several dead turtles around the margin of a ponded area near a manufacturing plant for arsenic-based cotton defoliants. The concentration of arsenic in the pondwater was 63 mg As/litre. The turtles appeared to have been blind before death, with the tissue of the eyelid and the nasal area appearing to be keratinized. Three of five live turtles collected downstream in the headwaters of Finfeather Lake (Bryan, Texas, USA) were blind; the arsenic concentration in the lake was 8.5 mg/litre. Sorensen et al. (1985) collected green sunfish (*L. cyanellus*) from Municipal Lake, Bryan, Texas (water concentration 13.6 mg As/litre). Cellular changes in hepatocytes of sunfish were compared with levels of arsenic in the liver. Significant increases were observed in the volumes occupied by necrotic and fibrous bodies as arsenic levels in the liver increased. The surface density of rough endoplasmic reticulum increased significantly with increasing arsenic concentrations.

9.2.3 Terrestrial organisms

9.2.3.1 Plants

Injury symptoms on crop plants resulting from toxic quantities of arsenic in soils were noted in the 1930s, when it was found that young trees planted in old orchard soils that had been treated with organic arsenicals grew slowly and were stunted (Snyder, 1935). In addition to being stunted, young apple trees had leaf symptoms that indicated water-deficiency stress, which implied injury to the roots (NAS, 1977). Trappe et al. (1973) report that apple trees growing in soil contaminated from the earlier use of lead arsenate insecticide (300 mg As(V)/kg) were stunted and had badly stunted rootlets that were sparsely mycorrhizal. Trees growing in the same area in soil at < 50 mg As/kg showed healthy growth and vigorous rootlets that were intensely mycorrhizal.

Symptoms similar to those previously attributed to arsenic toxicity from the use of lead arsenate insecticide were reported by Aggett & Aspell (1980) for apple trees growing near a timber mill processing wood treated with copper chrome arsenate. Total inorganic soil arsenic, available soil arsenic, arsenic residues in leaves and observed symptoms decreased with increasing distance from the source of the contamination.

Woolson et al. (1971) studied the effect of 29 arsenic-contaminated soils on the growth of corn plants (*Z. mays*). A significant correlation was found between growth reduction and increasing arsenic concentration with a much better correlation with available arsenic than with total arsenic.

Wild (1974) notes an extensive flora present on arsenical mine spoils in Africa (arsenic levels < 30 g/kg) but did not determine if these species exhibited resistance. A wide range of arsenate plant tolerances have been demonstrated from the arsenate-contaminated regions of the world. These include species such as *Andropogon scoparius* (Rocovich & West, 1975), *Agrostis capillaris* (Porter & Peterson, 1977), *Plantago laceolata* (Pollard, 1980), *Holcus lanatus* (Meharg & Macnair, 1990, 1991b, 1992) and *Silene vulgaris* (Paliouris & Hutchinson, 1991). The mechanistic basis of such adaptations in grasses is discussed in section 9.1.3.1.

9.2.3.2 Vertebrates

Buck et al. (1976) reported two cases of arsenic toxicosis in domesticated animals. Seven animals in a herd of 75 cattle died and when analysed were found to contain arsenic concentrations in the liver and kidney of 12.8–18.2 and 42.5–60 mg/kg respectively. The animals were being raised on poor-quality hay with little other forage available; however, they were also found to have access to a container of herbicide containing sodium arsenite. In another case five horses died when fed on grass cuttings from a field which had been treated with a crabgrass control formulation (47% As_2O_3; 3.5% lead arsenate). Liver and kidney tissue contained arsenic levels of 21.6 and 21–24 mg/kg respectively. Hullinger et al. (1998) reported a case of arsenic toxicosis attributable to ingestion of ashes from burned posts which had been treated with an arsenic-containing preservative. Lack of normal salt supplementation to the herd was conducive to pica-like behaviour and ingestion of toxic ashes. Seven cows from a herd of 37 developed diarrhoea, weakness, stumbling and prostration. Four of the affected cattle died within 48 h. Arsenic was detected in liver tissue at 4.2 mg/kg, in abomasal contents at 42 mg/kg and in rumen contents at 105 mg/kg.

The herbicidal properties of arsenic made its use as a tree-debarker an important factor in the wood pulp industry in the north-eastern USA in the 1940s and 1950s. Cook (1953) reported two cases in New York State (USA) in which approximately 10 white-tailed deer consumed fatal amounts of sodium arsenite that was used to debark pulp trees. Field studies by Boyce & Verme (1954) showed that 923–2770 mg of arsenic (as sodium arsenite) was lethal to deer when licked from the bark of treated trees. The authors also report that wildlife kills from arsenic poisoning in Michigan (USA) in 1952 amounted to five deer, four porcupines, and one rabbit on about 81 ha of commercially treated trees. The practice of debarking trees with arsenicals has been replaced by mechanical techniques.

Swiggart et al. (1972) reported that 11 white-tailed deer (*Odocoileus virginianus*) were found dead after the aerial application of arsenic acid (52.3% As_2O_3) for the control of Johnson grass (*Sorghum halapense*). Livers from five of the deer were analysed and found to contain arsenic concentrations ranging from 16.8 to 24.3 mg/kg. The authors state that the levels of arsenic found were sufficient to have killed the deer but that the herbicide was not registered for such a use and, therefore, the deaths were attributed to misuse of the pesticide.

10. EVALUATION OF HUMAN HEALTH RISKS AND EFFECTS ON THE ENVIRONMENT

10.1 Effects on human health

10.1.1 Acute effects

Inorganic arsenic is acutely toxic and ingestion of large doses leads to gastrointestinal symptoms, disturbances of cardiovascular and central nervous system functions, multiorgan failure and eventually death. In survivors, bone marrow depression, haemolysis, hepatomegaly, melanosis, polyneuropathy and encephalopathy may be observed.

10.1.2 Vascular effects

Several studies in Taiwan have demonstrated an association between arsenic ingestion and blackfoot disease (BFD), with clear exposure–response effects related to both the well-water arsenic levels and duration of use of arsenic-contaminated drinking-water. Preclinical cases have also been identified by Doppler ultrasound in formerly exposed individuals, suggesting that the effects of arsenic ingestion can persist after exposure has declined or ceased. Several other studies and case reports of subjects exposed to arsenic from many sources, in countries other than Taiwan, document an association with peripheral vascular alterations. However, the extreme form and high prevalence of BFD found in Taiwan has not been reported in other parts of the world, and contributing factors, such as malnutrition or other concurrent exposures, may play a role in the pathophysiology of the disease.

Hypertension is associated with long-term exposure to arsenic, but this evidence is limited to cross-sectional studies, one occupational and two environmental (Taiwan and Bangladesh), all three of which found elevations in blood pressure with arsenic exposure. The two environmental studies demonstrated exposure–response relationships. It should be noted that although hypertension is not a very important cause of death itself, it is a major risk factor for other vascular diseases.

Several studies in Taiwan show a relationship between arsenic exposure and mortality from cardiovascular diseases (CVD), including exposure–response relationships. Similar results have generally not been observed in other arsenic drinking-water studies or in a medicinal study, in all of which the exposure levels have been lower. In the occupational studies, mortality from arteriosclerosis and coronary heart disease was elevated in the latest report from the Tacoma cohort (USA), but no statistically significant increases for these effects have been found in the Ronnskar (Sweden) or Anaconda (USA) smelter cohorts. The study in Utah found an excess of mortality from hypertensive heart disease but there were only a small number of deaths.

Only very limited evidence exists for an association between arsenic exposure and cerebrovascular disease. Some of the Taiwanese studies have shown an elevated risk of death from cerebrovascular disease, but the data are inconsistent across studies and the elevations, where present, are small compared with those for CVD. Studies from other countries provide only very limited support for the Taiwanese findings, but exposure levels were considerably lower.

10.1.3 Diabetes mellitus

In Taiwan, the prevalence and mortality rates of diabetes mellitus were higher among the population of the BFD-endemic area. There was also an exposure–response relationship between cumulative arsenic exposure and the prevalence of diabetes mellitus. A similar exposure–response pattern was observed in a study in Bangladesh, where prevalence of keratosis was used as a surrogate for arsenic exposure. Two occupational studies found an association of borderline statistical significance between diabetes mellitus and exposure to arsenic.

10.1.4 Neurological effects

Although there is good evidence that acute arsenic poisoning causes neurological effects, especially in the peripheral nervous system, there is little evidence of neurological effects from long-term lower-level environmental or occupational exposure. The few published studies have suggested changes in peripheral nerve function after arsenic exposure, but the studies have been limited by small numbers, different methods used to assess the end-points and co-exposure to other known neurotoxins.

10.1.5 Cancer of the lung, bladder, and kidney

Studies in Taiwan, Chile and Argentina show consistently high mortality risks from lung, bladder and kidney cancer among populations exposed to arsenic via drinking-water. Where exposure–response relations have been studied, the risk of cancer for these sites increases with increasing exposure. Even when tobacco smoking has been considered, the exposure–response relationship remains.

Not all studies of populations exposed to arsenic have reported positive findings for increased lung, bladder and kidney cancer. Exposure in these studies have not been as high as those in Taiwan, Chile or Argentina, and the sample sizes of the study populations may not have provided the statistical power to detect increased risks.

Studies on populations occupationally exposed to arsenic, such as smelter workers, pesticide manufacturers and miners in many different countries, consistently demonstrate an excess lung cancer risk among the arsenic-exposed. Although all these groups are exposed to other chemicals in addition to arsenic, it is unlikely that some other common factor could explain the findings. The lung cancer risk increases with increasing arsenic exposure in all studies where exposure–response relationships have been investigated. Tobacco smoking has been investigated in several studies in two of the three main smelter cohorts, and was not found to be the cause of the increased lung cancer risk attributed to arsenic; however, it was found to be interactive with arsenic in increasing the lung cancer risk.

Risks of kidney or bladder cancer are not consistently elevated in studies among people occupationally exposed to arsenic. This difference between the occupational and environmental studies may reflect lower systemic concentrations of arsenic after inhalation exposure than after oral exposure.

It is difficult to determine the lowest arsenic drinking-water concentration at which increased risks of lung, bladder and kidney cancer may be found. Most of the studies where these effects have been observed were conducted in Taiwan. The exposure categories of studies conducted in the BFD-endemic area in Taiwan have historically been rather broad (e.g. < 300 µg/litre, 300–600 µg/litre, and > 600 µg/litre). A recent paper on the BFD-endemic area in Taiwan, however, reported increased risks of bladder and lung cancer mortality in persons consuming drinking-water with arsenic concentrations < 50 µg/litre.[1]

In Argentina, significantly elevated bladder, lung and kidney cancer mortality were found in the high-exposure group where over 75% of the measurements of arsenic in drinking-water were higher than the detection limit of 40 µg/litre. For the measurements over the detection limit, the average concentration was 178 µg/litre, which can be taken as the lowest exposure where these effects are observed. Exposure concentrations were not provided for the low- or intermediate-exposure groups although bladder, lung and kidney cancer mortality were significantly elevated for men and lung cancer mortality was significantly elevated for women in the intermediate-exposure group. Thus the lowest exposure where elevated kidney cancer risk could be observed would have to be considerably lower than 178 µg/litre.

In a case–control study conducted in Chile, there was an exposure–response relationship for the risk of lung cancer over all exposure categories, and the increased risk was statistically significant at exposure strata of 30–50 µg/litre and above. In a case–control study in Finland, a statistically significantly elevated risk of bladder cancer was observed at ≥0.5–64 µg As/litre drinking-water

---

[1]   While this EHC was in the press, a cohort study from north-eastern Taiwan (Chiou et al., 2001) also reported an exposure-dependent increase in the risk of bladder cancer in exposure categories 10–50, 50–100, and > 100 µg/litre, with relative risks of transitional cell carcinoma of 1.0, 1.9 (CI 0.1–32.5), 8.2 (0.7–99.1) and 15.3 (1.7–139.9). Unlike all earlier Taiwanese studies, this study used estimates of individual (rather than village average) drinking-water arsenic concentrations, and incidence rather than mortality as the end-point. Arsenic measurements in the well-water were performed using a hydride-generation atomic absorption method, and the results were adjusted for age, sex and cigarette smoking.

concentration but only when exposure was 3–9 years before diagnosis.

10.1.6 Cancer and precancerous lesions of the skin

Arsenic ingestion in drinking-water has been shown to be associated with a high risk of skin cancer. Well-documented studies on skin cancer after arsenic ingestion from drinking-water have been conducted in several populations in different countries, the largest of which were in Taiwan. Association of exposure to arsenic with skin cancer has also been observed in studies on patients treated with arsenicals. An early study found an excess of skin cancer mortality after occupational exposure to arsenic, but this was not observed in other occupational studies of arsenic exposure.

The lowest arsenic drinking-water concentration where an increased risk of skin cancer could be observed is in the lowest exposure group in the exposed Taiwan population (i.e. < 300 µg/litre). It should be noted that this is a very broad exposure category and the lowest concentration associated with skin cancer could have been considerably lower.

The lowest arsenic drinking-water concentration where an elevated risk of arsenic-associated skin lesions (hyperpigmentation and/or keratosis) has been found can be estimated from the study in West Bengal to be less than 50 µg/litre.

10.1.7 Cancer at other sites

In two partly overlapping studies in Taiwan, an elevated mortality from liver cancer was observed in relation to arsenic exposure from drinking-water. In one of the two studies in Chile, but not in the study in Argentina, such a relationship was observed.

Cancer at other sites in relation to arsenic exposure has been little studied outside Taiwan. The sites that have exhibited an elevated risk include oesophagus, stomach, small intestine, colon, nose, larynx, bone and prostate, as well as lymphoma and leukaemia. A study in the USA and another in Australia, neither of which showed a clear-cut increase in the risk of lung, bladder, or kidney cancer, showed moderately elevated mortality from cancer of the prostate. The studies on occupational exposure of arsenic have not found any consistent relationship between exposure to arsenic and cancer at sites other than lung.

It cannot be stated with certainty that arsenic exposure causes cancer at sites other than lung, skin, kidney and bladder. It is apparent that if such a causality exists, the relative risk of cancer at such sites must be lower than that for the sites for which the causality has been demonstrated.

10.1.8 Reproductive toxicity

Several studies have examined a number of reproductive endpoints in relation to arsenic exposure, and the results suggest elevations in fetal, neonatal and postnatal mortality, lowered birth weight, spontaneous abortions, stillbirths, pre-eclampsia and congenital malformations. However, there is no consistent evidence for any one particular end-point.

10.1.9 Genotoxicity

Genotoxicity studies in relation to arsenic exposure have included exposed and unexposed individuals from several populations and analyses have been based on various tissues, including blood, buccal and bladder cells as well as sections from tumour biopsies or Bowen's disease.

Even with some negative findings, the overall weight of evidence (Table 45) indicates that arsenic can cause clastogenic damage in different cell types, with different end-points, in exposed individuals. Clastogenic effects have also been observed in cells from cancer patients. Arsenic is thus clastogenic in humans *in vivo* However, no *HPRT* gene mutation was seen in the single study in lymphocytes or increases in *ras* or *p53* gene expression in cells from cancer or Bowen's disease patients with long-term exposure to arsenic, except for one study with increased *p53* expression in Bowen's disease patients with such exposure compared to patients without exposure.

Table 45. Genotoxicity studies after As exposure – weight of evidence approach

Micronuclei	Sister chromatid exchanges	Chromosome aberrations	Aneuploidy	HPRT mutation	p53/ras expression
+ (USA)	lymphocytes + (Argentina)	lymphocytes -,+ (Mexico)	lymphocytes + (Argentina)	lymphocytes - (Mexico)	- urothelial cancer patients (Taiwan)
+ (Northern Chile)	- (Mexico)	+ (Mexico)			- Bowen's disease (Taiwan)
+ (Argentina)	- (Argentina)	± (Finland)			+ Bowen's disease (Taiwan)
Buccal cells	+ Bowen's disease (Taiwan)	+ cancer patients (Taiwan)			
- (USA)	- cancer patients, BFD (Taiwan)				
+ (Mexico)					

10.1.10   *Supporting data from experimental studies*

Several animal carcinogenicity studies on arsenic have been carried out, but limitations such as high dose levels, limited time of exposure and limited number of animals make these inconclusive. However, in a recent study, exposure of female mice to arsenic in drinking-water was associated with increased incidence in tumours involving mainly lung, liver, gastrointestinal tract and skin. Inorganic arsenic did not induce point mutations in bacteria or in mammalian cells. However, arsenic can produce chromosomal aberrations *in-vitro*, affect methylation and repair of DNA, induce cell proliferation, transform cells, and promote tumours. One study has indicated that DMA may cause bladder cancer in male rats at high doses.

10.1.11   *Conclusions*

Arsenic exposure via drinking-water is causally related to cancer in the lungs, kidney, bladder and skin. Drinking-water arsenic concentrations of ≤50 µg/litre have been associated with increased risks of cancer in the bladder and lung. Precursors of skin cancer have been associated with drinking-water arsenic levels ≤ 50 µg/litre. Occupational exposure to airborne arsenic is causally related to cancer of the lung. Cumulative exposure to ≥0.75 mg/m^3 · year has been associated with an increased risk of cancer of the lung.

Arsenic is considered to be genotoxic in humans on the basis of clastogenicity in exposed individuals and findings *in vitro*.

Arsenic exposure via drinking-water induces PVD. Whether arsenic alone is sufficient to cause the extreme form of this disease, BFD, is not known.

Conclusions on the causality of the relationship between arsenic exposure and other health effects are less clear-cut. The evidence is strongest for hypertension and CVD, suggestive for diabetes and reproductive effects and weak for cerebrovascular disease, long-term neurological effects and cancer at sites other than lung, bladder, kidney and skin.

10.2 Evaluation of effects on the environment

10.2.1 Exposure

Arsenic is continually cycled through all environmental compartments. Arsenic can be elevated to high levels in water and soil because of the underlying geology or geothermal activity. In the aquatic environment arsenic concentrations can also become elevated in some estuaries and in waters near heavy industrial or mining and mineral-processing areas. The highest concentrations of arsenic in soil tend to be associated with mining waste.

Mean total arsenic concentrations in air from remote and rural areas range from 0.02 to 4 ng/m^3. Mean total arsenic concentrations in urban areas range from 3 to about 200 ng/m^3; much higher concentrations (> 1000 ng/m^3) have been measured in the vicinity of industrial sources.

Reported concentrations of arsenic in surface waters are summarized in Fig. 6. Concentrations of arsenic in open ocean seawater are typically 1–2 µg/litre. Arsenic is widely distributed in surface freshwaters, and background concentrations in rivers and lakes are generally < 2 µg/litre except in areas with volcanic rock and sulfide mineral deposits. Mean arsenic concentrations of 500 µg/litre and a maximum of 25 mg/litre have been reported for geothermal waters. Enhanced arsenic levels of < 10 mg/litre have been reported near anthropogenic sources such as mining and agrochemical manufacture. Mean sediment arsenic concentrations range from 5 to 3000 mg/kg, with the higher levels occurring in areas of contamination.

Reported concentrations of arsenic in soils are summarized in Fig. 7. Background concentrations in soil tend to range from 1 to 40 mg/kg, with a mean value of 5 mg/kg. Naturally elevated levels of arsenic in soils may be associated with geological substrata such as sulfide ores. Anthropogenically contaminated soils can have concentrations of arsenic up to several percent.

Three major modes of arsenic biotransformation have been found to occur in the environment: redox transformation between arsenite and arsenate, reduction and methylation of inorganic arsenic

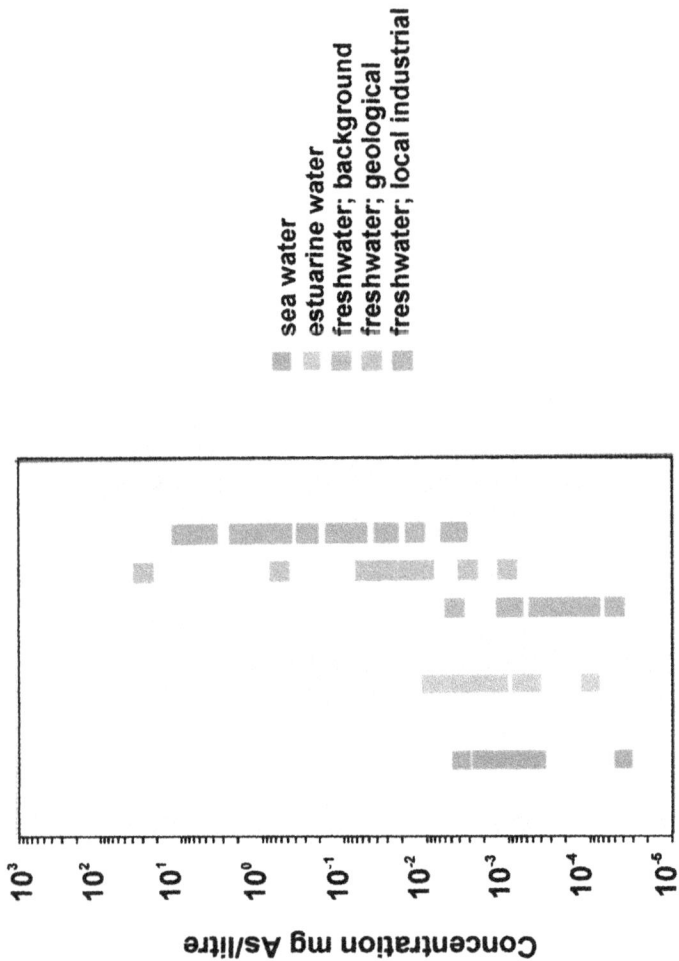

Fig. 6. Reported concentrations of arsenic in seawater, estuarine water and freshwater. The seawater values will be background levels; some anthropogenic component is possible in estuarine concentrations values. Freshwater values are separated into background (anthropogenic input unlikely), geological (volcanic/geothermal) concentrations local to industrial activity (mining/agrochemicalmanufacture).

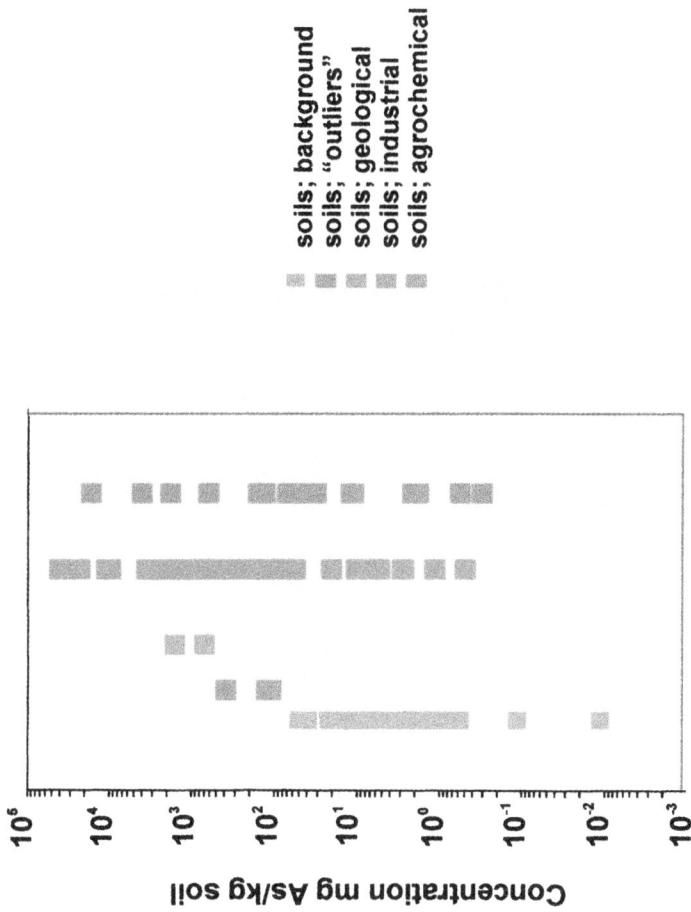

Fig. 7. Reported concentrations of arsenic in soils. Values plotted as "outliers" are the upper end of ranges where bedrock or freshwater sediments were thought to contribute to higher than normal arsenic levels. "Geological" values are for volcanic areas. "Industrial" values include mining, smelting and manufacture of agrochemicals; the "agrochemicals" category covers concentrations following use of pesticides, sheep dips, etc.

and biosynthesis of more complex organic arsenic compounds. There is biogeochemical cycling of compounds formed from these processes. Bioaccumulation of organic arsenic compounds, after their biogenesis from inorganic forms, occurs in aquatic organisms. Biomagnification in aquatic food chains has not been observed. Terrestrial plants may accumulate inorganic arsenic by root uptake from the soil or by adsorption of airborne arsenic deposited on the leaves. Background arsenic concentrations in living organisms are usually < 1 mg/kg (fresh weight) in freshwater and terrestrial biota. These levels are higher in biota collected near anthropogenic sources or areas with geothermal activity. Marine organisms can normally contain much higher arsenic concentrations (< 100 mg/kg fresh weight) and these are predominantly organic arsenic species such as arsenosugars (macroalgae) and arsenobetaine (invertebrates and fish).

10.2.2 *Effects*

Aquatic and terrestrial biota show a wide range of sensitivities to the few arsenic species (inorganic arsenic and the simple methylated acids) that have been tested. Phytoplankton have generally been the most sensitive to inorganic arsenic (both arsenate and arsenite). Toxicity of inorganic arsenic in freshwater is summarized in Fig. 8. The lowest EC_{50} value for growth in freshwater algae was at an arsenate concentration of 48 µg As/litre with a LOEC of 5 µg As/litre. NOECs for arsenite in 28-day tests on daphnids were between 600 and 1300 µg As/litre. In embryo-larval tests on fish and amphibians the lowest LC_{50}s for arsenate were 540 and 40 µg As/litre respectively. Toxicity of inorganic arsenic in seawater is summarized in Fig. 9. Adverse effects on marine periphyton communities were observed at arsenate concentrations of 15–60 µg/litre. For marine invertebrates the lowest acute LC_{50}s of arsenate for copepods were 10.9 (nauplius), 19.8 (copepodid) and 27.5 µg As/litre (ovigerous female). The toxicity of organic arsenic to aquatic organisms is summarized in Fig. 10.

Terrestrial toxicity of inorganic arsenic and organoarsenicals previously used as pesticides is summarized in Fig. 11. Levels of soil arsenic reported to be toxic to terrestrial plants span a broad range, with toxicity thresholds ranging from around 30 mg/kg to 300 mg/kg with toxicity tending to be greater in sandy than in clay soils. The

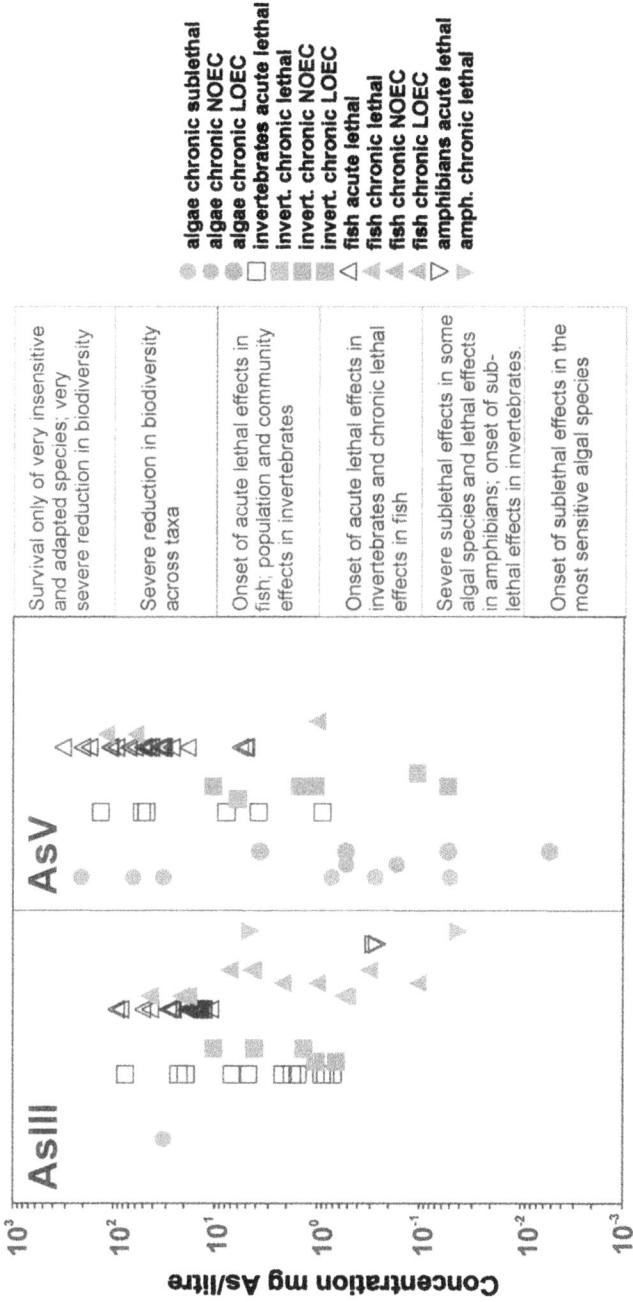

Fig. 8. Acute and chronic toxic effects of arsenite and arsenate in freshwater. Values are taken from the tables and text of Chapter 9. Likely effects are given for each order of magnitude increase in concentration. Effects assume no mitigation of toxicity in the environment.

397

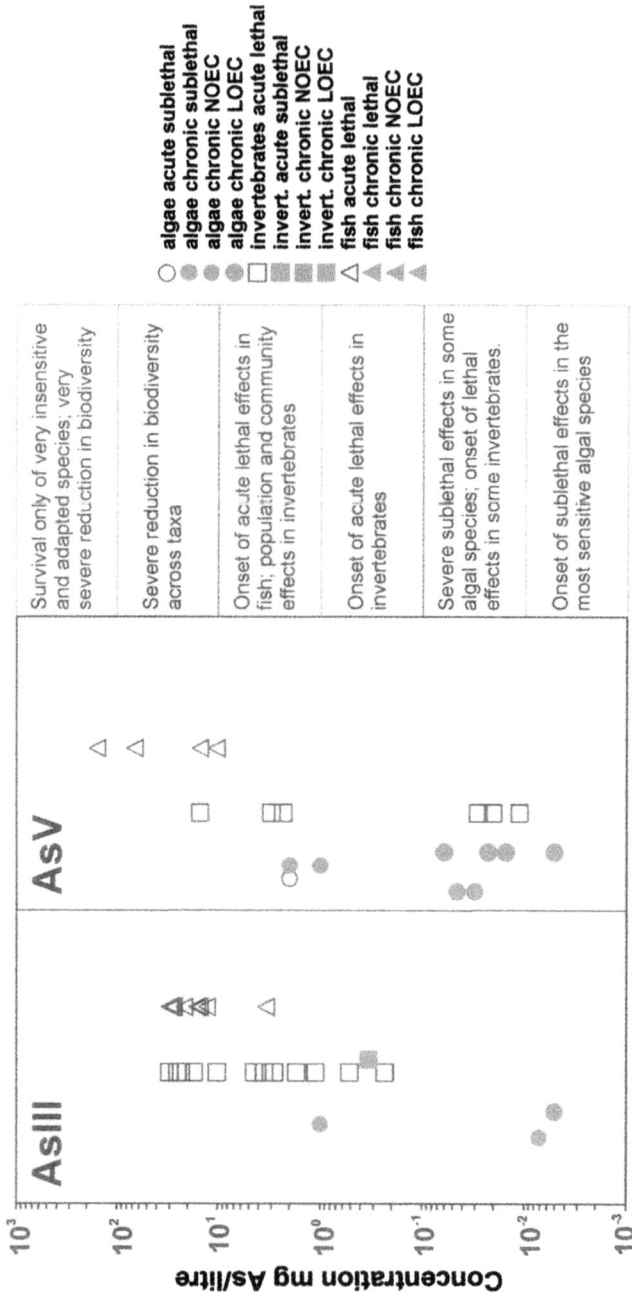

Fig. 9. Acute and chronic toxic effects of arsenite and arsenate in seawater. Values are taken from the tables and text of Chapter 9. Likely effects are given for each order of magnitude concentration. Effects assume no mitigation of toxicity in the environment.

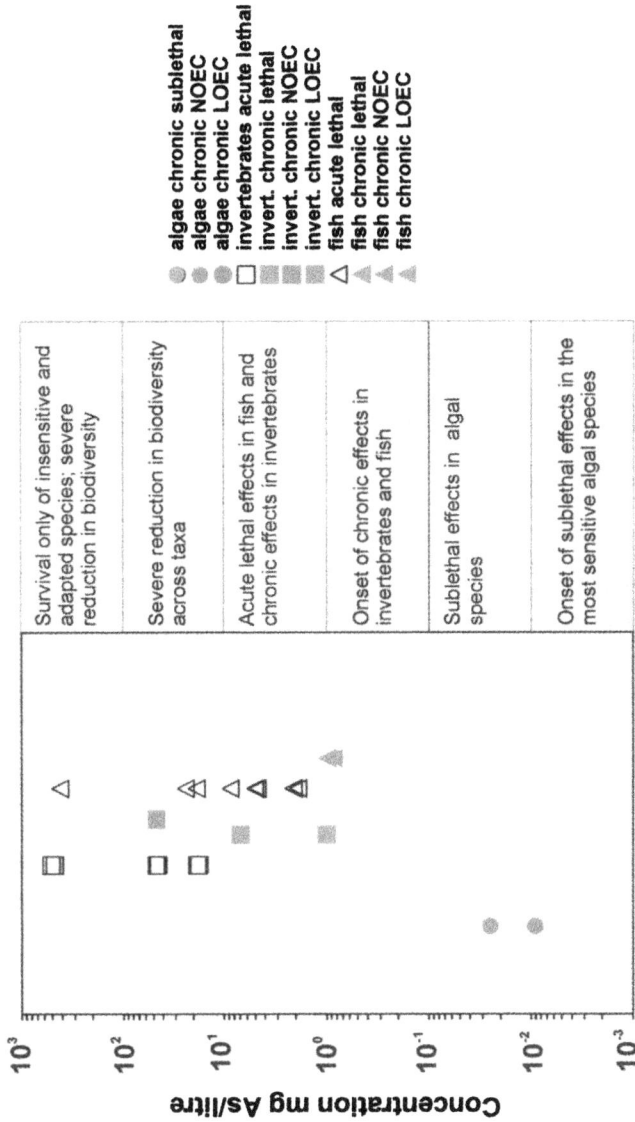

Fig. 10. Acute and chronic aquatic toxicity of organic arsenic (monomethyl arsonic acid; **MMA**). Values are taken from the tables and text of Chapter 9. Likely effects are given for each order of magnitude increase in concentration. Effects assume no mitigation of toxicity in the environment.

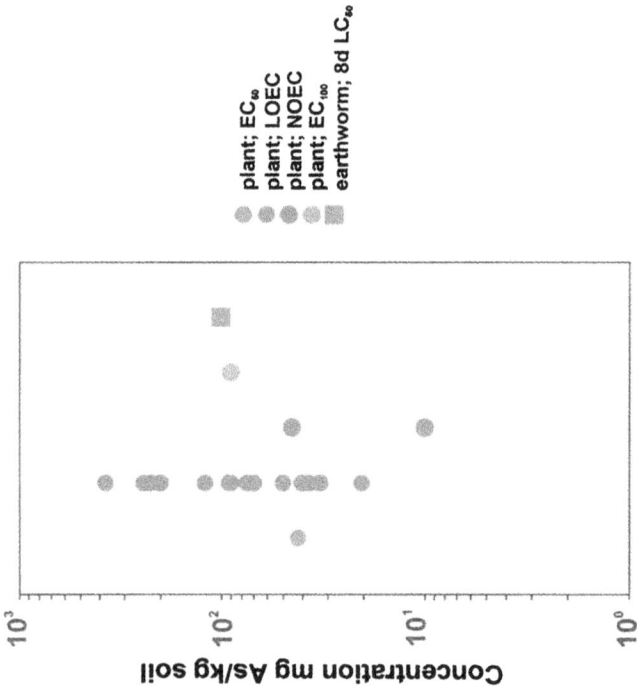

Fig. 11. Terrestrial toxicity of inorganic arsenic and organoarsenic compounds formerly used as pesticides. Values for plants are sub-lethal effects with the end-points as either growth or yield. Where values in original reports were quoted as application rates in kg/ha, these have been converted to mg/kg soil by assuming even incorporation into the top 5 cm of soil. For organic arsenicals, values have been recalculated as As.

lowest critical plant tissue concentration was around 1 mg/kg. Very limited data are available for soil invertebrates. The only toxicity data available for birds are from acute, single-dose oral exposure which cannot easily be related to feeding exposure in the environment. Longer-term feeding studies with birds have not used doses which led to significant adverse effects; since all reported NOECs are "greater than" the highest experimental dose, values cannot be used for assessing risk.

10.2.3 Environmental modification of toxicity

The figures summarizing aquatic toxicity list likely effects at differing concentrations of arsenic in surface waters. The suggested effects assume no mitigation of toxicity in the environment. In practice, the sensitivity of both aquatic and terrestrial organisms to arsenic is modified by biological and abiotic factors. In the environment arsenic toxicity is affected by temperature, pH, Eh, organic matter content, phosphate concentration, adsorption to solid matrices, and the presence of other substances and toxicants. In general, if the same organisms are compared under the same test conditions, then inorganic arsenicals are more toxic than the few pentavalent organic arsenic compounds likely to be encountered in the environment and arsenite is more toxic than arsenate. The mechanism of uptake of arsenate by organisms, and its mode of toxicity, differ considerably. This may explain some of the interspecies differences in response to different arsenic species. For example, the primary mechanism of arsenite toxicity is considered to result from its binding to protein sulfhydryl groups; arsenate is known to affect oxidative phosphorylation by competing with phosphate.

10.2.4 Risk evaluation

Conventional, generic risk assessment would apply uncertainty factors to the lowest reported chronic effects concentration. For arsenic in freshwaters, this would be 5 µg/litre for growth in algae. This concentration is similar to the upper limit of the natural range of arsenic concentrations in most surface freshwaters. It is almost four orders of magnitude lower than the highest natural concentrations of arsenic in geothermal regions. Since communities of organisms occur in surface waters across the whole natural range

(0.05–25 000 µg/litre), a single protective concentration target for arsenic is inappropriate. Although there is no direct evidence that populations of organisms living at the higher end of the range for most surface waters (around 2 µg/litre) are less sensitive to arsenic than those at the bottom end of the range (0.05 µg/litre), this might be inferred from laboratory and field evidence. There is clear laboratory and field evidence that populations living at much higher concentrations have adapted to high inorganic arsenic levels. In addition, factors outlined above (section 10.2.3) may mitigate arsenic toxicity in the environment. Realistically, risk assessment for inorganic arsenic can only be done on a site-by-site basis taking into account background arsenic concentrations, local population tolerance and other local mitigating factors.

For populations of organisms living in low inorganic arsenic environments with little possible mitigation (e.g. low phosphate levels), a concentration of around 5 µg/litre would be protective of all but the most sensitive algae. Adapted populations at high natural inorganic arsenic concentrations will be specialized communities, possibly of lower biodiversity but probably of high conservation interest. Areas polluted by anthropogenic activity, therefore, present the risk manager with different options based on both practicability and desirability of remediation; if adapted communities have developed over time, these might be destroyed by remediation. Clearly the contamination of pristine areas with arsenic to levels which cause adaptation and decreased biodiversity is unacceptable.

From the limited data available for the marine environment it seems that inorganic arsenic concentrations in the open ocean do not overlap with reported toxic thresholds, suggesting minimal risk to organisms. The extreme sensitivity of some marine algae to inorganic arsenic is consistent with general low exposure and, therefore, no development of tolerance. Absorption of high concentrations of arsenate from ambient seawater by algae (accumulated as arsenosugars) may be linked to low phosphate levels. The reason for the accumulation of high concentrations of arsenobetaine in marine fish and invertebrates is uncertain. In estuaries, anthropogenic and geological sources may contribute to higher concentrations of arsenic which pose some risk to sensitive populations not previously exposed.

Reported total arsenic concentrations in soil can be very high. However, total arsenic is a poor indicator of toxicity to plants. Bioavailable arsenic represents a small percentage of total soil arsenic (10% or less and usually < 2%). Even with this low level of bioavailability, sensitive plants will be eliminated at some of the reported soil concentrations. On severely contaminated mine wastes, specialized plant communities tolerant to arsenic have developed. Some tolerant plants grow on wastes with total arsenic levels of several percent by weight. Communities are likely to be low in biodiversity at high arsenic concentrations.

Available data on soil invertebrates are very limited. Results for earthworms from one study suggest risk in soils with greater than background concentrations; the worms for this study were taken from soil low in arsenic. There is no published information on development of tolerance in soil invertebrate populations.

11. RECOMMENDATIONS FOR FUTURE RESEARCH

The Task Group identified the following areas where additional research might improve the assessment of risks to human health and environment induced by exposure to arsenic.

11.1 Human health

Better exposure–response epidemiological studies are required to characterize the potential health effects of arsenic at low levels of exposure, notably:

- longitudinal studies on cardiovascular morbidity, hypertension and diabetes

- epidemiological studies on reproductive end-points for which associations have already been suggested.

- longitudinal studies of neurological effects

- identification of sensitive subpopulations

- Influence of nutrition on arsenic metabolism and arsenic-induced effects

- biomarkers for arsenic exposure and health effects

- development of animal models for studying the carcinogenic and non-carcinogenic effects of arsenic

- better characterizations of the metabolism of arsenic, including formation of reactive intermediate metabolites, and possible genetic polymorphisms of the enzyme activities involved, and other factors influencing the metabolism

- mechanisms of action of the carcinogenic and non-carcinogenic effects of arsenic

- characterization of exposure to and bioavailability and toxicity of different arsenic species in foods, as well as variation in the levels of arsenic species in the same food items from different geographical areas

- development of robust, sensitive and accurate analytical techniques suitable for field measurements of arsenic in water and air.

11.2 Environmental

Further studies are required on:

- Global cycling and relative contributions of natural and anthropogenic sources of As.

- Concentrations and speciation in as well as effects of arsenic on zooplankton and phytoplankton in estuarine and marine ecosystems and freshwater aquatic systems.

- Concentrations of and possible effects of arsenic on terrestrial species.

- Metabolism of inorganic arsenic in algae, fish and shellfish.

12. PREVIOUS EVALUATIONS BY INTERNATIONAL BODIES

Arsenic and arsenic compounds were evaluated by the International Agency for Research on Cancer (IARC, 1973) and the evaluation updated in supplement 7 (IARC, 1987). There was *sufficient evidence* for carcinogenicity to humans and *limited evidence* for carcinogenicity to animals, and the overall evaluation was that arsenic and arsenic compounds are carcinogenic to humans (Group 1). This evaluation applies to the group of chemicals (i.e. arsenic and arsenic compounds) as a whole and not necessarily to all individual chemicals within the group.

The World Health Organization has given a provisional guideline value of 10 µg/litre for arsenic in drinking-water as the practical quantification limit (WHO, 1996).

The World Health Organization (WHO, 2000) has estimated that the unit risk for arsenic-induced lung cancer (risk estimate for lifetime exposure to a concentration of 1 µg/m^3) is 1.5×10^{-3}.

REFERENCES

Abdel-Moati AR (1990) Speciation and behavior of arsenic in the Nile Delta lakes. Water Air Soil Pollut, **51**: 117–132.

Abdelghani AA, Reimers RS, Anderson AC, Englande AJ, Lo CP & Shariatpanahi M (1981) Transport and distribution of arsenic in sediments. Heavy metals in the environment. Proceedings of the 3rd International Conference. Amsterdam, September 1981, Geneva, WHO, pp 665–668.

Abdelghani AA, Anderson AC, Jaghabir M & Mather F (1986) Arsenic levels in blood, urine, and hair of workers applying monosodium methanearsonate (MSMA). Arch Environ Health, **41**: 163–169.

Abdo KM, Elwell MR, Montgomery CA, Thompson MB, Thompson RB & Prejean JD (1989) Toxic responses in F344 rats and B6C3F1 mice given roxarsone in their diets for up to 13 weeks. Toxicol Lett, **45**: 55–66.

Acharyya SK, Chakraborty P, Lahiri S, Raymahashay BC, Guha S & Bhowmik A (1999) Arsenic poisoning in the Ganges delta. *Nature* **401**(6753): 545.

Adkison DL & Sundberg JP (1991) 'Lipomatous' hamartomas and choristomas in inbred laboratory mice. Vet Pathol, **28**: 305–312.

Agahian B, Lee JS, Nelson JH & Johns RE (1990) Arsenic levels in fingernails as a biological indicator of exposure to arsenic. Am Ind Hyg Assoc J, **51**: 646–651.

Aggett J & Aspell AC (1980) Arsenic contamination in an apple orchard. Environ Pollut, **22**: 39–46.

Aggett J & Kriegman MR (1988) The extent of formation of arsenic(III) in sediment interstitial waters and its release to hypolimnetic waters in Lake Ohakuri. Water Res, **22**(4): 407–411.

Aggett J & O'Brien GA (1985) Detailed model for the mobility of arsenic in Lacustrine sediments based on measurements in Lake Ohakuri. Environ Sci Technol, **19**(3): 231–238.

Aggett J & Roberts LS (1986) Insight into the mechanism of accumulation of arsenate and phosphate in hydro lake sediments by measuring the rate of dissolution with ethylenediaminetetraacetic acid. Environ Sci Technol, **20**(2): 183–186.

Ahmann D, Roberts AL, Krumholz LR & Morel FMM (1994) Microbe grows by reducing arsenic. Nature, **371**: 750.

Ahmann D, Krumholz LR, Hemond HF, Lovley DR & Morel FMM (1997) Microbial mobilization of arsenic from sediments of the Aberjona watershed. Environ Sci Technol, **31**: 2923–2930.

Aichberger K & Hofer GF (1989) [Contents of arsenic, mercury and selenium in agricultural soils of Upper Austria]. (in German) Bodenkultur, 40(1): 1–11.

Akeredolu FA, Barrie LA, Olson MP, Oikawa KK, Pacyna JM & Keeler GJ (1994) The flux of anthropogenic trace metals into the Arctic from the mid-latitudes in 1979/80. Atmos Environ, 28(8): 1557–1572.

Akkari KH, Frans RE & Lavy TL (1986) Factors affecting degradation of MSMA in soil. Weed Sci, 34: 781–787.

Albores A, Koropatnick J, Cherian MG & Zelazowski AJ (1992) Arsenic induces and enhances rat hepatic metallothionein production in vivo. Chem Biol Interact, 85: 127–140.

Albores A, Sinal CJ, Cherian MG & Bend JR (1995) Selective increase of rat lung cytochrome P450 1A1 dependent monooxygenase activity after acute sodium arsenite administration. Can J Physiol Pharmacol, 73: 153–158.

Alexander J & Aaseth J (1985) Excretion of arsenic in rat bile: a role of complexing ligands containing sulphur and selenium. Nutr Res Suppl I: 515–519.

Amonoo-Neizer EH, Nyamah D & Bakiamoh SB (1996) Mercury and arsenic pollution in soil and biological samples around the mining town of Obuasi, Ghana. Water Air Soil Pollut 91(3/4): 363–373.

Amundsen CE, Hanssen JE, Semb A & Steinnes E (1992) Long-range atmospheric transport of trace elements to southern Norway. Atmos Environ 26A(7): 1309–1324.

Anastasia FB & Kender WJ (1973) Influence of soil arsenic on the growth of lowbush blueberry. J Environ Qual, 2(3): 335–337.

Anderson AC, Abdelghani AA, Smith PM, Mason JW & Englande AJ (1975) The acute toxicity of MSMA to black bass (*Micropterus dolomieu*), crayfish (*Procambarua* sp.) and channel catfish (*Ictalurus lacustris*). Bull Environ Contam Toxicol, 14(3): 330–333.

Anderson LCD & Bruland KW (1991) Biogeochemistry of arsenic in natural waters: the importance of methylated species. Environ Sci Technol, 25(3): 420–427.

Andreae MO (1978) Distribution and speciation of arsenic in natural waters and some marine algae. Deep Sea Res, 25: 391–402.

Andreae MO (1979) Arsenic speciation in seawater and interstitial waters: the influence of biological-chemical interactions on the chemistry of a trace element. Limnol Oceanogr, 24(3): 440–452.

Andreae MO (1980) Arsenic in rain and the atmospheric mass balance of arsenic. J Geophys Res 85(8C): 4512–4518.

Andreae MO (1983) Biotransformation of arsenic in the marine environment. In: Lederer WH & Fensterheim RJ ed. Arsenic: industrial, biochemical, environmental perspectives. New York, Van Nostrand Reinhold, pp 378–392.

Andreae MO & Andreae TW (1989) Dissolved arsenic species in the Schelde Estuary and watershed, Belgium. Estuar Coast Mar Sci, **29**: 421–433.

Andreae MO & Froelich PN (1984) Arsenic, antimony, and germanium biogeochemistry in the Baltic Sea. Tellus 36B: 101–117.

Andreae MO & Klumpp DW (1979) Biosynthesis and release of organoarsenic compounds by marine algae. Environ Sci Technol, **13**(6): 741.

Andren S, Schutz A, Vahter M, Attewell R, Johansson L, Wallers S & Skerfving S (1988) Environmental exposure to lead and arsenic among children living near a glassworks. Sci Total Environ, **77**(1): 25–34.

ANZFA (Australia New Zealand Food Authority) (1994) The 1994 Australian Market Basket Survey. A total diet survey of pesticides and contaminants. Canberra, Australia New Zealand Food Authority.

Apel M & Stoeppler M (1983) Speciation of arsenic in urine of occupationally non-exposed persons. International Conference Heavy Metals in the Environment, 1983, Heidelberg. Vol. 1. Edinburgh, CEP Consultants.

Aposhian HV (1989) Biochemical toxicology of arsenic. In: Hodgson E, Bend JR, Philpot RM eds. Reviews in biochemical toxicology, Vol. 10. New York, Elsevier Science.

Aposhian HV, Gurzau ES, Le CX, Gurzau A, Healy SM, Lu X, Ma M, Yip L, Zakharyan RA, Maiorino RM, Dart RC, Tircus MG, Gonzalez-Ramirez D, Morgan DL, Avram D & Aposhian MM (2000) Occurrence of monomethylarsonous acid in urine of humans exposed to inorganic arsenic. Chem Res Pharmacol, **13**: 693–697.

Arbouine MW & Wilson HK (1992) The effect of seafood consumption on the assessment of occupational exposure to arsenic by urinary arsenic speciation. J Trace Elem Electrolytes Health Dis, **6**: 153–160.

Arle HF & Hamilton KC (1971) Topical applications of DSMA and MSMA in irrigated cotton. Weed Sci, **19**(5): 545–547.

Armstrong BK, McNulty JC, Levitt LJ, Williams KA & Hobbs MST (1979) Mortality in gold and coal miners in Western Australia with special reference to lung cancer. Br J Ind Med, **36**: 199–205.

Armstrong CW, Stroube RB, Rubio T, Siudyla EA & Miller GB Jr (1984) Outbreak of fatal arsenic poisoning caused by contaminated drinking water. Arch Environ Health, **39**: 276–279.

Arrighi HM & Hertz-Picciotto I (1996) Controlling the healthy worker survivor effect: an example of arsenic exposure and respiratory cancer. Occup Environ Med, 53: 455–462.

Aschengrau A, Zierler S & Cohen A (1989) Quality of community drinking water and the occurrence of spontaneous abortion. Arch Environ Health, 44(5): 283–289.

Asher CJ & Reay PF (1979) Arsenic uptake by barley seedlings. Aust J Plant Physiol, 6 459–466.

Astolfi E, Maccagno A, Garcia Fernandez JC, Vaccaro R & Stimola R (1981) Relation between arsenic in drinking water and skin cancer. Biol Trace Elem Res, 3(2): 133–143.

ATSDR (1993) Toxicological profile for arsenic. TP-92/02. Atlanta, GA, Agency for Toxic Substances and Disease Registry, U.S. Department of Health & Human Services.

ATSDR (2000) Toxicological Profile for Arsenic (update). Atlanta, GA, Agency for Toxic Substances and Disease Registry, U.S. Department of Health & Human Services.

Attar KM, El Faer MZ, Rawdah TN & Tawabini BS (1992) Levels of arsenic in fish from the Arabian Gulf. Mar Pollut Bull, 24(2): 94–97.

Aurilio AC, Mason RP & Hemond HF (1994) Speciation and fate of arsenic in three lakes of the Aberjona watershed. Environ Sci Technol, 28(4): 577–585.

Axelson O, Dahlgren E, Jansson CD & Rehnlund SO (1978) Arsenic exposure and mortality: a case-referent study from a Swedish copper smelter. Br J Ind Med, 35 8–15.

Azcue JM, Mudroch A, Rosa F & Hall GEM (1994a) Effects of abandoned gold mine tailings on the arsenic concentrations in water and sediments of Jack of Clubs Lake, BC. Environ Technol, 15(7): 669–678.

Azcue JM, Nriagu JO & Schiff S (1994b) Role of sediment porewater in the cycling of arsenic in a mine-polluted lake. Environ Int, 20(4): 517–527.

Babich H, Martin-Alguacil N & Borenfreund E (1989) Arsenic-selenium interactions determined with cultured fish cells. Toxicol Lett, 45: 157–164.

Baker MD, Inniss WE, Mayfield CI, Wong PTS & Chau YK (1983a) Effect of pH on the methylation of mercury and arsenic by sediment microorganisms. Environ Technol Lett, 4: 89–100.

Baker MD, Wong PTS, Chau YK, Mayfield CI & Inniss WE (1983b) Methylation of arsenic by freshwater green algae. Can J Fish Aquat Sci, 40: 1254–1257.

Baker RS, Arle HF, Miller JH & Holstun JT (1969) Effects of organic arsenical herbicides on cotton response and chemical residues. Weed Sci, 17: 37–40.

Baldi F, Bianco MA & Pepi M (1995) Mercury, arsenic and boron resistant bacteria isolated from the phyllosphere as positive bioindicators of airborne pollution near geothermal plants. Sci Total Environ, **164**(2): 99–107.

Ball AL, Rom WN & Glenne B (1983) Arsenic distribution in soil surrounding the Utah copper smelter. Am Ind Hyg Assoc J, **44**(5): 341–348.

Barbaris B & Betterton EA (1996) Initial snow chemistry survey of the Mogollon Rim in Arizona. Atmos Environ, **30**(17) 3093–3103.

Barbaud A, Mougeolle JM & Schmutz JL (1995) Contact hypersensitivity to arsenic in a crystal factory worker. Contact Dermatitis, **33**: 272–273.

Barchowsky A, Dudek EJ, Treadwell MD & Wetterhahn KE (1996) Arsenic induces oxidant stress & NF-B activation in cultured aortic endothelial cells. Free Radical Biol Med, **21**: 783–790.

Barrows ME, Petrocelli SR, Macek KJ & Carroll JJ (1980) Bioconcentration and elimination of selected water pollutants by bluegill sunfish (*Lepomis macrochirus*). In: Haque R ed. Dynamics, exposure and hazard assessment of toxic chemicals. Ann Arbor, MI, Ann Arbor Science Publishers, pp 379–392.

Barry GA, Chudek PJ, Best EK & Moody PW (1995) Estimating sludge application rates to land based on heavy metal and phosphorus sorption characteristics of soil. Water Res, **29**(9): 2031–2034.

Bartolomé B, Cordoba S, Nieto S, Fernandez-Herrera J & Garcia-Diez A (1999) Acute arsenic poisoning: clinical and histopathological features. Br J Dermatol, **141**: 1106–1109.

Bates MN, Smith AH & Cantor KP (1995) Case-control study of bladder cancer and arsenic in drinking water. Am J Epidemiol, **141**: 523–530.

Baxley MN, Hood RD, Vedel GC, Harrison WP & Szczech GM (1981) Prenatal toxicity of orally administered sodium arsenite in mice. Bull Environ Contam Toxicol, **26**: 749–756.

Beard HC & Lyerly LA (1961) Separation of arsenic from antimony and bismuth by solvent extraction. Anal Chem, **33**(12) 1781–1782.

Beauchemin D, Siu KWM, Mclaren JW & Berman SS (1989) Determination of arsenic species by high-performance liquid chromatography-inductively coupled plasma mass spectrometry. J Anal At Spectrom, **4**: 285–289.

Beavington F & Cawse PA (1978) Comparative studies of trace elements in air particulate in northern Nigeria. Sci Total Environ, **10**: 239–244.

Beceiro-Gonzalez E, Gonzalez-Soto E, Lopez-Mahia P & Prada-Rodriguez D (1997)Total arsenic and selenium levels in atmospheric particulate matter of La Coruna (Spain). Sci Total Environ, **208**: 207–211.

Bech J, Poschenrieder C, Llugany M, Barceló J, Tume P, Tobias FJ, Barranzuela JL & Vásquez ER (1997) Arsenic and heavy metal contamination of soil and vegetation around a copper mine in Northern Peru. Sci Total Environ, **203**: 83–91.

Becket WAS, Moore JL, Keogh JP & Bleecker ML (1986) Acute encelopathy due to occupational exposure to arsenic. Br J Ind Med, **43**: 66-67.

Beckman L & Nordenson I (1986) Interaction between some common genotoxic agents. Hum Hered, **36**: 397–401.

Bekemeier H & Hirschelmann R (1989) Reactivity of resistance blood vessels ex vivo after administration of toxic chemicals to laboratory animals: arteriolotoxicity. Toxicol Lett, **49**: 49–54.

Bell E, Sher S, Hull B, Merrill C, Rose S, Chamson A, Asselinea D, Dubertwot L, Coulomb B & Lapiere C, Nusgens B, Neveux Y (1983) The reconstitution of living skin. J Invest Dermatol, **81**: 2s–10s.

Belzile N (1988) The fate of arsenic in sediments of the Laurentian Trough. Geochim Cosmochim Acta, **52**: 2293–2302.

Bencko V & Symon K (1969) Dynamics of arsenic accumulation in hairless mice after per oral administration. J Hyg Epidemiol Microbiol Immunol, **13**: 248–253.

Bencko V, Dvorak V & Symon K (1973) Organ retention of parenterally administered arsenic (labeled with ^{74As}) in mice preliminarily exposed to the element in drinking water: a study in arsenic tolerance. J Hyg Epidemiol Microbiol Immunol, **17**: 165–168.

Bencko V, Rossner P, Havrankova H, Puzanova A & Tucek M (1978) Effects of the combined action of selenuim & arsenic on mice versus suspension culture of mice fibroblasts. In: Fours JR, Gut I eds. Industrial and environmental xenobiotics. In vitro versus in vivo biotransformation and toxicity. Amsterdam, Excerpta Medica, pp 312–316.

Bengtsson BE & Bergstrom B (1987) A flowthrough fecundity test with *Nitocra spinipes* (Harpacticoidea, crustacea) for aquatic toxicity. Ecotoxicol Environ Saf,14(3): 260–268.

Benramdane L, Accominotti M, Fanton L, Malicier D & Vallon J-J (1999) Arsenic speciation in human organs following fatal arsenic trioxide poisoning: A case report. Clin Chem, **45**: 301–306.

Benson AA & Summons RE (1981) Arsenic accumulation in Great Barrier Reef invertebrates. Science, **211**: 482–483.

Benson AA, Katayama M & Knowles FC (1988) Arsenate metabolism in aquatic plants. Appl Organomet Chem, **2**: 349–352.

Benson LM, Porter EK & Peterson PJ (1981) Arsenic accumulation, tolerance, and geotypic variation in plants on arsenical mine wastes in S.W. England. J Plant Nutr, **3**(1–4): 655–666.

Berg T, Røyset O & Steinnes E (1995a) Moss (*Hylocomium splendens*) used as biomonitor of atmospheric trace element deposition: estimation of uptake efficiencies. Atmos Environ, **29**(3): 353–360.

Berg T, Røyset O, Steinnes E & Vadset M (1995b) Atmospheric trace element deposition: Principal component analysis of ICP-MS data from moss samples. Environ Pollut, **88**(1): 67–77.

Bertholf LM & Pilson JE (1941) Studies on toxicity to honeybees of acid lead arsenate, calcium arsenate, phenothiazine and cryolite. J Econ Entomol, **34**(1): 24–33.

Bertolero F, Marafante E, Rade JE, Pietra R & Sabbioni E (1981) Biotransformation and intracellular binding of arsenic in tissues of rabbits after intraperitoneal administration of ^{74}As labelled arsenite. Toxicology, **20**(1): 35–44.

Bertolero F, Pozzi G, Sabbioni E & Saffiotti U (1987) Cellular uptake and metabolic reduction of pentavalent to trivalent arsenic as determinants of cytotoxicity and morphological transformation. Carcinogenesis, **8**: 803–808.

Beyer WN & Cromartie EJ (1987) A survey of Pb, Cu, Zn, Cd, Cr, As, and Se in earthworms and soil from diverse sites. Environ Monit Assess, **8**(1): 27–36.

BGS (British Geological Survey) (2000) http://www.bgs.ac.uk/arsenic/bphase1/B_find.htm#

BGS & DPHE (2001). Arsenic contamination of groundwater in Bangladesh. Kinniburgh DG & Smedley PL ed. Vol 2: Final report. British Geological Survey Report WC/00/19 Keyworth, UK, British Geological Survey.

Bhumbla DK & Keefer RF (1994) Arsenic mobilization and bioavailability in soils. In: Nriagu JO ed. Arsenic in the environment: Part I: Cycling and characterization. New York, John Wiley & Sons, pp 51–82.

Biesinger KE & Christensen GM (1972) Effects of various metals on survival, growth, reproduction, and metabolism of *Daphnia magna*. J Fish Res Board Can, **29**: 1691–1700.

Binder S, Forney D, Kaye W & Paschal D (1987) Arsenic exposure in children living near a copper smelter. Bull Environ Contam Toxicol, **39**: 114–121.

Birge WJ (1978) Aquatic toxicology of trace elements of coal and fly ash. In: Thorp JH & Gibbons JW ed. Energy and stress in aquatic systems. Springfield, VA, NTIS, pp 219–240.

Birge WJ, Hudson JE, Black JA & Westerman AG (1978) Embryo-larval bioassays on inorganic coal elements and *in situ* biomonitoring of coal-waste effluents. In: Samuel DE, Stauffer JR, Hocutt CH & Mason WT ed. Surface mining and fish/wildlife needs in the Eastern United States (PB 298353). Springfield, VA, NTIS, pp 97–104.

Bishop RF & Chisholm D (1962) Arsenic accumulation in Annapolis valley orchard soils. Can J Soil Sci, **42**: 77–80.

Biswas BK, Dhar RK, Samanta G, Mandal BK, Chakraborti D, Faruk I, Islam KS, Chowdhury MM, Islam A & Roy S (1998) Detailed study report of Samta, one of the arsenic-affected villages of Jessore District, Bangladesh. Curr Sci, **74**(2): 134–145.

Blackwell RQ & Yang TH, Ai I (1961) Preliminary report on arsenic levels in water and food from the endemic blackfoot area. J Formos Med Assoc, **60**: 1139–1140.

Blakley BR, Sisodia CS & Mukkur TK (1980) The effect of methylmercury, tetraethyl lead, and sodium arsenite on the humoral immune response in mice. Toxicol Appl Pharmacol, **52**: 245–254.

Blanck H & Wängberg S-Å (1988a) Induced community tolerance in marine periphyton established under arsenate stress. Can J Fish Aquat Sci, **45**: 1816–1819.

Blanck H & Wängberg S-Å (1988b) Validity of an ecotoxicological test system: short-term and long-term effects of arsenate on marine periphyton communities in laboratory systems. Can J Fish Aquat Sci, **45**(10) 1807–1815.

Blanck H & Wängberg S-Å (1991) Pattern of cotolerance in marine periphyton communities established under arsenate stress. Aquat Toxicol, **21**(1/2): 1–14.

Blasco F (1975) Réduction partielle de l'arséniate en arsénite et exsorbtion de l'arsenic par *Chlorella pyrenoidosa*. Physiol Vég, **13**(2): 185–201.

Blot W & Fraumeni JF Jr (1975) Arsenical air pollution and lung cancer. Lancet, **2**: 142–144.

Blum JJ (1966) Phosphate uptake by phosphate-starved *Euglena*. J Gen Physiol, **49**: 1125–1136.

Blus LJ, Neely BS, Lamont TG & Mulhern B (1977) Residues of organochlorines and heavy metals in tissues and eggs of brown pelicans, 1969–1973. Pestic Monit J, **11**(1): 40–53.

Boampong C, Brindle ID, Le XC, Pidwerbesky L & Ceccarelli-Ponzoni CM (1988) Interference reduction by L-cysteine in the determination of arsenic by hydride generation. Anal Chem, **60**: 1185–1188.

Bodewig FG, Valenta P & Nürnberg HW (1982) Trace determination of As(III) and As(V) in natural waters by differential pulse anodic stripping voltammetry. Fresenius Z Anal Chem, **311**: 187–191.

Bohn A (1975) Arsenic in marine organisms from West Greenland. Mar Pollut Bull, **6**(6): 87–89.

Bolla-Wilson K & Bleecker ML (1987) Neuropsychological impairment following inorganic arsenic exposure. J Occup Med, **29**: 500–503.

Bollinger CT, van Zijl P & Louw JA (1992) Multiple organ failure with the adult respiratory distress syndrome in homicidal arsenic poisoning. Respiration, **59**: 57–61.

Borgmann U, Cove R & Loveridge C (1980) Effects of metals on the biomass production kinetics of freshwater copepods. Can J Fish Aquat Sci, **37**: 567–575.

Borgono JM, Vicent P, Venturino H & Infante A (1977) Arsenic in the drinking water of the city of Antofagasta: epidemiological and clinical study before and after the installation of a treatment plant. Environ Health Perspect, **19**: 103–105.

Borum DR & Abernathy CO (1994) Human oral exposure to inorganic arsenic. In: Chappell WR, Abernathy CO, & Cothern CR eds. Arsenic exposure and health. London, Science, pp 21–29.

Borzsonyi M, Bereczky A, Rudnai P, Csanady M & Horvath A (1992) Epidemiological studies on human subjects exposed to arsenic in drinking water in Southeast Hungary. Arch Toxicol, **66**: 77–78.

Bottino NR, Newman RD, Cox ER, Stockton R, Hoban M, Zingaro RA & Irgolic KJ (1978) The effects of arsenate and arsenite on the growth and morphology of the marine unicellular algae *Tetraselmis chuii* (Chlorophyta) and *Hymenomonas carterae* (Chrysophyta). J Exp Mar Biol Ecol, **33**: 153–168.

Bowen HJM (1979) Elemental chemistry of the elements. London, Academic Press.

Boyce AP & Verme IJ (1954) Toxicity of arsenite debarkers to deer in Michigan. Proceedings of the 16th Midwest Wildlife Conference, 1954, St Louis, Mo. Cited in NAS (1977).

Boyle RW & Jonasson IR (1973) The geochemistry of arsenic and its use as an indicator element in geochemical prospecting. J Geochem Explor, **2**: 251–296.

Braman RS & Foreback CC (1973) Methylated forms of arsenic in the environment. Science, **182**: 1247–1249.

Brannon JM & Patrick WH (1987) Fixation, transformation, and mobilization of arsenic in sediments. Environ Sci Technol, **21**(5): 450–459.

Brayer AF, Callahan CM, Wax PM (1997) Acute arsenic poisoning from ingestion of 'snakes'. Pediatr Emerg Care, **13**(6): 394–396.

Bright DA, Coedy B, Dushenko WT & Reimer KJ (1994) Arsenic transport in a watershed receiving gold mine effluent near Yellowknife, Northwest Territories, Canada. Sci Total Environ, **155**(3): 237–252.

415

Bright DA, Dodd M & Reimer KJ (1996) Arsenic in subArctic lakes influenced by gold mine effluent: the occurrence of organoarsenicals and 'hidden' arsenic. Sci Total Environ, 180: 165–182.

Brimblecombe P (1979) Atmospheric arsenic. Nature, 280: 104–105.

Bringmann G & Kühn R (1977) Grenzwerte der Schadwirkung wassergefahrdender Stoffe gegen Bakterien (Pseudomonas putida) und Grünalgen (Scenedesmus quadricauda) im Zellvermehrungshemmtest. Z Wasser Abwasser Forsch, 10: 87–98.

Bringmann G & Kühn R (1978) Testing of substances for their toxicity threshold: model organisms Microcystis (Diplocystis) aeruginosa and Scenedesmus quadricauda. Mitt Int Ver Limnol, 21: 275–284.

Brockbank CI & Batley GE, Low GC (1988) Photochemical decomposition of arsenic species in natural waters. Environ Technol Lett, 9(12): 1361–1366.

Brooke PJ & Evans WH (1981) Determination of total inorganic arsenic in fish, shellfish and fish products. Analyst 106(1262): 514–520.

Brouwer OF, Okenhout W, Edelbroek PM, de Kom JFM, de Wolff FA & Peters ACB (1992) Increased neurotoxicity of arsenic in methylenetetrahydrofolate reductase deficiency. Clin Neurol Neurosurg, 94: 307–310.

Brown CC & Chu KC (1983a) Implications of the multistage theory of carcinogenesis applied to occupational arsenic exposure. J Natl Cancer Inst, 70: 455–463.

Brown CC & Chu KC (1983b) A new method for the analysis of cohort studies: implications of the multistage theory of carcinogenesis aplied to occupational arsenic exposure. Environ Health Perspect, 50: 293–308.

Brown IR & Rush SJ (1984) Induction of a 'stress' protein in intact mammalian organs after the intravenous administration of sodium arsenite. Biochem Biophys Res Commun, 120: 150–155.

Brown JL & Kitchin KT (1996) Arsenite, but not cadmium, induces ornithine decarboxylase and heme oxygenase in rat liver: relevance to arsenic carcinogenesis. Cancer Lett, 98: 227–231.

Brown JL, Kitchin KT & George M (1997) Dimethylarsinic acid treatment alters six different rat biochemical parameters: relevance to arsenic carcinogenesis. Teratogenesis Carcinog Mutagen, 17: 71–84.

Brown LM, Pottern LM & Blot JW (1984) Lung cancer in relation to environmental pollutants emitted from industrial sources. Environ Res, 34: 250–261.

Brown RM, Newton D, Pickford CJ & Sherlock JC (1990) Human metabolism of arsenobetaine ingested with fish. Human Exp Toxicol, 9: 41–46.

Brumbaugh WG, Ingersoll CG, Kemble NE, May TW & Zajicek JL (1994) Chemical characterization of sediments and pore water from the upper Clark Fork River and Milltown Reservoir, Montana. Environ Toxicol Chem, 13(12) 1971–1983.

Bryant V, Newbery DM, McLusky DS & Campbell R (1985) Effect of temperature and salinity on the toxicity of arsenic to three estuarine invertebrates (*Corophium volutator, Macoma balthica, Tubifex costatus*). Mar Ecol Progr Ser, 24: 129–137.

Bröer S, Ji G, Bröer A & Silver S (1993) Arsenic efflux governed by the arsenic resistance determinant of *Staphylococcus aureus* plasmid pl258. J Bacteriol, 175(11): 3480–3485.

Buchancova J, Klimentova G, Knizkova M, Mesko D, Galikova E, Kubik J,Fabianova E & Jakubis M. (1998) Health status of workers of a thermal power station exposed for prolonged periods to arsenic and other elements from fuel. Cent Eur J Public Health, 6(1): 29–36.

Buchet J & Lison D (1998) Mortality by cancer in groups of the Belgian population with a moderately increased intake of arsenic. Int Arch Environ Health, 71: 125–130.

Buchet JP & Lauwerys R (1981) Evaluation of exposure of inorganic arsenic in man. In: Facchetti S ed. Analytical techniques for heavy metals in biological fluids. Amsterdam, Elsevier Science, pp 75–90.

Buchet JP & Lauwerys R (1985) Study of inorganic arsenic methylation by rat liver in vitro: relevance for the interpretation of observations in man. Arch Toxicol, 57: 125–129.

Buchet JP & Lauwerys R (1987) Study of factors influencing the in vivo methylation of inorganic arsenic in rats. Toxicol Appl Pharmacol, 91: 65–74.

Buchet JP & Lauwerys R (1988) Role of thiols in the in vitro methylation of inorganic arsenic by rat liver cytosol. Biochem Pharmacol, 37: 3149–3153.

Buchet JP, Roels H, Lauwerys R, Bruaux P, Thoreau-Claeys F, Lafontaine A & Verduyn G (1980) Repeated surveillance of exposure to cadmium, manganese, and arsenic in school-age children living in rural, urban, and nonferrous smelter areas in Belgium. Environ Res, 22: 95–108.

Buchet JP, Lauwerys R & Roels H (1981a) Comparison of the urinary excretion of arsenic metabolites after a single oral dose of sodium arsenite, monomethylarsonate, or dimethylarsinate in man. Int Arch Occup Environ Health, 48: 71–79.

Buchet JP, Lauwerys R & Roels H (1981b) Urinary excretion of inorganic arsenic and its metabolites after repeated ingestion of sodium metaarsenite by volunteers. Int Arch Occup Environ Health, 48: 111–118.

Buchet JP, Geubel A, Pauwels S, Mahieu P & Lauwerys R (1984) The influence of liver disease on the methylation of arsenite in humans. Arch Toxicol, 55: 151–154.

Buchet JP, Pauwels J & Lauwerys R (1994) Assessment of exposure to inorganic arsenic following ingestion of marine organisms by volunteers. Environ Res, 66: 44–51.

Buchet JP, Lauwerys RR & Yager JW (1995) Lung retention & bioavailability of arsenic after single intratracheal administration of sodium arsenite, sodium arsenate, fly ash & copper smelter dust in the hamster. Environ Geochem Health, **17**: 182–188.

Buchet JP, Lison D, Ruggeri M, Foa V & Elia G (1996) Assessment of exposure to inorganic arsenic, a human carcinogen, due to the consumption of seafood. Arch Toxicol, **70**: 773–778.

Buck WM, Osweiler GD & VanGelder GA (1976) Arsenic I: Inorganic, aliphatic and trivalent organic arsenicals. In: Clinical and diagnostic veterinary toxicology, 2nd ed. Dubuque, Kendall-Hunt, pp 281 288.

Budd K & Craig SR (1981) Resistance to arsenate toxicity in the bluegreen alga *Synechococcus leopoliensis*. Can J Bot, **59**: 1518–1521.

Budd K, Casey JR & MacArthur JD (1986) Arsenite toxicity and arsenite tolerance in the cyanobacterium *Synechococcus leopoliensis*. Can J Bot, **64**: 2433–2440.

Buhl KJ & Hamilton SJ (1990) Comparative toxicity of inorganic contaminants released by placer mining to early life stages of salmonids. Ecotoxicol Environ Saf,**20**(3): 325–342.

Buhl KJ & Hamilton SJ (1991) Relative sensitivity of early life stages of arctic grayling, coho salmon, and rainbow trout to nine inorganics. Ecotoxicol Environ Saf, **22**(2): 184–197.

Buiatti E, Kriebel D, Geddes M, Santucci M & Pucci N (1985) A case control study of lung cancer in Florence, Italy. I. Occupational risk factors. J Epidemiol Community Health,**39**: 244–250.

Bulbulyan MA, Jourenkova NJ, Boffetta P, Astashevsky SV, Mukeria AF & Zaridze DG (1996) Mortality in a cohort of Russian fertilizer workers. Scand J Work Environ Health, **22**(1): 27–33.

Burkhard EG, Dutkiewicz VA & Husain L (1994) A study of SO_2, SO_4^{2-} and trace elements in clear air and clouds above the midwestern United States. Atmos Environ, **28**(8): 1521–1533.

Burleson FG, Simeonova PP, Germolec DR & Luster MI (1996) Dermatotoxic chemical stimulate of c-jun and c-fos transcription and AP-1 binding in human keratinocytes. Res. Commun Molec Pathol Pharmacol, **93**: 131–148.

Burns LA, Sikorski EE, Saady JJ & Munson AE (1991) Evidence for arsenic as the immunosuppressive component of gallium arsenide. Toxicol Appl Pharmacol, **110**: 157–169.

Burns LA, McCay JA, Brown R & Munson AE (1993) Arsenic in the sera of gallium arsenide-exposed mice inhibits bacterial growth and increases host resistance. J Pharmacol Exp Ther, **265**(2): 795–800.

418

Burton GA (1987) Occurrence of bacterial resistance to arsenite, copper, and selenite in adverse habitats. Bull Environ Contam Toxicol, **39**(6): 990–997.

Burton GA, Lazorchak JM, Waller WT & Lanza GR (1987) Arsenic toxicity changes in the presence of sediment. Bull Environ Contam Toxicol, **38**(3): 491–499.

Bustamante J, Dock L, Vahter M, Fowler B & Orrenius S (1997) The semiconductor arsenic and indium induce apoptosis in rat thymocytes. Toxicology, **118**: 129–136.

Byrd JT (1988) The seasonal cycle of arsenic in estuarine and nearshore waters of the South Atlantic Bight. Mar Chem, **25**: 383–394.

Byrne AR & Tusek-Znidaric M (1983) Arsenic accumulation in the mushroom *Laccaria amethystina*. Chemosphere, **12**: 1113–1117.

Callahan MA, Slimak MW, Gabel NW, May IP, Fowler CF, Freed JR, Jennings P, Durfee RL, Whitmore FC, Maestri B, Mabey WR, Holt BR & Gould C (1979) Water-related environmental fate of 129 priority pollutants. Vol I. Introduction and technical background, metals and inorganics, pesticides and PCBs. EPA-440/4-79-029a. Washington, DC, U.S. Environmental Protection Agency, Office of Water Planning and Standards.

Caltabiano MM, Koesler TP, Poste G & Greig RG (1986) Induction of 32- and 34-kDa stress proteins by sodium arsenite, heavy metals, and thiol-reactive agents. J Biol Chem, **261**: 13381–13386.

Camardese MB, Hoffman DJ, LeCaptain LJ & Pendleton GW (1990) Effects of arsenate on growth and physiology in mallard ducklings. Environ Toxicol Chem, **9**(6): 785–795.

Capar SG & Yess NJ (1996) US Food and Drug administration survey of cadmium, lead and other elements in clams and oysters. Food Add Contam, **13**(5): 553–560.

Carbonell-Barrachina AA, Aarabi MA, DeLaune RD, Gambrell RP &Patrick WH (1998) The influence of arsenic chemical form and concentration on *Spartina patens* and *Spartina alterniflora* growth and tissue arsenic concentration. Plant Soil, **198**: 33–43.

Cardwell RD, Foreman DG, Payne TR & Wilbur DJ (1976) Acute toxicity of selected toxicants to six species of fish. EPA/600/3-76/008. Duluth, MN: U.S. Environmental Protection Agency.

Carey PL, McLaren RG & Adams JA (1996) Sorption of cupric, dichromate and arsenate ions in some New Zealand soils. Water Air Soil Pollut, **87**(1–4): 189–203.

Carlin A, Shi W, Dey S & Rosen BP (1995) The *ars* operon of *Escherichia coli* confers arsenical and antimonial resistance. J Bacteriol, **177**(4): 981–986.

Carlson CL & Carlson CA (1994) Impacts of coal pile leachate on a forested wetland in South Carolina. Water Air Soil Pollut, **72**: 89–109.

Carmignani M, Boscolo P & Iannaccone A (1983) Effects of chronic exposure to arsenate on the cardiovascular function of rats. Br J Ind Med, **40**: 280–284.

Carmignani M, Boscolo P & Castellino N (1985) Metabolic fate and cardiovascular effects of arsenic in rats and rabbits chronically exposed to trivalent and pentavalent arsenic. Arch Toxicol Suppl, **8**: 452–455.

Carpenter R, Peterson ML & Jahnke RA (1978) Sources, sinks, and cycling of arsenic in the Puget Sound region. In: Wiley ML ed. Estuarine interactions. New York, Academic Press, pp 459–480.

Carpenter SJ (1987) Developmental analysis of cephalic axial dysraphic disorders. Anat Embryol, **176**: 345–365.

Castilla MI, Cortazar Y & Buigues JP (1993) Determination of arsenic (III/V) by X-ray fluorescence spectroscopy in waters. Agrochimica, **37**(3): 211–218.

Cebrian ME, Albores A, Aguilar M & Blakely E (1983) Chronic arsenic poisoning in the North of Mexico. Human Toxicol, **2**: 121–133.

Cebrian ME, Albores A, Connelly JC & Bridges JW (1988) Assessment of arsenic effects on cytosolic heme status using tryptophan pyrrolase as an index. J Biochem Toxicol, **3**: 77–86.

Cervantes C (1995) Bacterial resistance to arsenic compounds. Rev Latinoam Microbiol, **37**(4): 387–395.

Ch'i IC & Blackwell RQ (1968) A controlled retrospective study on blackfoot disease, an endemic peripheral gangrene disease in Taiwan. Am J Epidemiol, **88**: 7–24.

Chaineau E, Binet S, Pol D, Chatellier G & Meininger V (1990) Embryotoxic effects of sodium arsenite and sodium arsenate on mouse embryos in culture. Teratology, **41**: 105–112.

Chakraborti D, Adams F & Irgolic KJ (1986) Compound-specific determination of arsenite at sub-nanogram concentrations in freshwater and seawater. Fresenius Z Anal Chem, **323**(4): 340–342.

Chakraborti D, Das D, Chatterjee A, Jin Z & Jiang SG (1992) Direct determination of some heavy metals in urban air particulates by electrothermal atomic absorption spectrometry using Zeeman background correction after simple acid decomposition. Part IV: Application to Calcutta air particulates. Environ Technol, **13**(1): 95–100.

Chakraborti D, Samanta G, Mandal BK, Roy Chowdhury T, Chanda CR, Biswas BK, Dhar RK, Basu GK & Saha KC (1998) Calcutta's industrial pollution: Groundwater arsenic contamination in a residential area and sufferings of people due to industrial effluent discharge – an eight-year study report. Curr Sci, **74**(4): 346–355.

Challenger F (1945) Biological methylation. Chem Rev, **36**: 315–361.

Chan TYK (1994) The prevalence use and harmful potential of some Chinese herbal medicines in babies and children. Vet Human Toxicol, **36**(3): 238–240.

Chang TK, Shyu GS, Lin YP & Chang NC (1999) Geostatistical analysis of soil arsenic content in Taiwan. J Environ Sci Health A: Toxic Hazardous Subst Environ Eng, **34**(7): 1485–1501.

Chang WC, Chen SH, Wu HL, Shi GY, Murota S & Morita I (1991) Cytoprotective effect of reduced glutathione in arsenical-induced endothelial cell injury. Toxicology, **69**: 101–110.

Chappell J, Chiswell B & Olszowy H (1995) Speciation of arsenic in a contaminated soil by solvent extraction. Talanta, **42**(3): 323–329.

Chatterjee A & Mukherjee A (1999) Hydrogeological investigation of ground water arsenic contamination in South Calcutta. Sci Total Environ, **225**(3): 249–262.

Chatterjee A, Das D & Chakraborti D (1993) A study of ground water contamination by arsenic in the residential area of Behala, Calcutta due to industrial pollution. Environ Pollut, **80**: 57–65.

Chatterjee A, Das D, Mandal BK, Chowdhury TR, Samanta G & Chakraborti D (1995) Arsenic in ground water in six districts of West Bengal, India: the biggest arsenic calamity in the world. Part 1. Arsenic species in drinking water and urine of the affected people. Analyst, **120**: 643–650.

Chen C-J & Wang C-J (1990) Ecological correlation between arsenic level in well water and age-adjusted mortality from malignant neoplasms. Cancer Res, **50**: 5470–5474.

Chen C-J, Chuang Y-C, Lin T-M & Wu H-Y (1985) Malignant neoplasms among residents of a blackfoot disease endemic area in Taiwan: high-arsenic artesian well water and cancers. Cancer Res, **45**: 5895–5899.

Chen C-J, Chuang Y-C, You S-L, Lin T-M & Wu H-Y (1986) A retrospective study on malignant neoplasms of bladder, lung and liver in blackfoot disease endemic area in Taiwan. Br J Cancer, **53**: 399–405.

Chen C-J, Kuo TL & Wu MM (1988a) Arsenic and cancers. *Lancet*, **1**: 414–415.

Chen C-J, Wu M-M, Lee S-S, Wang J-D, Cheng S-H & Wu H-Y (1988b) Atherogenicity and carcinogenicity of high-arsenic artesian well water. Multiple risk-factors and related malignant neoplasms of blackfoot disease. Arteriosclerosis, **8**: 452–460.

Chen C-J, Chen CW, Wu MM & Kuo TL (1992) Cancer potential in liver, lung, bladder and kidney due to ingested inorganic arsenic in drinking water. *Br J Cancer*, **66**: 888–892.

Chen C-J, Hsueh Y-M, Lai M-S, Shyu M-P, Chen S-Y, Wu M-M, Kuo T-L & Tai T-Y (1995) Increased prevalence of hypertension and long-term arsenic exposure. Hypertension, **25**: 53–60.

Chen C-J, Chiou H-Y, Chiang M-H, Lin L-J & Tai T-Y (1996) Dose-response relationship between ischemic heart disease mortality and long-term arsenic exposure. Arterioscler Thromb Vasc Biol, **16**(4): 504–510.

Chen GS, Asai T, Suzuki Y, Nishioka K & Nishiyama S (1990) A possible pathogenesis for Blackfoot disease. J Dermatol, **17**: 599–608.

Chen KP & Wu HY (1962) Epidemiologic studies on blackfoot disease 2. A study of source of drinking water in relation to the disease. J Formos Med Assoc, **61**: 611–618.

Chen KP, Wu HY & Wu TC (1962) Epidemiologic studies on blackfoot disease in Taiwan: III. Physicochemical characteristics of drinking water in endemic blackfoot disease areas. Mem Coll Med Natl Taiwan Univ, **8**: 115–129.

Chen SL, Dzeng SR, Yang MH, Chiu KH, Shieh GM & Wai CM (1994) Arsenic species in groundwaters of the blackfoot disease area, Taiwan. Environ Sci Technol, **28**(5): 877–881.

Cheng CN & Focht DD (1979) Production of arsine and methylarsines in soil and in culture. Appl Environ Microbiol, **38**(3): 494–498.

Chiang H-S, Guo H-R, Hong C-L, Lin S-M & Lee E-F (1993) The incidence of bladder cancer in the blackfoot disease endemic area in Taiwan. Br J Urol, **71**: 274–8.

Chiang HS, Hong CL, Guo HR, Lee EF & Chen TY (1988) Comparative study on the high prevalence of bladder cancer in the blackfoot disease endemic area in Taiwan. Taiwan I Hsueh Hui Tsa Chih, **87**: 1074–1080.

Chilvers DC & Peterson PJ (1987) Global cycling of arsenic. In: Hutchinson TC & Meema KM ed. Lead, mercury, cadmium and arsenic in the environment. Chichester, John Wiley & Sons, pp 279–303.

Chin KV, Tanaka S, Darlington G, Pastan I & Gottesman MM (1990) Heat shock and arsenite increase expression of the multidrug resistance (MDR1) gene in human renal carcinoma cells. J Biol Chem, **265**: 221–226.

Chiou H-Y, Hsueh Y-M, Liaw K-F, Horng S-F, Chiang M-H, Pu Y-S, Lin JS-N, Huang C-H & Chen C-J (1995) Incidence of internal cancers and ingested inorganic arsenic: a seven-year follow-up study in Taiwan. Cancer Res, **55**: 1296–1300.

Chiou H-Y, Huang W-I, Su C-L, Chang S-F, Hsu Y-H & Chen C-J (1997a) Dose-response relationship between prevalence of cerebrovascular disease and ingested inorganic arsenic. Stroke, **28**: 1717–1723.

Chiou PW-S, Chen K-L & Yu B (1997b) Effects of roxarsone on performance, toxicity, tissue accumulation and residue of eggs and excreta in laying hens. J Sci Food Agric,**74**: 229–236.

Chiou H-Y, Chiou S-T, Hsu Y-H, Tseng C-H, Wei M-L & Chen C-J (2001) Incidence of transitional cell carcinoma and arsenic in drinking water: A follow-up study of 8102 residents in an arseniasis-endemic area in Northeastern Taiwan. Am J Epidemiol, 153: 411–418.

Chisholm D & MacPhee AW (1972) Persistence and effects of some pesticides in soil. J Econ Entomol, 65(4): 1010–1013.

Choucair AK & Ajox ET (1988) Hair and nails in arsenical neuropathy. Ann Neurol, 23: 628–629.

Chowdhury TR, Mandal BK, Samanta G, Basu GK, Chowdhury PP, Chanda CR, Karan NK, Lodh D, Dhar RK, Das D, Saha KC & Chakraborti D (1997) Arsenic in groundwater in six districts of West Bengal, India: Arsenic in groundwater in six districts of West Bengal, India: the biggest arsenic calamity in the world: the status report up to August, 1995. In: Abernathy CO, Calderon RL, & Chappell WR ed. Arsenic. Exposure and health effects. London, Chapman & Hall, pp 93–112.

Chrostek WJ, Elesh E & Taylor JS (1980) Health Hazard Evaluation Report. Jeannette Glass Company, Jeannette, PA. HE 80-19-765. Cincinnati, OH, National Institute for Occupational Health.

Chunguo C & Zihui L (1988) Chemical speciation and distribution of arsenic in water, suspended solids and sediment of Xiangjiang River, China. Sci Total Environ, 77(1): 69–82.

Chutke NL, Ambulkar MN & Weginwar RG, Garg AN (1994) Substoichiometric isotope dilution analysis of arsenic in biological and environmental standard reference materials by solvent extraction using toluene-3,4-dithiol in benzene. J Radioanal Nucl Chem, 185(1): 145–156.

Chutke NL, Ambulkar MN & Garg AN (1995) An environmental pollution study from multielemental analysis of pedestrian dust in Nagpur city, Central India. Sci Total Environ, 164(3): 185–194.

Civantos DP, Lopez Rodriguez A, Aguado-Borruey JM & Julia Narvaez JA (1995) Fulminant malignant arrhythmia and multiorgan failure in acute arsenic poisoning. Chest, 108(6): 1774–1775.

Clark DR, Cantu R, Cowman DF & Maxson DJ (1998) Uptake of arsenic and metals by tadpoles at an historically contaminated Texas site. Ecotoxicology, 7: 61–67.

Clement WH & Faust SD (1981) The release of arsenic from contaminated sediments and muds. J Environ Sci Health, A16(1): 87–122.

Cockell KA & Hilton JW (1985) Chronic toxicity of dietary inorganic and organic arsenicals to rainbow trout (Salmo gairdneri R.). Fed Proc, 44(4): 938.

Cockell KA & Hilton JW (1988) Preliminary investigations on the comparative chronic toxicity of four dietary arsenicals to juvenile rainbow trout (*Salmo gairdneri* R.). Aquat Toxicol, **12**: 73–82.

Cockell KA, Hilton JW & Bettger WJ (1991) Chronic toxicity of dietary disodium arsenate heptahydrate to juvenile rainbow trout (*Oncorhynchus mykiss*). Arch Environ Contam Toxicol, **21**(4): 518–527.

Coles DG, Ragaini RC, Ondov JM, Fisher GL, Silberman D, Prentice BA (1979) Chemical studies of stack fly ash from a coal-fired power plant. Environ Sci Technol, **13**(4): 455–459.

Collecchi P, Esposito M, Brera S, Mora E, Mazzucotelli A & Oddone M (1985) The distribution of arsenic and cobalt in patients with laryngeal carcinoma. J Appl Toxicol, **6**: 287–289.

Collinson LP & Dawes IW (1995) Isolation, characterization & overexpression of the yeast gene, GLR1, encoding glutathione reductase. Gene, **156**: 123–127.

Concha G, Nermell B & Vahter M (1998a) Metabolism of inorganic arsenic in children with chronic high arsenic exposure in Northern Argentina. Environ Health Perspec, **106**: 355–359.

Concha G, Nermell B & Vahter M (1998b) Exposure to inorganic arsenic metabolites during early human development. Toxicolog Sci, **44**(2): 185–90.

Concha G, Vogler G, Nermell B & Vahter M (1998c) Low-level arsenic excretion in breast milk of native Andean women exposed to high levels of arsenic in the drinking water. Int Arch Occup Environ Health, **71**: 42–46.

Conway HL (1978) Sorption of arsenic and cadmium and their effects on growth, micronutrient utilization, and photosynthetic pigment composition of *Asterionella formosa*. J Fish Res Board Can, **35**: 286–294.

Cook D (1953) Chemi-peeling and wildlife. NY State Conserv, **7**(6): 8.

Cordier S, Thériault G & Iturra H (1983) Mortality pattern in a population living near a copper smelter. Environ Res, **31**: 311–322.

Corr JJ (1997) Measurement of molecular species of arsenic and tin using elemental and molecular dual mode analysis by ionspray mass spectrometry. J Anal At Spectrom, **12**: 537–546.

Corr JJ & Larsen EH (1996) Arsenic speciation by liquid chromatography coupled with ionspray tandem mass spectrometry. J Anal At Spectrom, **11**: 1215–1224.

Counts JL & Goodman JI (1995) Alterations in DNA methylation may play a variety of roles in carcinogenesis. Cell, **83**: 13–15.

Cowell BC (1965) The effects of sodium arsenite and silvex on the plankton populations in farm ponds. Trans Am Fish Soc, **94**: 371–377.

Cox CD & Ghosh MM (1994) Surface complexation of methylated arsenates by hydrous oxides. Water Res, **28**(5): 1181–1188.

Cox DP & Alexander M (1973) Production of trimethylarsine gas from various arsenic compounds by three sewage fungi. Bull Environ Contam Toxicol, **9**: 84–88.

Crane RK & Lipmann F (1953) The effect of arsenate on aerobic phosphorylation. J Biol Chem, **201**: 235–243.

Crearley JE (1973) Arsenic contamination of Finfeather and Municipal Lakes in the City of Bryan. Austin, TX, Texas Water Quality Board.

Crecelius EA (1975) The geochemical cycle of arsenic in Lake Washington and its relation to other elements. Limnol Oceanogr, **20**: 441–451.

Crecelius EA (1978) Modification of the arsenic speciation technique using hydride generation. Anal Chem, **50**(6): 826–827.

Crecelius EA, Bothner MH & Carpenter R (1975) Geochemistries of arsenic, antimony, mercury, and related elements in sediments of Puget Sound. Environ Sci Technol, 9(4): 325–333.

Crecelius EA, Bloom NS, Cowan CE & Jenne EA (1986) Speciation of selenium and arsenic in natural waters and sediments: arsenic speciation. Research Report, EA-4641, Vol 2. Palo Alto, CA, EPRI (Electrical Power Research Institute).

Crecelius EA, Apts CW, Bingler LS & Cotter OA (1994) The cycling of arsenic species in coal-fired power plant cooling reservoirs. In: Nriagu JO ed. Arsenic in the environment: Part I: Cycling and characterization. New York, John Wiley & Sons, pp 83–97.

Crossen PE (1983) Arsenic and SCE in human lymphocytes. Mutat Res, **119**: 415–419.

Cruz AC, Fomsgaard IS & Lacayo J (1994) Lead, arsenic, cadmium and copper in Lake Asososca, Nicaragua. Sci Total Environ, **155**(3): 229–236.

Cullen NM, Wolf LR & St Clair D (1995) Pediatric arsenic ingestion. Am J Emerg Med, **13**: 432–435.

Cullen WR & Reimer KJ (1989) Arsenic speciation in the environment. Chem Rev, **89**(4): 713–764.

Cullen WR, Froese CL, Lui A, McBride BC, Patmore DJ & Reimer M (1977) The aerobic methylation of arsenic by microorganisms in the presence of L-methionine-methyl-d3. J Organomet Chem, **139**: 61–69.

Cullen WR, McBride BC & Pikett AW (1979a) The transformation of arsenicals by *Candida humicola*. Can J Microbiol, **25**: 1201–1205.

Cullen WR, McBride BC & Reimer M (1979b) Induction of the aerobic methylation of arsenic by *Candida humicola*. Bull Environ Contam Toxicol, **21**: 157–161.

Cullen WR, McBride BC & Reglinski J (1984a) The reaction of methylarsenicals with thiols: some biological implications. J Inorg Biochem, **21**: 179–194.

Cullen WR, McBride BC & Reglinski J (1984b) The reduction of trimethylarsine oxide to trimethylarsine by thiols: a mechanistic model for the biological reduction of arsenicals. J Inorg Biochem, **21**: 45–60.

Cullen WR, Li H, Pergantis SA, Eigendorf GK & Harrison LG (1994) The methylation of arsenate by a marine alga *Polyphysa peniculus* in the presence of L-methionine-methyl-d-3. Chemosphere, **28**(5): 1009–1019.

Curtis MW, Copeland TL & Ward CH (1979) Acute toxicity of 12 industrial chemicals to freshwater and saltwater organisms. Water Res, **13**: 137–141.

Custer TW & Hohman WL (1994) Trace elements in canvasbacks (*Aythya valisineria*) wintering in Louisiana, USA, 1987–1988. Environ Pollut, **84**(3): 253–259.

Cuzick J, Sasieni P & Evans S (1992) Ingested arsenic, keratoses, and bladder cancer. Am J Epidemiol, **136**: 417–421.

Da Costa EWB (1972) Variation in the toxicity of arsenic compounds to microorganisms and the suppression of the inhibitory effects by phosphate. Appl Microbiol, **23**(1): 46–53.

Dabeka RW, McKenzie AD, Lacroix GMA, Cleroux C, Bowe S, Graham RA & Conacher HBS (1993) Survey of arsenic in total diet food composites and estimation of the dietary intake of arsenic by Canadian adults and children. J AOAC Int, **76**: 14–25.

Daghir NJ & Hariri NN (1977) Determination of total arsenic residues in chicken eggs. J Agric Food Chem, **25**(5): 1009–1010.

Dang HS, Jaiswal DD & Somasundaram S (1983) Distribution of arsenic in humans tissues and milk. Sci Total Environ, **29**: 171–175.

Darland JE & Inskeep WP (1997) Effects of pH and phosphate competition on the transport of arsenate. J Environ Qual, **26**: 1133–1139.

Das D, Chatterjee A, Mandal BK, Samanta G & Chakraborti D (1995) Arsenic in ground water in six districts of West Bengal, India: the biggest arsenic calamity in the world. Part 2. Arsenic concentration in drinking water, hair, nails, urine, skin-scale and liver tissue (biopsy) of the affected people. Analyst, **120**: 917–924.

Das D, Samanta G, Mandal BK, Chowdhury TR, Chanda CR, Chowdhury PP, Basu GK & Chakraborti D (1996) Arsenic in groundwater in six districts of West Bengal, India. Environ Geochem Health, **18**: 5–15.

Das T, Roy Choudhury A, Sharma A & Talukder G (1993) Modification of clastogenicity of three known clastogens by garlic extract in mice in vivo. Environ Mol Mutagen, 21: 383–388.

Davenport JR & Peryea FJ (1991) Phosphate fertilizers influence leaching of lead and arsenic in a soil contaminated with lead arsenate. Water Air Soil Pollut, 58: 101–110.

Davidson CI, Goold WD, Mathison TP, Wiersma GB, Brown KW & Reilly MT (1985) Airborne trace elements in Great Smoky Mountains, Olympic, and Glacier National Parks. Environ Sci Technol, 19(1): 27–35.

Davies PJ (1974) Arsenic in sediments on the continental shelf of southeast Australia. Search, 5(8): 394–397.

Davis A, Ruby MV & Bergstrom PD (1992) Bioavailability of arsenic and lead in soils from the Butte, Montana, mining district. Environ Sci Technol, 26: 461–468.

Davis RD, Beckett PHT & Wollan E (1978) Critical levels of twenty potentially toxic elements in young spring barley. Plant Soil, 49: 395–408.

de Bettencourt AMM (1988) On arsenic speciation in the Tagus Estuary. Neth J Sea Res, 22(3): 205–212.

De Koe T (1994) *Agrostis castellana* and *Agrostis delicatula* on heavy metal and arsenic enriched sites in NE Portugal. Sci Total Environ, 145(1/2): 103–109.

De Sastre MSR, Varillas A, Kirschbaum P (1992) Arsenic content in water in the northwest area of Argentina,. Arsenic in the environment and its incidence on health (International seminar proceedings), 1992, Universidad de Chile (Santiago). pp 91–99.

DeKimpe J, Cornelis R, Mees L & Vanholder R (1996) Basal metabolism of intraperitoneally injected carrier-free [74]As-labeled arsenate in rabbits. Fundam Appl Toxicol, 34: 240–248.

Deknudt G, Leonard A, Arany J, Jenar-Du Boisson G & Delavignette E (1986) In vivo studies in male mice on the mutagenic effects of inorganic arsenic. Mutagenesis, 1: 33–34.

Del Razo LM, Arellano MA, Cebrián ME (1990) The oxidation states of arsenic in well-water from a chronic arsenicism area of northern Mexico. Environ Pollut, 64(2): 143–153.

Delnomdedieu M, Basti MM, Otvos JD & Thomas DJ (1993) Transfer of arsenite from glutathione to dithiols: a model of interaction. Chem Res Toxicol, 6(5): 598–602.

Delnomdedieu M, Basti MM, Otvos JD & Thomas DJ (1994a) Reduction of binding of arsenate and dimethylarsinate by glutathione: a magnetic resonance study. Chem Biol Interact, 90: 139–155.

Delnomdedieu M, Basti MM, Otvos JD & Thomas DJ (1994b) Complexation of arsenic species in rabbit erythrocytes. Chem Res Toxicol, **7**: 621–627.

Deuel LE & Swoboda AR (1972a) Arsenic solubility in a reduced environment. Soil Sci Soc Am Proc, **36**: 276–278.

Deuel LE & Swoboda AR (1972b) Arsenic toxicity to cotton and soybeans. J Environ Qual, **1**(3): 317–320.

DFG (Deutsche Forschungsgemeinschaft) (1999) List of MAK and BAT Values 1999. Commission for the Investigation of Health Hazards of Chemical Compounds in the Work Area, Report No 35. Weinheim, Wiley-VCH.

DG Environment (2000) Ambient air pollution by As, Cd and Ni compounds. Position paper, Final version, October 2000. Brussels, European Commission DG Environment.

Dhar RK, Biswas BK, Samanta G, Mandal BK, Chakraborti D, Roy S, Fafar A, Islam A, Ara G, Kabir S, Khan AW, Ahmed SA & Hadi SA (1997) Groundwater arsenic calamity in Bangladesh. Curr Sci, **73**(1): 48–59.

Diamadopoulos E, Ioannidis S & Sakellaropoulos GP (1993) As(v) removal from aqueous solutions by fly ash. Water Res, **27**(12) 1773–1777.

Diamond ML (1995) Application of a mass balance model to assess in-place arsenic pollution. Environ Sci Technol, **29**(1): 29–42.

Diaz-Barriga F, Llamas E, Mejia JJ, Carrizales L, Santoyo ME, Vega-Vega L & Yanez L (1990) Arsenic-cadmium interaction in rats. Toxicology, **64**: 191–203.

Diaz-Barriga F, Santos MA, Mejia JD, Batres L, Yanez L, Carrizales L, Vera E, Razo LM Del & Cebrian ME (1993) Arsenic and cadmium exposure in children living near a smelter complex in San Luis Pososi, Mexico. Environ Res, **62**: 242–250.

Dickens R & Hiltbold AE (1967) Movement and persistence of methanearsonates in soil. Weeds, **15**: 299–304.

Ding H, Wang J, Dorsey JG & Caruso JA (1995) Arsenic speciation by micellar liquid chromatogaphy with inductively coupled plasma mass spectrometric detection. J Chrom A, **694**: 425–431.

Dmuchowski W & Bytnerowicz A (1995) Monitoring environmental pollution in Poland by chemical analysis of Scots pine (*Pinus sylvestris* L.) needles. Environ Pollut, **87**(1): 87–104.

Dolan R, Loon J Van, Templeton D & Paudyn A (1990) Assessment of ICP-MS for routine multielement analysis of soil samples in environmental trace element studies. Fresenius J Anal Chem, **336**(2): 99–105.

Domingo JL, Bosque MA & Piera V (1991) *meso*-2,3-dimercaptosuccinic acid and prevention of arsenite embryotoxicity and teratogenicity in the mouse. Fundam Appl Toxicol, **17**: 314–320.

Done AK & Peart AJ (1971) Acute toxicities of arsenical herbicides. Clin Toxicol, **4**: 343–355.

Dong JT & Luo XM (1994) Effects of arsenic on DNA damage and repair in human fetal lung fibroblasts. Mutat Res, **315**: 11–15.

Doyle MO & Otte ML (1997) Organism-induced accumulation of iron, zinc and arsenic in wetland soils. Environ Pollut, **96**(1): 1–11.

Dubreuilh W (1910) Kératose arsenicale et cancer arsenical. Ann Derm Syph, **1**: 65–83.

Dudas MJ (1984) Enriched levels of arsenic in post-active acid sulfate soils in Alberta. Soil Sci Soc Am J, **48**: 1451–1452.

Dudas MT (1987) Accumulation of native arsenic in acid sulphate soils in Alberta, Canada. J Soil Sci, **67**: 317–331.

Dudka S & Markert B (1992) Baseline concentrations of AS, Ba, Be, Li, Nb, Sr and V in surface soils of Poland. Sci Total Environ, **122**(3): 279–290.

Dulout FN, Grill CA, Seoane AI, Maderna CR, Nilsson R, Vahter M, Darroudi F & Natarajan AT (1996) Chromosomal aberrations in peripheral blood lymphocytes from native Andean women and children from Northwestern Argentina exposed to arsenic in drinking water. Mutat Res, **370**: 151–158.

Dwyer FJ, Burch SA, Ingersoll CG & Hunn JB (1992) Toxicity of trace element and salinity mixtures to striped bass (*Morone saxatilis*) and *Daphnia magna*. Environ Toxicol Chem, **11**: 513–520.

Ebdon L, Walton AP, Millward GE & Whitfield M (1987) Methylated arsenic species in estuarine porewaters. Appl Organomet Chem, **1**: 427–433.

Ebdon L, Hill S, Walton AP & Ward RW (1988) Coupled chromatography–atomic spectrometry for arsenic speciation – a comparative study. Analyst, **113**: 1159–1165.

Edmonds JS & Francesconi KA (1977) Methylated arsenic from marine fauna. Nature, **265**: 436.

Edmonds JS & Francesconi KA (1981a) Arseno-sugars from brown kelp (*Ecklonia radiata*) as intermediates in cycling of arsenic in a marine ecosystem. Nature, **289**: 602–604.

Edmonds JS & Francesconi KA (1981b) Isolation and identification of arsenobetaine from the American lobster *Homarus americanus*. Chemosphere, **10**(9): 1041–1044.

Edmonds JS & Francesconi KA (1981c) The origin and chemical form of arsenic in the school whiting. Mar Pollut Bull, **12**(3): 92–96.

Edmonds JS & Francesconi KA (1987a) Transformations of arsenic in the marine environment. Experientia, **43**: 553–557.

Edmonds JS & Francesconi KA (1987b) Trimethylarsine oxide in estuary catfish (*Cnidoglanis macrocephalus*) and school whiting (*Sillago bassensis*) after oral administration of sodium arsenate; and as a natural component of estuary catfish. Sci Total Environ, **64**: 317–323.

Edmonds JS & Francesconi KA (1988a) The methylation of arsenic by marine macro-algae. In: Craig PJ & Glockling F ed. The biological alkylation of heavy elements. Proceedings of a conference, 17–18 September 1987, London. Special Publication No. 66. Cambridge, Royal Society of Chemistry, pp 138–141.

Edmonds JS & Francesconi KA (1988b) The origin of arsenobetaine in marine animals. Appl Organomet Chem, **2**: 297–302.

Edmonds JS & Francesconi KA (1993) Arsenic in seafoods: human health aspects and regulations. Marine Pollut Bull, **26**(12) 665–674.

Edmonds JS, Francesconi KA & Hansen JA (1982) Dimethyloxarsylethanol from anaerobic decomposition of brown kelp (*Ecklonia radiata*): a likely precursor of arsenobetaine in marine fauna. Experientia, **38**: 643–644.

Edmonds JS, Shibata Y, Francesconi KA, Yoshinaga J & Morita M (1992) Arsenic lipids in the digestive gland of the western rock lobster *Panulirus cygnus*: An investigation by HPLC ICP-MS. Sci Total Environ, **122**(3): 321–335.

Eguchi N, Kuroda K & Endo G (1997) Metabolites of arsenic induced tetraploids and mitotic arrest in cultured cells. Arch Environ Contam Toxicol, **32**: 141–145.

Eisler R (1988) Arsenic hazards to fish, wildlife, and invertebrates: a synoptic review. Biological Report 85(1.12). Laurel, MD, U.S. Department of the Interior,

Elfving DC, Stehn RA, Pakkala IS & Lisk DJ (1979) Arsenic content of small mammals indigenous to old orchard soils. Bull Environ Contam Toxicol, **21**: 62–64.

Elfving DC, Wilson KR, Ebel JG, Manzell KL, Gutenmann WH & Lisk DJ (1994) Migration of lead and arsenic in old orchard soils in the Georgian Bay region of Ontario. Chemosphere, **29**(2): 407–413.

Elkhatib EA, Bennett OL & Wright RJ (1984) Arsenite sorption and desorption in soils. Soil Sci Soc Am J, **48**(5): 1025–1030.

Endo G, Kuroda K, Okamoto A & Horiguchi S (1992) Dimethylarsenic acid induces tetraploids in Chinese hamster cells. Bull Environ Contam Toxicol, **48**: 131–137.

Engel R & Smith A (1994) Arsenic in drinking water and mortality from vascular disease: an ecologic analysis in 30 counties in the United States. Arch Environ Health, **49**: 418–427.

Engman J & Jorhem L (1998) Toxic and essential elements in fish from Nordic waters with the results seen from the perspective of analytical quality assurance. Food Add Contam, **15**: 884–892.

Enserink EL, Maas-Diepeveen JL & Van Leeuwen CJ (1991) Combined effects of metals; an ecotoxicological evaluation. Water Res, **25**(6): 679–687.

Enterline PE (1983) Sorting out multiple causal factors in individual cases. In: Chiazze L, Lundin FE, Watkins D eds. Methods and issues in occupational and environmental epidemiology. Ann Arbor, MI, Ann Arbor Science, pp 177–182.

Enterline PE & Marsh GM (1980) Mortality studies of smelter workers. Am J Ind Med, **1**: 251–259.

Enterline PE & Marsh GM (1982) Cancer among workers exposed to arsenic and other substances in a copper smelter. Am J Epidemiol, **116**(6): 895–911.

Enterline PE, Henderson VL & Marsh GM (1987a) Exposure to arsenic and respiratory cancer. A reanalysis. Am J Epidemiol, **125**: 929–938.

Enterline PE, Marsh GM, Esmen NA, Henderson VL, Callahan CM & Paik M (1987b) Some effects of cigarette smoking, arsenic, and SO2 on mortality among US copper smelter workers. J Occup Med, **29**: 831–841.

Enterline PE, Day R & Marsh GM (1995) Cancers related to exposure to arsenic at a copper smelter. Occup Environ Med, **52**: 28–32.

ERG (1997) Report on the Expert panel on arsenic carcinogenicity: Review and workshop, Prepared by Eastern Research Group, Inc. 110 Hartwell Avenue, Lexington, MA 02173, EPA contract No. 68-C6-0041. Washington, DC, National Center for Environmental Assessment, U.S. Environmental Protection Agency.

Erry BV, Macnair MR, Meharg AA, Shore RF & Newton I (1999) Arsenic residues in predatory birds from an area of Britain with naturally and anthropogenically elevated arsenic levels. Environ Pollut, **106**(1): 91–95.

Esser KB (1996) Reference concentrations for heavy metals in mineral soils, oat, and orchard grass (*Dactylis glomerata*) from three agricultural regions in Norway. Water Air Soil Pollut, **89**(3/4): 375–397.

Esteban M, Arino C, Ruisanchez I, Larrechi MS & Rius FX (1994) Expert system for the voltammetric determination of trace metals, Part IV. Methods for speciation of chromium and arsenic. Anal Chim Acta, **285**: 193–208.

Ettajani H, Amiard-Triquet C, Jeantet AY & Amiard JC (1996) Fate and effects of soluble or sediment-bound arsenic in oysters (*Crassostrea gigas* Thun.). Arch Environ Contam Toxicol, **31**(1): 38–46.

Falk H, Caldwell GG, Ishak KG, Thomas LB & Popper H (1981a) Arsenic-related hepatic angiosarcoma. Am J Ind Med, **2**: 43–50.

Falk H, Herbert J, Crowley S, Ishak KG, Thomas LB, Popper H & Caldwell GG (1981b) Epidemiology of hepatic angiosarcoma in the United States: 1964–1974. Environ Health Perspect, **41**: 107–113.

Falkner KC, McCallum GP & Bend JR (1993a) Effects of arsenite treatment on NAD(P)H:quinone acceptor oxidoreductase activity in liver, lung, kidney and heart of the rat. Drug Metab Dispos, **21**: 334–337.

Falkner KC, McCallum GP, Cherian MG & Bend JR (1993b) Effects of acute sodium arsenite administration on the pulmonary chemical metabolizing enzymes, cytochrome P-450 monooxygenase, NAD(P)H:quinone acceptor oxidoreductase and glutathione *S*-transferase in guinea pig: comparison with effects inliver and kidney. Chem Biol Interactions, **86**: 51–68.

Fargasová A (1994a) Comparative toxicity of five metals on various biological subjects. Bull Environ Contam Toxicol, **53**(2): 317–324.

Fargasová A (1994b) Effect of Pb, Cd, Hg, As, and Cr on germination and root growth of *Sinapis alba* seeds. Bull Environ Contam Toxicol, **52**(3): 452–456.

Farmer JG & Lovell MA (1986) Natural enrichment of arsenic in Loch Lomond sediments. Geochim Cosmochim Acta, **50**: 2059–2067.

Farmer JG & Johnson LR (1990) Assessment of occupational exposure to inorganic arsenic based on urinary concentrations and speciation of arsenic. Br J Ind Med, **47**(5): 342–348.

Faust SD, Winka A, Belton T & Tucker R (1983) Assessment of the chemical and biological significance of arsenical compounds in a heavily contaminated watershed. Part II. Analysis and distribution of several arsenical species. J Environ Sci Health, **A18**(3): 389–411.

Faust SD, Winka AJ & Belton T (1987a) An assessment of chemical and biological significance of arsenical species in the Maurice River drainage basin (N.J.). Part I. Distribution in water and river and lake sediments. J Environ Sci Health, **22A**(3): 209–237.

Faust SD, Winka AJ & Belton T (1987b) An assessment of chemical and biological significance of arsenical species in the Maurice River drainage basin (N.J.). Part II. Partitioning of arsenic into bottom sediments. J Environ Sci Health, **22A**(3): 239–262.

Featherstone AM, Butler ECV, O'Grady BV & Michel P (1998) Determination of arsenic species in sea water by hydride generation atomic fluorescence spectroscopy. J Anal Energy Spectrom, **13**(12) 1355–1360.

Feldman RG, Niles CA, Kelly-Hayes M, Sax DS, Dixon WJ, Thompson DJ & Landau E (1979) Peripheral neuropathy in arsenic smelter workers. Neurology, **29**: 939–944.

Ferguson JF & Gavis J (1972) A review of the arsenic cycle in natural waters. Water Res, **6**: 1259–1274.

Ferm VH & Hanlon DP (1985) Constant rate exposure of pregnant hamsters to arsenate during early gestation. Environ Res, **37**: 425–432.

Ferreccio C, Gonzalez C, Solari J & Noder C (1996) Bronchopulmonary cancer in workers exposed to arsenic: a case control study. Rev Med Chil, **124**(1): 119–123.

Ferreccio C, Psych CG, Stat VM, Gredis GM & Sancha AM (1998) Lung cancer and arsenic exposure in drinking water: a case-control study in northern Chile. Cad Saude Publica **14**(Suppl 3): 193–198.

Ferreccio C, Gonzalez C, Milosavjlevic V, Marshall G, Sancha AM & Smith AH (2000) Lung cancer and arsenic concentrations in drinking water in Chile. Epidemiology, **11**: 673–679.

Fesmire FM, Schauben JL & Roberge RJ (1988) Survival following massive arsenic ingestion. Am J Emerg Med, **6**: 602–606.

Fierz U (1965) Catamnestic research into the side effects of inorganic arsenotherapy in skin diseases. Dermatologica, **131**: 41–58.

Fincher RME & Koerker RM (1987) Long term survival in acute arsenic encephalopathy follow-up using newer measures of electrophysiolooogic parameters. Am J Med, **82**: 549–552.

Fisher DL (1982) Cultured rat embryo accumulation of DNA, RNA, and protein following maternal administration of sodium arsenate. Environ Res, **28**: 1–9.

Flanjak J (1982) Inorganic and organic arsenic in some commercial East Australian crustacea. J Sci Food Agric, **33**(6): 579–83.

Florence TM, Stauber JL & Ahsanullah M (1994) Toxicity of nickel ores to marine organisms. Sci Total Environ, **148**: 139–156.

Foa V, Colombi A, Maroni M, Buratti M & Calzaferri G (1984) The speciation of the chemical forms of arsenic in the biological monitoring of exposure to inorganic arsenic. Sci Total Environ, **34**: 241–259.

Fordyce FM, Williams TM, Palittpapapon A & Charoenchaisei P (1995) Hydrogeochemistry of arsenic in an area of chronic mining-related arsenism, Ron Phibun District. Keyworth, UK: British Geological Survey.

Forget J, Pavillon JF, Menasria MR & Bocquené G (1998) Mortality and LC$_{50}$ values for several stages of the marine copepod *Tigriopus brevicornis* (Müller) exposed to the metals arsenic and cadmium and the pesticides atrazine, carbofuran, dichlorvos, and malathion. Ecotoxicol Environ Saf, **40**: 239–244.

Fowler SW & Ünlü MY (1978) Factors affecting bioaccumulation and elimination of arsenic in the shrimp *Lysmata seticaudata*. Chemosphere, **7**(9): 711–720.

Francesconi KA, Micks P, Stockton RA & Irgolic KJ (1985) Quantitative determination of arsenobetaine, the major water-soluble arsenical in three species of crab, using high pressure liquid chromatographic and an inductively coupled argon plasma emission spectrometer as the arsenic-specific detector. Chemosphere, **14**(10) 1443–1453.

Francesconi KA, Edmonds JS & Stick RV (1989) Accumulation of arsenic in yelloweye mullet (*Aldrichetta forsteri*) following oral administration of organoarsenic compounds and arsenate. Sci Total Environ, **79**(1): 59–67.

Franke KW & Moxon AL (1936) A comparison of the minimum fatal doses of selenium, tellurium, arsenic and vanadium. J Pharm Exp Ther, **58**: 454–459.

Freeman GB, Johnson JD, Killinger JM, Liao SC, Davis OA, Ruby MV, Chaney RL, Lovre SC & Bergstrom PD (1993) Bioavailability of arsenic in soil impacted by smelter activities following oral administration in rabbits. Fundam Appl Toxicol, **21**: 83–88.

Freeman GB, Schoof RA, Ruby MV, Davis OA, Dill JA, Liao SC, Lapin CA & Bergstrom PD (1995) Bioavailability of arsenic in soil and house dust impacted by smelter activities following oral administration in cynomolgus monkeys. Fundam Appl Toxicol, **28**: 215–222.

Freeman HC, Uthe JF, Fleming RB, Odense PH, Ackman RG, Landry G & Musial C (1979) Clearance of arsenic ingested by man from arsenic contaminated fish. Bull Environ Contam Toxicol, **22**: 224–229.

Freeman JW & Couch JR (1956) Prolonged encophalopahty with arsenic poisoning. Neurology, **28**: 853–855.

Freeman MC (1985) The reduction of arsenate to arsenite by an *Anabaena*-bacteria assemblage isolated from the Waikato River. N Z J Mar Freshw Res, **19**: 277–282.

Freeman MC, Aggett J & O'Brien G (1986) Microbial transformations of arsenic in Lake Ohakuri, New Zealand. Water Res, **20**(3): 283–294.

Frost F, Harter L, Milham S, Royce R, Smith AH, Hartley J & Enterline P (1987) Lung cancer among women residing close to an arsenic emitting copper smelter. Arch Environ Health, **42**: 148–152.

Gailer J & Irgolic KJ (1996) Retention behavior of arsenobetaine, arsenocholine, trimethylarsine oxide and tetramethylarsonium iodide on a styrene-divinylbenzene column with benzenesulfonates as ion-pairing reagents. J Chromatogr A, **730**: 219–229.

Gaines TB (1960) The acute toxicity of pesticides to rats. Toxicol Appl Pharm, **2**: 88–99.

Gaines TB & Linder RE (1986) Acute toxicity of pesticides in adult and weanling rats. Fundam Appl Toxicol, **7**: 299–308.

Galba J (1972) [Desorption of arsenates from soil by water]. (in Slovakian) Pol 'nohospodarstvo, **18**(11) 945–952.

Galy P, Touraine R, Brune J, Gallois P, Roudier P, Loire R, Lheureux P & Wiesendanger T (1963a) Les cancers broncho-pulmonaires de l'intoxication arsenicale chronique chez les viticulterus du Beaujolais. Lyon Méd, **43**: 735–744.

Galy P, Touraine R, Brune J, Roudier R & Gallois P (1963b) Le cancer pulmonaire d'origine arsenical des vignerons du Beaujolais. J Fr Med Chir Thorac, **17**: 303–311.

Gao S & Burau RG (1997) Environmental factors affecting rates of arsine evolution from and mineralization of arsenicals in soil. J Environ Qual, **26**: 753–763.

Garcia-Vargas G, Cebrian ME, Albores A, Lim CK & De Matteis F (1995) Time-dependent porphyric response in mice subchronically exposed to arsenic. Hum Exp Toxicol, **14**(6): 475–483.

Garland M, Morris JS, Rosner BA, Stampfer MJ, Spate VL, Baskett CJ, Willett WC & Hunter DJ (1993) Toenail trace element levels as biomarkers: reproducibility over a 6-year period. Cancer Epidemiol Biomarkers Prev, **2**: 493–497.

Geiszinger A, Goessler W, Kuehnelt D, Francesconi K & Kosmus W (1998) Determination of arsenic compounds in earthworms. Environ Sci Technol, **32**(15) 2238–2243.

George GM, Frahm LJ & McDonnell JP (1973) Dry ashing method for determination of total arsenic in animal tissues: collaborative study. J AOAC Int, **56**(4): 793–797.

George RK & Roscoe RS (1951) Microdetermination of arsenic and its application to biological material. Anal Chem, **23**(6): 914–919.

Georis B, Cardenas A, Buchet JP & Lauwerys R (1990) Inorganic arsenic methylation by rat tissue slices. Toxicology, **63**: 73–84.

Gerhardsson L, Dahlgren E, Eriksson A, Lagerkvist BEA, Lundstrom J & Nordberg GF (1988) Fatal arsenic poisoning – a case report. Scand J Work Environ Health, **14**: 130–133.

German J (1984) The embryonic stress hypothesis of teratogenesis. Am J Med, **76**: 293–301.

German J & Louie E, Banerjee D (1986) The heat-shock response in vivo: experimental induction during mammalian organogenesis. Teratogen Carcinog Mutagen, **6**: 555–562.

Germolec DR, Spalding J, Boorman GA, Wilmer JL, Yoshida T, Simeonova PP, Bruccoleri A, Kayama F, Gaido K, Tennant R, Burleson F, Dong W, Lang RW & Luster MI (1997) Arsenic can mediate skin neoplasia by chronic stimulation of keratinocyte-derived growth factors. Mutat Res, **386**: 209–281.

Geubel AP, Mairlot MC, Buchet JP, Dive C & Lauwerys R (1988) Abnormal methylation capacity in human liver cirrhosis. Int J Clin Pharm Res, VIII: 117–122.

Geyer L (1898) Über die chronischen Hautveränderungen beim Arsenicismus und Betrachtungen über die Masserkrankungen in Reichenstein in Schlesien. Arch Derm Syphilol (Berlin), **43**: 221–280.

Gibbs PE, Langston W, Burt GR & Pascoe PL (1983) *Tharyx marioni* (Polychaeta): a remarkable accumulator of arsenic. J Mar Biol Assoc UK, **63**: 313–325.

Gibson DP, Brauninger R, Shaffi HS, Kerckaert GA, LeBoeuf RA, Isfort RJ & Aardema MJ (1997) Induction of micronuclei in Syrian hamster embryo cells: comparison to results in the SHE cell transformation assay for national toxicology program test chemicals. Mutat Res, **392**: 61–70.

Giddings JM & Eddlemon GK (1977) The effects of microcosm size and substrate type on aquatic microcosm behavior and arsenic transport. Arch Environ Contam Toxicol, **6**: 491–505.

Ginsburg JM & Lotspeich WD (1963) Interrelations of arsenate and phosphate in the dog kidney. Am J Physiol, **205**: 707–714.

Gjerde DT, Wiederin DR, Smith FG & Mattson BM (1993) Metal speciation by means of microbore columns with direct-injection nebulization by inductively coupled plasma atomic emission spectroscopy. J Chromatogr, **640**: 73–78.

Glooschenko WA & Arafat N (1988) Atmospheric deposition of arsenic and selenium across Canada using Sphagnum moss as a biomonitor. Sci Total Environ, **73**(3): 269–275.

Goebel HH, Schmidt PF, Bohl J, Tettenborn B, Kramer G & Gutmann L (1990) Polyneuropathy due to acute arsenic intoxication: biopsy studies. J Neuropathol Exp Neurol, **49**: 137–149.

Goede AA (1985) Mercury, selenium, arsenic and zinc in waders from the Dutch Wadden Sea. Environ Pollut, **37**: 287–309.

Goering PL, Maronpot RR & Fowler BA (1988) Effect of intratracheal gallium arsenide administration on □-aminolevulinic acid dehydratase in rats: relationship to urinary excretion of aminolevulinic acid. Toxicol Appl Pharmacol, **92**: 179–193.

Goldberg S & Glaubig RA (1988) Anion sorption on a calcareous, montmorillonitic soil – arsenic. Soil Sci Soc Am J, **52**(5): 1297–1300.

Golding LA, Timperley MH & Evans CW (1997) Non-lethal responses of the freshwater snail *Potamopyrgus antipodarum* to dissolved arsenic. Environ Monit Assess, **47**(3): 239–254.

Goldstein SH & Babich H (1989) Differential effects of arsenite and arsenate to *Drosophila melanogaster* in a combined adult/developmental toxicity assay. Bull Environ Contam Toxicol, **42**: 276–282.

Gomez-Ariza JL, Sanchez Rodas D & Giraldez I (1998) Selective extraction of iron oxide associated arsenic species from sediments for speciation with coupled HPLC HG AAS. J Anal Energy Spectrom, **13**(12) 1375–1379.

Goncalo S, Silva MS, Goncalo M & Baptista AP (1980) Occupational contact dermatitis to arsenic trioxide. In: Forsch PJ & Dooms-Goossens A ed. Current topics in contact dermatitis. Berlin. Springer-Verlag, pp 333–336.

Gonsebatt ME, Vega L, Montero R, Garcia-Vargas G, Razo LM Del, Albores A, Cebrian ME & Ostrosky-Wegman P (1994) Lymphocyte replicating ability in individuals exposed to arsenic via drinking water. Mutat Res, **313**: 293–99.

Gonsebatt ME, Vega L, Salazar AM, Montero R, Guzmán P, Blas J, Del Razo LM, García-Vargas G, Albores A, Cebrian ME, Kelsh M & Ostrosky-Wegman P (1997) Cytogenetic effects in human exposure to arsenic. Mutat Res, **386**: 219–28.

Gonzalez MJ, Aguilar MV & Para MCC (1995) Gastrointestinal absorption of inorganic arsenic (V): the effect of concentration and interactions with phosphate and dichromate. Vet Hum Toxicol, **37**: 131–136.

Gottlieb K, Koehler JR & Tessari J (1993) Non-analytic problems in detecting arsenic and cadmium in children living near a cadmium refinery in Denver, Colorado. J Exposure Anal Environ Epidemiol, **3**: 139–153.

Grandjean P, Weihe P, Needham LL, Burse VW, Patterson DGJ, Sampson EJ, Jorgensen PJ & Vahter M (1995) Relation of a seafood diet to mercury, selenium, arsenic, and polychlorinated biphenyl and organochlorine concentrations in human milk. J Environ Sci Health A: Environ Sci Eng Toxic Hazardous Substance Contr, **30**(4): 921–938.

Greaves W, Rom W, Lyon J, Varley G, Wright DD & Chiu G (1981) Relationship between lung cancer and distance of residence from non-ferrous smelter stack effluent. Am J Ind Med, **2**: 15–23.

Gregus Z, Gyurasics A & Csanaky I (2000) Biliary and urinary excretion of inorganic arsenic: Monomethylarsonous acid as a major biliary metabolite in rats. Toxicol Sci, **56**: 18–25.

Greshonig H & Irgolic KJ (1997) The mercuric-bromide-stain method and the Natelson method for the determination of arsenic: implications for assessment of risks from exposure to arsenic in Taiwan. In: Abernathy CO, Calderon RL & Chappell WR ed. Arsenic. Exposure and health effects. London, Chapman & Hall, pp 17–32.

Gresser MJ (1981) ADP-arsenate, formation by submitochondrial particles under phosphorylating conditions. J Biol Chem, **256**: 5981–5983.

Greulach U & Henze G (1995) Analysis of arsenic(V) by cathodic stripping voltammetry. Anal Chim Acta, **306**: 217–223.

Groen K, Vaessen HAMG, Kliest JJG, de Boer JLM, van Ooik T, Timmerman A & Vlug RF (1994) Bioavailability of inorganic arsenic from bog ore containing soil in the dog. Environ Health Perspect, **102**: 182–184.

Guitart R, Torra M, Cerradelo S, Puig Casado P, Mateo R & To Figueras J (1994) Pb, Cd, As, and Se concentrations in livers of dead wild birds from the Ebro Delta, Spain. Bull Environ Contam Toxicol, **52**(4): 523–529.

Gulens J, Champ DR & Jackson RE (1979) Influence of redox environments on the mobility of arsenic in ground water. In: Jenne EA ed. Chemical modeling in aqueous systems, ACS Symposium Series 93. Washington, D.C, American Chemical Society, pp 81–95.

Gunderson EL (1995) FDA total diet study-1986–1991-dietary intakes of pesticides, selected elements, and other chemicals. J AOAC Int, **78**: 1353–1363.

Guo H-R, Chiang H-S, Hu H, Lipsitz SR & Monson RR (1994) Arsenic in drinking water and urinary cancers: a preliminary report. In: Chappell WR, Abernathy CO & Cothern CR ed. Arsenic exposure and health. Northwood, UK, Science and Technology Letters, pp 119–128.

Guo HR, Chiang HS, Hu H, Lipsitz SR & Monson RR (1997) Arsenic in drinking water and incidence of urinary cancers. Epidemiology, **8**: 545–550.

Guo HR, Lipsitz SR, Hu H & Monson RR (1998) Using ecological data to estimate a regression model for individual data: the association between arsenic in drinking water and incidence of skin cancer. Environ Res, **79**: 82–93.

Gupta SK & Chen KY (1978) Arsenic removal by adsorption. J Water Pollut Control Fed, **50**: 493–506.

Gustafsson JP & Tin NT (1994) Arsenic and selenium in some Vietnamese acid sulphate soils. Sci Total Environ, **151**(2): 153–158.

Gyurasics A, Varga F & Gregus Z (1991) Glutathione-dependent biliary excretion of arsenic. Biochem Pharm, **42**: 465–468.

Haddad E & Zikovsky L (1985) Determination of Al, As, Co, Cr, Cs, Fe, Mn, Sb, Sc, W and Zn in the workroom air by instrumental neutron activation analysis. J Radioanal Nucl Chem Lett, **93**(6): 371–378.

Haegele MA & Tucker RK (1974) Effects of 15 common environmental pollutants on eggshell thickness in mallards and *Coturnix*. Bull Environ Contam Toxicol, **11**(1): 98–102.

Hakala E & Pyy L (1992) Selection determination of toxicologically important arsenic species in urine by high-performance liquid chromatography-hydride generation atomic absorption spectrometry. J Anal Energy Spectrom, **7**: 191–195.

Hakala E & Pyy L (1995) Assessment of exposure to inorganic arsenic by determining the arsenic species excreted in urine. Toxicol Lett, **77**: 249–258.

Hall LL, George SE, Kohan MJ, Styblo J & Thomas DJ (1997) In vitro methylation of inorganic arsenic in mouse intestinal cecum. Toxicol Appl Pharm, **147**: 101–109.

Hamamoto E (1955) [Infant arsenic poisoning by powdered milk.] (in Japanese) Nihon Iji Shimpo, **1649**: 3–12.

Hamilton SJ & Buhl KJ (1990) Safety assessment of selected inorganic elements to fry of chinook salmon (*Oncorhynchus tshawytscha*). Ecotoxicol Environ Saf, **20**(3): 307–324.

Hamilton SJ & Buhl KJ (1997) Hazard assessment of inorganics, individually and in mixtures, to two endangered fish in the San Juan River, New Mexico. Environ Toxicol Water Qual, **12**(3): 195–209.

Hanaoka K, Matsumoto T, Tagawa S & Kaise T (1987) Microbial degradation of arsenobetaine, the major water soluble organoarsenic compound occurring in marine animals. Chemosphere, **16**(10–12): 2545–2550.

Hanaoka K, Hasegawa S, Kawabe N, Tagawa S & Kaise T (1990) Aerobic and anaerobic degradation of several arsenicals by sedimentary microorganisms. Appl Organomet Chem, **4**: 239–243.

Hanaoka K, Koga H, Tagawa S & Kaise T (1992) Degradation of arsenobetaine to inorganic arsenic by the microorganisms occurring in the suspended substances. Comp Biochem Physiol, **101B**(4): 595–599.

Hansen SH, Larsen EH, Pritzl G & Cornett C (1992) Speciation of seven arsenic compounds by high performance liquid chromatography with on-line detection by hydrogen-argon flame atomic absorption spectrometry and inductively coupled plasma mass spectrometry. J Anal Spectrom, **7**: 629–634.

Harako A (1986) Studies on arsenic in environment around the Osorezan Volcano region (Part 3). Arsenic in soil. Hirosaki Med J, **38**: 232–243.

Harrington JM, Middaugh JP, Morse DL & Housworth J (1978) A survey of a population exposed to high concentrations of arsenic in well water in Fairbanks, Alaska. Am J Epidemiol, **108**(5): 377–385.

Harrison JWE, Packman EW & Abbott DD (1958) Acute oral toxicity and chemical and physical properties of arsenic trioxides. AMA Arch Ind Health, **17**: 118–123.

Hartwell SI, Jin JH, Cherry DS & Cairns J (1989) Toxicity versus avoidance response of golden shiner, *Notemigonus crysoleucas*, to five metals. J Fish Biol, **35**(3): 447–456.

Hartwig A, Groblinghoff UD, Beyersmann D, Natarajan AT, Filon R & Mullenders LHF (1997) Interaction of arsenic (III) with nucleotide excision repair in UV irradiated human fibroblasts. Carcinog Oxf, **18**(2): 399–405.

Häsänen E, Lipponen M, Kattainen R, Markkanen K, Minkkinen P & Brjukhanov P (1990) Elemental concentrations of aerosol samples from the Baltic Sea area. Chemosphere, **21**(3): 339–347.

Hasegawa H, Sohrin Y, Matsui M, Hojo M & Kawashima M (1994) Speciation of arsenic in natural waters by solvent extraction and hydride generation atomic absorption spectrometry. Anal Chem, **66**: 3247–3252.

Haswell SJ, O'Neill P & Bancroft KC (1985) Arsenic speciation in soil-pore waters from mineralized and unmineralized areas of south-west England. Talanta, **32**: 69–72.

Hayakawa O & Watanabe N (1982) Characteristics of arsenite and arsenate adsorption on soils. Mem Hokkaido Inst Technol, **10**: 179–187.

Healy SM, Zakharyan RA & Aposhian HV (1997) Enzymatic methylation of arsenic compounds: IV. in vitro and in vivo deficiency of the methylation of arsenite and monomethylarsonic acid in the guinea pig. Mutat Res, **386**: 229–239.

Healy SM, Casarez EA, Ayala-Fierro F & Aposhian HV (1998) Enzymatic methylation of arsenic compounds, V. arsenite methyltransferase activity in tissues of mice. Toxicol Appl Pharm, **148**: 65–70.

Hei TK, Liu SX & Waldren C (1998) Mutagenicity of arsenic in mammalian cells: role of reactive oxygen species. Proc Natl Acad Sci U S A, **95**: 8103–8107.

Heitkemper D, Creed J & Caruso J (1989) Speciation of arsenic in urine using high-performance liquid chromatography with inductively coupled plasma mass spectrometric detection. J Anal Spectrom, **4**: 279–284.

Helgesen H & Larsen EH (1998) Bioavailability and speciation of arsenic in carrots grown in contaminated soil. Analyst, **123**(5): 791–6.

Hellou J, Fancey LL & Payne JF (1992) Concentrations of twenty-four elements in bluefin tuna, *Thunnus thynnus* from the Northwest Atlantic. Chemosphere, **24**(2): 211–218.

Hertz-Picciotto I & Smith A (1993) Observations on the dose–response curve for arsenic exposure and lung cancer. Scand J Work Environ Health, **19**: 217–26.

Hertz-Picciotto I, Smith AH, Holtzman D, Lipsett M & Alexeeff G (1992) Synergism between occupational arsenic exposure and smoking in the induction of lung cancer. Epidemiology, **3**: 23–31.

Hertz-Picciotto I, Arrighi HM & Hu SW (2000) Does arsenic exposure increase the risk for circulatory disease? Am J Epidemiol, **151**: 174–181.

Heyman A, Preiffer JB, Willett RW & Talor HM (1956) Peripheral neuropathy caused by arsenical intoxiction. A study of 41 cases with observations on the effects of BAL (2,3-dimercapto-propanol). N Engl J Med, **254**: 401–409.

Higham AM & Tomkin RPT (1993) Determination of trace quantities of selenium and arsenic in canned tuna fish by using electroanalytical techniques. Food Chem, **48**: 85–93.

Hill AB & Faning EL (1948) Studies on the incidence of cancer in a factory handling inorganic compounds of arsenic: mortality experience in the factory. Br J Ind Med, **5**: 1–6.

Hill EF & Camardese MB (1986) Lethal dietary toxicities of environmental contaminants and pesticides to *Coturnix*. Fish and Wildlife Technical Report 2. Washington, DC, U.S. Department of the Interior Fish and Wildlife Service.

Hiltbold AE, Hajek BF & Buchanan GA (1974) Distribution of arsenic in soil profiles after repeated applications of MSMA. Weed Sci, **22**(3): 272–275.

Hindmarsh J, McLetchie OR, Heffernan LPM, Hayne OA, Ellenberger HA, McCurdy RF & Thiebaux HJ (1977) Electromyographic abnormalities in chronic environmental arsenicalism. In: Brown SS, ed. Clinical chemistry and chemical toxicology of metals, 1. Amsterdam, Elsevier/North Holland Biomedical Press, pp 287–293.

Hinwood AL, Jolley DJ & Sim MR (1999) Cancer incidence and high environmental arsenic concentrations in rural populations: Results of an ecological study. J Environ Health Res, **9**: 131–141.

Hirata M, Hisanaga A, Tanaka A & Ishinishi N (1988) Glutathione and methylation of inorganic arsenic in hamsters. Appl Organomet Chem, **2**: 315–321.

Hirata M, Mohri T, Hisanaga A & Ishinishi N (1989) Conversion of arsenite and arsenate to methylarsenic and dimethylarsenic compounds by homogenates prepared from livers and kidneys of rats and mice. Appl Organomet Chem, **3**: 335–341.

Hirata M, Tanaka A, Hisanaga A & Ishinishi N (1990) Effects of glutathione depletion on the acute nephrotoxic potential of arsenite and on arsenic metabolism in hamsters. Toxicol Appl Pharm, **106**: 469–481.

Hoag WG (1963) Spontaneous cancer in mice. Ann N Y Acad Sci, **108**: 805–831.

Hochadel JF & Waalkes MP (1997) Sequence of exposure to cadmium and arsenic determines the extent of toxic effects in male Fischer rats. Toxicology, **116**: 89–98.

Hodgson JT & Jones RD (1990) Mortality of a cohort of tin miners 1941–86. Br J Ind Med, **47**: 665–676.

Hoffman DJ, Sanderson CJ, LeCaptain LJ, Cromartie E & Pendleton GW (1992) Interactive effects of arsenate, selenium, and dietary protein on survival, growth, and physiology in mallard ducklings. Arch Environ Contam Toxicol, **22**(1): 55–62.

Holak W & Specchio JJ (1991) Determination of total arsenic, As(III) and As(V), in foods by atomic absorption spectrophotometry. At Spectrosc, **12**(4): 105–108.

Holcman A & Stibilj V (1997) Arsenic residues in eggs from laying hens fed with a diet containing arsenic (III) oxide. Arch Environ Contam Toxicol, **32**: 407–410.

Holcombe GW, Phipps GL & Fiandt JT (1983) Toxicity of selected priority pollutants to various aquatic organisms. Ecotoxicol Environ Saf, **7**: 400–409.

Hollibaugh JT, Seibert DLR & Thomas WH (1980) A comparison of the acute toxicities of ten heavy metals to phytoplankton from Saanich Inlet, B.C., Canada. Estuar Coast Mar Sci, **10**: 93–105.

Holm TR, Anderson MA, Iverson DG & Stanforth RS (1979) Heterogeneous interactions of arsenic in aquatic systems. In: Jenne EA ed. Chemical modeling in aqueous systems, ACS Symposium Series 93. Washington, DC, American Chemical Society, pp 711–736.

Honda K, Hatayama T, Takahashi K & Yukioka M (1992) Heat shock proteins in human and mouse embryonic cells after exposure to heat shock or teratogenic agents. Teratogen Carcinog Mutagen, **11**: 235–244.

Hood RD & Harrison WP (1982) Effects of prenatal arsenite exposure in the hamster. Bull Environ Contam Toxicol, **29**: 671–678.

Hood RD, Harrison WP & Vedel GC (1982) Evaluation of arsenic metabolites for prenatal effects in the hamster. Bull Environ Contam Toxicol, **29**: 679–687.

Hood RD & Vedel-Macrander GC (1984) Evaluation of the effect of BAL (2,3-dimercaptopropanol) on arsenite-induced teratogenesis in mice. Toxicol Appl Pharm, **73**: 1–7.

Hood RD, Vedel-Macrander GC, Zaworotko MJ, Tatum FM & Meeks RG (1987) Distribution, metabolism, and fetal uptake of pentavalent arsenic in pregnant mice following oral or intraperitoneal administration. Teratology, **35**: 19–25.

Hood RD, Vedel GC, Zaworotko MJ, Tatum FM & Meeks RG (1988) Uptake, distribution, and metabolism of trivalent arsenic in the pregnant mouse. J Toxicol Environ Health, **25**: 423–434.

Hopenhayn-Rich C, Smith AH & Goeden HM (1993) Human studies do not support the methylation threshold hypothesis for the toxicity of inorganic arsenic. Environ Res, 60: 161–177.

Hopenhayn-Rich C, Biggs ML, Smith AH, Kalman DA & Moore LE (1996a) Methylation study of a population environmentally exposed to arsenic in drinking water. Environ Health Perspect, 104: 620–628.

Hopenhayn-Rich C, Biggs ML, Kalman DA, Moore LE & Smith AH (1996b) Arsenic methylation patterns before and after changing from high to lower concentrations of arsenic in drinking water. Environ Health Perspect, 104: 1200–1207.

Hopenhayn-Rich C, Biggs ML, Fuchs A, Bergoglio R, Tello EE, Nicolli H & Smith AH (1996c) Bladder cancer mortality associated with arsenic in drinking water in Argentina. Epidemiology, 7: 117–124.

Hopenhayn-Rich C, Biggs ML & Smith AH (1996d) Arsenic and bladder cancer mortality – Reply. Epidemiology 7: 558.

Hopenhayn-Rich C, Biggs ML & Smith AH (1998) Lung and kidney cancer mortality associated with arsenic in drinking water in Cordoba, Argentina. Int J Epidemiol, 27: 561–569.

Hopenhayn-Rich C, Hertz-Picciotto I, Browning S, Ferreccio C & Peralta C (1999) Reproductive and developmental effects associated with chronic arsenic exposure. Arsenic exposure and health effects. Oxford, Elsevier Science, pp 151–164.

Hopenhayn-Rich C, Browning S, Hertz-Picciotto I, Ferreccio C, Peralta C & Gibb H (2000) Chronic arsenic exposureand risk of infant mortality in two areas in Chile. Environ Health Perspect, 108: 667–673.

Hörnström E (1990) Toxicity test with algae – a discussion on the batch method. Ecotoxicol Environ Saf, 20(3): 343–353.

Hothem RL & Welsh D (1994) Contaminants in eggs of aquatic birds from the grasslands of central California. Arch Environ Contam Toxicol, 27(2): 180–185.

Howard AG, Arbab-Zavar MH & Apte S (1984) The behaviour of dissolved arsenic in the estuary of the river Beaulieu. Estuar Coast Mar Sci, 19: 493–504.

Howard AG, Apte SC, Comber SDW & Morris RJ (1988) Biogeochemical control of the summer distribution and speciation of arsenic in the Tamar Estuary. Estuar Coast Mar Sci, 27(4): 427–443.

Howard AG & Apte SC (1989) Seasonal control of arsenic speciation in an estuarine ecosystem. Appl Organomet Chem, 3(6): 499–507.

Hsieh LL, Chen HJ, Hsieh JT, Jee SH, Chen GS & Chen CJ (1994) Arsenic-related Bowen's disease and paraquat-related skin cancerous lesions show no detectable ras and p53 gene alterations. Cancer Lett, **86**: 59–65.

Hsu YH, Li SY, Chiou HY, Yeh PM, Liou JC & Hsueh YM, Chang SH & Chen CJ (1997) Spontaneous and induced sister chromatid exchanges and delayed cell proliferation in peripheral lymphocytes of Bowen's disease patients and matched controls of arseniasis-hyperendemic villages in Taiwan. Mutat Res, **386**(3): 241–251.

Hsu CH, Yang SA, Wang JY, Yu HS & Lin SR (1999) Mutational spectrum of p53 gene in arsenic-related skin cancers from the blackfoot disease endemic area of Taiwan. Br J Cancer, **80**(7): 1080–1086.

Hsueh YM, Cheng GS, Wu MM, Yu HS, Kuo TL & Chen CJ (1995) Multiple risk factors associated with arsenic-induced skin cancer: effects of chronic liver disease and malnutritional status. Br J Cancer, **71**: 109–114.

Hsueh YM, Wu WL, Huang YL, Chiou HY, Tseng CH & Chen CJ (1998) Low serum carotene level and increased risk of ischemic heart disease related to long-term arsenic exposure. Atherosclerosis, **141**: 249–257.

Hu Y, Su L & Snow E (1998) Arsenic toxicity is enzyme specific and its affects on ligation are not caused by the direct inhibition of DNA repair enzymes. Mutat Res, **401**: 203–218.

Hua C, Jagner D & Renman L (1987) Automated determination of total arsenic in sea water by flow constant-current stripping analysis with gold fibre electrodes. Anal Chim Acta, **201**: 263–268.

Huang RN & Lee TC (1996) Cellular uptake of trivalent arsenite and pentavalent arsenate in KB cells cultured in phosphate-free medium. Toxicol Appl Pharm, **136**(2): 243–249.

Hudson RH, Tucker RK & Haegele MA (1984) Handbook of toxicity of pesticides to wildlife. Resource Publication 153. Washington, DC, U.S. Department of the Interior Fish and Wildlife Service.

Huffman GP, Huggins FE, Shah N & Zhao J (1994) Speciation of arsenic and chromium in coal and combustion ash by XAFS spectroscopy. Fuel Process Technol, **39**: 47–62.

Hughes JP, Polissar L & van Belle G (1988) Evaluation and synthesis of health effects studies of communities surrounding arsenic producing industries. Int J Epidemiol, **17**: 407–13.

Hughes MF & Kenyon EM (1998) Dose-dependent effects on the disposition of monomethylarsonic acid and dimethylarsinic acid in the mouse after intravenous administration. J Toxicol Environ Health, **53**: 101–118.

Hughes MF & Thompson DJ (1996) Subchronic dispositional and toxicological effects of arsenate administered in drinking water to mice. J Toxicol Environ Health, **49**: 177–196.

Hughes MF, Menache M & Thompson DJ (1994) Dose-dependent disposition of sodium arsenate in mice following acute oral exposure. Fundam Appl Toxicol, **22**: 80–89.

Hughes MF, Mitchell CT, Edwards BC & Rahman MS (1995) In vitro percutaneous absorption of dimethylarsenic acid in mice. J Toxicol Environ Health, **45**: 101–112.

Huiliang H, Jagner D & Renman L (1988) Flow potentiometric and constant-current stripping analysis for arsenic (V) without prior chemical reduction to arsenic (III) – application to the determination of total arsenic in seawater and urine. Anal Chim Acta, **207**: 37–46.

Hullinger G, Sangster L, Colvin B & Frazier K (1998) Bovine arsenic toxicosis from ingestion of ashed copper chrome arsenate treated timber. Vet Hum Toxicol, **40**: 147–148.

Hunder F, Nguyen P-T, Schumann K & Fichtl B (1993) Influence of inorganic and organic arsenicals on intestinal transfer of nutrients. Res Comm Chem Pathol Pharm, **80**: 83–92.

Hutchinson J (1887) Arsenic cancer. Br Med J, **2**: 1280.

Hutton M & Symon C (1986) The quantities of cadmium, lead, mercury and arsenic entering the U.K. environment from human activities. Sci Total Environ, **57**: 129–150.

Huysmans KD & Frankenberger WT (1990) Arsenic resistant microorganisms isolated from agricultural drainage water and evaporation pond sediments. Water Air Soil Pollut, **53**(1/2): 159–168.

Huysmans KD & Frankenberger WT (1991) Evolution of trimethylarsine by a *Penicillium* sp. isolated from agricultural evaporation pond water. Sci Total Environ, **105**: 13–28.

IARC (International Agency for Research on Cancer) (1973) Arsenic and inorganic arsenic compounds. IARC Monographs on the Evaluation of Carcinogenic Risk of Chemicals to Man. Vol. 2. Some inorganic and organometallic compounds. Lyon, International Agency for Research on Cancer, pp 48–73.

IARC (International Agency for Research on Cancer) (1980) Arsenic and arsenic compounds. IARC Monographs on the Evaluation of Carcinogenic Risk of Chemicals to Humans. Vol. 23. Some metals and metal compounds. Lyon, International Agency for Research on Cancer, pp 39–142.

IARC (International Agency for Research on Cancer) (1987) IARC Monographs on the Evaluation of the Carcinogenic Risk of Chemicals to Humans. Overall evaluations of carcinogenicity: An Updating of IARC Monographs Vols 1–42, Supplement 7. Lyon, International Agency for Research on Cancer.

IARC (International Agency for Research on Cancer) (1993) Exposures in the glass manufacturing industry. IARC Monographs on the Evaluation of the Carcinogenic Risk of Chemicals to Humans, Vol. 58. Beryllium, cadmium, mercury, and exposure in the glass manufacturing industry. Lyon, International Agency for Research on Cancer, pp 347–375.

Ihrig MM, Shalat SL & Baynes C (1998) A hospital-based case-control study of stillbirths and environmental exposure to arsenic using an atmospheric dispersion model linked to a geographical information system. Epidemiolology, 9: 290–294.

ILO (International Labour Office)(1991) Occupational exposure limits for airborne toxic substances, 3rd ed. Geneva, International Labour Office.

Inglis A & Davis EL (1972) Effects of water hardness on the toxicity of several organic and inorganic herbicides to fish. Technical Paper 67. Washington, DC, U.S. Department of the Interior, Bureau of Sport Fisheries and Wildlife.

Ip C & Ganther H (1988) Efficacy of trimethylselenonium versus selenite in cancer chemoprevention and its modulation by arsenite. Carcinogenesis, 9: 1481–1484.

Ip C & Ganther H (1992) Biological activities of trimethylselenonium as influenced by arsenite. J Inorg Biochem, 46: 215–222.

IPCS (International Programme on Chemical Safety) (1981) Arsenic. Geneva, World Health Organization.

Ishinishi N, Yamamoto A, Hisanaga A & Inamasu T (1983) Tumorigenicity of arsenic trioxide to the lung in Syrian golden hamsters by intermittent instillations. Cancer Lett, 21: 141–147.

Ismael A & Roberts RD (1992) Arsenic in small mammals. Environ Technol, 13(11): 1091–1095.

Itoh T, Zhang YF, Murai S, Saito H, Nagahama H, Miyate H, Saito Y & Abe E (1990) The effect of arsenic trioxide on brain monoamine metabolism and locomotor activity of mice. Toxicol Lett, 54: 345–353.

Jacobs LW, Syers JK & Keeney DR (1970a) Arsenic sorption by soils. Soil Sci Soc Am Proc, 34: 750–754.

Jacobs LW, Walsh LW & Keeney DR (1970b) Arsenic residue toxicity to vegetable crops grown on plainfield sand. Agron J, 62: 588–591.

Jaghabir MTW, Abdelghani A & Anderson AL (1988) Oral and dermal toxicity of MSMA to New Zealand white rabbits. Oryctalagus cuniculus. Bull Environ Contam Toxicol, 40: 119–122.

Jaghabir MTW, Abdelghani A & Anderson AL (1989) Histopathological effect of monosodiummethanearsonate (MSMA) on New Zealand white rabbits (Oryctalagus cuniculus). Bull Environ Contam Toxicol, 42: 289–293.

Jaghabir MW, Abdelghani AA & Anderson AC (1994) Absorption, distribution and elimination of arsenic in New Zealand white rabbits (Oryctalagus cuniculus) following multiple oral doses of monosodium methane arsonate (MSMA). Dirasat, 21B: 137–145.

Jakubowski M, Trzcinka-Ochocka M, Razniewska G & Matczak W (1998) Biological monitoring of occupational exposure to arsenic by determining urinary content of inorganic arsenic and its methylated metabolites. Int Arch Occup Environ Health, 71: S29–S32.

Järup L (1992) Dose response relations for occupational exposure to arsenic and cadmium. PhD thesis. Karolinska Institutet, Stockholm.

Järup L & Pershagen G (1991) Arsenic exposure, smoking, and lung cancer in smelter workers – a case-control study. Am J Epidemiol, 134: 545–551.

Järup L, Pershagen G & Wall S (1989) Cumulative arsenic exposure and lung cancer in smelter workers: a dose-response study. Am J Ind Med, 15: 31–41.

Jenner HA & Janssen-Mommen JPM (1993) Duckweed *Lemna minor* as a tool for testing toxicity of coal residues and polluted sediments. Arch Environ Contam Toxicol, 25(1): 3–11.

Jensen GE & Hansen ML (1998) Occupational arsenic exposure and glycosylated haemoglobin. Analyst, 123: 77–80.

Jensen GE, Christensen JM & Poulsen OM (1991) Occupational and environmental exposure to arsenic – incrase urinary arsenic in children. Sci Total Environ, 107: 169–177.

Jha AN, Noditi M, Nilsson R & Natarajan AT (1992) Genotoxic effects of sodium arsenite on human cells. Mutat Res, 284: 215–221.

Ji G & Silver S (1995) Bacterial resistance mechanisms for heavy metals of environmental concern. J Ind Microbiol, 14(2): 61–75.

Jiang QQ & Singh BR (1994) Effect of different forms and sources of arsenic crop yield and arsenic concentration. Water Air Soil Pollut, 74(3/4): 321–343.

Johns C & Luoma SN (1990) Arsenic in benthic bivalves of San Francisco Bay and the Sacramento/San Joaquin River Delta. Sci Total Environ, 97/98: 673–684.

Johnson CA & Thornton I (1987) Hydrological and chemical factors controlling the concentration of Fe, Cu, Zn and As in a river system contaminated by acid mine drainage. Water Res, 21(3): 359–365.

Johnson DL (1972) Bacterial reduction of arsenate in sea water. Nature, 240: 44–45.

Johnson DL & Burke RM (1978) Biological mediation of chemical speciation. II. Arsenate reduction during marine phytoplankton blooms. Chemosphere, 7(8): 645–648.

Johnson RK, Eriksson L & Wiederholm T (1992) Ordination of profundal zoobenthos along a trace metal pollution gradient in northern Sweden. Water Air Soil Pollut, 65(3/4): 339–351.

Johnston D, Oppermann H, Jackson J & Levinson W (1980) Induction of four proteins in chick embryo cells by sodium arsenite. J Biol Chem, **255**: 6975–6980.

Jones CA, Inskeep WP & Neuman DR (1997) Arsenic transport in contaminated mine tailings following liming. J Environ Qual, **26**: 433–439.

Jonnalagadda SB & Nenzou G (1996a) Studies on arsenic rich mine dumps: I. Effect on the surface soil. J Environ Sci Health, **A31**(8): 1909–1915.

Jonnalagadda SB & Nenzou G (1996b) Studies on arsenic rich mine dumps: III. Effect on the river water. J Environ Sci Health, **A31**(10) 2547–2555.

Jonnalagadda SB & Nenzou G (1997) Studies on arsenic rich mine dumps: II. The heavy element uptake by vegetation. J Environ Sci Health, **A32**(2): 455–464.

Julshamn K, Andersen A, Ringdal O & Mørkøre J (1987) Trace elements intake in the Faroe Islands. I. Element levels in edible parts of pilot whales (*Globicephalus meleanus*). Sci Total Environ, **65**: 53–62.

Julshamn K, Maage A & Larsen EH (1996) Studies of critical factors in the determination of arsenic in standard reference materials of marine origin by ETAAS NMKL interlaboratory study. Fresenius J Anal Chem 355(3/4): 304–307.

Jurewicz S & Buikema AL (1980) Effects of arsenate on algae, *Daphnia*, and mosquito fish. Va J Sci, **31**: 124.

Kabir H & Bilgi C (1993) Ontario gold miners with lung cancer. Occupational exposure assessment in establishing work-relatedness. J Occup Med, **35**: 1203–1207.

Kachinskas DJ, Phillips MA, Qin Q, Stokes JD & Rice RH (1994) Arsenate perturbation of human keratinocyte differentiation. Cell Growth Differ, **5**: 1235–1241.

Kachinskas DJ, Qin Q, Phillips MA & Rice RH (1997) Arsenate suppression of human keratinocyte programming. Mutat Res, **386**(3): 253–261.

Kagey BT, Bumgarner JE & Creason JP (1977) Arsenic levels in maternal-fetal tissue sets. In: Hemphill DD ed. Trace substances in environmental health XI. A symposium, Columbia, University of Missouri Press, pp 252–256.

Kaise T & Fukui S (1992) The chemical form and acute toxicity of arsenic compounds in marine organisms. Appl Organomet Chem, **6**: 155–160.

Kaise T, Watanabe S & Itoh K (1985) The acute toxicity of arsenobetaine. Chemosphere, **14**: 1327–1332.

Kaise T, Hanaoka K & Tagawa S (1987) The formation of trimethylarsine oxide from arsenobetaine by biodegradation with marine microorganisms. Chemosphere, 16(10–12): 2551–2558.

448

Kaise T, Yamauchi H, Horiguchi Y, Tani T, Watanabe S, Hirayama T & Fukui S (1989) A comparative study on acute toxicity of methylarsonic acid, dimethylarsinic acid and trimethylarsine oxide in mice. Appl Organomet Chem, **3**: 273–277.

Kalman DA, Hughes J, Belle G van, Burbacher T, Dolgiano D, Coble K, Mottet NK & Polissar L (1990) The effect of variable environmental arsenic contamination on urinary concentrations of arsenic species. Environ Health Perspect, **89**: 145–151.

Karagas MR, Morris JS, Weiss JE, Spate V, Baskett C & Greenberg ER (1996) Toenail samples as an indicator of drinking water arsenic exposure. Cancer Epidemiol Biomarkers Prev, **5**: 849–852.

Kato K, Ito H & Okamoto K (1997) Modulation of the arsenite-induced expression of stress proteins by reducing agents. Cell Stress Chaperones, **2**(3): 199–209.

Kavanagh PJ, Farago ME, Thornton I & Braman RS (1997) Bioavailability of arsenic in soil and mine wastes of the Tamar valley, SW England. Chem Speciation Bioavailability, **9**(3): 77–81.

Kavanagh P, Farago ME, Thornton I, Goessler W, Kuehnelt D, Schlagenhaufen C & Irgolic KJ (1998) Urinary arsenic species in Devon and Cornwall residents, UK. A pilot study. Analyst, **123**(1): 27–29.

Keeley PE & Thullen RJ (1971) Cotton response to temperature and organic arsenicals. Weed Sci, **19**(3): 297–300.

Kelley JA, Jaffe DA, Baklanov A & Mahura A (1995) Heavy metals on the Kola Peninsula: aerosol size distribution. Sci Total Environ, **160/161**: 135–138.

Kenney LJ & Kaplan JH (1988) Arsenate substitutes for phosphate in the human red cell sodium pump and anion exchanger. J Biol Chem, **263**(17): 7954–7960.

Kenyon EM, Hughes MF & Levander OA (1997) Influence of dietary selenium on the disposition of arsenate in the female B6C3F1 mouse. J Toxicol Environ Health, **51**: 279–299.

Kerkvliet NI, Steppan LB, Koller LD & Exon JH (1980) Immunotoxicology studies of sodium arsenate-effects of exposure on tumor growth and cell-mediated tumor immunity. J Environ Pathol Toxicol, **4**: 65–79.

Keyse SM & Tyrrell RM (1989) Heme oxygenase is the major 32-kDa stress protein induced in human skin fibroblasts by UVA radiation, hydrogen peroxide and sodium arsenite. Proc Natl Acad Sci U S A, **86**: 99–103.

Khangarot BS & Ray PK (1989) Sensitivity of midge larvae of *Chironomus tentans* Fabricius (Diptera Chironomidae) to heavy metals. Bull Environ Contam Toxicol, **42**(3): 325–330.

Khangarot BS, Sehgal A & Bhasin MK (1985) Man and biosphere – studies on the Sikkim Himalayas. Part 5: Acute toxicity of selected heavy metals on the tadpoles of *Rana hexadactyla*. Acta Hydrochim Hydrobiol, **13**(2): 259–263.

Kipling MD (1977) Arsenic. In: Lenihan J, Fletcher WW eds. The chemical environment. Glasgow, Blackie, pp 93–120.

Klaassen CD (1974) Biliary excretion of arsenic in rats, rabbits, and dogs. Toxicol Appl Pharm, **29**: 447–457.

Klemperer NS & Pickart CM (1989) Arsenite inhibits two step in the ubiquitin-dependent proteolytic pathway. J Biol Chem, **32**: 19245–19252.

Klimecki WT, Borchers AH, Egbert RE, Nagle RB, Carter DE & Bowden GT (1997) Effects of acute and chronic arsenic exposure of human-derived keratinocytes in an in vitro human skin equivalent system: a novel model of human arsenicism. Toxicol Vitro, **11**: 89–98.

Klumpp DW (1980) Characteristics of arsenic accumulation by the seaweeds *Fucus spiralis* and *Ascophyllum nodosum*. Mar Biol, **58**: 257–264.

Klumpp DW & Peterson PJ (1979) Arsenic and other trace elements in the waters and organisms of an estuary in SW England. Environ Pollut, **19**: 11–20.

Knauer K, Behra R & Hemond H (1999) Toxicity of inorganic and methylated arsenic to algal communities from lakes along an arsenic contamination gradient. Aquatic Toxicology, **46**(34): 221–230.

Knox S, Langston WJ, Whitfield M, Turner DR & Liddicoat MI (1984) Statistical analysis of estuarine profiles. II. Application to arsenic in the Tamar Estuary (S.W.England). Estuar Coast Mar Sci, **18**: 623–638.

Kochhar TC, Howard W, Hoffmann S & Brammer-Carleton L (1996) Effect of trivalent and pentavalent arsenic in causing chromosome alterations in cultured Chinese hamster ovary (CHO) cells. Toxicol Lett, **84**: 37–42.

Kondo M & Ichikawa I (1994) Different acute effects of oral and intratracheal administration of disodium arsenate and gallium arsenide on heme synthesis in rats. Appl Organomet Chem, **8**(3): 215–221.

Koons RD & Peters CA (1994) Axial distribution of arsenic in individual human hairs by solid sampling graphite furnace AAS. J Anal Toxicol, **18**: 36.

Korte NE & Fernando Q (1991) A review of arsenic (III) in groundwater. CRC Crit Rev Environ Control, **21**(1): 1–39.

Kosnett MJ & Becker CE (1988) Dimercaptosuccinic acid: Utility in acute and chronic Arsenic poisoning. Vet Hum Toxicol, **30**(4): 369.

Kraus RJ & Ganther HE (1989) Synergistic toxicity between arsenic and methylated selenium compounds. Biol Trace Elem Res, **20**: 105–113.

Kreppel H, Bauman JW, Liu J & Klaassen CD (1990) Arsenite induction of metallothionein in mice. Toxicologist, **10**: 81.

Kreppel H, Bauman JW, Liu J, McKim JM & Klaassen CD (1993) Induction of metallothionein by arsenicals in mice. Fundam Appl Toxicol, **20**: 184–189.

Krishnaja AP, Rege MS & Joshi AG (1987) Toxic effects of certain heavy metals (Hg, Cd, Pb, As and Se) on the intertidal crab *Scylla serrata*. Mar Environ Res, **21**(2): 109–119.

Krishnakumari L, Varshney PK, Gajbhiye SN, Govindan K & Nair VR (1983) Toxicity of some metals on the fish *Therapon jarbua* (Forsskal, 1775). India J Mar Sci, **12**: 64–66.

Kristiansen J, Christensen JM, Iversen BS & Sabbioni E (1997) Toxic trace element reference levels in blood and urine: influence of gender and lifestyle factors. Sci Total Environ, **204**(2): 147–60.

Kuehnelt D, Goessler W & Irgolic KJ (1997) Arsenic compounds in terrestrial organisms II: arsenocholine in the mushroom *Amanita muscaria*. Appl Organom Chem, **11**: 459–470.

Kuo TL (1968) Arsenic content of artesian well water in endemic area of chronic arsenic poisoning. Rep Inst Pathol Natl Taiwan Univ, **20**: 7–13.

Kuo TT, Hu S, Lo SK & Chan HL (1997) p53 expression and proliferative activity in Bowen's disease with or without chronic arsenic exposure. Hum Pathol, **28**(7): 786–790.

Kuratsune M, Tokudome S, Shirakusa T, Yoshida M & Tokumitsu Y (1974) Occupational lung cancer among copper smelters. Int J Cancer **15**(13) 552–8.

Kurttio P, Komulainen H, Hakala E, Kahelin H & Pekkanen J (1998) Urinary excretion of arsenic species after exposure to arsenic present in drinking water. Arch Environ Contam Toxicol, **34**: 297–305.

Kurttio P, Pukkala E, Kahelin H, Auvinen A & Pekkanen J (1999) Arsenic concentrations in well water and risk of bladder and kidney cancer in Finland. Environ Health Perspect, **107**: 705–710.

Kusiak RA, Springer J, Ritchie AC & Muller J (1991) Carcinoma of the lung in Ontario gold miners: possible aetiological factors. Br J Ind Med, **48**: 808–817.

Kusiak RA, Ritchie AC, Muller J & Springer J (1993) Mortality from lung cancer in Ontario uranium miners. Br J Ind Med, **50**: 920–928.

Lacayo ML, Cruz A, Calero S, Lacayo J, Fomsgaard I (1992) Total arsenic in water, fish, and sediments from Lake Xolotlán, Managua, Nicaragua. Bull Environ Contam Toxicol, **49**(3): 463–470.

Lagerkvist B, Linderholm H & Nordberg GF (1986) Vasospastic tendency and Raynaud's phenomenon in smelter workers exposed to arsenic. Environ Res, **39**: 465–474.

Lagerkvist BEA, Linderholm H & Nordberg G (1988) Arsenic and Raynaud's phenomenon. Int Arch Occup Environ Health, **60**: 361–364.

Lagunas R (1980) Sugar-arsenate esters: thermodynamics and biochemical behavior. Arch Biochem Biophys, **205**(1): 67–75.

Lai M-S, Hsueh Y-M, Chen C-J, Shyu M-P, Chen S-Y, Kuo T-L, Wu M-M & Tai T-Y (1994) Ingested inorganic arsenic and prevalence of diabetes mellitus. Am J Epidemiol, **139**: 484–492.

Lai VWM, Cullen WR, Harrington CF & Reimer KJ (1998) Seasonal changes in arsenic speciation in *Fucus* species. Appl Organomet Chem, **12**: 243–251.

Laintz KE, Yu JJ & Wai CM (1992) Separation of metal ions with sodium bis(trifluoroethyl)dithio-carbamate chelation and supercritical fluid chromatography. Anal Chem, **64**: 311–315.

Landsberger S & Wu D (1995) The impact of heavy metals from environmental tobacco smoke on indoor air quality as determined by compton suppression neutron activation analysis. Sci Total Environ, **173**(174): 323–337.

Langlois C & Langis R (1995) Presence of airborne contaminants in the wildlife of northern Québec. Sci Total Environ, **160/161**: 391–402.

Langston WJ (1980) Arsenic in U.K. estuarine sediments and its availability to benthic organisms. J Mar Biol Assoc UK, **60**: 869–881.

Langston WJ (1983) The behavior of arsenic in selected United Kingdom estuaries. Can J Fish Aquat Sci, **40** (Suppl 2): 143–150.

Larramendy ML, Popescu NC & DiPaolo JA (1981) Induction by inorganic metal salts of sister chromatid exchanges and chromosome aberrations in human and Syrian hamster cell strains. Environ Mutagen, **3**: 597–606.

Larsen EH (1991) Electrothermal atomic absorption spectrometry of inorganic and organic arsenic species using conventional and fast furnace programmes. J Anal Spectrom, **6**: 375–377.

Larsen EH, Pritzl G & Hansen SH (1993) Arsenic speciation in seafood samples with emphasis on minor constituents – an investigation by high-performance liquid chromatography with inductively coupled plasma mass spectrometric detection. J Anal At Spectrom, **8**: 97–116.

Latarjet R, Galy P, Maret G & Gallois P (1964) Cancers broncho-pulmonaires et intoxication arsenicale chez les vignerons du Beaujolais (Bronchopulmonary cancers and arsenical poisoning among Beaujolais vine-dressers). Mem Acad Chir, **90**: 384–390.

Le XC, Ma M (1997) Speciation of arsenic compounds by using ion-pair chromatography with atomic spectrometry and mass spectrometry detection. J Chromatogr A, **764**: 55–64.

Le XC, Cullen WR & Reimer KJ (1993) Determination of urinary arsenic and impact of dietary arsenic intake.Talanta, **40**(2): 185–193.

Le XC, Cullen WR & Reimer KJ (1994a) Effect of cysteine on the speciation of arsenic by using hydride generation atomic absorption spectrometry. Anal Chim Acta, **285**: 277–285.

Le XC, Cullen WR & Reimer KJ (1994b) Speciation of arsenic compounds in some marine organisms. Environ Sci Technol, **28**: 1598–1604.

Le XC, Cullen WR & Reimer KJ (1994c) Human urinary arsenic excretion after one-time ingestion of seaweed, crab and shrimp. Clin Chem, **40**: 617–624.

Leder A, Kuo A, Cardiff RD, Sinn E & Leder P (1990) v-Ha-*ras* transgene abrogates the initiation step I mouse skin tumorigenesis: Effects of phorbol esters and retinoic acid. Proc Natl Acad Sci U S A, **87**: 9178–9182.

Lee AM & Fraumeni JF Jr (1969) Arsenic and respiratory cancer in man: an occupational study. J Natl Cancer Inst, **42**: 1045–1052.

Lee CK, Low KS & Hew NS (1991) Accumulation of arsenic by aquatic plants. Sci Total Environ, **103**(2/3): 215–227.

Lee TC & Ho IC (1994) Differential cytotoxic effects of arsenic on human and animal cells. Environ Health Perspect, **102**: 101–105.

Lee TC, Huang RY & Jan KY (1985a) Sodium arsenite enhances the cytotoxicity, clastogenicity and 6-thioguanine-resistant mutagenicity of ultraviolet light in Chinese hamster ovary cells. Mutat Res, **148**: 83–89.

Lee TC, Oshimura M & Barrett JC (1985b) Comparison of arsenic-induced cell transformation, cytotoxicity, mutation and cytogenetic effects in Syrian hamster embryo cells in culture. Carcinogenesis, **6**: 1421–1426.

Lee TC, Lee KC, Tzeng YJ, Huang RY & Jan KY (1986a) Sodium arsenite potentiates the clastogenecity and mutagencity of DNA crosslinking agents. Environ Mutagen, **8**: 119–128.

Lee TC, Wang-Wuu S, Huang RY, Lee KCC & Jan KY (1986b) Differential effects of pre- and posttreatment of sodium arsenite on the genotoxicity of methyl methanesulfonate in Chinese hamster ovary cells. Cancer Res, **46**: 1854–1857.

Lee TC, Tanaka N, Lamb PW, Gilmer TM & Barrett JC (1988) Induction of gene amplification by arsenic. Science, **241**: 79–81.

Lee-Feldstein A (1983) Arsenic and respiratory cancer in humans: Follow-up of copper smelter employees in Montana. J Natl Cancer Inst, **70**: 601–609.

Lee-Feldstein A (1986) Cumulative exposure to arsenic and its relationship to respiratory cancer among copper smelter employees. J Occup Med, **28**: 296–302.

Lee-Feldstein A (1989) A comparison of several measures of exposure to arsenic. Matched case-control study of copper smelter employees. Am J Epidemiol, **129**: 112–124.

Leivuori M & Niemistö L (1995) Sedimentation of trace metals in the Gulf of Bothnia. Chemosphere, **31**(8): 3839–3856.

Leland HV & Scudder BC (1990) Trace elements in *Corbicula fluminea* from the San Joaquin River, California. Sci Total Environ, **97/98**: 641–672.

Leoni L & Sartori F (1996) Heavy metals and arsenic in sediments from the continental shelf of the Northern Tyrrhenian/Eastern Ligurian seas. Mar Environ Res, **41**(1): 73–98.

Lerda D (1994) Sister-chromatid exchange (SCE) among individuals chronically exposed to arsenic in drinking water. Mutat Res, **312**: 111–120.

Lerman S & Clarkson TW (1983) The metabolism of arsenite and arsenate by the rat. Fundam Appl Toxicol, **3**: 309–314.

Lerman SA, Clarkson TW & Gerson RJ (1983) Arsenic uptake and metabolism by liver cells is dependent on arsenic oxidation state. Chem Biol Interactions, **45**: 401–406.

Levin-Scherz JK, Patrick JD, Weber FH & Garabedian CJ (1987) Acute arsenic ingestion. Ann Emerg Med, **16**: 702–704.

Levine AJ (1997) p53, the cellular gatekeeper for growth and division. Cell, **88**: 323–331.

Lewis DR, Southwick JW, Ouellet-Hellstrom R, Rench J & Calderon RL (1999) Drinking water arsenic in Utah: a cohort mortality study. Environ Health Perspect, **107**: 359–365.

Li J, Chen SZ, Yang SH, Zhang MX, Yu Z & Sun X (1989) [Distribution and speciation of arsenic in the seawater of the Huanghe River estuary and adjacent areas] (in Chinese) J Ocean Univ Qingdao, **19**(4): 87–96.

Li JH & Rossman TG (1989) Inhibition of DNA ligase activity by arsenite: a possible mechanism of its comutagenesis. Mol Toxicol, **2**: 1–9.

Li JH & Rossman TG (1991) Comutagenesis of sodium arsenite with ultraviolet radiation in Chinese hamster V79 cells. Biol Metals, **4**: 197–200.

Li YM & Broome JD (1999) Arsenic targets tubulins to induce apoptosis in myeloid leukemia cells. Cancer Res, **59**(4): 776–780.

Lima AR, Curtis C, Hammermeister DE, Markee TP, Northcott CE & Brooke LT (1984) Acute and chronic toxicities of arsenic(III) to fathead minnows, flagfish, daphnids, and an amphipod. Arch Environ Contam Toxicol, **13**: 595–601.

Lin T-H, Huang Y-L & Wang M-Y (1998) Arsenic species in drinking water, hair, fingernails, and urine of patients with blackfoot disease. J Toxicol Environ Health A, **53**: 85–93.

Lin ZQ, Schuepp PH, Schemenauer RS & Kennedy GG (1995) Trace metal contamination in and on Balsam fir (*Abies balsamea* (L) Mill.) foliage in Southern Quebec, Canada. Water Air Soil Pollut, **81**(1/2): 175–191.

Lindgren A, Vahter M & Dencker L (1982) Autoradiographic studies on the distribution of arsenic in mice and hamsters administered [74]As-arsenite or -arsenate. Acta Pharm Toxicol, **51**: 253–265.

Lindgren A, Danielsson BRG, Dencker L & Vahter M (1984) Embryotoxicity of arsenite and arsenate: distribution in pregnant mice and monkeys and effects on embryonic cells *in vitro*. Acta Pharm Toxicol, **54**: 311–320.

Lindsay DM & Sanders JG (1990) Arsenic uptake and transfer in a simplified estuarine food chain. Environ Toxicol Chem, **9**(3): 391–395.

Liou SH, Gu TL & Chen CJ (1996) Hypersensitivity to mitomycin C-induced sister chromatid exchange as a biomarker of past exposure to arsenic. Epidemiol Biomarkers Prev, **5**: 103–107.

Liou SH, Lung JC, Chen YH, Yang T, Hsieh LL, Chen CJ & Wu TN (1999) Increased chromosome-type chromosome aberration frequencies as biomarkers of cancer risk in a blackfoot endemic area. Cancer Res, **59**: 1481–1484.

Liu Y, Lopez-Avila V, Zhu JJ, Wiederin DR & Beckert WF (1995) Capillary electrophoresis coupled on-line with inductively coupled plasma mass spectrometry for elemental speciation. Anal Chem, **67**: 2020–2025.

Liu YT & Chen Z (1996) A retrospective lung cancer mortality study of people exposed to insoluble arsenic and radon. Inst Occup Med, **140**: 137–148.

Livesey NT & Huang PM (1981) Adsorption of arsenate by soil and its relation to selected chemical properties and anions. Soil Sci, **131**(2): 88–94.

Lo MC (1975) Report in the investigation of arsenic content of well water in the province of Taiwan. Nan-Tour, Taiwan Provincial Institute of Sanitary Department.

Lo MC, Hsen YC & Lin BK (1977) The second report on the investigation of arsenic content in underground water in Taiwan province. Taichung, Provincial Institute of Environmental Sanitation.

Lopez S, Miyashita Y & Simons SS Jr (1990) Structurally based, selective interaction of arsenite with steroid receptors. J Biol Chem, **265**(27) 16039–16042.

Loppi S & Bargagli R (1996) Lichen biomonitoring of trace elements in a geothermal area (central Italy). Water Air Soil Pollut, **88**(1/2): 177–187.

Lowe TP, May TW, Brumbaugh WG & Kane DA (1985) National contaminant biomonitoring program: Concentrations of seven elements in freshwater fish, 1978–1981. Arch Environ Contam Toxicol, **14**: 363–388.

Lubin JH & Fraumeni JF Jr (2000) Re: 'Does arsenic exposure increase the risk for circulatory disease?'. Am J Epidemiol, **152**: 290–293.

Lubin JH, Pottern LM, Blot WJ, Tokudome S, Stone BJ & Fraumeni JF Jr (1981) Respiratory cancer among copper smelter workers: recent mortality statistics. J Occup Med, **23**: 779–784.

Lubin JH, Pottern LM, Blot WJ, Stone BJ & Fraumeni JF Jr (2000) Respiratory cancer in a cohort of copper smelter workers: results from more than 50 years of follow-up. Am J Epidemiol, **151**: 554–565.

Lüchtrath H (1983) The consequences of chronic arsenic poisoning among Moselle wine growers. Pathoanatomical investigations of post-mortem examinations between 1960 and 1977. J Cancer Res Clin Oncol, **105**: 173–182.

Lugo G, Cassady G & Palmisano P (1969) Acute maternal arsenic intoxication with neonatal death. Ann J Dis Child, **117**: 328–330.

Lunde G (1977) Occurrence and transformation of arsenic in the marine environment. Environ Health Perspect, **19**: 47–52.

Luten JB, Riekwel-Booy G & Rauchbaar A (1982) Occurrence of arsenic in plaice (*Pleuronectes platessa*), nature of organo-arsenic compound present and its excretion by man. Environ Health Perspect, **45**: 165–170.

MAFF UK (1997) Survey of arsenic in food. The eighth report of the steering group on food surveillance, the working party on the monitoring of foodstuffs for heavy metals. Food surveillance paper No. 8. MAFF, UK.

Ma M & Le SC (1998) Effect of arsenosugar ingestion on urinary arsenic speciation. Clin Chem, **44**: 539–550.

Mabuchi K, Lilienfeld AM & Snell LM (1979) Lung cancer among pesticide workers exposed to inorganic arsenicals. Arch Environ Health, **34**: 312–320.

Mabuchi K, Lilienfeld AM & Snell LM (1980) Cancer and occupational exposure to arsenic: a study of pesticide workers. Prev Med, **9**: 51–77.

Macnair MR & Cumbes Q (1987) Evidence that arsenic tolerance in *Holcus lanatus* L. is caused by an altered phosphate uptake system. New Phytol, **107**: 387–394.

Macy JM, Nunan K, Hagen KD, Dixon DR, Harbour PJ, Cahill M & Sly LI (1996) *Chrysiogenes arsenatis* gen. nov., sp. nov., a new arsenate-respiring bacterium isolated from gold mine wastewater. Int J Syst Bacteriol, **46**(4): 1153–1157.

Madany IM, Wahab AAA & Al-Alawi Z (1996) Trace metals concentrations in marine organisms from the coastal areas of Bahrain, Arabian Gulf. Water Air Soil Pollut, **91**(3/4): 233–248.

Maeda S, Nakashima S, Takeshita T & Higashi S (1985) Bioaccumulation of arsenic by freshwater algae and the application to the removal of inorganic arsenic from an aqueous phase. Part II. *Chlorella vulgaris* isolated from arsenic-polluted environment. Sep Sci Technol, **20**(2/3): 153–161.

Maeda S, Kumeda K, Maeda M, Higashi S & Takeshita T (1987a) Bioaccumulation of arsenic by freshwater algae (*Nostoc* sp.) and the application to the removal of inorganic arsenic from an aqueous phase. Appl Organomet Chem, **1**: 363–370.

Maeda S, Wada H, Kumeda K, Onoue M, Ohki A, Higashi S & Takeshita T (1987b) Methylation of inorganic arsenic by arsenic-tolerant freshwater algae. Appl Organomet Chem, **1**: 465–472.

Maeda S, Fujita S, Ohki A, Yoshifuku I, Higashi S & Takeshita T (1988) Arsenic accumulation by arsenic-tolerant freshwater blue-green alga (*Phormidium* sp.). Appl Organomet Chem, **2**: 353–357.

Maeda S, Inoue R, Kozono T, Tokuda T, Ohki A & Takeshita T (1990a) Arsenic metabolism in a freshwater food chain. Chemosphere, **20**(1/2): 101–108.

Maeda S, Ohki A, Miyahara K, Takeshita T & Higashi S (1990b) Growth characteristics and arsenic metabolism of two species of arsenic-tolerant bacteria. Appl Organomet Chem, **4**: 245–250.

Maeda S, Ohki A, Tokuda T & Ohmine M (1990c) Transformation of arsenic compounds in a freshwater food chain. Appl Organomet Chem, **4**: 251–254.

Maeda S, Kusadome K, Arima H, Ohki A & Naka K (1992a) Uptake and excretion of total inorganic arsenic by the freshwater alga *Chlorella vulgaris*. Appl Organomet Chem, **6**: 399–405.

Maeda S, Kusadome K, Arima H, Ohki A & Naka K (1992b) Biomethylation of arsenic and its excretion by the alga *Chlorella vulgaris*. Appl Organomet Chem, **6**: 407–413.

Maeda S, Ohki A, Kusadome K, Kuroiwa T, Yoshifuku I & Naka K (1992c) Bioaccumulation of arsenic and its fate in a freshwater food chain. Appl Organomet Chem, **6**: 213–219.

Maekawa A, Takizawa H & Hayashi Y (1981) Spontaneous tumors of mice, rats and hamsters used for carcinogenicity studies. Mutagens Toxicol, **4**: 22–41.

Maenhaut W (1987) Particle-induced X-ray emission spectrometry: an accurate technique in the analysis of biological environmental and geological samples. Anal Chim Acta, **195**: 125–140.

Magnuson ML, Creed JT & Brockhoff CA (1996) Speciation of arsenic compounds by ion chromatography with inductively coupled plasma mass spectrometry detection utilizing hydride generation with a membrane separator. J Anal Spectrom, **11**: 893–898.

Maher WA (1983) Inorganic arsenic in marine organisms. Mar Pollut Bull, **14**(8): 308–310.

Maher WA (1985a) Arsenic in coastal waters of South Australia. Water Res, **19**(7): 933–934.

Maher WA (1985b) The presence of arsenobetaine in marine animals. Comp Biochem Physiol 80C(1): 199–201.

Maher WA (1988) Arsenic in the marine environment of south Australia. In: Craig PJ & Glocking F ed. The biological alkylation of heavy elements. Proceedings of a conference, 17–18 September 1987, London. Special Publication no.66. Royal Society of Chemistry, Cambridge, pp 120–126.

Maher W & Butler E (1988) Arsenic in the marine environment. Appl Organomet Chem, **2**: 191–214.

Mahieu B, Buchet JP & Lauwerys R (1987) Evolution clinique et biologique d'une intoxication orale aiguë par l'anhydride arsénieux et considérations sur l'attitude thérapeutique. J Toxicol Clin Exp, **7**: 273–278 (in French).

Maitani T, Saito N, Abe M, Uchiyama S & Saito Y (1987) Chemical form-dependent induction of hepatic zinc-thionein by arsenic administration and effect of co-administered selenium in mice. Toxicol Lett, **39**: 63–70.

Mäki-Paakkanen J, Kurttio P, Paldy A & Pekkanen J (1998) Association between the clastogenic effect in peripheral lymphocytes and human exposure to arsenic through drinking water. Environ Mol Mutagenesis, **32**: 301–313.

Mandal BK, Chowdhury TR, Samanta G, Basu GK, Chowdhury PP, Chanda CR, Lodh D, Karan NK, Dhar RK, Tamili DK, Das D, Saha KC & Chakraborti D (1996) Arsenic in groundwater in seven districts of West Bengal, India -- The biggest arsenic calamity in the world. Curr Sci, **70**(11): 976–986.

Mankovska B (1986) Accumulation of As, Sb, S, and Pb in soil and pine forest. Ekológia, **5**(1): 71–79.

Manning BA & Goldberg S (1997) Arsenic(III) and arsenic(V) adsorption on three California soils. Soil Sci, **162**(12) 886–895.

Mannio J, Järvinen O, Tuominen R & Verta M (1995) Survey of trace elements in lake waters of Finnish Lapland using the ICP-MS technique. Sci Total Environ, **160/161**: 433–439.

Marafante E & Vahter M (1987) Solubility, retention, and metabolism of intratracheally and orally administered inorganic arsenic compounds in the hamster. Environ Res, **42**: 72–82.

Marafante E, Bertolero F, Edel J, Pietra R & Sabbioni E (1982) Intracellular interaction and biotransformation of arsenite in rats and rabbits. Sci Total Environ, **24**: 27–39.

Marafante E, Vahter M & Dencker L (1984) Metabolism of arsenocholine in mice, rats and rabbits. Sci Total Environ, **34**: 223–240.

Marafante E, Vahter M & Envall J (1985) The role of the methylation in the detoxication of arsenate in the rabbit. Chem Biol Interact, **56**: 225–238.

Marafante E, Vahter M, Norin H, Envall J, Sandstrom M, Christakopoulos A & Ryhage R (1987) Biotransformation of dimethylarsinic acid in mouse, hamster and man. J Appl Toxicol, **7**: 111–117.

Marsh GM, Stone RA, Esmen NA, Gula MJ, Gause CK, Petersen NJ, Meaney FJ, Rodney S & Prybylski D (1997) A case-control study of lung cancer mortality in six Gila Basin, Arizona smelter towns. Environ Res, **75**: 56–72.

Marsh GM, Stone RA, Esmen NA, Gula MJ, Gause CK, Petersen NJ, Meaney FJ, Rodney S & Prybylski D (1998) A case-control study of lung cancer mortality in four rural Arizona smelter towns. Arch Environ Health, **53**: 15–28.

Martin M, Osborn KE, Billig P & Glickstein N (1981) Toxicities of ten metals to *Crassostrea gigas* and *Mytilus edulis* embryos and *Cancer magister* larvae. Mar Pollut Bull, **12**(9): 305–308.

Martin WE & Nickerson PR (1973) Mercury, lead, cadmium, and arsenic residues in starlings – 1971. Pestic Monit J, **7**(1): 67–72.

Martinez G, Cebrian M & Chamorro G, Jauge P (1983) Urinary uroporphyrin as an indicator of arsenic exposure in rats. Proc West Pharm Soc, **26**: 171–174.

Mass MJ & Wang L (1997) Arsenic alters cytosine methylation patterns of the promoter of the tumor suppressor gene *p53* in human lung cells: a model for a mechanism of carcinogenesis. Mutat Res, **386**: 263–277.

Masscheleyn PH, Delaune RD & Patrick WH (1991a) Effect of redox potential and pH on arsenic speciation and solubility in a contaminated soil. Environ Sci Technol, **25**(8): 1414–1419.

Masscheleyn PH, Delaune RD & Patrick WH (1991b) Heavy metals in the environment: arsenic and selenium chemistry as affected by sediment redox potential and pH. J Environ Qual, **20**(3): 522–527.

Mast TJ, Greenspan BJ, Dill JA, Stoney KH, Evanoff JJ & Rommereim RL (1990) Inhalation developmental toxicology studies: gallium arsenide in mice and rats: final report. Technical Report NTIS/DE91 005300. Springfield, VA, National Technical Information Service.

Mast TJ, Dill JA, Greenspan BJ, Evfanoff JJ, Morrissey RE & Schwetz BA (1991) The developmental toxicity of inhaled gallium arsenide in rodents. Teratology, **43**: 455–461.

Matanoski G, Landau E, Tonascia J, Lazar C, Elliott EA, McEnroe W & King K (1981) Cancer mortality in an industrial area of Baltimore. Environ Res, **25**: 8–28.

Mathieu D, Mathieu-Nolf M, Germain-Alonso M, Neviere R, Furon D & Wattel F (1992) Massive arsenic poisoning – effect of hemodialysis and dimercaprol on arsenic kinetics. Intensive Care Med, **18**: 47–50.

Matisoff G, Khourey CJ, Hall JF, Varnes AW & Strain WH (1982) The nature and source of arsenic in Northeastern Ohio ground water. Ground Water, **20**(4): 446–456.

Matsuto S, Kasuga H, Okumoto H & Takahashi A (1984) Accumulation of arsenic in blue-green alga, *Phormidium* sp. Comp Biochem Physiol, **78C**(2): 377–382.

Matsuto S, Stockton RA & Irgolic KJ (1986) Arsenobetaine in the red crab, *Chionoecetes opilio*. Sci Total Environ, **48**: 133–140.

May TW & McKinney GL (1981) Cadmium, lead, mercury, arsenic, and selenium concentrations in freshwater fish, 1976–77 – National Pesticide Monitoring Program. Pestic Monit J, **15**(1): 14–38.

Mayer FL (1987) Acute toxicity handbook of chemicals to estuarine organisms. PB87–188686. Gulf Breeze, FL, Environmental Research Laboratory, US EPA.

Mayer FL & Ellersieck MR (1986) Manual of acute toxicity: interpretation and data base for 410 chemicals and 66 species of freshwater animals. Resource Publication 160. Washington, DC, U.S. Department of the Interior Fish and Wildlife Service.

Mayes PA (1983) In: Martin, DW, Mayes, PA & Rodwell VW ed. Harper's review of biochemistry, 19th ed. Los Altos, CA, Lange Medical Publishers, p 165.

Mazumder DNG, Haque R, Ghosh N, De BK, Santra A, Chakraborty D & Smith AH (1998) Arsenic levels in drinking water and the prevalence of skin lesions in West Bengal, India. Int J Epidemiol, **27**: 871–877.

McBride BC & Wolfe RS (1971) Biosynthesis of dimethylarsine by methanobacterium. Biogeochemistry, **10**: 4312–4317.

McBride BC, Merilees H, Cullen WR & Pickett W (1978) Anaerobic and aerobic alkylation of arsenic. ACS Symp Ser, **82**: 94–115.

McGeachy SM & Dixon DG (1989) The impact of temperature on the acute toxicity of arsenate and arsenite to rainbow trout (*Salmo gairdneri*). Ecotoxicol Environ Saf, **17**(1): 86–93.

McGeehan SL (1996) Arsenic sorption and redox reactions: Relevance to transport and remediation. J Environ Sci Health, **A31**(9): 2319–2336.

McLaren RG, Naidu R, Smith J & Tiller KG (1998) Fractionation and distribution of arsenic in soils contaminated by cattle dip. J Environ Qual, **27**: 348–354.

McLaren SJ & Kim ND (1995) Evidence for a seasonal fluctuation of arsenic in New Zealand's longest river and the effect of treatment on concentrations in drinking water. Environ Pollut, **90**(1): 67–73.

Mealey J Jr, Brownell GL & Sweet WH (1959) Radioarsenic in plasma, urine, normal tissues, and intracranial neoplasma. Arch Neurol Psychiatr, **81**: 310–320.

Meharg AA & Macnair MR (1990) An altered phosphate uptake system in arsenate-tolerant *Holcus lanatus* L. New Phytol, **116**: 29–35.

Meharg AA & Macnair MR (1991a) The mechanisms of arsenate tolerance in *Deschampsia cespitosa* (L.) Beauv. and *Agrostis capillaris* L. Adaptation of the arsenate uptake system. New Phytol, **119**: 291–297.

Meharg AA & Macnair MR (1991b) Uptake, accumulation and translocation of arsenate in arsenate-tolerant and non-tolerant *Holcus lanatus* L. New Phytol, **117**: 225–231.

Meharg AA & Macnair MR (1992) Suppression of the high affinity phosphate uptake system: a mechanism of arsenate tolerance in *Holcus lanatus* L. J Exp Bot, **43**(249): 519–524.

Meharg AA, Naylor J & Macnair MR (1994) Phosphorus nutrition of arsenate-tolerant and nontolerant phenotypes of velvetgrass. J Environ Qual, **23**(2): 234–238.

Meharg AA, Shore RF & Broadgate K (1998) Edaphic factors affecting the toxicity and accumulation of arsenate in the earthworm *Lumbricus terrestris*. Environ Toxicol Chem, **17**(6): 1124–1131.

Merry RH, Tiller KG & Alston AM (1983) Accumulation of copper, lead and arsenic in some Australian orchard soils. Aust J Soil Res, **21**(4): 549–561.

Merry RH, Tiller KG & Alston AM (1986) The effects of contamination of soil with copper, lead and arsenic on the growth and composition of plants. I. Effects of season, genotype, soil temperature and fertilizers. Plant Soil, **91**(1): 115–128.

Merwin I, Pruyne PT, Ebel JG, Manzell KL & Lisk DJ (1994) Persistence, phytotoxicity, and management of arsenic, lead and mercury residues in old orchard soils of New York State. Chemosphere, 29(6): 1361–1367.

Metcalfe-Smith JL (1994) Influence of species and sex on metal residues in freshwater mussels (Family Unionidae) from the St. Lawrence River, with implications for biomonitoring programs. Environ Toxicol Chem, 13(9): 1433–1443.

Michiels L, Rauwelaert E Van der, Hasselt F Van, Kas K & Merregaert J (1993) fau cDNA encodes a ubiquitin-like-S30 fusion protein and is expressed as an antisense sequence in the Finkel–Biskis–Reilly murine sarcoma virus. Oncogene, 8(9): 2537–2546.

Michnowicz CJ & Weaks TE (1984) Effects of pH on toxicity of As. Cr, Cu. Ni, and Zn to *Selenastrum capricornutum* Printz. Hydrobiologia, 118: 299–305.

Miles AK, Calkins DG & Coon NC (1992) Toxic elements and organochlorines in harbor seals (*Phoca vitulina richardsi*), Kodiak, Alaska, USA. Bull Environ Contam Toxicol, 48(5): 727–732.

Millward GE, Kitts HJ, Ebdon L, Allen JI & Morris AW (1997) Arsenic in the Thames plume, UK. Mar Environ Res, 44(1): 51–67.

Ministerie van Sociale Zaken en Werkgelegenheid (2000) [National list of maximal accepted concentrations 2000] (in Dutch). Den Haag, Sdu Uitgevers.

Minkkinen P & Yliruokanen I (1978) The arsenic distribution in Finnish peat bogs. Kem Kemi, 7/8: 331–335.

Minkowitz S (1964) Multiple carcinomata following ingestion of medicinal arsenic. Ann Intern Med, 61: 296–299.

Mirkes PE & Cornel L (1992) A comparison of sodium arsenite-hyperthermia-induced stress responses and abnormal development in cultured postimplantation rat embryos. Teratology, 46: 251–259.

Mirkes PE, Dogget B & Cornel L (1994) Induction of a heat shock response (HSP 72) in rat embryos exposed to selected chemical teratogens. Teratology, 49: 135–142.

Misaelides P, Samara C, Noli F, Kouimtzis T & Anousis I (1993) Toxic element concentrations in airborne particulate matter in the area of Thessaloniki, Greece. Sci Total Environ, 130/131: 139–146.

Mizuta N, Mizuta M, Ito F, Ito T, Uchida H, Watanabe Y, Akama H, Murakami T, Hayashi F, Nakamura K, Yamaguchi T, Mizuia W, Oishi S & Matsumura H (1956) An outbreak of acute arsenic poisoning caused by arsenic contaminated soy soyce (shoyu): A clinical report of 220 cases. Bull Yamaguchi Med Sch, 4: 131–150.

Mohri T, Hisanaga A & Ishinishi N (1990) Arsenic intake & excretion by Japanese adults: A 7-day duplicate diet study. Food Chem Toxicol, 28: 521–529.

Mok WM & Wai CM (1989) Distribution and mobilization of arsenic species in the creeks around the Blackbird mining district, Idaho. Water Res, **23**(1): 7–13.

Mok WM & Wai CM (1990) Distribution and mobilization of arsenic and antimony species in the Coeur d'Alene River, Idaho. Environ Sci Technol, **24**(1): 102–108.

Mok WM & Wai CM (1994) Mobilization of arsenic in contaminated river waters. In: Nriagu JO ed. Arsenic in the environment: Part I: Cycling and characterization. New York, John Wiley & Sons, pp 99–117.

Moore DF, O'Callaghan CA, Berlyne G, Ogg CS, Davies HA, House IM & Henry JA (1994) Acute arsenic poisoning: Absence of polyneuropathy after treatment with 2,3-dimercaptopropanesulphonate (DMPS). J Neurol Neurosurg Psychiatry, **57**: 1133–1135.

Moore JN, Ficklin WH & Johns C (1988) Partitioning of arsenic and metals in reducing sulfidic sediments. Environ Sci Technol, **22**(4): 432–437.

Moore LE, Warner ML, Smith AH, Kalman D & Smith MT (1996) Use of the fluorescent micronucleus assay to detect the genotoxic effects of radiation and arsenic exposure in exfoliated human epithelial cells. Environ Mol Mutagen, **27**: 176–184.

Moore LE, Smith AH, Hopenhayn-Rich C, Biggs ML, Kalman DA & Smith MT (1997b) Micronuclei in exfoliated bladder cells among individuals chronically exposed to arsenic in drinking water. Cancer Epidemiol Biomarkers Prev, **6**: 31–6.

Moore LE, Smith AH, Hopenhayn-Rich C, Biggs ML, Kalman DA & Smith MT (1997c) Decrease in bladder cell micronucleus prevalence after intervention to lower the concentration of arsenic in drinking water. Cancer Epidemiol Biomarkers Prev, **6**: 1051–6.

Moore MM, Harrington-Brock K & Doerr CL (1997a) Relative genotoxic potency of arsenic and its methylated metabolites. Mutat Res, **386**: 279–290.

Morales KH, Ryan L, Kuo TL, Wu MM & Chen CJ (2000) Risk of internal cancers from arsenic in drinking water. Environ Health Perspect, **108**: 655–661.

Morrissey RE & Mottet NK (1983) Arsenic-induced exencephaly in the mouse and associated lesions occurring during neurulation. Teratology, **28**: 399–411.

Morse DL, Harrington JM, Housworth J, Landrigan PJ & Kelter A (1979) Arsenic exposure in multiple environmental media in children near a smelter. Clin Toxicol, **14**(4): 389–399.

Morton WE & Caron GA (1989) Encephalopathy: an uncommon manifestation of workplace arsenic poisoning? Am J Ind Med, **15**: 1–5.

Mudroch A & Capobianco JA (1979) Effects of mine effluent on uptake of Co, Ni, Cu, As, Zn, Cd, Cr and Pb by aquatic macrophytes. Hydrobiologia, **64**(3): 223–231.

Muir DCG, Wagemann R, Grift NP, Norstrom RJ, Simon M & Lien J (1988) Organochlorine chemical and heavy metal contaminants in white-beaked dolphins (*Lagenorhynchus albirostris*) and pilot whales (*Globicephala melaena*) from the coast of Newfoundland, Canada. Arch Environ Contam Toxicol, **17**(5): 613–629.

Muller WU, Streffer C & Fischer-Lahdo C (1986) Toxicity of sodium arsenite in mouse embryos in vitro and its influence on radiation risk. Arch Toxicol, **59**: 172–175.

Murai T, Iwata H, Otoshi T, Endo G, Horiguchi S & Fukushima S (1993) Renal lesions induced in F344/DuCrj rats by 4-weeks oral administration of dimethylarsinic acid. Toxicol Lett, **66**: 53–61.

Murphy MJ, Lyon LW & Taylor JW (1981) Subacute arsenic neuropathy: clinical and electrophysiological observations. J Neurol Neurosurg Psychiatry, **44**: 896–900.

Muse JO, Tudino MB, d'Huicque L, Troccoli OE & Carducci CN (1989) Atomic absorption spectrometric determination of inorganic and organic arsenic in some marine benthic algae of the southern Atlantic coasts. Environ Pollut, **58**(4): 303–312.

NAS (National Academy of Sciences) (1977) Medical and biological effects of environmental pollutants: arsenic. Washington, DC, National Academy of Sciences.

Naddy RB, La Point TW & Klaine SJ (1995) Toxicity of arsenic, molybdenum and selenium combinations to *Ceriodaphnia dubia*. Environ Toxicol Chem, **14**(2): 329–336.

Nagymajtenyi L, Selypes A & Berencsi G (1985) Chromosomal aberrations and fetotoxic effects of atmospheric arsenic exposure in mice. J Appl Toxicol, **5**: 61–63.

Nakadaira H, Yamamoto M & Katoh K (1995) Arsenic levels in soil of a town polluted 35 years ago (Nakajo, Japan). Bull Environ Contam Toxicol, **55**(5): 650–657.

Nakamura M, Matsuzono Y, Tanaka S & Hashimoto Y (1990) Chemical form of arsenic compounds and distribution of their concentrations in the atmosphere. Appl Organomet Chem, **4**: 223–230.

Naqvi SM & Flagge CT (1990) Chronic effects of arsenic on American red crayfish, *Procambarus clarkii*, exposed to monosodium methanearsonate (MSMA) herbicide. Bull Environ Contam Toxicol, **45**(1): 101–106.

Naqvi SM, Davis VO & Hawkins RM (1985) Percent mortalities and LC$_{50}$ values for selected microcrustaceans exposed to Treflan, Cutrine-plus, and MSMA herbicides. Bull Environ Contam Toxicol, **35**: 127–132.

Naqvi SM, Hawkins R & Naqvi NH (1987) Mortality response and LC$_{50}$ values for juvenile and adult crayfish, *Procambarus clarkii* exposed to Thiodan (insecticide), Treflan, MSMA, Oust (herbicides) and Cutrine-plus (Algicide). Environ Pollut, **48**: 275–283.

Naqvi SM, Flagge CT & Hawkins RL (1990) Arsenic uptake and depuration by red crayfish, *Procambarus clarkii*, exposed to various concentrations of monosodium methanearsonate (MSMA) herbicide. Bull Environ Contam Toxicol, **45**(1): 94–100.

Narang APS & Datta DV (1983) Brain arsenic concentrations in fulminant hepatitis. J Assoc Phys India, **31**: 518–519.

Navarro M, Sánchez M, López H & López MC (1993) Arsenic contamination levels in waters, soils, and sludges in southeast Spain. Bull Environ Contam Toxicol, **50**(3): 356–362.

Neff JM (1997) Ecotoxicology of arsenic in the marine environment. Environ Toxicol Chem, **16**(5): 917–927.

Neiger RD & Osweiler GD (1989) Effect of subacute low level dietary sodium arsenite on dogs. Fundam Appl Toxicol, **13**: 439–451.

Nelson DA, Calabrese A, Nelson BA, MacInnes JR & Wenzloff DR (1976) Biological effects of heavy metals on juvenile bay scallops, *Argopecten irradians*, in short-term exposures. Bull Environ Contam Toxicol, **16**: 275–282.

Nelson WC, Lykins MH, Mackey J, Newill VA, Finklea JF & Hammer DI (1973) Mortality among orchard workers exposed to lead arsenate spray: a cohort study. J Chronic Dis, **26**: 105–118.

Nemec MD, Holson JF, Farr CH & Hood RD (1998) Developmental toxicity assessment of arsenic acid in mice and rabbits. Reprod Toxicol, **12**: 647–658.

Neubauer O (1947) Arsenical cancer: a review. Br J Cancer, **1**: 192–251.

Neumann H, Bode-Kirchhoff A, Madeheim A & Wetzel A (1998) Toxicity testing of heavy metals with the *Rhizobium*-legume symbiosis: High sensitivity to cadmium and arsenic compounds. Environ Sci Pollut Res, **5**(1): 28–36.

Ng JC (1999) Speciation, bioavailability and toxicology of arsenic in the environment. PhD Thesis. University of Queensland,.

Ng JC & Moore MR (1996) Bioavailability of arsenic in soils from contaminated sites using a 96 hour rat blood model. In: Langley A, Markey B & Hill H eds. The health risk assessment and management of contaminated sites. Contaminated Sites Monograph Series. No. 5. South Australia, Commonwealth Department of Human Services and Health and the Environmental Protection Agency, pp 355–363.

Ng JC, Johnson D, Imray P, Chiswell B & Moore M (1998a) Speciation of arsenic metabolites in the urine of occupational workers and experimental rats using an optimised hydride cold-trapping method. Analyst, **123**: 929–933.

Ng JC, Kratzmann SM, Qi L, Crawley H, Chiswell B & Moore MR (1998b) Speciation and absolute bioavailability: risk assessment of arsenic-contaminated sites in a residential suburb in Canberra. Analyst, **123**: 889–892.

Ng JC, Seawright AA, Qi L, Garnett CM, Moore MR & Chiswell B (1998c) Tumours in mice induced by chronic exposure of high arsenic concentration in drinking water. Third International Conference on Arsenic Exposure and Health Effects, 12–15 July 1998, San Diego, CA. p 28.

Ng JC, Seawright AA, Qi L, Garnett CM, Chiswell B & Moore MR (1999) Tumours in mice induced by exposure to sodium arsenate in drinking water. In: Abernathy C, Caldron R & Chappell W eds. Arsenic exposure and health effects. Oxford, Elsevier Science, pp 217–223.

NHMRC (National Health and Medical Research Council) (1996) Australian drinking water guidelines. Canberra, Commonwealth of Australia.

NHW/DOE (National Health & Welfare/Department of the Environment) (1993) Arsenic and its compounds – supporting documentation. Canadian Environmental Protection Act Priority Substances List. Ottawa, National Health & Welfare/Department of the Environment.

Nichols JW, Wedemeyer GA, Mayer FL, Dickhoff WW, Gregory SV, Yasutake WT & Smith SD (1984) Effects of freshwater exposure to arsenic trioxide on the parr-smolt transformation of coho salmon (*Oncorhynchus kisutch*). Environ Toxicol Chem, **3**: 142–149.

Nickson R, McArthur J, Burgess W, Ahmed KM, Ravenscroft P & Rahman M (1998) Arsenic poisoning of Bangladesh groundwater [letter]. Nature, **395**(6700): 338.

Nicolli HB, Suriano JM, Gomez Peral MA, Ferpozzi LH & Baleani OA (1989) Groundwater contamination with arsenic and other trace elements in an area of the Pampa province of Cordoba, Argentina. Environ Geol Water Sci, **14**: 3–16.

Nordenson I & Beckman L (1991) Is the genotoxic effect of arsenic mediated by oxygen free radicals? Hum Hered, **41**: 71–73.

Nordstrom S, Beckman L & Nordenson I (1978a) Occupational and environmental risks in and around a smelter in northern Sweden; I. Variations in birth weight. *Hereditas*, **88**: 43–46.

Nordstrom S, Beckman L & Nordenson I (1978b) Occupational and environmental risks in and around a smelter in northern Sweden; III. Frequencies of spontaneous abortion. *Hereditas*, **88**: 51–54.

Nordstrom S, Beckman L & Nordenson I (1979a) Occupational and environmental risks in and around a smelter in northern Sweden; V. Spontaneous abortion among female employees and decrease birth weight in their offspring. *Hereditas*, **90**: 291–296.

Nordstrom S, Beckman L & Nordenson I (1979b) Occupational and environmental risks in and around a smelter in northern Sweden; VI. Congenital malformations. *Hereditas*, **90**: 297–302.

Norheim G, Skaare JU & Wiig Ø (1992) Some heavy metals, essential elements, and chlorinated hydrocarbons in polar bear (*Ursus maritimus*) at Svalbard. Environ Pollut, **77**(1): 51–57.

Norin H & Christakopoulos A (1982) Evidence for the presence of arsenobetaine and another organoarsenical in shrimps. Chemosphere, **11**(3): 287–298.

Norin H, Vahter M, Christakopoulos A & Sandström M (1985) Concentration of inorganic and total arsenic in fish from industrially polluted water. Chemosphere, **14**(3/4): 325–334.

Norstrom RJ, Schweinsberg RE & Collins BT (1986) Heavy metals and essential elements in livers of the polar bear (*Ursus maritimus*) in the Canadian Arctic. Sci Total Environ, **48**: 195–212.

NRC (National Research Council) (1999) Arsenic in drinking water. Subcommittee on Arsenic in Drinking Water, Committee on Toxicology, Board on Environmental Studies and Toxicology Commission on Life Science. Washington, DC, National Academy Press.

NRCC (National Research Council of Canada) (1978) Effects of arsenic in the Canadian environment. Ottawa, NRCC.

Nunoshiba T & Nishioka H (1987) Sodium arsenite inhibits spontaneous and induced mutations in *Escherichia coli*. Mutat Res, **184**: 99–105.

Nygren O, Nilsson CA & Lindahl R (1992) Occupational exposure to chromium, copper & arsenic during work with impregnated wood in joinery shops. Ann Occup Hyg, **36**: 509–517.

O'Neil P (1990) Arsenic. In: Alloway BJ ed. Heavy metals in soils. Glasgow, Blackie and Sons, pp 83–99.

OSHA (2000) Occupational Safety and Health Agency. http://www.osha-slc.gov/OshStd_data/1910_1018.html.

Oberly TJ, Piper CE & McDonald DS (1982) Mutagenicity of metal salts in the L5178Y mouse lymphoma assay. J Toxicol Environ Health, **9**: 367–376.

Ochi T, Nakajima F, Sakurai T, Kaise T & Oya-Ohta Y (1996) Dimethylarsinic acid causes apoptosis in HL-60 cells via interaction with glutathione. Arch Toxicol, **70**: 815–821.

Odanaka Y, Matano O & Goto S (1980) Biomethylation of inorganic arsenic by the rat and some laboratory animals. Bull Environ Contam Toxicol, **24**: 452–459.

Offergelt JA, Roels H, Buchet JP, Boeckx M & Lauwerys R (1992) Relation between airborne arsenic trioxide and urinary excretion of inorganic arsenic and its methylated metabolites. Br J Ind Med, **49**: 387–393.

Oh SJ (1991) Electrophysiological profile in arsenic neuropathy. J Neurol Neurosurg Psychiatr, **54**: 1103–1105.

Ohlendorf HM, Marois KC, Lowe RW, Harvey TE & Kelly PR (1991) Trace elements and organochlorines in surf scoters from San Francisco Bay, 1985. Environ Monit Assess, **18**(2): 105–122.

Ohyama S, Ishinishi N, Hisanaga A & Yamamoto A (1988) Comparative chronic toxicity, including tumorigenicity, of gallium arsenide & arsenic trioxide intratracheally instilled into hamsters. Appl Organomet Chem, **2**: 333–337.

Okui T & Fujiwara Y (1986) Inhibition of human excision DNA repair by inorganic arsenic and the co-mutagenic effect in V79 Chinese hamster cells. Mutat Res, **172**: 69–76.

Oladimeji AA, Qadri SU & deFreitas ASW (1984) Long-term effects of arsenic accumulation in rainbow trout, *Salmo gairdneri*. Bull Environ Contam Toxicol, **32**: 732–741.

Oladimeji AA, Qadri SU, Tam GKH & DeFreitas ASW (1979) Metabolism of inorganic arsenic to organoarsenicals in rainbow trout (*Salmo gairdneri*). Ecotoxicol Environ Saf, **3**: 394–400.

Olesik JW, Kinzer JA & Olesik SV (1995) Capillary electrophoreses – inductively coupled plasma spectrometry for rapid elemental speciation. Anal Chem, **67**: 1–12.

Omura M, Hirata M, Tanaka A, Zhao M, Makitar Y, Inoue N, Gotoh K & Ishinishi N (1996a) Testicular toxicity evaluation of arsenic-containing binary compound semiconductors, gallium arsenide and indium arsenide, in hamsters. Toxicol Lett, **89**: 123–129.

Omura M, Tanaka A, Hirata M, Zhao M, Makita Y, Inoue N, Gotoh K & Ishinishi N (1996b) Testicular toxicity of gallium arsenide, indium arsenide, and arsenic oxide in rats by repetitive intratracheal instillation. Fundam Appl Toxicol, **32**: 72–78.

Onishi H (1969) Arsenic. In: Wedepohl KH ed. Handbook of Geochemistry, Vol. II-2. New York, Springer.

Onken BM & Hossner LR (1995) Plant uptake and determination of arsenic species in soil solution under flooded conditions. J Environ Qual, **24**(2): 373–381.

Orvini E, Gills TE & LaFleur PD (1974) Method for determination of selenium, arsenic, zinc, cadmium, and mercury in environmental matrices by neutron activaion analysis. Anal Chem, **46**(9): 1294–1297.

Osburn HS (1957) Cancer of the lung in Gwanda. Centr Afr J Med, **3**: 215–223.

Osburn HS (1969) Lung cancer in a mining district in Rhodesia. S Afr Med J, **43**: 1307–1312.

Oscarson DW, Huang PM & Liaw WK (1980) The oxidation of arsenite by aquatic sediments. J Environ Qual, 9(4): 700–703.

Osowski SL, Brewer LW, Baker OE & Cobb GP (1995) The decline of mink in Georgia, North Carolina, and South Carolina: the role of contaminants. Arch Environ Contam Toxicol, **29**(3): 418–423.

Ostrosky-Wegman P, Gonsebatt ME, Montero R, Vega L, Barba H, Espinosa J, Palao A, Cortinas C, Garcia-Vargas G, Del Razo LM & Cebrian M (1991) Lymphocyte proliferation kinetics and genotoxic findings in a pilot study on individuals chronically exposed to arsenic in Mexico. Mutat Res, **250**: 477–482.

Ott MG, Holder BB & Gordon HL (1974) Respiratory cancer and occupational exposure to arsenicals. Arch Environ Health, **29**: 250–255.

Otte ML, Rozema J, Beek MA, Kater BJ & Broekman RA (1990) Uptake of arsenic by estuarine plants and interactions with phosphate, in the field (Rhine Estuary) and under outdoor experimental conditions. Sci Total Environ, **97/98**: 839–854.

Outridge PM & Noller BN (1991) Accumulation of toxic trace elements by freshwater vascular plants. Rev Environ Contam Toxicol, **121**: 1–63.

Oya-Ohta Y, Kaise T & Ochi T (1996) Induction of chromosomal aberrations in cultured human fibroblasts by inorganic and organic arsenic compounds and the different roles of glutathione in such induction. Mutat Res, **357**: 123–129.

Pacyna JM, Bartonova A, Cornille P & Maenhaut W (1989) Modelling of long-range transport of trace elements. A case study. Atmos Environ, **23**(1): 107–114.

Page GW (1981) Comparison of groundwater and surface water for patterns and levels of contamination by toxic substances. Environ Sci Technol, **15**(12): 1475–1481.

Pain DJ, Amiard-Triquet C & Sylvestre C (1992) Tissue lead concentrations and shot ingestion in nine species of waterbirds from the Camargue (France). Ecotoxicol Environ Saf, **24**(2): 217–233.

Palawski D, Hunn JB & Dwyer FJ (1985) Sensitivity of young striped bass to organic and inorganic contaminants in fresh and saline waters. Trans Am Fish Soc, **114**: 748–753.

Paliouris G & Hutchinson TC (1991) Arsenic, cobalt and nickel tolerances in two populations of *Silene vulgaris* (Moench) Garcke from Ontario, Canada. New Phytol, **117**: 449–459.

Parris GE & Brinckman FE (1976) Reactions which relate to environmental mobility of arsenic and antimony. II. Oxidation of trimethylarsine and trimethylstibine. Environ Sci Technol, **10**(12): 1128–1134.

Paschal DC, DiPietro ES, Phillips DL & Gunter EW (1989) Age dependence of metals in hair in a selected U.S. population. Environ Res, **48**: 17–28.

Passino DRM & KrAm JM (1980) Toxicity of arsenic and PCBs to fry of deepwater ciscoes (*Coregonus*). Bull Environ Contam Toxicol, **24**: 527–534.

Passino DRM & Novak AJ (1984) Toxicity of arsenate and DDT to the cladoceran *Bosmina longirostris*. Bull Environ Contam Toxicol, **33**: 325–329.

Pedersen GA, Mortensen GK & Larsen EH (1994) Beverages as a source of toxic trace element intake. Food Addit Contaminants, **14**(3): 351–363.

Peirson DH, Cawse PA & Cambray RS (1974) Chemical uniformity of airborne particulate material, and a maritime effect. Nature, **251**: 675–679.

Penrose WR (1975) Biosynthesis of organic arsenic compounds in brown trout (*Salmo trutta*). J Fish Res Board Can, **32**: 2385–2390.

Penrose WR, Black R & Hayward MJ (1975) Limited arsenic dispersion in sea water, sediments and biota near a continuous source. J Fish Res Board Can, **32**: 1275–1281.

Penrose WR, Conacher HBS, Black R, Méranger JC, Miles W, Cunningham HM & Squires WR (1977) Implications of inorganic/organic interconversion on fluxes of arsenic in marine food webs. Environ Health Perspect, **19**: 53–59.

Peoples SA (1975) Review of arsenical pesticides. In: Woolson EA ed. Arsenical pesticides. Washington, DC, American Chemical Society, pp 1–12.

Pergantis S, Heithmar EM & Hinners TA (1997) Speciation of arsenic animal feed additives by microbore high-performance liquid chromatography with inductively coupled plasma mass spectrometry. Analyst, **122**: 1063–1068.

Pershagen G (1985) Lung cancer mortality among men living near an arsenic-emitting smelter. Am J Epidemiol, **122**: 684–694.

Pershagen G & Bjorklund N-E (1985) On the pulmonary tumorigenicity of arsenic trisulfide and calcium arsenate in hamsters. Cancer Lett, **27**: 99–104.

Pershagen G, Elinder CG & Bolander AM (1977) Mortality in a region surrounding an arsenic emitting plant. Environ Health Perspect, **19**: 133–137.

Pershagen G, Wall S, Taube A & Linnman L (1981) On the interaction between occupational arsenic exposure and smoking and its relationship to lung cancer. Scand J Work Environ Health, **7**: 302–309.

Pershagen G, Lind B & Bjorklund NE (1982) Lung retention and toxicity of some inorganic arsenic compounds. Environ Res, **29**: 425–434.

Pershagen G, Nordberg G & Bjorklund NE (1984) Carcinomas of the respiratory tract in hamsters given arsenic trioxide and/or benzo(a)pyrene by the pulmonary route. Environ Res, **34**: 227–241.

Pershagen G, Bergman F, Klominek J, Damber L & Wall S (1987) Histological types of lung cancer among smelter workers exposed to arsenic. Br J Ind Med, **44**: 454–458.

Peryea FJ & Creger TL (1994) Vertical distribution of lead and arsenic in soils contaminated with lead arsenate pesticide residues. Water Air Soil Pollut, **78**(3/4): 297–306.

Peters RA (1955) Biochemisty of some toxic agents. I. Present state of knowledge of biochemical lesions induced by trivalent arsenical poisoning. Bull Johns Hopkins Hosp, **97**: 1–20.

Peterson ML & Carpenter R (1983) Biogeochemical processes affecting total arsenic and arsenic species distributions in an intermittently anoxic fjord. Mar Chem, **12**: 295–321.

Petrick JS, Ayala-Fierro F, Cullen WR, Carter DE & Aposhian VH (2000) Monomethylarsonous acid (MMAIII) is more toxic than arsenite in Chang human hepatocytes. Toxicol Appl Pharmacol, **168**: 203–207.

Pettine M, Camusso M & Martinotti W (1992) Dissolved and particulate transport of arsenic and chromium in the Po River (Italy). Sci Total Environ, **119**: 253–280.

Phillips DJH (1990) Arsenic in aquatic organisms: A review, emphasizing chemical speciation. Aquat Toxicol, **16**(3): 151–186.

Phillips DJH & Depledge MH (1985) Metabolic pathways involving arsenic in marine organisms: a unifying hypothesis. Mar Environ Res, **17**: 1–12.

Phillips DJH & Depledge MH (1986) Chemical forms of arsenic in marine organisms, with emphasis on *Hemifusus* species. Water Sci Technol, **18**(4/5): 213–222.

Pietenpol JA, Holt JT, Stein RW & Moses HL (1990) Transforming growth factor 1 suppression of c-*myc* gene transcription: role in inhibition of keratinocyte proliferation. Proc Natl Acad Sci U S A, **87**: 9178–9182.

Pinto SS & Bennett BM (1963) Effect of arsenic trioxide exposure on mortality. Arch Environ Health, **7**: 583–591.

Pinto SS, Varner MO, Nelson KW, Labbe AL & White LD (1976) Arsenic trioxide absorption & excretion in industry. J Occup Med, **18**: 677–680.

Pinto SS, Enterline PE, Henderson V & Varner MO (1977) Mortality experience in relation to a measured arsenic trioxide exposure. Environ Health Perspect, **19**: 127–130.

Pinto SS, Henderson V & Enterline PE (1978) Mortality experience of arsenic-exposed workers. Arch Environ Health, **33**: 325–331.

Pollard AJ (1980) Diversity of metal tolerances in *Plantago lanceolata* L. from the southeastern United States. New Phytol, **86**: 109.

471

Pollisar L, Lowry-Coble K, Kalman DA, Hughes JP, Belle G Van, Covert DS, Burvacher TM, Bolgiano D & Mottet NK (1990) Pathways of human exposure to arsenic in a community surrounding a copper smelter. Environ Res, **53**: 29–47.

Pomroy C, Charbonneau SM & McCullough RS, Tam GKH (1980) Human retention studies with ^{74}As. Toxicol Appl Pharm, **53**: 550–556.

Porter EK & Peterson PJ (1975) Arsenic accumulation by plants on mine waste (United Kingdom). Sci Total Environ, **4**: 365–371.

Porter EK & Peterson PJ (1977) Biogeochemistry of arsenic on polluted sites in S. W. England. In: Hemphill DD ed. Trace substances in environmental health. Proceedings of the University of Missouri's 11th annual conference. Columbia, Mo, 7–9 June 1977, Missouri, University of Missouri, pp 89–99.

Prasad G (1994) Removal of arsenic(V) from aqueous systems by adsorption onto some geological materials. In: Nriagu JO ed. Arsenic in the environment: Part I: Cycling and characterization. New York, John Wiley & Sons, pp 133–154.

Presley BJ, Taylor RJ & Boothe PN (1992) Trace metal concentrations in sediments of the Eastern Mississippi Bight. Mar Environ Res, **33**(4): 267–282.

Pretty JR, Bludbaugh EA & Caruso JA (1993) Determination of arsenic (III) and selenium (IV) using an on-line anodic stripping voltammetry flow cell with detection by inductively coupled plasma atomic emission spectrometry and inductively coupled plasma mass spectrometry. Anal Chem, **65**: 3396–3403.

Proudfoot FG, Jackson ED, Hulan HW & Salisbury CDC (1991) Arsanilic acid as a growth promoter for chicken broilers when administered via either feed or drinking water. Can J Anim Sci, **71**: 221–226.

Prukop JA & Savage NL (1986) Some effects of multiple sublethal doses of monosodium methanearsonate (MSMA) herbicide on hematology, growth and reproduction of laboratory mice. Bull Environ Contam Toxicol, **36**: 337–341.

Qiao YL, Taylor PR, Yao SX, Erozan YS, Luo XC, Barrett MJ, Yan QY, Giffen CA, Huang SQ, Maher MM, Forman MR & Tockman MS (1997) Risk factors and early detection of lung cancer in a cohort of Chinese tin miners. Ann Epidemiol, **7**: 533–541.

Rabano ES, Castillo NT, Torre KJ, Solomon PA (1989) Speciation of arsenic in ambient aerosols collected in Los Angeles. J Air Waste Manage Assoc, **39**(1): 76–80.

Rahman M & Axelson O (1995) Diabetes mellitus and arsenic exposure: a second look at case-control data from a Swedish copper smelter. Occup Environ Med, **52**: 773–774.

Rahman M, Wingren G & Axelson O (1996) Diabetes mellitus among Swedish glass workers – an effect of arsenic exposure? Scand J Work Environ Health, **22**: 146–149.

Rahman M, Tondel M, Ahmad SA & Axelson O (1998) Diabetes mellitus associated with arsenic exposure in Bangladesh. Am J Epidemiol, **148**: 198–203.

Rahman M, Tondel M, Ahmad SA, Chowdhury IA, Faruquee MH & Axelson O (1999a) Hypertension and arsenic exposure in Bangladesh. Hypertension, **33**: 74–78.

Rahman M, Tondel M, Chowdhury IA & Axelson O (1999b) Relations between exposure to arsenic, skin lesions, and glucosuria. Occup Environ Med, **56**: 277–281.

Rahman MS & Hughes MF (1994) In vitro percutaneous absorption of monosodium methanarsonate and disodium methanearsonate in female B6C3F1 mice. J Toxicol Environ Health, **41**: 421–433.

Rahman MS, Hall LL & Hughes MF (1994) In vitro percutaneous absorption of sodium arsenate B6C3F1 mice. Toxicol Vitro, **8**: 441–448.

Raie RM (1996) Regional variation in As, Cu, HG, and Se and interaction between them. Ecotoxicol Environ Saf, **35**: 248–252.

Ramirez P, Eastmond DA, Laclette JP & Ostrosky-Wegman P (1997) Disruption of microtubule assembly and spindle formation as a mechanism for the induction of aneuploid cells by sodium arsenite and vanadium pentoxide. Mutat Res, **386**(3): 291–298.

Ramirez-Campos J, Ramos-Peek J, Martinez-Barros M, Zamora-Peralta M & Martinez-Cerrato J (1998) [Peripheral neuropathy caused by acute arsenic poisoning]. (in Spanish) Gac Med Mex, **134**: 241–246.

Rasmussen RE & Menzel DB (1997) Variation in arsenic-induced sister chromatid exchange in human lymphocytes and lymphoblastoid cell lines. Mutat Res, **386**: 299–306.

Ratnam KV, Espy MJ, Muller SA, Smith TF & Su WPD (1992) Clinicopathologic study of arsenic-induced skin lesions: no definite association with human papillomavirus. J Am Acad Derm, **27**: 120–122.

Reay PF (1972) The accumulation of arsenic from arsenic-rich natural waters by aquatic plants. J Appl Ecol, **9**: 557–565.

Reichl FX, Szinicz L, Kreppel H & Forth W (1988) Effect of arsenic on carbohydrate metabolism after single or repeated injection in guinea pigs. Arch Toxicol, **62**: 473–475.

Reichl FX, Szinicz L, Kreppel H, Fichtl B & Forth W (1990) Effect of glucose in mice after acute experimental poisoning with arsenic trioxide (As_2O_3). Arch Toxicol, **64**: 336–338.

Reimann C, Caritat P De, Halleraker JH, Volden T, Äyräs M, Niskavaara H, Chekushin VA & Pavlov VA (1997) Rainwater composition in eight arctic catchments in northern Europe (Finland, Norway and Russia). Atmos Environ, **31**(2): 159–170.

Reimer KJ (1989) The methylation of arsenic in marine sediments. Appl Organomet Chem, **3**(6): 475–490.

Reimer KJ & Thompson JAJ (1988) Arsenic speciation in marine interstitial water. The occurrence of organoarsenicals. Biogeochemistry, 6: 211–237.

Reish DJ (1993) Effects of metals and organic compounds on survival and bioaccumulation in two species of marine gammaridean amphipod, together with a summary of toxicological research on this group. J Nat Hist, 27: 781–794.

Rencher AC, Carter MW & McKee DW (1977) A retrospective epidemiological study of mortality at a large western copper smelter. J Occup Med, 19: 754–758.

Repetto G, Sanz P & Repetto M (1994) Comparative in vitro effects of sodium arsenite and sodium arsenate on neuroblastoma cells. Toxicology, 92: 143–153.

Reuther R (1992) Arsenic introduced into a littoral freshwater model ecosystem. Sci Total Environ, 115(3): 219–237.

Richardson CW, Price JD & Burnett E (1978) Arsenic concentrations in surface runoff from small watersheds in Texas. J Environ Qual, 7(2): 189–192.

Riedel GF (1993) The annual cycle of arsenic in a temperate estuary. Estuaries, 16(3A): 533–540.

Riedel GF, Sanders JG & Osman RW (1987) The effect of biological and physical disturbance on the transport of arsenic from contaminated estuarine sediments. Estuar Coast Mar Sci, 25: 693–706.

Riedel GF, Sanders JG & Osman RW (1989) The role of three species of benthic invertebrates in the transport of arsenic from contaminated estuarine sediment. J Exp Mar Biol Ecol, 134: 143–155.

Rin K, Kawaguchi K, Yamanaka K, Tezuka M, Oku N & Okada S (1995) DNA-strand breaks induced by dimethylarsinic acid a metabolite of inorganic arsenics, are strongly enhanced by superoxide anion radicals. Biol Pharm Bull, 18: 45–48.

Rivara MI, Cebrian M, Corey G, Hernandez M & Romieu I (1997) Cancer risk in an arsenic-contaminated area of Chile. Toxicol Ind Health, 13: 321–338.

Robertson FN (1989) Arsenic in ground-water under oxidizing conditions, south-west United States. Environ Geochem Health, 11: 171–185.

Robertson JL & McLean JA (1985) Correspondence of the LC_{50} for arsenic trioxide in a diet-incorporation experiment with the quantity of arsenic ingested as measured by X-ray, energy-dispersive spectrometry. J Econ Entomol, 78: 1035–1036.

Rocovich SE & West DA (1975) Arsenic tolerance in a population of the grass *Andropogon scoparius* Michx. Science, 188: 263–264.

Roels H, Buchet J-P, Truc J, Croquet F & Lawerys R (1982) The possible role of direct ingestion on the overall absorption of cadmium or arsenic in workers exposed to CdO and As₂O₃ dust. Am J Ind Med, 3: 53–65.

Rogers EH, Chernoff N & Kavlock RJ (1981) The teratogenic potential of cacodylic acid in the rat and mouse. Drug Chem Toxicol, 4: 49–61.

Rom WN, Varley G, Lyon JL & Shopkow S (1982) Lung cancer mortality among residents living near the El Paso smelter. Br J Ind Med, 39: 269–272.

Romo-Kröger CM & Llona F (1993) A case of atmospheric contamination at the slopes of the Los Andes mountain range. Atmos Environ, 27A(3): 401–404.

Romo-Kröger CM, Morales JR, Dinator MI, Llona F & Eaton LC (1994) Heavy metals in the atmosphere coming from a copper smelter in Chile. Atmos Environ, 28(4): 705–711.

Rosemarin A, Notini M & Holmgren K (1985) The fate of arsenic in the Baltic Sea *Fucus vesiculosus* ecosystem. Ambio, 14(6): 342–345.

Rosen BP (1995) Resistance mechanisms to arsenicals and antimonials. J Basic Clin Physiol Pharm, 6: 251–263.

Rosenstein R, Peschel A, Wieland B & Gotz F (1992) Expression and regulation of the antimonite, arsenite, and arsenate resistance operon of *Staphylococcus xylosus* plasmid pSX267. J Bacteriol, 174(11) 3676–3683.

Rosner MH & Carter DE (1987) Metabolism and excretion of gallium arsenide and arsenic oxides by hamsters following intratracheal instillation. Fundam Appl Toxicol, 9: 730–737.

Rossman TG (1981) Enhancement of uv-mutagenesis by low concentrations of arsenic in *E. coli.* Mutat Res, 91: 207–211.

Rossman TG & Wolosin D (1992) Differential susceptibility to carcinogen-induced amplification of SV40 and dhfr sequences in SV40-transformed human keratinocytes. Mol Carcinog, 6: 203–213.

Rossman TG & Wang ZL (1999) Expression cloning for arsenite resistance resulted in isolation of tumor suppressor fau cDNA: possible involvement of the ubiquitin system in arsenic carcinogenesis. Carcinogenesis, 20(2): 311–316.

Rossman TG, Stone D, Molina M & Troll W (1980) Absence of arsenite mutagenicity in *E. coli* and Chinese hamster cells. Environ Mutagen, 2: 371–379.

Rossman TG, Goncharova EJ, Rajah & Wang Z (1997) Human cells lack the inducible tolerance to arsenite seen in hamsters. Mutat Res, 386: 307–314.

Roth F (1955) Haemangioendothelioma of the liver after chronic arsenic intoxication. Zent Allg Pathol, 93: 424–425.

Roth F (1957a) Arsenic-liver tumours (haemangioendothelioma). Z Krebsforsch, **61**: 468–503.

Roth F (1957b) [Late consequencies of chronic arsenicism in Moselle vine-dressers] (in German). Dtsch Med Wochenschr, **82**: 211–217.

Roth F (1958) [On bronchial carcinoma in arsenic-exposed vineyard workers] (in German). Virchows Arch A Pathol Anat Histopathol, **331**: 119–137.

Rowland IR & Davies MJ (1981) In vitro metabolism of inorganic arsenic by the gastrointestinal microflora of the rat. J Appl Toxicol, **1**: 278–283.

Roy WR, Hassett JJ & Griffin RA (1986) Competitive coefficients for the adsorption of arsenate, molybdate, and phosphate mixtures by soils. Soil Sci Soc Am J, **50**(5): 1176–1182.

RoyChoudhury A, Das T, Sharma A & Talukder G (1996) Dietary garlic extract in modifying clastogenic effects of inorganic arsenic in mice: two-generation studies. Mutat Res, **359**: 165–170.

Rutishauser U, Acheson A, Hall AK & Sunchine J (1988) N-CAM as a regulator of cell–cell interactions. Science, **240**: 53–57.

Sabbioni E, Fischbach M, Possi G, Pietra R, Gallorini M & Piette JL (1991) Cellular retention, toxicity and carcinogenic potential of seafood arsenic. I. Lack of cytotoxicity and transforming activity of arsenobetaine in the BALB/3T3 cell line. Carcinogenesis, **12**: 1287–1291.

Sadana RS (1983) Determination of arsenic in the presence of copper by differential pulse cathodic stripping voltammetry at a hanging mercury drop electrode. Anal Chem, **55**: 304–307.

Sadiq M (1990) Arsenic chemistry in marine environments: a comparison between theoretical and field observations. Mar Chem, **31**: 285–297.

Sadiq M (1997) Arsenic chemistry in soils: an overview of thermodynamic predictions and field observations. Water Air Soil Pollut, **93**: 117–136.

Sadiq M, Zaidi TH & Mian AA (1983) Environmental behavior of arsenic in soils: theoretical. Water Air Soil Pollut, **20**(4): 369–377.

Sadler R, Olszowy H, Shaw G, Biltoft R & Connell D (1994) Soil and water contamination by arsenic from a tannery waste. Water Air Soil Pollut, **78**(1/2): 189–198.

Saffiotti U & Bertolero F (1989) Neoplastic transformation of BALB/3T3 cells by metals and the quest for induction of a metastatic phenotype. Biol Trace Elem Res, **21**: 475–482.

Saiki MK & May TW (1988) Trace element residues in bluegills and common carp from the lower San Joaquin River, California, and its tributaries. Sci Total Environ, **74**: 199–217.

Saiki MK & Palawski DU (1990) Selenium and other elements in juvenile striped bass from the San Joaquin Valley and San Francisco Estuary, California. Arch Environ Contam Toxicol, **19**(5): 717–730.

Sakata M (1987) Relationship between adsorption of arsenic(III) and boron by soil and soil properties. Environ Sci Technol, **21**(11): 1126–1130.

Sandberg GR & Allen IK (1975) A proposed arsenic cycle in an agronomic ecosystem. ACS Symp Ser **7**. Washington, DC, American Chemical Society, pp 124–147.

Sanders JG (1979a) The concentration and speciation of arsenic in marine macro-algae. Estuar Coast Mar Sci, **9**: 95–99.

Sanders JG (1979b) Effects of arsenic speciation and phosphate concentration on arsenic inhibition of *Skeletonema costatum* (Bacillariophyceae). J Phycol, **15**: 424–428.

Sanders JG (1980) Arsenic cycling in marine systems. Mar Environ Res, **3**: 257–266.

Sanders JG (1986) Direct and indirect effects of arsenic on the survival and fecundity of estuarine zooplankton. Can J Fish Aquat Sci, **43**(3): 694–699.

Sanders JG & Cibik SJ (1985) Adaptive behavior of euryhaline phytoplankton communities to arsenic stress. Mar Ecol Progr Ser, **22**: 199–205.

Sanders JG & Cibik SJ (1988) Response of Chesapeake Bay phytoplankton communities to low levels of toxic substances. Mar Pollut Bull, **19**(9): 439–444.

Sanders JG & Osman RW (1985) Arsenic incorporation in a salt marsh ecosystem. Estuar Coast Mar Sci, **20**: 387–392.

Sanders JG & Riedel GF (1987) Control of trace element toxicity by phytoplankton. Recent Adv Phytochem, **21**: 131–149.

Sanders JG & Vermersch PS (1982) Response of marine phytoplankton to low levels of arsenate. J Plankton Res, **4**(4): 881–893.

Sanders JG & Windom HL (1980) The uptake and reduction of arsenic species by marine algae. Estuar Coast Mar Sci, **10**: 555–567.

Sanders JG, Osman RW & Riedel GF (1989) Pathways of arsenic uptake and incorporation in estuarine phytoplankton and the filter-feeding invertebrates *Eurytemora affinis*, *Balanus improvisus* and *Crassostrea virginica*. Mar Biol, **103**(3): 319–325.

Sanderson KV (1963) Arsenic and skin cancer. Trans St John's Hosp Dermatol Soc, **49**: 115–122.

Sandström A & Wall S (1993) Cancer incidence among male salaried employees at a smeltery in northern Sweden. Acta Oncol, **32**: 9–14.

Sandström AIM, Wall SGI & Taube A (1989) Cancer incidence and mortality among Swedish smelter workers. Br J Ind Med, **46**: 82–89.

Sandström AM & Wall SG (1992) Continued surveillance of cancer incidence among Swedish smelter workers. Acta Oncol, **31**: 11–17.

Sanford RA & Klein DA (1988) Environmental bioremediation for organometallic compounds: microbial growth and arsenic volatilization from soil and retorted shale. Appl Organomet Chem, **2**(2): 159–169.

Sanok WJ, Ebel JG, Manzell KL, Gutenmann WH & Lisk DJ (1995) Residues of arsenic and lead in potato soils on Long Island. Chemosphere, **30**(4): 803–806.

Sapunar-Postruznik J, Bazulic D & Kubala H (1996) Estimation of dietary intake of arsenic in the general population of the Republic of Croatia. Sci Total Environ, **191**: 119–123.

Sarkar A & Jana S (1986) Heavy metal pollutant tolerance of *Azolla pinnata*. Water Air Soil Pollut, **27**(1/2): 15–18.

Savabieasfahani M, Lochmiller RL, Rafferty DP & Sinclair JA (1998) Sensitivity of wild cotton rats (Sigmodon hispidus) to the immunotoxic effects of low-level arsenic exposure. Arch Environ Contam Toxicol, **34**: 289–296.

Schaefer ED & Pipes WO (1973) Temperature and the toxicity of chromate and arsenate to the rotifer *Philodina roseola*. Water Res, **7**: 1781–1790.

Schaller KH, Schiele R & Valentin H (1982) [Questions on the action of arsenic in mixing, melting and charging in the lead crystal industry] (in German). Arbeitsmedizinische Untersuchungen zur Einwirkung von Fremdstoffen und Schadstoffen in der keramischen und Glas-Industrie [Medical study at work of the action of impurities and hazardous substances in the ceramics and glass industry] No. 31. Würzburg, Berufsgenossenschaft der keramischen und Glas-Industrie.

Schaller KH, Schiele R & Valentin H (1982) [Questions on the action of arsenic in mixing, melting and charging in the lead crystal industry] (in German). No. 31 [Professional Association of the Ceramics and Glass Industry], Würzburg, pp.

Schmitt CJ & Brumbaugh WG (1990) National Contaminant Biomonitoring Program: concentrations of arsenic, cadmium, copper, lead, mercury, selenium, and zinc in U.S. freshwater fish, 1976–1984. Arch Environ Contam Toxicol, **19**(5): 731–747.

Schrauzer GN, Seck JA, Holland RJ, Beckham TM, Rubin EM & Sibert JW (1972) Reductive dealkylation of alkylcobaloximes, alkylcobalamins, and related compounds: Stimulation of corrin dependent reductase and methyl group transfer reactions. Bioinorg Chem, **2**: 93–124.

Schroeder WH, Dobson M, Kane DM & Johnson ND (1987) Toxic trace elements associated with airborne particulate matter: a review. J Air Pollut Control Assoc 37(11) 1267–1285.

Schults DW, Ferraro SP, Ditsworth GR & Sercu KA (1987) Selected chemical contaminants in surface sediments of Commencement Bay and the Tacoma Waterways, Washington, USA. Mar Environ Res, 22(4): 271–295.

Schweizer EE (1967) Toxicity of DSMA soil residues to cotton and rotational crops. Weeds, 15: 72–76.

Scott N, Hatelid KM, MacKenzie NE & Carter DE (1993) Reactions of arsenic(III) and arsenic(V) species with glutathione. Chem Res Toxicol, 6: 102–106.

Scudlark JR & Church TM (1988) The atmospheric deposition of arsenic and association with acid precipitation. Atmos Environ, 22(5): 937–943.

Scudlark JR & Johnson DL (1982) Biological oxidation of arsenite in seawater. Estuar Coast Mar Sci, 14: 693–706.

Scudlark JR, Conko KM & Church TM (1994) Atmospheric wet deposition of trace elements to Chesapeake Bay: CBAD study year 1 results. Atmos Environ, 28(8): 1487–1498.

Seddon HR (1951) Diseases of domestic animals in Australia, part 3, tick and mite infestations (also animals, miscellaneous insects, etc. harmful to stock). Commonwealth of Australia, Department of Health, Service Publication (Division of Veterinary Hygiene), 7: 1–200. Canberra, Commonwealth Government Printer.

Sen AK & De AK (1987) Adsorption of arsenic on coal fly ash. Indian J Technol, 25(6): 259–261.

Seyler P & Martin JM (1989) Biogeochemical processes affecting arsenic species distribution in a permanently stratified lake. Environ Sci Technol, 23(10): 1258–1263.

Seyler P & Martin JM (1990) Distribution of arsenite and total dissolved arsenic in major French estuaries: dependence on biogeochemical processes and anthropogenic inputs. Mar Chem, 29(2/3): 277–294.

Shacklette HT & Boerngen JG (1984) Element concentrations in soils and other surficial materials of the conterminous United States. Professional Paper 1270. Washington, DC, US Geological Survey.

Shah PV, Fisher HL, Sumler MR, Monroe RJ, Chernoff N & Hall LL (1987) Comparison of the penetration of 14 pesticides through the skin of young and adult rats. J Toxicol Environ Health, 21: 353–366.

Sharif AKM, Alamgir M, Krishnamoorthy KR & Mustafa AI (1993) Determination of arsenic, chromium, mercuryl, selenium and zinc in tropical marine fish by neutron activation. J Radioanal Nucl Chem, **170**(2): 299–307.

Sharples JM, Meharg AA, Chambers SM & Cairney JWG (1999) Arsenate sensitivity in ericoid and ectomycorrhizal fungi. Environ Toxicol Chem, **18**(8): 1848–1855.

Sheely JW & Jones JH (1993) Assessment of arsenic exposures and congtrols in gallium arsenide production. Am Ind Hyg Assoc J, **54**: 61–69.

Sheppard SC (1992) Summary of phytotoxic levels of soil arsenic. Water Air Soil Pollut, **64**(3/4): 539–550.

Shibata A, Ohneseit PF, Tsai YC, Spruck CH, Nichols PW, Chiang H, Lai MK & Jones PA (1994) Mutational spectrum in the p53 gene in bladder tumours from the endemic area of black foot disease in Taiwan. Carcinogenesis, **15**(6): 1085–7.

Shibata Y, Sekiguchi M, Otsukit A & Morita M (1996) Arsenic compounds in zoo- and phytoplankton of marine origin. Appl Organomet Chem, **10**: 713–719.

Shinagawa A, Shiomi K, Yamanaka H & Kikuchi T (1983) Selective determination of inorganic arsenic (III), (V) and organic arsenic in marine organisms. Bull Jap Soc Sci Fish, **49**(1): 75–78.

Shiomi K, Shinagawa A, Igarashi T, Yamanaka H & Kikuchi T (1984) Evidence for the presence of arsenobetaine as a major arsenic compound in the shrimp *Sergestes lucens*. Experientia, **40**: 1247–1248.

Shiomi K, Kakehashi Y, Yamanaka H & Kikuchi T (1987) Identification of arsenobetaine and a tetramethylarsonium salt in the clam *Meretrix luseria*. Appl Organomet Chem, **1**: 177–183.

Shiomi K, Aoyama M, Yamanaka H & Kikuchi T (1988a) Chemical forms of arsenic in sponges, sea anemones and sea hare. Comp Biochem Physiol, **90**: 361–366.

Shiomi K, Horiguchi Y & Kaise T (1988b) Acute toxicity and rapid excretion in urine of tetramethylarsonium salts found in some marine animals. Appl Organomet Chem, **2**: 385–389.

Shotyk W (1996) Natural and anthropogenic enrichments of As, Cu, Pb, Sb, and Zn in ombrotrophic versus minerotrophic peat bog profiles, Jura Mountains, Switzerland. Water Air Soil Pollut, **90**(3/4): 375–405.

Shukla JP, Shukla KN & Dwivedi UN (1987) Survivality and impaired growth in arsenic treated fingerlings of *Channa punctatus*, a fresh water murrel. Acta Hydrochim Hydrobiol, **15**(3): 307–311.

Shum SCK, Neddersen R & Houk RS (1992) Elemental speciation by liquid chromatography-inductively coupled plasma mass spectrometry with direct injection nebulization. Analyst, **117**: 577–582.

Sikorski EE, McCay JA, White KL, Bradley SG & Munson AE (1989) Immunotoxicity of the semiconductor gallium arsenide in female B6C3F1 mice. Fundam Appl Toxicol, **13**: 843–858.

Sikorski EE, Burns LA, McCoy KL, Stern M & Munson AE (1991a) Suppression of splenic accessory cell function in mice exposed to gallium arsenide. Toxicol Appl Pharm, **110**: 143–156.

Sikorski EE, Burns LA, Stern ML, Luster MI & Munson AE (1991b) Splenic cell targets in galium arsenide-induced suppression of the primary antibody response. Toxicol Appl Pharm, **110**: 129–142.

Simonato L, Moulin JJ, Javelaud B, Ferro G, Wild P, Winkelman R & Saracci R (1994) A retrospective mortality study of workers exposed to arsenic in a gold mine and refinery in France. Am J Ind Med, **25**: 625–633.

Simons SS Jr, Chakraborti PK & Cavanaugh AH (1990) Arsenite and cadmium(II) as probes of glucocorticoid receptor structure and function. J Biol Chem, **265**(4): 1938–45.

Singh DB, Prasad G, Rupainwar DC & Singh VN (1988) As(III) removal from aqueous solution by adsorption. Water Air Soil Pollut, **42**(3/4): 373–386.

Siu KWM, Guevremont R, Le Blanc JCY, Gardner GJ & Berman SS (1991) Electrospray interfacing for the coupling of ion-exchange and ion-pairing chromatography to mass spectrometry. J Chromatogr, **554**: 27–38.

Skaare JU, Markussen NH, Norheim G, Haugen S & Holt G (1990) [Levels of polychlorinated biphenyls, organochlorine pesticides, mercury, cadmium, copper, selenium, arsenic, and zinc in the harbour seal, *Phoca vitulina*, in Norwegian waters]. (in Norwegian) Environ Pollut, **66**(4): 309–324.

Slejkovec Z, vanElteren JT & Byrne AR (1998) A dual arsenic speciation system combining liquid chromatographic and purge and trap gas chromatographic separation with atomic fluorescence spectrometric detection. Anal Chim Acta, **20**(358): 51–60.

Smith A, Goycolea M, Haque R & Biggs ML (1998) Marked increase in bladder and lung cancer mortality in a region of Northern Chile due to arsenic in drinking water. Am J Epidemiol, **147**: 660–669.

Smith AH, Arroyo AP, Mazumder DN, Kosnett MJ, Hernandez AL, Beeris M, Smith MM & Moore LE (2000) Arsenic-induced skin lesions among Atacameno people in Northern Chile despite good nutrition and centuries of exposure. Environ Health Perspect, **108**: 617–620.

Smith RA, Alexander RB & Wolman MG (1987) Water-quality trends in the nation's rivers. Science, **235**: 1607–1615.

Smith TJ, Crecelius EA & Reading JC (1977) Airborne arsenic exposure and excretion of methylated arsenic compounds. Environ Health Perspect, **19**: 89–93.

Snyder JC (1935) Crops planted in pulled orchards. Proc Wash State Hortic Assoc, **31**: 48–54.

Sobel W, Bond GG, Baldwin CL & Ducommun DJ (1988) An update of respiratory cancer and occupational exposure to arsenicals. Am J Ind Med, **13**: 263–270.

Soignet SL, Maslak P, Wang ZG, Jhanwar S, Calleja E, Dardashti LJ, Corso D, DeBlasio A, Gabrilove J, Scheinberg DA, Pandolfi PP & Warrell RP Jr (1998) Complete remission after treatment of acute promyelocytic leukemia with arsenic trioxide. N Engl J Med, **339**(19): 1341–1348.

Solomon PA, Altshuler SL & Keller ML (1993) Arsenic speciation in atmospheric aerosols at The Geysers. J Air Waste Manage Assoc, **43**(5): 765–768.

Sommers SC & McManus RG (1953) Multiple arsenical cancers of skin and internal organs. Cancer, **6**: 347–359.

Sonderegger JL & Ohguchi T (1988) Irrigation related arsenic contamination of a thin, alluvial aquifer, Madison River Valley, Montana, USA. Environ Geol Water Sci, **11**(2): 153–161.

Sorensen EMB (1976) Toxicity and accumulation of arsenic in green sunfish (*Lepomis cyanellus*) exposed to arsenate in water. Bull Environ Contam Toxicol, **15**(6): 756–761.

Sorensen EMB, Henry RE & Ramirez-Mitchell R (1979) Arsenic accumulation, tissue distribution and cytotoxicity in teleosts following indirect aqueous exposures. Bull Environ Contam Toxicol, **21**: 162–169.

Sorensen EMB, Ramirez-Mitchell R, Pradzynski A, Bayer TL & Wenz LL (1985) Stereological analyses of hepatocyte changes parallel arsenic accumulation in the livers of green sunfish. J Environ Pathol Toxicol Oncol, **6**(2): 195–210.

Spehar RL & Fiandt JT (1986) Acute and chronic effects of water quality criteria-based metal mixtures on three aquatic species. Environ Toxicol Chem, **5**: 917–931.

Spehar RL, Fiandt JT, Anderson RL & DeFoe DL (1980) Comparative toxicity of arsenic compounds and their accumulation in invertebrates and fish. Arch Environ Contam Toxicol, **9**: 53–63.

Stanley TR, Spann JW, Smith GJ & Rosscoe R (1994) Main and interactive effects of arsenic and selenium on mallard reproduction and duckling growth and survival. Arch Environ Contam Toxicol, **26**(4): 444–451.

Steevans DR, Walsh LM & Keeney DR (1972) Arsenic phytotoxicity on a Plainfield sand as affected by ferric sulfate or aluminum sulfate. J Environ Qual, 1(3): 301–303.

Steinnes E, Solberg W, Petersen HM & Wren CD (1989) Heavy metal pollution by long range atmospheric transport in natural soils of Southern Norway. Water Air Soil Pollut, 45(3/4): 207–218.

Stevens JT, Hall LL, Farmer JD, DiPasquale LC, Chernoff N & Durham WF (1977) Disposition of cacodylic acid in rats after intravenous, intratracheal, or per oral administration. Environ Health Perspect, 19: 151–157.

Stevens JT, DiPasquale LC & Farmer JD (1979) The acute inhalation toxicology of the technical grade organoarsenical herbicides, cacodylic acid and disodium methanearsonic acid; a route comparison. Bull Environ Contam Toxicol, 21: 304–311.

Stilwell DE & Gorny KD (1997) Contamination of soil with copper, chromium, and arsenic under decks built from pressure treated wood. Bull Environ Contam Toxicol, 58: 22–29.

Stocker R, Glazer AN & Ames BN (1987) Antioxidant activity of albumin-bound bilirubin. Proc Natl Acad Sci USA, 84: 5918–5922.

Stoeppler M, Burow M, Backhaus F, Schramm W & Nürnberg HW (1986) Arsenic in seawater and brown algae of the Baltic and the North Sea. Mar Chem, 18(2–4): 321–334.

Stone M & Marsalek J (1996) Trace metal composition and speciation in street sediment: Sault Ste. Marie. Can Water Air Soil Pollut, 87(1–4): 149–169.

Stronkhorst J (1992) Trends in pollutants in blue mussel *Mytilus edulis* and flounder *Platichthys flesus* from two Dutch estuaries, 1985–1990. Mar Pollut Bull, 24(5): 250–258.

Styblo M & Thomas DJ (1995) In vitro inhibition of glutathione reductase by arsenotriglutathione. Biochem Pharm, 49: 971–977.

Styblo M & Thomas DJ (1997) Binding of arsenicals to proteins in an in vitro methylation system. Toxicol Appl Pharm, 147: 1–8.

Styblo M, Delnomdedieu M & Thomas DJ (1995) Biological mechanisms and toxicological consequences of the methylation of arsenic. In: Goyer RA & Cherian MG ed. Toxicology of metals – biochemical aspects. Berlin, Springer, pp 407–433.

Styblo M, Delnomdedieu M & Thomas DJ (1996) Mono- and dimethylation of arsenic in rat liver cytosol in vitro. Chem Biol Interact, 99: 147–164.

Styblo M, Serves SV, Cullen WR & Thomas DJ (1997) Comparative inhibition of yeast glutathione reductase by arsenicals and arsenothiols. Chem Res Toxicol, 10: 27–33.

Styblo M, Del Razo LM, LeCluyse EL, Hamilton GA, Wang C, Cullen WR & Thomas DJ (1999) Metabolism of arsenic in primary cultures of human and rat hepatocytes. Chem Res Toxicol, 12(7): 560–565.

Styblo M, Razo LM Del, Vega L, Germolec DR, LeCluyse EL, Hamilton GA, Reed W, Wang C, Cullen WR & Thomas DJ (2000) Comparative toxicity of trivalent and pentavalent inorganic and methylated arsenicals in rat and human cells. Arch Toxicol, 74: 289–299.

Subra I, Hubert G, Aubert S, Héry M & Elcsbache JM (1999) Exposition professionnelle aux métaux lors de l'usinage des bois traités au cuivre, chrome, arsenic. INRS Cahiers de notes documentaire – Hygiene et sécurité du travail No. 175. Vandoeuvre, France, Centre de recherche de l'INRS.

Suzuki K, Wakao N, Sakurai Y, Kimura T, Sakka K & Ohmiya K (1997) Transformation of *Escherichia coli* with a large plasmid of *Acidiphilium multivorum* AIU 301 encoding arsenic resistance. Appl Environ Microbiol, 63(5): 2089–2091.

Swiggart RC, Whitehead CJ, Curley A & Kellogg FE (1972) Wildlife kill resulting from the misuse of arsenic acid herbicide. Bull Environ Contam Toxicol, 8(2): 122–128.

Szinicz L & Forth W (1988) Effects of As_0O_3 on gluconeogenesis. Arch Toxicol, 61: 444–449.

Tabacova S, Little RE, Balabaeva L, Pavlova S & Petrov I (1994a) Complications of pregnancy in relation to maternal lipid peroxides, glutathione, and exposure to metals. Reprod Toxicol, 8: 217–224.

Tabacova S, Baird DD, Balabaeva L, Lolova D & Petrov I (1994b) Placental arsenic and cadmium in relation to lipid peroxides and glutathione levels in maternal-infant pairs from a copper smelter area. *Placenta*, 15: 873–881.

Tabacova S, Hunter ES III & Gladen BC (1996) Developmental toxicity of inorganic arsenic in whole embryo culture: oxidation state, dose, time, and gestational age dependence. Toxicol Appl Pharmacol, 138: 298–307.

Takahashi A, Kawakami H, Bada A, Okonogi Y & Matsuto S (1990) Effects of phosphate on arsenate inhibition in a marine cyanobacterium, *Phormidium* sp. Appl Organomet Chem, 4: 269–279.

Takamatsu T, Aoki H & Yoshida T (1982) Determination of arsenate, arsenite, monomethylarsonate, and dimethylarsinate in soil polluted with arsenic. Soil Sci, 133(4): 239–246.

Takatsu A & Uchiumi A (1998) Abnormal arsenic accumulation by fish living in a naturally acidified lake. Analyst, 123: 73–75.

Takeuchi IK (1979) Embryotoxicity of arsenic acid: light and electron microscopy of its effect on neurulation-stage rat embryo. J Toxicol Sci, 4: 405–416.

Tam GKH, Charbonneau SM, Bryce F, Pomroy C & Sandi E (1979) Metabolism of inorganic arsenic ([74]As) in humans following oral ingestion. Toxicol Appl Pharm, 50: 319–322.

Tam GKH, Charbonneau SM, Bryce F & Sandi E (1982) Excretion of a single oral dose of fish-arsenic in man. Bull Environ Contam Toxicol, **28**: 669–673.

Tamaki S & Frankenberger WT (1992) Environmental biochemistry of arsenic. Rev Environ Contam Toxicol, **124**: 79–110.

Tammes PM & Lint de MM (1969) Leaching of arsenic from soil. Neth J Agric Sci, **17**(2): 128–132.

Tan AS, Su HQ, Li XR & Wang TH (1983) [Distribution of arsenic through waters, sediments and benthonic organisms of Bohai Bay]. (in Chinese) Marine Sci Quingdao, **4**: 28–30.

Tanaka T (1990) Arsenic in the natural environment. Part II: Arsenic concentrations in thermal waters from Japan. Appl Organomet Chem, **4**: 197–203.

Tanner CC & Clayton JS (1990) Persistence of arsenic 24 years after sodium arsenite herbicide application to Lake Rotoroa, Hamilton, New Zealand. N Z J Mar Freshw Res, **24**: 173–179.

Tao SSH & Bolger PM (1998) Dietary intakes of arsenic in the United States. 3rd International Conference on arsenic exposure and health effects, 12–15 July 1998, San Diego, CA.

Tariq J, Jaffar M & Moazzam M (1991) Concentration correlations between major cations and heavy metals in fish from the Arabian sea. Mar Pollut Bull, **22**(11): 562–565.

Taylor D, Maddock BG & Mance G (1985) The acute toxicity of nine 'grey list' metals (arsenic, boron, chromium, copper, lead, nickel, tin, vanadium and zinc) to two marine fish species: dab (*Limanda limanda*) and grey mullet (*Chelon labrosus*). Aquat Toxicol, **7**: 135–144.

Taylor PR, Qiao YL, Schatzkin A, Yao SX, Lubin J, Mao BL, Rao JY, McAdams M, Xuan XZ & Li JY (1989) Relation of arsenic exposure to lung cancer among tin miners in Yunnan Province, China. Br J Ind Med, **46**: 881–886.

Temple PJ, Linzon SN & Chai BL (1977) Contamination of vegetation and soil by arsenic emissions from secondary lead smelters. Environ Pollut, **12**: 311–320.

Tessier A, Campbell PGC & Bisson M (1979) Sequential extraction procedure for the speciation of particulate trace metals. Anal Chem, **51**(7): 844–850.

Tezuka M, Hanioka K, Yamanaka K & Okada S (1993) Gene damage induced in human alveolar type II (L-132) cells by exposure to dimethylarsinic acid. Biochem Biophys Res Comm, **191**: 1178–1183.

Thanabalasingam P & Pickering WF (1986) Arsenic sorption by humic acids. Environ Pollut, **12B**(3): 233–246.

Thiel T (1988) Phosphate transport and arsenate resistance in the cyanobacterium *Anabaena variabilis*. J Bacteriol, 170(3): 1143–1147.

Thiers H, Colomb D, Moulin G & Colin L (1967) Le cancer cutané arsenical des viticulteurs du Beaujolais. Ann Dermatol Syphilol, **94**: 133–158.

Thomas P, Finnie JK & Williams JG (1997) Feasibility of identification and monitoring of arsenic species in soil and sediment samples by coupled high-performance liquid chromatography inductively coupled plasma mass spectrometry. J Anal Spectrom, 12: 1367–1372.

Thompson D (1993) A chemical hypothesis for arsenic methylation in mammals. Chem Biol Interact, **88**: 89–114.

Thorgeirsson UP, Dalgard DW, Reeves J & Adamson RH (1994) Tumor incidence in a chemical carcinogenesis study of nonhuman primates. Regul Toxicol Pharm, **19**: 130–151.

Thursby GB & Steele RL (1984) Toxicity of arsenite and arsenate to the marine macroalga *Champia parvula* (Rhodophyta). Environ Toxicol Chem, **3**: 391–397.

Tice RR, Yager JW, Andrews P & Crecelius E (1997) Effect of hepatic methyl donor status on urinary excretion and DNA damage in B6C3F1 mice treated with sodium arsenite. Mutat Res, **386**: 315–334.

Tiller KG, Merry RH, Zarcinas BA & Ward TJ (1989) Regional geochemistry of metal-contaminated surficial sediments and seagrasses in upper Spencer Gulf, South Australia. Estuar Coast Mar Sci, **28**(5): 473–493.

Tinwell H, Stephens SC & Ashby J (1991) Arsenite as the probable active species in the human carcinogenicity of arsenic: mouse micronucleus assays on Na and K arsenite, orpiment, and Fowler's solution. Environ Health Perspect, **95**: 205–210.

Tokudome S & Kuratsune M (1976) A cohort study on mortality from cancer and other causes among workers at a metal refinery. Int J Cancer, **17**: 310–317.

Tollestrup K, Daling JR & Allard J (1995) Mortality in a cohort of orchard workers exposed to lead arsenate pesticide spray. Arch Environ Health, **50**: 221–229.

Tondel M, Rahman M, Magnuson A, Chowdhury IA, Faruquee MH & Ahmad SA (1999) The relationship of arsenic levels in drinking water and the prevalence rate of skin lesions in Bangladesh. Environ Health Perspect, **107**: 727–729.

Trappe JM, Stahly EA, Benson NR & Duff DM (1973) Mycorrhizal deficiency of apple trees in high arsenic soils. HortScience, **8**(1): 52–53.

Tremblay GH & Gobeil C (1990) Dissolved arsenic in the St Lawrence Estuary and the Saguenay Fjord, Canada. Mar Pollut Bull **21**(10) 465–469.

Trepka MJ, Heinrich J, Schulz C, Krause C, Popescu M, Wjst M & Wichmann H-E (1996) Arsenic burden among children in industrial areas of eastern Germany. Sci Total Environ, **180**: 95–105.

Tripathi RM, Raghunath R & Krishnamoorthy TM (1997) Arsenic intake by the adult population in Bombay City. Sci Total Environ, **208**: 89–95.

Tsai SM, Wang TN & Ko YC (1998) Cancer mortality trends in a blackfoot disease endemic community of Taiwan following water source replacement. J Toxicol Environ Health, **55**: 389–404.

Tsai SM, Wang TN & Ko YC (1999) Mortality for certain diseases in areas with high levels of arsenic in drinking water. Arch Environ Health, **54**: 186–193.

Tseng C, Chong C, Heng L, Tseng C & Tai T (2000a) The incidence of type 2 diabetes mellitus in Taiwan. Diabetes Res Clin Pr, **50** (Suppl 2): S61–S64.

Tseng CH, Chong CK, Chen CJ, Lin BJ & Tai TY (1995) Abnormal peripheral microcirculation in seemingly normal subjects living in blackfoot-disease-hyperendemic villages in Taiwan. Int J Microcirc Clin Exp, **15**: 21–27.

Tseng CH, Chong CK, Chen CJ & Tai TY (1996) Dose-response relationship between peripheral vascular disease and ingested inorganic arsenic among residents in blackfoot disease endemic villages in Taiwan. Atherosclerosis, **120**: 125–133.

Tseng CH, Chong CK, Chen CJ & Tai TY (1997) Lipid profile and peripheral vascular disease in arseniasis-hyperendemic villages in Taiwan. Angiology, **48**: 321–335.

Tseng CH, Tai TY, Chong CK, Tseng CP, Lai MS, Lin B, Chiou HY, Hsueh YM, Hsu KH & Chen CJ (2000b) Long-term arsenic exposure and incidence of non-insulin-dependent diabetes mellitus: a cohort study in arseniasis-hyperendemic villages in Taiwan. Environ Health Perspect, **108**: 847–851.

Tseng W-P (1977) Effects and dose-response relationship of skin cancer and blackfoot disease with arsenic. Environ Health Perspect, **19**: 109–119.

Tseng W-P, Chu H-M, How S-W, Fong J-M, Lin C-S & Yeh S (1968) Prevalence of skin cancer in an endemic area of chronic arsenicism in Taiwan. J Natl Cancer Inst, **40**: 453–463.

Tsuda T, Kume Y, Yamamoto M, Nagira T & Aoyama H (1987) An epidemiological study on cancer in certified arsenic poisoning patients in Toroku. Jpn J Ind Health, **29**: 496–497.

Tsuda T, Nagira T, Yamamoto M, Kurumatani N, Hotta N, Harada M & Aoyama H (1989) [Malignant neoplasms among residents who drank well water contaminated by arsenic from a King's Yellow factory.] (in Japanese) Sangyo Ika Daigaku Zasshi, **11** (Suppl 1): 289–301.

Tsuda T, Nagira T, Yamamoto M & Kume Y (1990) An epidemiological study on cancer in certified arsenic poisoning patients in Toroku. Ind Health, **28**: 53–62.

Tsuda T, Babazono A, Yamamoto E, Kurumatani N, Mino Y, Ogawa T, Kishi Y & Aoyama H (1995) Ingested arsenic and internal cancer: a historical cohort study followed for 33 years. Am J Epidemiol, **141**: 198–209.

Tsukamoto H, Parker HR, Gribble DH, Mariassy A & Peoples SA (1983) Nephrotoxicity of sodium arsenate in dogs. Am J Vet Res, **44**: 2324–2330.

Turpeinen R, Pantsar Kallio M, Haggblom M & Kairesalo T (1999) Influence of microbes on the mobilization, toxicity and biomethylation of arsenic in soil. Science Total Environ, **15**(236): 173–180.

Ünlü MY (1979) Chemical transformation and flux of different forms of arsenic in the crab *Carcinus maenas*. Chemosphere, **8**(5): 269–275.

Ünlü MY & Fowler SW (1979) Factors affecting the flux of arsenic through the mussel *Mytilus galloprovincialis*. Mar Biol, **51**: 209–219.

Umweltbundesamt (1997) [Data about the environment. The condition of the environment in Germany]. (in German) Federal Ministry for the Environment, Berlin. Germany.

Urieta I, Jalon M & Eguilero I (1996) Food surveillance in the Basque Country (Spain), II: Estimation of the dietary intake of organochlorine pesticides, heavy metals, arsenic, aflatoxin M1, iron and zinc through the total diet study, 1990/91. Food Addit Contam, **13**: 29–52.

US DOI (US Department of the Interior) (1991) http://ntp-db.niehs.gov/htdocs/ARC/ARC_KC/Arsenic_Cmpds.htlm.

US EPA (US Environmental Protection Agency) (1982) An exposure and risk assessment for arsenic. EPA 440/4-85-005. Washington, DC, US Environmental Protection Agency.

US EPA (US Environmental Protection Agency) (1985) Ambient water quality criteria for arsenic – 1984. EPA/440/5-84/033. Washington, DC, US Environmental Protection Agency.

US EPA (US Environmental Protection Agency) (1992) Second draft for the drinking water criteria document on arsenic. Prepared Under ICAIR Program No. 1524, for EPA Contract 68-C8-0033, ERG Subcontract No. LSI-8700, ERG Work Assignment no. 2-19, Life Systems. Washington, DC, US Environmental Protection Agency.

US EPA (US Environmental Protection Agency) (1996) Bioavailability of arsenic and lead in environmental substrates. 1. results of an oral dosing study of immature swine. Superfund/Office of Environmental Assessment, EPA 910/R-96-002. Washington, DC, US Environmental Protection Agency.

US EPA (US Environmental Protection Agency) (1997) Peer review of EPA's research plan for arsenic in drinking water. Draft report. Ad Hoc Subcommittee On Arsenic Research, Board Of Scientific Counselors (BOSC), Office Of Research and Development. Washington, DC, US Environmental Protection Agency.

Uthus EO (1992) Evidence for arsenic essentiality. Environ Geochem Health, **14**: 55–58.

Vaessen HA & van Ooik A (1989) Speciation of arsenic in Dutch total diets: methodology and results. Z Lebensm Unters Forsch, **189**(3): 232–5.

Vahter M (1981) Biotransformation of trivalent and pentavalent inorganic arsenic in mice and rats. Environ Res, **25**: 286–293.

Vahter M (1988) Biological availability and toxicity of arsenic contaminated soil. Report No. 3. Stockholm, National Institute of Environmental Medicine.

Vahter M (1999) Methylation of inorganic arsenic in different mammalian species and population groups Sci Progr, **82**: 69–88.

Vahter M & Envall J (1983) In vivo reduction of arsenate in mice and rabbits. Environ Res, **32**: 14–24.

Vahter M &Gustafsson B (1980) Biotransformation of arsenic in germfree and conventional mice. In: Anke M, Schneider H-J, Bruckner C ed. Proceedings of 3rd Symposium on Trace Elements. Arsenic, 7–11 July 1980, Jena. Jena, Abteilung Wissenschaftliche Publikationen der Friedrich-Schiller-Universität, pp 123–129.

Vahter M & Lind B (1986) Concentrations of arsenic in urine of the general population in Sweden. Sci Total Environ, **54**: 1–12.

Vahter M & Marafante E (1983) Intracellular distribution and metabolic fate of arsenite and arsenate in mice and rabbits. Chem Biol Interact, **47**: 29–44.

Vahter M & Marafante E (1985) Reduction and binding of arsenate in marmoset monkeys. Arch Toxicol, **57**: 119–124.

Vahter M & Marafante E (1987) Effects of low dietary intake of methionine, choline or proteins on the biotransformation of arsenite in the rabbit. Toxicol Lett, **37**: 41–46.

Vahter M & Norin H (1980) Metabolism of [74]As-labeled trivalent and pentavalent inorganic arsenic in mice. Environ Res, **21**: 446–457.

Vahter M, Marafante E, Lindgren A & Dencker L (1982) Tissue distribution and subcellular binding of arsenic in marmoset monkeys after injection of [74]As-arsenite. Arch Toxicol, **51**: 65–77.

Vahter M, Marafante E & Dencker L (1983) Metabolism of arsenobetaine in mice, rats and rabbits. Sci Total Environ, **30**: 197–211.

Vahter M, Marafante E, Dencker L (1984) Tissue distribution and retention of [74]As-dimethylarsinic acid in mice and rats. Arch Environ Contam Toxicol, **13**: 259–264.

Vahter M, Friberg L, Rahnster B, Nygren A, Nolinder P (1986) Airborne arsenic and urinary excretion of metabolites of inorganic arsenic among smelter workers. Int Arch Occup Environ Health, **57**: 79–91.

Vahter M, Concha G, Nermell B, Nilsson R, Dulout F & Natarajan AT (1995a) A unique metabolism of inorganic arsenic in native Andean women. Eur J Pharmacol Environ Toxicol Pharmacol Sect, **293**: 455–462.

Vahter M, Couch R, Nermell B & Nilsson R (1995b) Lack of methylation of inorganic arsenic in the chimpanzee. Toxicol Appl Pharm, **133**: 262–268.

Valentine JL (1994) Chapter 14 – Review of health assessments for US/Canada populations exposed to arsenic in drinking water. In: Chappell WR, Abernathy CO & Cothern CR eds. Arsenic – Exposure and health. Northwood, UK, Science and Technology Letters, pp 139–152.

Valentine JL, Kang HK & Spivey G (1979) Arsenic levels in human blood, urine, and hair in response to exposure via drinking water. Environ Res, **20**: 24–32.

Valkonen S, Savolainen H & Jarvisalo J (1983) Arsenic distribution and neurochemical effects in peroral sodium arsenite exposure of rats. Bull Environ Contam Toxicol, **30**(3): 303–308.

Van Cleuvenbergen RJA, Van Mol WE & Adams FC (1988) Arsenic speciation in water by hydride cold trapping – quartz furnace atomic absorption spectrometry: An evaluation. J Anal Spectrom, **3**: 169–176.

Van der Hoek EE & Comans RNJ (1996) Modeling arsenic and selenium leaching from acidic fly ash by sorption on iron (hydr)oxide in the fly ash matrix. Environ Sci Technol, **30**(2): 517–523.

van der Sloot HA, Hoede D, Wijkstra J, Duinker JC & Nolting RF (1985) Anionic species of V, As, Se, Mo, Sb, Te, and W in the Scheldt and Rhine estuaries and the southern bight (North Sea). Estuar Coast Mar Sci, **21**: 633–651.

VanderKop PA & MacNeil JD (1989) Effects of arsanilic acid and monensin when given simultaneously in the diet of broiler chicks. Vet Hum Toxicol, **31**(3): 209–213.

Varanasi U, Stein JE, Tilbury KL, Meador JP, Sloan CA, Clark RC & Chan SL (1994) Chemical contaminants in gray whales (*Eschrichtius robustus*) stranded along the west coast of North America. Sci Total Environ, **145**(1/2): 29–53.

Varsanyi I (1989) Arsenic in deep groundwater. In: Miles DL ed. Water–rock interaction (WRI-6). Rotterdam, A. A. Balkema, pp 715–718.

Vaughan GT & Greenslade PM (1998) Sensitive bioassays for risk assessment of contaminated soils. Investigation Report CET/IR 55. Environmental Research Trust Final Report. Sydney, Australia, CSIRO Division of Coal and Energy Technology.

Velez D, Ybanez N & Montoro R (1996) Monomethylarsonic and dimethylarsinic acid contents in seafood products. J Agric Food Chem, **44**: 859–864.

Vermeer K & Thompson JAJ (1992) Arsenic and copper residues in waterbirds and their food down inlet from the Island Copper Mill. Bull Environ Contam Toxicol, **48**(5): 733–738.

Vermette SJ, Peden ME, Willoughby TC, Lindberg SE & Weiss AD (1995) Methodology for the sampling of metals in precipitation: Results of the National Atmospheric Deposition Program (NADP) pilot network. Atmos Environ **29**(11) 1221–1229.

Vianna NJ, Brady JA & Cardamone AT (1981) Epidemiology of angiosarcoma of liver in New York State. N Y State J Med, **81**: 895–899.

Villanueva C & Kogevinas M (1999) Comments on 'Drinking water arsenic in Utah: a cohort mortality study'. Environ Health Perspect, **107**: A544.

Viren J & Silvers A (1999) Nonlinearity in the lung cancer dose-response for airborne arsenic: Apparent confounding by year of hire in evaluating lung cancer risks from arsenic exposure in Tacoma smelter workers. Regul Toxicol Pharm, **30**(21): 117–129.

Vocke RW, Sears KL, O'Toole JJ & Wildman RB (1980) Growth responses of selected freshwater algae to trace elements and scrubber ash slurry generated by coal-fired power plants. Water Res, **14**: 141–150.

Von Endt DW, Kearney PC & Kaufman DD (1968) Degradation of monosodium methanearsonic acid by soil microorganisms. J Agric Food Chem, **16**(1): 17–20.

Vogel AE (1954) Special tests for small amounts of arsenic. In: Vogel AE (ed) A textbook of macro and semi-micro qualitative inorganic analysis, 4th ed. London, Longmans, pp 242–247.

Wagemann R, Snow NB, Rosenberg DM & Lutz A (1978) Arsenic in sediments, water, and aquatic biota from lakes in the vicinity of Yellowknife, Northwest Territories, Canada. Arch Environ Contam Toxicol, **7**: 169–191.

Wahlberg JE & Boman A (1986) Contact sensitivity to arsenical compounds. Dermatosen, **34**: 10–12.

Wakao N, Koyatsu H, Komai Y, Shimokawara H, Sakurai Y & Shiota H (1988) Microbial oxidation of arsenite and occurrence of arsenite-oxidizing bacteria in acid mine water from a sulfur-pyrite mine. Geomicrobiol J, **6**: 11–24.

Waldman JM, Lioy PJ, Zelenka M, Jing L, Lin YN, He QC, Qian ZM, Chapman R & Wilson WE (1991) Wintertime measurements of aerosol acidity and trace elements in

Wuhan, a city in central China. Atmos Environ, **25B**(1): 113–120.

Wall S (1980) Survival and mortality pattern among Swedish smelter workers. Int J Epidemiol, **9**: 73–87.

Walsh LM & Keeney DR (1975) Behavior and phytotoxicity of inorganic arsenicals in soils. ACS Symp Ser, **7**: 35–52.

Walsh PR, Duce RA & Fasching JL (1979) Considerations of the enrichment, sources, and flux of arsenic in the troposphere. J Geophys Res, **84**(4C): 1719–1726.

Wang C & Lazarides E (1984) Arsenite-induced changes in methylation of the 70,000 dalton heat shock proteins in chicken embryo fibroblasts. Biochem Biophys Res Commun, **119**: 735–743.

Wang DS, Weaver RW & Melton JR (1984) Microbial decomposition of plant tissue contaminated with arsenic and mercury. Environ Pollut, **34**: 275–282.

Wang HF (1993) Glutathione *S*-transferase facilitates the excretion of arsenic from arsenic-resistant Chinese hamster ovary cells. Biochem Biophys Res Commun, **192**: 1093–1099.

Wang LF, Liu HD, Lin FF, Su MY, Xu XF, Sun XZ, Aihaidi, Ma L, Yang Y & Huang ZW (1993) Endemic arsenism in a village of Xinjiang: epidemiology, clinical and preventive studies for 9 years. Endemic Dis Bull, **8**: 71–78.

Wang TS & Huang H (1994) Active oxygen species are involved in the induction of micronuclei by arsenite in XRS-5 cells. Mutagenesis, **9**(3): 253–257.

Wang TS, Kuo CF, Jan KY & Huang H (1996a) Arsenite induces apoptosis in Chinese hamser ovary cells by generation of reactive oxygen species. J Cell Physiol, **169**: 256–268.

Wang Z & Rossman TG (1993) Stable and inducible arsenite resistance in Chinese hamster cells. Toxicol Appl Pharmacol, **118**: 80–86.

Wang Z, Dey S, Rosen BP & Rossman TG (1996b) Efflux mediated resistance to arsenicals in arsenic resistant and hypersensitive Chinese hamster cells. Toxicol Appl Pharmacol, **137**(1): 112–119.

Wängberg S-Å & Blanck H (1990) Arsenate sensitivity in marine periphyton communities established under various nutrient regimes. J Exp Mar Biol Ecol, **139**: 119–134.

Wanibuchi H, Yamamoto S, Chen H, Yoshida K, Endo G, Hori T & Fukushima S (1996) Promoting effects of dimethylarsinic acid on *N*-butyl-*N*-(4-hydroxybutyl)nitrosamine-induced urinary bladder carcinogenesis in rats. Carcinogenesis, **17**: 2435–2439.

Wanibuchi H, Hori T, Meenakshi V, Ichihara T, Yamamoto S, Yano Y, Otani S, Nakae D, Konishi Y & Fukushima S (1997) Promotion of rat hepatocarcinogenesis by dimethylarsinic acid: association with elevated ornithine decarboxylase activity and formation of 8 hydroxydeoxyguanosine in the liver. Jpn J Cancer Res **88**(12) 1149–1154.

Warner ML, Moore LE, Smith MT, Kalman DA, Fanning E & Smith AH (1994) Increased micronuclei in exfoliated bladder cells of individuals who chronically ingest arsenic-contaminated water in Nevada. Cancer Epidemiol Biomarkers Prev, **3**: 583–590.

Warren HV, Jorksy SJ & Gould CE (1983) Quantitative analysis of zinc, copper, lead, molybdenum, bismuth, mercury and arsenic in brain and other tissues from multiple sclerosis and non-multiple sclerosis cases. Sci Total Environ, **29**: 163–169.

Waslenchuk DG (1978) The budget and geochemistry of arsenic in a continental shelf environment. Mar Chem, **7**: 39–52.

Waslenchuk DG & Windom HL (1978) Factors controlling the estuarine chemistry of arsenic. Estuar Coast Mar Sci, **7**: 455–464.

Watson AP, Van Hook RI & Reichle DE (1976) Toxicity of organic and inorganic arsenicals to an insect herbivore. Environ Sci Technol, **10**(4): 356–359.

Watson WA, Veltri JC & Metcalf TJ (1981) Acute arsenic exposure treated with oral d-penicillamine. Vet Hum Toxicol, **23**: 164–166.

Wauchope RD (1975) Fixation of arsenical herbicides, phosphate and arsenate in alluvial soils. J Environ Qual, 4(3): 355–358.

Wauchope RD & McDowell LL (1984) Adsorption of phosphate, arsenate, methanearsenate and cacodylate by lake and stream sediments: comparison with soils. J Environ Qual, **13**(3): 499–504.

Weaver RW, Melton JR, Wang D & Duble RL (1984) Uptake of arsenic and mercury from soil by Bermuda grass *Cynodon dactylon*. Environ Pollut, **33**: 133–142.

Webb DR, Wilson SE & Carter DE (1986) Comparative pulmonary toxicity of gallium arsenide, gallium (III) oxide or arsenic (III) oxide intratracheally instilled into rats. Toxicol Appl Pharm, **82**: 405–416.

Webb DR, Wilson SE & Carter DE (1987) Pulmonary clearance & toxicity of respirable gallium arsenide particulates intratracheally instilled into rats. Am Ind Hyg Assoc, **48**: 660–667.

Wei M, Wanibuchi H, Yamamoto S, Li W & Fukushima S (1999) Urinary bladder carcinogenicity of dimethylarsinic acid in male F344 rats. Carcinogenesis, **20**(9): 1873–1876.

Weinshilboum RM (1992) Methylation pharmacogenetics: thiopurine methyltransferase as a model system. Xenobiotica, **22**: 1055–1071.

Weir PA & Hine CH (1970) Effects of various metals on behavior of conditioned goldfish. Arch Environ Health, **20**: 45–51.

Welch AH, Lico MS & Hughes JL (1988) Arsenic in groundwater of the western United States. Ground Water, **26**(3): 333–347.

Welch K, Higgins I, Oh M & Burchfiel C (1982) Arsenic exposure, smoking, and respiratory cancer in copper smelter workers. Arch Environ Health, **37**(6): 325–335.

Wells JM & Richardson DHS (1985) Anion accumulation by the moss *Hylocomium splendens*: uptake and competition studies involving arsenate, selenate, selenite, phosphate, sulphate and sulphite. New Phytol, **101**: 571–583.

Welz B, Schubert-Jacobs M & Sperling M (1990) Investigation of reactions and atomization of arsine in heated quartz tube using atomic absorption and masss spectrometry. Spectrochem Acta, **45B**(11): 1235–1256.

Wenclawiak BW & Krah M (1995) Reactive supercritical fluid extraction and chromatography of arsenic species. Fresenius J Anal Chem, **351**: 134–138.

Wester RC, Maibach HI, Sedik L, Melendres J & Wade M (1993) In vitro and in vivo percutaneous absorption and skin decontamination of arsenic from water and soil. Fundam Appl Toxicol, **20**: 336–340.

Wewerka EM, Bertino JPL, Wagner P, Williams JM, Wanek PL & Wangen LE (1978) Trace element characterisation of coal wastes, second annual progress report. DOE LA-7360-PR; EPA-600/7-78-028a, Washington DC.

White DH, King KA & Prouty RM (1980) Significance of organochlorine and heavy metal residues in wintering shorebirds at Corpus Christi, Texas, 1976–77. Pestic Monit J, **14**(2): 58–63.

Whitworth MR, Pendleton GW, Hoffman DJ & Camardese MB (1991) Effects of dietary boron and arsenic on the behavior of mallard ducklings. Environ Toxicol Chem, **10**(7): 911–916.

WHO (1996) Guidelines for drinking-water quality, 2nd edition, Vol 2. Health criteria and other supporting information. Geneva, World Health Organization.

WHO (2000) WHO Air quality guidelines for Europe, 2nd edition. Copenhagen, WHO Regional Office for Europe.

Wicklund KG, Daling JR, Allard J & Weiss NS (1988) Respiratory cancer among orchardists in Washington State, 1968 to 1980. J Occup Med, **30**: 561–564.

Wiemayer SN, Lamont TG & Locke LN (1980) Residues of environmental pollutants and necropsy data for eastern United States ospreys, 1964–1973. Estuaries, **3**(3): 155–167.

Wiencke JK & Yager JW (1992) Specificity of arsenite in potentiating cytogenetic damage induced by the DNA crosslinking agent diepoxybutane. Environ Mol Mutagen, 19: 195–200.

Wiersma D, van Goor BJ & van der Veen NG (1986) Cadmium, lead, mercury, and arsenic concentrations in crops and corresponding soils in The Netherlands. J Agric Food Chem, 34: 1067–1074.

Wild H (1974) Arsenic tolerant plant species established on arsenical mine dumps in Rhodesia. Kirkia, 9: 265–278.

Willhite CC (1981) Arsenic-induced axial skeletal (dysraphic) disorders. Exp Mol Pathol, 34: 145–158.

Williams M, Fordyce F, Paijitprapapon A & Charoenchaisri P (1996) Arsenic contamination in surface drainage and groundwater in part of the southeast Asian tin belt, Nakhon Si Thammarat Province, southern Thailand. Environ Geol, 27: 16–33.

Windebank AJ (1986) Specific inhibition of myelination by lead in vitro; comparison with arsenic, thallium, and mercury. Exp Neurol, 94: 203–212.

Wlodarczyk B, Bennett GD, Calvin JA, Craig JC & Finnell RH (1996) Arsenic-induced alterations in embryonic transcription factor gene expression: implications for abnormal neural development. Develop Genet, 18: 306–315.

Wolfsperger M, Hauser G, Gobler W & Schlagenhaufen C (1994) Heavy metals in human hair samples from Austria and Italy: influence of sex and smoking habits. Sci Total Environ, 156: 235–242.

Woller A, Mester Z & Fodor P (1995) Determination of arsenic species by high-performance liquid chromatography ultrasonic nebulization atomic fluorescence spectrometry. J Anal Energy Spectrom, 10: 609–613.

Wong O, Whorton MD, Foliari DE & Lowengart R (1992) An ecologic study of skin cancer and environmental arsenic exposure. Int Arch Occup Environ Health, 64: 235–241.

Wong PTS, Chau YK, Luxon L & Bengert GA (1977) Methylation of arsenic in the aquatic environment. In: Hemphill DD ed. Trace substances in environmental health. Proceedings of the University of Missouri's 11th Annual Conference, 7–9 June 1977, Columbia, MO, University of Missouri, pp 100–106.

Wood JM (1974) Biological cycles for toxic elements in the environment. Science, 183: 1049–1052.

Woods JS & Fowler BA (1978) Altered regulation of mammalian hepatic heme biosynthesis and urinary porphyrin excretion during prolonged exposure to sodium arsenate. Toxicol Appl Pharm, 43(2): 361–71.

Woods JS & Southern MR (1989) Studies on the etiology of trace metal-induced porphyria: effects of porphyrinogenic metals on coproporphyrinogen oxidase in rat liver and kidney. Toxicol Appl Pharm, **97**(1): 183–90.

Woolson EA (1973) Arsenic phytotoxicity and uptake in six vegetable crops. Weed Sci, **21**: 524–527.

Woolson EA (1977a) Generation of alkylarsines from soil. Weed Sci, **25**(5): 412–416.

Woolson EA (1977b) Fate of arsenicals in different environmental substrates. Environ Health Perspect, **19**: 73–81.

Woolson EA (1983) In: Fowler BA ed. Biological and environmental effects of arsenic. Amsterdam, Elsevier Science, pp 51–139.

Woolson EA & Kearney PC (1973) Persistence and reactions of 14C-cacodylic acid in soils. Environ Sci Technol, **7**: 47–50.

Woolson EA & Isensee AR (1981) Soil residue accumulation from three arsenic sources. Weed Sci, **29**(1): 17–21.

Woolson EA, Axley JH & Kearney PC (1971) Correlation between available soil arsenic, estimated by six methods, and response of corn (*Zea mays* L). Soil Sci Soc Am Proc, **35**(1): 101–105.

Woolson EA, Axley JH & Kearney PC (1973) The chemistry and phytotoxicity of arsenic in soil: 2. Effects of time and phosphorus. Soil Sci Soc Am Proc, **37**: 254–259.

Wrench JJ & Addison RF (1981) Reduction, methylation, and incorporation of arsenic into lipids by the marine phytoplankton *Dunaliella tertiolecta*. Can J Fish Aquat Sci, **38**: 518–523.

Wu HY, Chen KP, Tseng WP & Hsu JL (1961) Epidemiologic studies on Blackfoot disease. 1. Prevalence and incidence of the disease by age, sex, year, occupation and geographic distribution. Mem Coll Med Nat Taiwan Univ, **7**: 33–50.

Wu M-M, Kuo T-L, Hwang Y-H & Chen C-J (1989) Dose-response relation between arsenic concentration in well water and mortality from cancers and vascular disease. Am J Epidemiol, **130**: 1123–1132.

Wu ZY, Han M, Lin ZC & Ondov JM (1994) Chesapeake Bay atmospheric deposition study, year, 1: sources and dry deposition of selected elements in aerosol particles. Atmos Environ, **28**(8): 1471–1486.

Wyttenbach A, Bajo S, Furrer V, Langenauer M & Tobler L (1997) The accumulation of arsenic, bromine and iodine in needles of Norway spruce (*Picea abies* [L.] Karst.) at sites with low pollution. Water Air Soil Pollut, **94**: 417–430.

Xu H, Allard B & Grimvall A (1988) Influence of pH and organic substance on the adsorption of As(V) on geologic materials. Water Air Soil Pollut, 40(3/4): 293–305.

Xu H, Allard B & Grimvall A (1991) Effects of acidification and natural organic materials on the mobility of arsenic in the environment. Water Air Soil Pollut, 57/58: 269–278.

Xu JL & Thornton I (1985) Arsenic in garden soils and vegetable crops in Cornwall, England: implications for human health. Environ Geochem Health, 7(4): 131–133.

Xu ZY, Blot WJ, Xiao HP, Wu A, Feng YP, Stone BJ, Sun J, Ershow AG, Henderson BE & Fraumeni JF Jr (1989) Smoking, air pollution, and the high rates of lung cancer in Shenyang, China. J Natl Cancer Inst, 81: 1800–1806.

Xu ZY, Blot WJ, Fraumeni JF Jr, Zhao DZ, Stone BJ, Yin Q, Wu A, Henderson BE & Guan BP (1991) Environmental determinants of lung cancer in Shenyang, China. In O'Neill IK, Chen J, Bartsch H ed. Relevance to human cancer of N-nitroso compounds, tobacco smoke and mycotoxins. IARC Scientific Publication 105. Lyon, International Agency for Research on Cancer, pp 460-465.

Yadava KP, Tyagi BS & Singh VN (1988) Removal of arsenic(III) from aqueous solution by China clay. Environ Technol Lett, 6(11): 1233–1244.

Yager JW & Wiencke JK (1998) Inhibition of poly(ADP-ribose)polymerase by arsenite. Mutat Res, 386: 345–351.

Yager JW, Hicks JB & Fabianova E (1997) Airborne arsenic and urine excretion of arsenic metabolites during boiler cleaning operations in a Slovak coal-fired power plant. Environ Health Perspec, 105: 836–842.

Yamamoto A, Hisanaga A & Ishinishi N (1987) Tumorigenicity of inorganic arsenic compounds following intratracheal instillations to the lungs of hamsters. Int J Cancer, 40: 220–223.

Yamamoto S, Konishi Y, Matsuda T, Murai T, Shibata M, Matsui-Yuasa I, Otani S, Kuroda K, Endo G & Fukushima S (1995) Cancer induction by an organic arsenic compound, dimethylarsinic acid (cacodylic acid), in F344/DuCrj rats after pretreatment with five carcinogens. Cancer Res, 55: 1271–1276.

Yamanaka K & Okada S (1994) Induction of lung-specific DNA damage by metabolically methylated arsenics via the production of free radicals. Environ Health Perspect, 102: 37–40.

Yamanaka K, Hasegawa A, Sawamura R & Okada S (1989a) DNA strand breaks in mammalian tissues induced by methylarsenics. Biol Trace Elem Res, 21: 413–417.

Yamanaka K, Hasegawa A, Sawamura R & Okada S (1989b) Dimethylated arsenics induce DNA strand breaks in lung via the production of active oxygen in mice. Biochem Biophys Res Comm, 165: 43–50.

Yamanaka K, Ohba H, Hasegawa A, Sawamura R & Okada S (1989c) Mutagenicity of dimethylated metabolites of inorganic arsenics. Chem Pharm Bull, **37**: 2753–2756.

Yamanaka K, Hoshino M, Okamoto M, Sawamura R, Hasegawa A & Okada S (1990) Induction of DNA damage by dimethylarsine, a metabolite of inorganic arsenics is for the major part likely due to its peroxyl radical. Biochem Biophys Res Comm, **168**: 58–63.

Yamanaka K, Hasegawa A, Sawamura R & Okada S (1991) Cellular response to oxidative damage in lung induced by the administration of dimethylarsinic acid, a major metabolite of inorganic arsenics, in mice. Toxicol Appl Pharm, **108**: 205–213.

Yamanaka K, Tezuka M, Kato K, Hasegawa A & Okada S (1993) Crosslink formation between DNA and nuclear proteins by in vivo and in vitro exposure of cells to dimethylarsinic acid. Biochem Biophys Res Comm, **191**: 1184–1191.

Yamanaka K, Hayashi H, Kato K, Hasegaw A & Okada S (1995) Involvement of preferential formation of apurinic/apyrimidinic sites in dimethylarsenic-induced DNA strand breaks and DNA-protein crosslinks in cultured alveolar epithelial cells. Biochem Biophys Res Comm, **207**: 244–249.

Yamanaka K, Ohtsubo K, Hasegawa A, Hayashi H, Ohgi H, Kanisawa M & Okada S (1996) Exposure to dimethylarsinic acid, a main metabolite of inorganic arsenics, strongly promotes tumorigenesis initiated by 4-nitroquinoline 1-oxide in the lungs of mice. Carcinogenesis, **17**: 767–770.

Yamanaka K, Hayashi H, Kato K, Hasegawa A, Oku N & Okada S (1997) DNA single-strand breaks in L-132 cells resulting from inhibition of repair polymerization shortly after exposure to dimethylarsinic acid. Biol Pharm Bull, **20**: 163–167.

Yamaoka Y & Takimura O (1986) Marine algae resistant to inorganic arsenic. Agric Biol Chem, **50**(1): 185–186.

Yamaoka Y, Takimura O & Fuse H (1988) Environmental factors relating to arsenic accumulation by *Dunaliella* sp. Appl Organomet Chem, **2**: 359–364.

Yamaoka Y, Takimura O, Fuse H & Kamimura K (1992) Effects of arsenic on the organic component of the alga *Dunaliella salina*. Appl Organomet Chem, **6**: 357–362.

Yamauchi H & Yamamura Y (1979) Dynamic change of inorganic arsenic and methylarsenic compounds in human urine after oral intake as arsenic trioxide. Ind Health, **17**: 79–83.

Yamauchi H & Yamamura Y (1983) Concentration and chemical species of arsenic in human tissue. Bull Environ Contam Toxicol, **31**: 267–277.

Yamauchi H & Yamamura Y (1984a) Metabolism and excretion of orally administered dimethylarsinic acid in the hamster. Toxicol Appl Pharm, **74**: 134–140.

Yamauchi H & Yamamura Y (1985) Metabolism and excretion of orally administered arsenic trioxide in the hamster. Toxicology, **34**: 113–121.

Yamauchi H, Kaise T & Yamamura Y (1986a) Metabolism and excretion of orally administered arsenobetaine in the hamster. Bull Environ Contam Toxicol, **36**: 350–355.

Yamauchi H, Takahashi K & Yamamura Y (1986b) Metabolism and excretion of orally and intraperitoneally administered gallium arsenide in the hamster. Toxicology, **40**: 237–246.

Yamauchi H, Yamato N & Yamamura Y (1988) Metabolism and excretion of orally and intraperitoneally administered methylarsenic acid in the hamster. Bull Environ Contam Toxicol, **40**: 280–286.

Yamauchi H, Takahashi K, Mashiko M & Yamamura Y (1989a) Biological monitoring of arsenic exposure of gallium arsenide- and inorganic arsenic-exposed workers by determination of inorganic arsenic and its metabolites in urine and hair. Am Ind Hyg Assoc J, **50**: 606–612.

Yamauchi H, Takahashi K, Yamamura Y & Kaise T (1989b) Metabolism and excretion of orally and intraperitoneally administered trimethylarsine oxide in the hamster. Toxicol Environ Chem, **22**: 69–76.

Yamauchi H, Kaise T, Takahashi K & Yamamura Y (1990) Toxicity and metabolism of trimethylarsine in mice and hamster. Fundam Appl Toxicol, **14**: 399–407.

Yamauchi H, Takahashi K, Mashiko M, Saitoh J & Yamamura Y (1992) Intake of different chemical species of dietary arsenic by Japanese, and their blood and urinary arsenic levels. Appl Organomet Chem, **6**: 383–388.

Yamauchi H & Yamamura Y (1984b) Metabolism and excretion of orally ingested trimethylarsenic in man. Bull Environ Contam Toxicol, **32**: 682–687.

Yanez L, Carrizales L, Zanatta MT, Mejia JJ, Batres L & Diaz-Barriga F (1991) Arsenic-cadmium interaction in rats: toxic effects in the heart and tissue metal shifts. Toxicology, **67**: 227–234.

Ybanez N, Cervera ML & Montoro R (1992) Determination of arsenic in dry ahed seafood products by hydride generation atomic absorption spectrometry and a critical comparative study with platform furnace Zeeman-effect atomic absorption spectrometry and inductively coupled plasma atomic emission spectrometry. Anal Chim Acta, **258**: 61–71.

Yen HT, Chiang LC, Wen KH, Chang SF, Tsai CC, Yu CL & Yu HS (1996) Arsenic induces interleukin-8 expression in cultured keratinocytes. Arch Derm Res, **288**: 716–717.

Yoshida K, Chen H, Inoue Y, Wanibuchi H, Fukushima S, Kuroda K & Endo G (1997) The urinary excretion of arsenic metabolites after a single oral administration of dimethylarsinic acid to rats. Arch Environ Contam Toxicol, **32**: 416–421.

Yoshida K, Inoue Y, Kuroda K, Chen H, Wanibuchi H, Fukushima S & Endo G (1998) Urinary excretion of arsenic metabolites after long term oral administration of various arsenic compounds to rats. J Toxicol Environ Health, **54**(3): 179192.

Yost LJ, Schoof RA & Aucoin R (1998) Intake of inorganic arsenic in the North American diet. Hum Ecol Risk Assess, **4**: 137–152.

Yu JJ & Wai CM (1991) Chromatographic seperation of arsenic species with sodium bis(trifluoroethyl)dithiocarbamate chelation. Anal Chem, **63**: 842–845.

Yusof AM, Ikhsan ZB & Wood AKH (1994) The speciation of arsenic in seawater and marine species. J Radioanal Nucl Chem Artic, **179**(2): 277–283.

Zakharyan RA, Wildfang E & Aposhian HV (1996) Enzymatic methylation of arsenic compounds: III. the marmoset and tamarin, but not the rhesus, monkeys are deficient in methyltransferases that methylate inorganic arsenic. Toxicol Appl Pharm, **140**: 77–84.

Zaldivar R (1974) Arsenic contamination of drinking water and foodstuffs causing endemic chronic poisoning. Beitr Path Bd, **151**: 384–400.

Zaman K & Pardini RS (1995) An insect model for assessing arsenic toxicity: arsenic elevated glutathione content in the *Musca domestica* and *Trichoplusia ni*. Bull Environ Contam Toxicol, **55**: 845–852.

Zaroogian GE & Hoffman GL (1982) Arsenic uptake and loss in the American oyster, *Crassostrea virginica*. Environ Monit Assess, **1**: 345–358.

Zeiger E, Anderson B, Haworth S, Lawlor T & Mortelmans K (1992) Salmonella mutagenicity tests: V. Results from the testing of 311 chemicals. Environ Mol Mutagen, **19**(Suppl 21): 2–141.

Zhang L & Zhou K (1992) Background values of trace elements in the source area of the Yangtze River. Sci Total Environ, **125**: 391–404.

Zhang X, Cornelis R, Dekimpe J & Mees L (1996a) Arsenic speciation in serum of uraemic patients based on liquid chromatography with hydride generation atomic absorption spectrometry and on-line uv photo-oxidation digestion. Anal Chim Acta, **319**: 177–185.

Zhang X, Cornelis R, Kimpe J De, Mees L, Vanderbiesen V, De Cubber A & Vanholder R (1996b) Accumulation of arsenic species in serum of patients with chronic renal disease. Clin Chem, **42**: 1231–1237.

Zhang XR, Cornelis R, Dekimpe J, Mees L & Lameire N (1997) Speciation of arsenic in serum, urine, & dialysate of patients on continuous ambulatory peritoneal dialysis. Clin Chem, **43**: 406–408.

Zhang X, Cornelis R, Kimpe J De, Mees L & Lameire N (1998a) Study of arsenic-protein binding in serum of patients on continuous ambulatory peritoneal dialysis. Clin Chem, **44**: 141–147.

Zhang X, Cornelis R, Mees L, Vanholder R & Lameire N (1998b) Chemical speciation of arsenic in serum of uraemic patients. Analyst, **123**: 13–17.

Zhao CQ, Young MR, Diwan BA, Coogan TP & Waalkes MP (1997) Association of arsenic-induced malignant transformation with DNA hypomethylation and aberrant gene expression. Proc Natl Acad Sci U S A, **94**: 10907–10912.

Zhu B & Tabatabai MA (1995) An alkaline oxidation method for determination of total arsenic and selenium in sewage sludges. J Environ Qual, **24**(4): 622–626.

Zhuang GS, Wang YS, Tan MG, Zhi M, Pan WQ & Cheng YD (1990) Preliminary study of the distribution of the toxic elements As, Cd, and Hg in human hair and tissues by RNAA. Biol Trace Elem Res, **12**: 729–736.

Zierler S, Theodore M, Cohen A & Rothman K (1988) Chemical quality of maternal drinking water and congenital heart disease. Int J Epidemiol, **17**: 589–594.

Zima J & van den Berg CMG (1994) Determination of arsenic in sea water by cathodic stripping voltammetry in the presence of pyrrolidine dithiocarbamate. Anal Chim Acta, **289**: 291–298.

Zoltai SC (1988) Distribution of base metals in peat near a smelter at Flin Flon, Manitoba. Water Air Soil Pollut, **37**(1/2): 217–228.

RESUME

1. Propriétés et méthodes d'analyse

L'arsenic est un métalloïde largement répandu dans l'écorce terrestre où sa concentration moyenne est de 2 mg/kg. Il est présent à l'état de traces dans l'ensemble des roches, des sols et des eaux ainsi que dans l'air. L'arsenic présente quatre degrés d'oxydation : −3, 0, +3 et +5. En milieu réducteur, c'est l'arsénite qui constitue la forme prédominante (As III); en milieu oxygéné, c'est généralement l'arséniate (As V) qui constitue la forme stable. L'arsenic élémentaire n'est pas soluble dans l'eau. La solubilité des sels d'arsenic varie dans de larges proportions selon la valeur du pH et la nature des ions présents.

Il existe diverses méthodes instrumentales pour la recherche et le dosage de l'arsenic, notamment la spectrométrie d'absorption atomique (AAS), la spectrométrie de fluorescence atomique (AFS), la spectrométrie d'émission atomique à source plasma à couplage inductif, la spectrométrie d'émission Auger à source plasma à couplage inductif (ICP-AES) la spectrométrie de masse à source plasma à couplage inductif (ICP-MS) et la voltamétrie. Certaines d'entre elles peuvent être utilisées pour la détection spécifique d'un élément donné (par ex. l'ICP-MS) en les couplant avec à une technique de séparation chromatographique (chromatographie en phase liquide à haute performance ou chromatographie en phase gazeuse). Ces méthodes couplées sont utilisées pour le dosage des différentes espèces arsenicales. Pour quelques dérivés l'arsenic, on peut souvent accroître la sensibilité des dosages en utilisant des techniques basées sur la génération d'hydrure. Au Bangladesh, on utilise actuellement une trousse d'analyse basée sur la réaction colorée que l'arsine donne avec le bromure mercurique pour contrôler les eaux souterraines. Sur le terrain, la limite de détection autorisée par cette technique est de 50-100 µg/litre.

2. Etat naturel et sources environnementales diverses

L'arsenic est présent dans plus de 200 minéraux différents, dont le plus courant est l'arsénopyrite.

502

On estime qu'environ un tiers du flux arsenical atmosphérique est d'origine naturelle. La source naturelle la plus importante d'arsenic est l'activité volcanique, la volatilisation à basse température venant au second rang.

Dans de nombreuses régions du monde, et notamment au Bangladesh, les eaux souterraines utilisées pour la consommation contiennent de l'arsenic minéral géogène.

Des organoarsenicaux comme d'arsénobétaïne, l'arsénocholine, les sels de tétramééthylarsonium ou encore des arsénoglucides ou arsénolipides sont surtout présents dans les organismes marins mais on trouve également certains d'entre eux dans quelques espèces terrestres.

On prépare l'arsenic élémentaire en réduisant le trioxyde d'arsenic (As_2O_3) par le charbon de bois. Le trioxyde d'arsenic est un sous-produit des opérations de fonderie. On estime que 70% de la production mondiale d'arsenic est utilisée pour le traitement du bois d'oeuvre par l'arséniate double de cuivre et de chrome, 22% pour la préparation de produits chimiques à usage agricole, le reste entrant dans la composition de certains verres, produits pharmceutiques et alliages non ferreux.

Les exploitations minières, la fusion des métaux non ferreux et l'utilisation des combustibles fossiles sont les principaux processus industriels qui contribuent à la pollution anthropogénique de l'air, des sols et de l'eau par l'arsenic. De grandes étendues de terres agricoles sont depuis toujours contaminées par l'épandage de pesticides à base d'arsenic. La protection du bois d'oeuvre au moyen de dérivés arsenicaux contribue également à la pollution de l'environnement.

3. Transport et distribution dans l'environnement

De l'arsenic est libéré dans l'atmosphère lors de processus impliquant des températures élevées comme la combustion du charbon dans les centrales thermiques, le brûlage de la végétation et le volcanisme. La biométhylation a basse température et la réduction naturelles en arsines libère également de l'arsenic dans l'atmosphère. C'est essentiellement sous forme d'As_2O_3 que l'arsenic est libéré dans l'atmosphère où il est en majeure partie adsorbé à des particules en suspension. Ces particules sont dispersées par le vent et regagnent sol par en se déposant par voie humide ou par voie sèche. Les arsines qui résultent de l'activité microbienne se dégagent du sol ou des sédiments où elles ont pris naissance pour s'oxyder à l'air en composés arsenicaux non volatils qui se redéposent. L'arsenic est présent dans la colonne d'eau sous la forme d'arséniates, d' arsénites ou encore d'acide méthylarsonique ou diméthylarsinique. Dans le sol et les sédiments bien oxygénés, la presque totalité de l'arsenic est présent sous sa forme pentavalente (arséniate), qui est la plus stable thermodynamiquement. Certains arsénites ou arséniates peuvent alterner leur degré d'oxydation sous l'influence du potentiel rédox (Eh), du pH et de certains processus biologiques. Quelques dérivés de l'arsenic présentent une certaine affinité pour les surfaces argileuses et les matières organiques, ce qui peut influer sur leur comportement dans l'environnement. Les fluctuations du potentiel rédox, du pH, de la concentration en dérivés arsenicaux solubles et de la teneur en matières organiques en suspension peuvent avoir pour conséquence la libération d'arsenic. Des débris résultant de l'érosion des rochers et des sols peuvent être transportés par le vent ou entraînés par les eaux. De nombreux dérivés de l'arsenic sont adorbés sur les particules de terre et le lessivage des sols ne les transporte habituellement que sur de courtes distances.

On observe trois types principaux de biotransformation de l'arsenic dans l'environnement : interconversion rédox entre arsénites et arséniates, réduction et méthylation de l'arsenic ou biosynthèse d'organoarsenicaux. Les composés formés selon ces divers processus participent à un cycle biogéochimique.

4. Concentrations dans l'environnement et exposition humaine

Dans les zones reculées et en milieu rural, la concentration moyenne de l'arsenic dans l'air va de 0,02 à 4 ng/m^3. Dans l'air des villes, la concentration moyenne va de 3 à environ 200 ng/m^3; des concentrations beaucoup plus fortes (> 1000 ng/m^3) ont été relevées à proximité de sources industrielles, encore que la tendance soit à la réduction de ces teneurs du fait des mesures de lutte contre la pollution. Les eaux pélagiques ont le plus souvent une teneur de l'ordre de 1 à 2 µg/litre. L'arsenic est très répandu dans les eaux douces superficielles et sa concentration dans les cours d'eau et les lacs est généralement inférieure à 10 µg/litre, encore que l'on puisse trouver des valeurs pouvant aller jusqu'à 5 mg/litre dans des échantillons prélevés à proximité de sources anthropogéniques. Dans les eaux souterraines, la concentration de l'arsenic est d'environ 1 à 2 µg/litre sauf en présence de roches volcaniques ou de dépôts de sulfures, auquel cas la teneur en arsenic peut aller jusqu'à 3 mg/litre. La concentration moyenne dans les sédiments va de 5 à 3000 mg/kg, les valeurs les plus fortes se retrouvant dans les zones polluées. Dans le sol, la concentration de fond est comprise entre 1 et 40 mg/kg, la valeur moyenne se situant à 5 mg/kg. Une forte concentration naturelle d'arsenic peut être attribuée à certains substrats géologiques, par exemple à des minerais sulfurés. Dans les sols pollués par les activités humaines, la concentration de l'arsenic peut atteindre plusieurs grammes pour 100 ml.

Les résidus d'arsenic que l'on retrouve dans les organismes marins peuvent aller de < 1 à plus de 100 mg/kg, principalement sous la forme d'organoarsénicaux tels que les arsénoglucides (algues macroscopiques) et l'arsénobétaïne (invertébrés et poissons). Après biogénèse à partir des dérivés minéraux, les organoarsenicaux peuvent être accumulés par les organismes aquatiques. Le facteur de bioconcentration des dérivés arsenicaux est plus faible chez les invertébrés et les poissons d'eau douce que chez les organismes marins. On n'observe pas de bioamplification dans les différentes chaînes alimentaires aquatiques. La concentration de fond dans les biotes dulçaquicoles et terrestres est généralement inférieure à 1 mg/kg de poids frais. Les végétaux terrestres peuvent accumuler l'arsenic par captage radiculaire ou par adsorption foliaire des dépôts de particules aéroportées. Dans les échantillons biologiques prélevés à proximité de sources anthropogéniques où de zones d'activité géothermique, la concentration de l'arsenic est plus élevée. Certaines espèces en accumulent des quantités importantes, avec des concentrations qui peuvent atteindre 3000 mg/kg sur les sites miniers.

L'exposition à l'arsenic en dehors du cadre professionnel est essentiellement due à l'ingestion d'aliments et d'eau contaminés. La dose journalière totale est principalement constituée par l'apport alimentaire. Dans certaines régions, c'est l'eau de boisson qui constitue une source importante d'exposition à l'arsenic minéral. En pareil cas, elle intervient souvent de façon prépondérante dans l'apport journalier d'arsenic. Les résidus des industries extractives constituent également une source potentielle d'exposition à l'arsenic. L'apport journalier d'arsenic provenant de l'alimentation et des boissons se situe en général entre 20 et 300 µg. Selon des données limitées, environ 25% de l'arsenic présent dans l'alimentation s'y trouve sous la forme de dérivés minéraux, mais cela dépend en grande partie de la nature des aliments ingérés. Les poissons et les fruits de mer ont une faible teneur en arsenic minéral (< 1%). En revanche, la concentration en arsenic minéral est plus élevée dans la viande, la volaille, les produits laitiers et les céréales. L'exposition par la voie respiratoire peut correspondre à 10 µg/jour environ chez un fumeur et à environ 1 µg/jour chez un non fumeur et même davantage dans les zones polluées. La concentration des métabolites de l'arsenic minéral dans les urines (arsenic minéral, acides monométhylarsonique et diméthylarsinique) est révélatrice de la dose d'arsenic minéral absorbée par un individu. En général, elle va de 5 à 20 µg d'arsenic par litre, mais elle peut dépasser 1000 µg par litre.

Sur les lieux de travail qui bénéficient d'une pratique de pointe en matière d'hygiène et sécurité, l'exposition ne dépasse générale-ment pas 10 µg/m^3 (en moyenne pondérée par rapport au temps calculée sur 8 h). Cependant, des concentrations pouvant atteindre plusieurs milligrammes par mètre cube ont été relevées dans l'air de certains lieux de travail.

5. Cinétique et métabolisme

Lorsque des particules aéroportées sont inhalées, l'absorption de l'arsenic qu'elles renferment dépend dans une très large mesure de la solubilité et de la taille des particules. Les dérivés solubles de l'arsenic III et de l'arsenic V sont rapidement et largement résorbés au niveau des voies digestives. Pour beaucoup de dérivés arsenicaux, le métabolisme se caractérise par deux types principaux de réactions: 1) des réactions de réduction de l'arsenic V en arsenic III et 2) des réactions de méthylation oxydatives dans lesquelles les dérivés de l'arsenic III subissent des méthylations successives qui les transfor-ment en dérivés mono- di- et triméthylés, la *S*-adénosylméthionine (SAM) jouant le rôle de donneur de méthyle avec le glutathion(GSH) comme co-facteur essentiel. La méthylation de l'arsenic minéral en facilite l'excrétion car les produits qui en résultent, à savoir l'acide monométhylarsonique et l'acide diméthylarsinique, sont facilement éliminés par la voie urinaire. Il existe des différences interspécifiques quantitatives et qualitatives très importantes en ce qui concerne ces réactions de méthylation, puisque chez certaines espèces, la méthylation de l'arsenic est minime, voire inexistante (par ex. chez le ouistiti, le cobaye ou le chimpanzé). Cependant, chez l'Homme comme chez la plupart des animaux de laboratoire, l'arsenic subit une méthylation importante conduisant à des métabolites qui sont essentiellement excrétés par voie urinaire. Des facteurs comme la dose, l'âge, le sexe ou le tabagisme ne jouent qu'un rôle minime dans les importantes variations interindividuelles observées chez l'Homme en ce qui concerne la méthylation métabolique de l'arsenic. Sur trois études consacrées à ce problème, une seule a révélé une moins bonne méthylation de l'arsenic chez l'enfant. Les études effectuées sur des sujets humains donnent à penser que l'activité des méthyltransférases varie dans d'importantes proportions et on a avancé l'hypothèse d'un polymorphisme pour expliquer cet état de choses. L'expérimentation humaine et animale indique que la méthylation de l'arsenic pourrait être inhibée en cas d'exposition aiguë. Le métabolisme et la distribution de l'arsenic minéral peuvent dépendre du degré d'oxydation de cet élément, notamment à forte dose. Les études effectuées sur des animaux de laboratoire indiquent que l'administration d'arsenic minéral trivalent sous la forme d'As_2O_3 ou d'arsénite, par exemple, peut initialement donner lieu à de plus fortes concentrations, dans la plupart des tissus, que l'administration d'arsenic pentavalent. Il reste que la forme trivalente est plus largement méthylée, ce qui conduit en définitive à une excrétion similaire à long terme. Chez l'Homme et chez l'animal, on constate qu'après ingestion, des organoarsenicaux tels que l'acide monométhylarsonique, l'acide diméthylarsinique ou la bétaïne subissent une moindre métabolisation et sont plus rapidement éliminés dans l'urine que les dérivés minéraux.

La concentration de l'arsenic ou de ses métabolites dans le sang, les cheveux, les ongles ou les urines sert de biomarqueur de l'exposition à l'arsenic. Le taux sanguin n'est utile comme marqueur qu'en cas d'intoxication aiguë ou d'exposition chronique constante et de longue durée. L'arsenic est rapidement éliminé du sang et il est difficile de déterminer sous quelle forme chimique il s'y trouve.

L'arsenic présent dans les cheveux et les ongles peut constituer un indicateur d'une exposition ancienne, dans la mesure où l'on prend la précaution d'éliminer toute contamination externe. On peut également se servir de la teneur des cheveux en arsenic pour évaluer le temps écoulé depuis une exposition à l'arsenic. Le dosage des divers métabolites urinaires exprimés soit en arsenic minéral, soit par la somme de ces métabolites (arsenic minéral, acide monométhyl-arsonique, acide diméthylarsinique) fournit la meilleure évaluation quantitative d'une dose d'arsenic récemment absorbée. Toutefois, la consommation de certains produits de la mer, notamment des algues et quelques bivalves, peut constituer un facteur de confusion dans l'estimation de l'exposition à l'arsenic, en raison de la métabolisation des arsénoglucides en acide diméthylarsinique par l'organisme du consommateur ou de la présence même de cet acide arsinique dans les produits consommés. Pendant les 2 à 3 jours précédant un prélèvement d'urine destiné à la recherche d'une exposition à des dérivés minéraux de l'arsenic, il faut s'abstenir de consommer tout produit de ce genre.

6. Effets sur les animaux de laboratoire et les systèmes d'épreuve *in vitro*

Qu'il soit sous forme minérale ou organique, l'arsenic peut provoquer des effets indésirables chez les animaux de laboratoire. Cette action toxique va de l'intoxication aiguë mortelle à des effets chroniques comme le cancer. Le degré de toxicité de l'arsenic dépend essentiellement de la forme (organique ou minérale) sous laquelle il se trouve et de son degré d'oxydation dans l'espèce chimique en cause. On estime en général que les dérivés minéraux de l'arsenic sont plus toxiques que les organoarsenicaux et qu'à l'intérieur de ces deux catégories de composés, les dérivés de l'arsenic III sont plus toxiques que ceux de l'arsenic V, tout au moins à forte dose. L'arsenic s'attaque à plusieurs systèmes et organes comme les téguments, l'appareil respiratoire, le système cardiovasculaire, le système immunitaire, l'appareil génito-urinaire, les gonades, les voies digestives et le système nerveux.

Plusieurs études de cancérogénicité ont été effectuées sur l'animal, mais un certain nombre de limites relatives à la dose administrée, à la durée d'exposition et au nombre d'animaux utilisés n'ont pas permis d'obtenir des résultats concluants. Toutefois, on a récemment proposé un modèle animal qui pourrait se révéler intéressant pour de futures études de cancérogénicité. On a ainsi effectué une étude sur des souris femelles C57B1/6J, qui a consisté à donner aux animaux pendant 2 ans une eau de boisson contenant 500 µg d'As V par litre. Les résultats obtenus montrent que l'incidence des tumeurs a augmenté, les localisations principales étant le poumon, le foie, les voies digestives, et la peau. L'arsenic minéral ne provoque pas de mutations ponctuelles. L'arsenic peut toutefois causer des aberrations chromosomiques *in vitro*, affecter la méthylation et la réparation de l'ADN, provoquer la prolifération et la transformation des cellules et jouer le rôle de promoteur tumoral. Selon une étude, de fortes doses d'acide diméthylarsinique pourraient provoquer des cancers de la vessie chez le rat mâle.

7. Effets sur la santé humaine

Les dérivés minéraux solubles de l'arsenic sont fortement toxiques et l'ingestion de doses importantes entraîne l'apparition de symptômes gastrointestinaux et de troubles cardiovasculaires et neurologiques pouvant aboutir à la mort. Chez les survivants, on peut observer les anomalies suivantes: dépression médullaire, hémolyse, hépatomégalie, mélanose, polynévrite et encéphalopathie.

Il existe une relation de cause à effet entre une exposition de longue durée à l'arsenic présent dans l'eau de boisson et l'accroissement du risque de cancer de la peau, du poumon, de la vessie et du rein, ainsi que d'autres anomalies cutanées telles qu'une hyperkératose ou une modification de la pigmentation. Ces effets ont été mis en évidence dans de nombreuses études utilisant divers protocoles expérimentaux. Pour chacun de ces points d'aboutissement de l'action toxique de l'arsenic, on a pu établir une relation exposition-réponse et l'existence d'un risque important. C'est à Taiwan que ces effets ont été le plus étudiés, mais ils sont très largement attestés par des enquêtes menées dans les populations d'autres pays. On estime qu'il y a augmentation du risque de cancer du poumon ou de la vessie et de lésions cutanées d'origine arsenicale en cas de consommation d'eau de boisson dont la teneur est ≤ 50 µg/litre.

Il existe également une relation de cause à effet entre l'exposition, principalement respiratoire, à l'arsenic et le cancer du poumon. Le risque est important et des relations exposition-réponse ont été établies. On observe une augmentation du risque lorsque l'exposition cumulée est ≥ 0.75 mg/m^3.année (c'est-à-dire, par exemple, 15 ans d'exposition à une concentration dans l'air de 50 µg/m^3). Dans deux des trois principales cohortes de fondeurs étudiées notamment sous l'angle du tabagisme, on a constaté que cette habitude n'était pas responsable de l'accroissement du risque de cancer du poumon attribué à l'arsenic; en revanche, il apparaît que l'interaction entre tabagisme et arsenic intervient dans l'augmentation de ce risque.

Malgré quelques résultats négatifs, tout semble indiquer que chez les individus et les cancéreux, l'arsenic a des effets clastogènes sur différents types de cellules et avec différents points d'aboutissement. En ce qui concerne les mutations ponctuelles, les résultats sont largement négatifs.

A Taiwan, on a montré qu'une exposition chronique à l'arsenic peut provoquer une affection appelée "maladie du pied noir". Il s'agit d'une vasculite périphérique grave pouvant aboutir à la gangrène. Cette affection n'est pas attestée dans d'autres régions du monde et il est possible que d'autres facteurs soient à prendre en compte dans la pathologie observée à Taiwan. Quoi qu'il en soit, on a de bonnes raisons de penser, compte tenu des études effectuées dans plusieurs pays, que l'exposition à l'arsenic peut être à l'origine d'autres formes de vasculite périphérique.

Les conclusions relatives à la relation de cause à effet entre l'exposition à l'arsenic et d'autres pathologies sont moins tranchées. C'est dans le cas de l'hypertension et des cardiopathies que les éléments d'appréciation sont les plus convaincants; ils sont révélateurs pour ce qui est du diabète et des effets sur la reproduction et faibles en ce qui concerne les accidents vasculaires cérébraux, les effets neurologiques à long terme et les cancers de localisation autre que le poumon, la vessie, le rein et la peau.

8. Effets sur les autres êtres vivants dans leur milieu naturel

La sensibilité des organismes aquatiques et terrestres aux divers dérivés de l'arsenic est très contrastée. Elle peut varier sous l'influence de divers facteurs biologiques ou abiotiques. En général les composés minéraux sont plus toxiques que les organoarsenicaux et les arsénites davantage que les arséniates. Le type de toxicité et le mécanisme s'absorption des arséniates par les différents organismes varie considérablement. Ces considérations peuvent expliquer les différences de sensibilité aux arsénites et aux arséniates selon les espèces. On pense que la toxicité des arsénites résulte essentiellement de leur fixation aux groupements thiols des protéines. En ce qui concerne les arséniates, on sait qu'ils perturbent la phosphorylation oxydative par compétition avec les phosphates. Dans un environnement où la concentration en phosphates est élevée, les effets toxiques des arséniates se font généralement moins sentir. En raison de l'analogie entre phosphates et arséniates, les organismes qui vivent dans un milieu riche en arséniates, doivent fixer le phosphore tout en évitant l'action toxique de l'arsenic.

Les composés arsenicaux provoquent des intoxications aiguës et des effets chroniques au niveau individuel ou collectif à des concentrations allant de quelques microgrammes à plusieurs milligrammes par litre, selon la nature de l'espèce en cause, de la durée de l'exposition et du point d'aboutissement de l'effet toxique. Ces effets peuvent être mortels, se traduire par l'inhibition de la croissance ou de la photosynthèse, perturber la reproduction ou le comportement. Les milieux fortement pollués par l'arsenic se caractérisent par une limitation de l'abondance et de la diversité des espèces. Lorsque la concentration en arséniates atteint une valeur élevée, seules les espèces résistantes sont à même de survivre.

RESUMEN

1. Propiedades y procedimientos analíticos

El arsénico es un metaloide que se encuentra ampliamente distribuido en la corteza terrestre con una concentración media de 2 mg/kg. Está presente en cantidades ínfimas en todo tipo de rocas, suelos, agua y aire. El arsénico puede existir en cuatro estados de valencia: −3, 0, +3 y +5. En condiciones de reducción, la forma predominante es el arsenito (As(III)); el arseniato (As (V)) suele ser la forma estable en condiciones oxigenadas. El arsénico elemental no es soluble en agua. Las sales de arsénico tienen una amplia gama de solubilidades en función del pH y de las condiciones iónicas.

Hay diversas técnicas instrumentales para la determinación del arsénico. Entre ellas figuran la espectrometría de absorción atómica (AAS), la espectrometría de fluorescencia atómica (AFS), la espectrometría electrónica de Auger de plasma con acoplamiento inductivo (ICP-AES), la espectrometría de masas de plasma con acoplamiento inductivo (ICP-MS) y la voltametría. Algunas de éstas (por ejemplo, la espectrometría de masas de plasma con acoplamiento inductivo) pueden servir como detectores específicos del elemento cuando se combinan con técnicas de separación cromatográficas (por ejemplo, la cromatografía líquida de alto rendimiento y la cromatografía de gases). Estos métodos que combinan la cromatografía con la espectrometría se utilizan para determinar especies concretas de arsénico. Con frecuencia se puede conseguir una mayor sensibilidad para una serie limitada de compuestos de arsénico mediante el uso de técnicas de generación de hidruros. En Bangladesh se está utilizando actualmente un estuche de pruebas basado en la reacción coloreada de la arsina con el bromuro mercúrico para el análisis de las aguas freáticas, con un límite de detección de 50-100 µg/litro en las condiciones que se encuentran sobre el terreno.

2. Fuentes y presencia de arsénico en el medio ambiente

El arsénico está presente en más de 200 especies minerales, de las cuales la más común es la arsenopirita.

512

Se ha estimado que alrededor de un tercio del flujo atmosférico de arsénico es de origen natural. La actividad volcánica es la fuente natural más importante de arsénico, seguida de la volatilización a baja temperatura.

En el agua freática utilizada como agua de bebida en varias partes del mundo, por ejemplo en Bangladesh, hay arsénico inorgánico de origen geológico.

Los compuestos orgánicos de arsénico, por ejemplo la arsenobetaína, la arsenocolina, las sales de tetrametilarsonio, los arsenoazúcares y los lípidos con arsénico se encuentran fundamentalmente en organismos marinos, aunque también se han detectado algunos de estos compuestos en especies terrestres.

El arsénico elemental se obtiene por reacción del trióxido de arsénico (As_2O_3) con carbón vegetal. El As_2O_3 es un subproducto de las operaciones de fundición de metales. Se ha estimado que el 70% de la producción mundial de arsénico se utiliza en el tratamiento de la madera como arseniato de cobre y cromo, el 22% en productos químicos de uso agrícola y el resto en la obtención de vidrio, productos farmacéuticos y aleaciones no ferrosas.

La extracción y fundición de metales no ferrosos y la utilización de combustibles fósiles son los principales procesos industriales que contribuyen a la contaminación antropogénica del aire, el agua y el suelo con arsénico. El empleo de plaguicidas con arsénico a lo largo del tiempo ha dejado contaminadas amplias zonas de tierras agrícolas. El uso del arsénico en la conservación de la madera también ha llevado a la contaminación del medio ambiente.

3. Transporte y distribución en el medio ambiente

El arsénico se libera en la atmósfera mediante procesos de alta temperatura, como los de las centrales eléctricas alimentadas con carbón, la combustión de vegetación y los volcanes. El proceso natural de biometilación y reducción a arsinas a baja temperatura también libera arsénico en la atmósfera. La mayor parte del arsénico se libera en la atmósfera como As_2O_3, y se mantiene sobre todo adsorbido sobre la materia particulada. El viento dispersa estas partículas, que vuelven a la tierra mediante deposición húmeda o seca. Las arsinas que se liberan de fuentes microbianas en el suelo o los sedimentos se oxidan en el aire, reconvirtiendo el arsénico en formas no volátiles que vuelven a la tierra. Entre las formas de arsénico disueltas en la columna de agua figuran el arseniato, el arsenito, el ácido metilarsónico y el ácido dimetilarsínico. En aguas y sedimentos bien oxigenados, casi todo el arsénico presente se encuentra en estado pentavalente, termodinámicamente más estable (arseniato). Algunas especies de arsenito y arseniato pueden intercambiar el estado de oxidación en función del potencial de oxidación-reducción, el pH y los procesos biológicos. Algunas especies de arsénico tienen afinidad por las superficies con minerales de la arcilla y la materia orgánica y esto puede afectar a su comportamiento en el medio ambiente. Es posible la emisión de arsénico cuando se producen fluctuaciones en el potencial de oxidación-reducción, el pH, la concentración de arsénico soluble y el contenido orgánico de los sedimentos. La erosión del viento o el agua puede transportar rocas y suelo meteorizados. Muchos compuestos de arsénico tienden a adsorberse en el suelo, y con la lixiviación suelen recorrer distancias cortas en este medio.

En el medio ambiente se han observado tres sistemas principales de biotransformación del arsénico: la transformación de oxidación-reducción entre el arsenito y el arseniato, la reducción y metilación del arsénico y la biosíntesis de compuestos orgánicos de arsénico. Hay un ciclo biogeoquímico de los compuestos formados a partir de estos procesos.

4. Niveles en el medio ambiente y exposición humana

Las concentraciones medias de arsénico total en el aire de zonas lejanas y rurales oscilan entre 0,02 y 4 ng/m^3. Las concentraciones medias de arsénico total en las zonas urbanas oscilan entre 3 y unos 200 ng/m^3; se han medido concentraciones mucho más altas (> 1000 ng/m^3) en las proximidades de fuentes industriales, aunque en algunas zonas están disminuyendo gracias a las medidas de reducción de la contaminación. Las concentraciones de arsénico en alta mar suelen ser de 1-2 µg/litro. El arsénico está ampliamente distribuido en el agua dulce superficial, siendo normalmente las concentraciones en ríos y lagos inferiores a 10 µg/litro, aunque en muestras aisladas se puedan alcanzar hasta 5 mg/litro cerca de fuentes antropogénicas. Los niveles de arsénico en el agua freática son como promedio de alrededor de 1-2 µg/litro, excepto en las zonas con rocas volcánicas y depósitos de minerales de sulfuro, donde los niveles de arsénico pueden llegar a 3 mg/litro. Las concentraciones medias de arsénico en los sedimentos oscilan entre 5 y 3000 mg/kg, correspondiendo los niveles más altos a zonas contaminadas. La concentración de fondo en el suelo varía de 1 a 40 mg/kg, con valores medios frecuentes de alrededor de 5 mg/kg. Las concentraciones naturales elevadas de arsénico pueden estar asociadas con determinados sustratos geológicos, tales como las menas de sulfuro. Los suelos contaminados por actividades humanas pueden tener concentraciones de arsénico de hasta varios gramos por 100 ml.

Los organismos marinos suelen contener residuos de arsénico que oscilan entre < 1 y más de 100 mg/kg, predominantemente en forma de especies de arsénico orgánicas, como arsenoazúcares (macroalgas) y arsenobetaína (invertebrados y peces). En los organismos acuáticos se produce una bioacumulación de compuestos de arsénico orgánicos, tras su biogénesis a partir de las formas inorgánicas. Los factores de bioconcentración para los compuestos de arsénico son más bajos en los invertebrados y los peces de agua dulce que en los organismos marinos. No se ha observado bioamplificación en las cadenas alimentarias acuáticas. Las concentraciones de fondo de arsénico en la biota de agua dulce y terrestre son normalmente inferiores a 1 mg/kg (peso fresco). Las plantas terrestres pueden acumular arsénico por absorción radicular del suelo o mediante la adsorción del arsénico que deposita el aire en las hojas. Las concentraciones de arsénico son más elevadas en las muestras de biota recogidas en fuentes antropogénicas o en zonas con actividad geotérmica. Algunas especies acumulan niveles importantes, con concentraciones medias de hasta 3000 mg/kg en zonas con minas arsenicales.

La exposición humana no ocupacional al arsénico en el medio ambiente se produce fundamentalmente a través de la ingestión de alimentos y de agua. De éstos, suelen ser los alimentos los que más contribuyen a la ingesta diaria de arsénico total. En algunas zonas, el arsénico que contiene el agua de bebida es una fuente importante de exposición al arsénico inorgánico. En estos casos, el agua de bebida es con frecuencia lo que más contribuye a la ingesta de arsénico diaria. Los suelos contaminados, por ejemplo con desechos de minas, son también una fuente potencial de exposición al arsénico. La ingesta diaria de arsénico total a partir de los alimentos y las bebidas oscila generalmente entre 20 y 300 µg/día. Los limitados datos disponibles indican que alrededor del 25% del arsénico presente en los alimentos es inorgánico, pero esto depende sobre todo del tipo de alimentos ingeridos. Los niveles de arsénico inorgánico en los peces y los mariscos son bajos (< 1%). Productos alimenticios como la carne, los productos lácteos y los cereales tienen niveles más elevados de arsénico inorgánico. De la exposición pulmonar pueden proceder hasta alrededor de 10 µg/día en un fumador y en torno a 1 µg/día en una persona no fumadora, siendo más elevados los valores en las zonas contaminadas. La concentración de metabolitos de arsénico inorgánico en la orina (arsénico inorgánico, ácido metilarsónico y ácido dimetilarsínico) refleja la dosis absorbida de arsénico inorgánico con carácter individual. En general, oscila entre 5 y 20 µg de arsénico/litro, pero puede incluso superar los 1000 µg/litro.

En lugares de trabajo con prácticas de higiene ocupacional actualizadas, la exposición generalmente no supera los 10 µg/m^3 (promedio ponderado por el tiempo de ocho horas). Sin embargo, en algunos lugares se han notificado concentraciones de arsénico atmosférico en los recintos de trabajo de hasta varios mg por metro cúbico.

5. Cinética y metabolismo

La absorción de arsénico en las partículas suspendidas en el aire inhaladas depende fundamentalmente de la solubilidad y del tamaño de las partículas. Los compuestos de arsénico solubles, tanto pentavalentes como trivalentes, se absorben en gran medida del tracto gastrointestinal con rapidez. En muchas especies, el metabolismo del arsénico se caracteriza por dos tipos principales de reacciones: 1) reacciones de reducción del arsénico pentavalente a trivalente, y 2) reacciones de metilación oxidativa, en las cuales las formas de arsénico trivalente experimentan una metilación secuencial que da lugar a productos mono, di y trimetilados, utilizando *S*-adenosilmetionina como donante de metilo y glutatión como un cofactor esencial. La metilación del arsénico inorgánico facilita su excreción del organismo, puesto que los productos finales, el ácido metilarsónico y el ácido dimetilarsínico, se excretan fácilmente en la orina. Hay importantes diferencias interespecíficas cualitativas y cuantitativas en la metilación, hasta tal punto que algunas especies muestran una metilación mínima o nula (por ejemplo, el tití, el cobaya y el chimpancé). Sin embargo, el arsénico inorgánico se metila en gran parte en las personas y en los animales de laboratorio más comunes y los metabolitos se excretan fundamentalmente en la orina. Factores como la dosis, la edad, el sexo y el hábito de fumar contribuyen sólo de manera mínima a la importante variación que se observa en la metilación del arsénico en las personas. Sin embargo, sólo en un estudio de tres se ha observado una eficacia de la metilación más baja en los niños. Los estudios en las personas parecen indicar que existe una amplia diferencia en la actividad de las metiltransferasas y se ha planteado la hipótesis de la existencia de polimorfismo. De los estudios realizados en animales y personas parece deducirse que la metilación del arsénico se puede inhibir con niveles de exposición aguda elevados. El metabolismo y la eliminación del arsénico inorgánico pueden depender de su valencia, sobre todo cuando se trata de dosis elevadas. Los estudios en animales de laboratorio indican que la administración de arsénico inorgánico trivalente, como el As_2O_3 y el arsenito, produce inicialmente en la mayoría de los tejidos niveles más altos que la administración de arsénico pentavalente. Sin embargo, la forma trivalente se metila en mayor medida, dando lugar a una excreción prolongada semejante. Los compuestos organoarsenicales ingeridos, como los ácidos metilarsónico y dimetilarsínico y la arsenobetaína, se metabolizan mucho menos y se eliminan con mayor rapidez en la orina que el arsénico inorgánico, tanto en animales de laboratorio como en el ser humano.

Los niveles de arsénico o de sus metabolitos en la sangre, el pelo, las uñas y la orina se utilizan como biomarcadores de la exposición al arsénico. El arsénico en sangre es un biomarcador útil solamente en el caso de intoxicación aguda por arsénico o de exposición crónica estable elevada. El arsénico se elimina de la sangre con rapidez y la especiación de su formas químicas en la sangre es difícil. Su presencia en el pelo y las uñas puede ser un indicador de una exposición anterior al arsénico, siempre que se tenga la precaución de impedir la contaminación de las muestras con arsénico externo. El arsénico en el pelo se puede utilizar también para estimar el período de tiempo relativo desde una exposición aguda. La especiación de los metabolitos en la orina expresados como arsénico inorgánico o bien como la suma de los metabolitos (arsénico inorgánico + ácido metilarsónico + ácido dimetilarsínico) proporciona la mejor estimación cuantitativa de la dosis de arsénico absorbida recientemente. Sin embargo, el consumo de ciertos alimentos marinos, principalmente algas y algunos bivalvos, puede confundir en la estimación de la exposición al arsénico inorgánico, debido al metabolismo de los arsenoazúcares para dar lugar a ácido dimetilarsínico en el organismo o a la presencia de ácido dimetilarsínico en los alimentos marinos. Dichos alimentos deberían evitarse durante dos o tres días antes del muestreo de la orina para vigilar la exposición al arsénico inorgánico.

6. Efectos en los animales de laboratorio y en los sistemas *in vitro*

Tanto las formas orgánicas como las inorgánicas del arsénico pueden causar efectos adversos en los animales de laboratorio. Los efectos inducidos por el arsénico van desde la letalidad aguda hasta los efectos crónicos, como el cáncer. El grado de toxicidad del arsénico depende básicamente de la forma (por ejemplo, inorgánico u orgánico) y de su estado de oxidación. En general, se considera que los compuestos arsenicales inorgánicos son más tóxicos que los orgánicos, y dentro de estas dos categorías las formas trivalentes son más tóxicas que las pentavalentes, por lo menos en dosis elevadas. El arsénico afecta a varios sistemas diferentes de órganos, como el cutáneo, el respiratorio, el cardiovascular, el genitourinario, el reproductivo, el gastrointestinal y el nervioso.

Se han realizado varios estudios de carcinogenicidad en animales, pero no se ha llegado a ninguna conclusión debido a limitaciones como las dosis elevadas, el tiempo limitado de exposición y el escaso número de animales. Sin embargo, se ha descrito recientemente un modelo en animales que podría ser un instrumento útil para futuros estudios de carcinogenicidad. En ese estudio, se relacionó la exposición de ratones C57B1/6J hembras a concentraciones de arsénico en el agua de bebida de 500 µg de As(V)/litro durante dos años con una mayor incidencia de tumores que afectaban principalmente al pulmón, el hígado, el tracto gastrointestinal y la piel. El arsénico inorgánico no induce mutaciones puntuales. Sin embargo, el arsénico puede producir aberraciones cromosómicas *in vitro,* afectar a la metilación y reparación del ADN, inducir proliferación celular, transformar las células y favorecer la formación de tumores. En un estudio se ha indicado que en dosis elevadas el ácido dimetilarsínico puede provocar en ratas machos cáncer de la vejiga urinaria.

7. Efectos en el ser humano

El arsénico inorgánico soluble produce toxicidad aguda y la ingestión de dosis altas provoca síntomas gastrointestinales, trastornos de las funciones de los sistemas cardiovascular y nervioso y en último término la muerte. En los supervivientes se ha observado depresión de la médula ósea, hemólisis, hepatomegalia, melanosis, polineuropatía y encefalopatía.

La exposición prolongada al arsénico en el agua de bebida tiene una relación causal con un aumento de los riesgos de cáncer de piel, de pulmón, de vejiga y de riñón, así como con otros cambios cutáneos, por ejemplo hiperqueratosis y cambios de pigmentación. Estos efectos se han puesto de manifiesto en numerosos estudios utilizando diferentes diseños. Se han observado relación exposición-respuesta y riesgo alto para cada uno de estos efectos finales. Los efectos se han estudiado más detalladamente en Taiwan, pero también hay un gran número de pruebas de estudios sobre poblaciones en otros países. Se ha notificado un riesgo mayor de cáncer de pulmón y de vejiga y de lesiones cutáneas asociadas con el arsénico en relación con la ingestión de agua de bebida con concentraciones ≤ 50 µg de arsénico/litro.

Se ha encontrado una relación causal entre la exposición ocupacional al arsénico, principalmente por inhalación, y el cáncer de pulmón. Se han observado relaciones exposición-respuesta y riesgos altos. Se ha detectado un aumento del riesgo a niveles de exposición acumulativa $\geq 0,75$ (mg/m^3) por año (por ejemplo, 15 años de exposición a una concentración en el aire del recinto de trabajo de 50 µg/m^3). El humo del tabaco se ha investigado en dos de las tres cohortes principales de fundidores y no se observó que fuera la causa del aumento del riesgo de cáncer de pulmón atribuido al arsénico; sin embargo, se comprobó que había una interacción con el arsénico que daba lugar a un aumento del riesgo de cáncer de pulmón.

A pesar de que algunos resultados han sido negativos, el valor demostrativo en conjunto indica que el arsénico puede provocar daños clastogénicos en diferentes tipos de células con distintos efectos finales en las personas expuestas y en los enfermos de cáncer. Para las mutaciones puntuales, los resultados son fundamentalmente negativos.

Se ha demostrado que la exposición crónica al arsénico en Taiwan provocó "blackfoot", una forma grave de enfermedad vascular periférica que da lugar a cambios gangrenosos. Esta enfermedad no se ha documentado en otras partes del mundo y los resultados de Taiwan pueden depender de la contribución de otros factores. Sin embargo, hay pruebas convincentes de estudios en varios países de que la exposición al arsénico produce otras formas de enfermedad vascular periférica.

Las conclusiones sobre la causalidad de la relación entre la exposición al arsénico y otros efectos en la salud son menos claras. Las pruebas más convincentes son las relativas a la hipertensión y las enfermedades cardiovasculares, son dudosas para la diabetes e insuficientes para la enfermedad cerebrovascular, los efectos neurológicos prolongados y el cáncer en lugares distintos del pulmón, la vejiga, el riñón y la piel.

8. Efectos en otros organismos en el medio ambiente

La sensibilidad de la biota acuática y terrestre para las distintas especies de arsénico es muy variable. Depende de factores biológicos y abióticos. En general, los compuestos arsenicales inorgánicos son más tóxicos que los orgánicos, y el arsenito es más tóxico que el arseniato. El mecanismo de la toxicidad y el sistema de absorción del arseniato por los organismos varían considerablemente de unos a otros. Esto puede explicar las diferencias interespecíficas en la respuesta de los organismos al arseniato y el arsenito. Se considera que el mecanismo primario de la toxicidad del arsenito se deriva de su unión a los grupos sulfhidrilo de las proteínas. Se sabe que el arseniato afecta a la fosforilación oxidativa, porque compite con el fosfato. En condiciones en las cuales la concentración de fosfato es alta, la toxicidad del arseniato para la biota es generalmente reducida. Como el arseniato es análogo al fosfato, los organismos que viven en presencia de una concentración elevada de arseniato deben adquirir el fósforo nutritivo, pero evitando la toxicidad del arsénico.

Los compuestos de arsénico provocan efectos agudos y crónicos en las personas, las poblaciones y las comunidades a concentraciones que oscilan entre unos microgramos y miligramos por litro, dependiendo de la especie, el tiempo de exposición y los efectos finales medidos. Estos efectos incluyen la letalidad, la inhibición del crecimiento, de la fotosíntesis y de la reproducción y efectos de comportamiento. Las zonas contaminadas con arsénico se caracterizan por una abundancia y diversidad limitadas de especies. Si los niveles de arseniato son suficientemente altos, sólo es posible la presencia de especies resistentes.

THE ENVIRONMENTAL HEALTH CRITERIA SERIES(continued)

Flame retardants: tris(chloropropyl)
phosphate and tris(2-chloroethyl)
phosphate (No. 209, 1998)
Flame retardants: tris(2-butoxyethyl)
phosphate, tris(2-ethylhexyl) phosphate
and tetrakis(hydroxymethyl) phosphonium
salts (No. 218, 2000)
Fluorine and fluorides (No. 36, 1984)
Food additives and contaminants in food,
principles for the safety assessment of
(No. 70, 1987)
Formaldehyde (No. 89, 1989)
Fumonisin B$_1$ (No. 219, 2000)
Genetic effects in human populations,
guidelines for the study of (No. 46, 1985)
Glyphosate (No. 159, 1994)
Guidance values for human
exposure limits (No. 170, 1994)
Heptachlor (No. 38, 1984)
Hexachlorobenzene (No. 195, 1997)
Hexachlorobutadiene (No. 156, 1994)
Alpha- and beta-hexachlorocyclohexanes
(No. 123, 1992)
Hexachlorocyclopentadiene
(No. 120, 1991)
n-Hexane (No. 122, 1991)
Human exposure assessment
(No. 214, 2000)
Hydrazine (No. 68, 1987)
Hydrogen sulfide (No. 19, 1981)
Hydroquinone (No. 157, 1994)
Immunotoxicity associated with exposure
to chemicals, principles and methods for
assessment (No. 180, 1996)
Infancy and early childhood, principles for
evaluating health risks from chemicals
during (No. 59, 1986)
Isobenzan (No. 129, 1991)
Isophorone (No. 174, 1995)
Kelevan (No. 66, 1986)
Lasers and optical radiation (No. 23, 1982)
Lead (No. 3, 1977)[a]
Lead, inorganic (No. 165, 1995)
Lead – environmental aspects
(No. 85, 1989)
Lindane (No. 124, 1991)
Linear alkylbenzene sulfonates and related
compounds (No. 169, 1996)
Magnetic fields (No. 69, 1987)
Man-made mineral fibres (No. 77, 1988)
Manganese (No. 17, 1981)
Mercury (No. 1, 1976)[a]
Mercury – environmental aspects
(No. 86, 1989)
Mercury, inorganic (No. 118, 1991)
Methanol (No. 196, 1997)
Methomyl (No. 178, 1996)
2-Methoxyethanol, 2-ethoxyethanol, and
their acetates (No. 115, 1990)
Methyl bromide (No. 166, 1995)
Methylene chloride
(No. 32, 1984, 1st edition)
(No. 164, 1996, 2nd edition)
Methyl ethyl ketone (No. 143, 1992)
Methyl isobutyl ketone (No. 117, 1990)
Methylmercury (No. 101, 1990)
Methyl parathion (No. 145, 1992)

Methyl *tertiary*-butyl ether (No. 206, 1998)
Mirex (No. 44, 1984)
Morpholine (No. 179, 1996)
Mutagenic and carcinogenic chemicals,
guide to short-term tests for detecting
(No. 51, 1985)
Mycotoxins (No. 11, 1979)
Mycotoxins, selected: ochratoxins,
trichothecenes, ergot (No. 105, 1990)
Nephrotoxicity associated with exposure
to chemicals, principles and methods for
the assessment of (No. 119, 1991)
Neurotoxicity associated with exposure to
chemicals, principles and methods for the
assessment of (No. 60, 1986)
Neurotoxicity risk assessment for human
health, principles and approaches
(No. 223, 2001)
Nickel (No. 108, 1991)
Nitrates, nitrites, and N-nitroso compounds
(No. 5, 1978)[a]
Nitrogen oxides
(No. 4, 1977, 1st edition)[a]
(No. 188, 1997, 2nd edition)
2-Nitropropane (No. 138, 1992)
Noise (No. 12, 1980)[a]
Organophosphorus insecticides:
a general introduction (No. 63, 1986)
Paraquat and diquat (No. 39, 1984)
Pentachlorophenol (No. 71, 1987)
Permethrin (No. 94, 1990)
Pesticide residues in food, principles for
the toxicological assessment of
(No. 104, 1990)
Petroleum products, selected
(No. 20, 1982)
Phenol (No. 161, 1994)
d-Phenothrin (No. 96, 1990)
Phosgene (No. 193, 1997)
Phosphine and selected metal phosphides
(No. 73, 1988)
Photochemical oxidants (No. 7, 1978)
Platinum (No. 125, 1991)
Polybrominated biphenyls (No. 152, 1994)
Polybrominated dibenzo-p-dioxins and
dibenzofurans (No. 205, 1998)
Polychlorinated biphenyls and terphenyls
(No. 2, 1976, 1st edition)[a]
(No. 140, 1992, 2nd edition)
Polychlorinated dibenzo-p-dioxins and
dibenzofurans (No. 88, 1989)
Polycyclic aromatic hydrocarbons,
selected non-heterocyclic (No. 202, 1998)
Progeny, principles for evaluating health
risks associated with exposure to
chemicals during pregnancy
(No. 30, 1984)
1-Propanol (No. 102, 1990)
2-Propanol (No. 103, 1990)
Propachlor (No. 147, 1993)
Propylene oxide (No. 56, 1985)
Pyrrolizidine alkaloids (No. 80, 1988)
Quintozene (No. 41, 1984)
Quality management for chemical
safety testing (No. 141, 1992)
Radiofrequency and microwaves
(No. 16, 1981)

[a]   Out of print

WORLD HEALTH ORGANIZATION

CORRIGENDUM

ENVIRONMENTAL HEALTH CRITERIA NO. 224

ARSENIC AND ARSENIC COMPOUNDS (2ND EDITION)

Page 13, Table 2:

The CAS number for arsenenous acid should read **13768-07-5,** rather than 13768-07-05.

WORLD HEALTH ORGANIZATION

CORRIGENDUM

ENVIRONMENTAL HEALTH CRITERIA

NO. 224

ARSENIC AND ARSENIC COMPOUNDS

Page 34, Section 4.1.1, 2nd paragraph: The last four lines should read:

... the Olympic National Park, Washington (USA) to range from 76.7 to 208 $\mu g/m^2$. The average annual wet deposition of arsenic at Chesapeake bay (Maryland, USA) was found to be 49 μg As/m^2 (Scudlark et al., 1994).

Page 261, Table 31:

4th column:
Delete the words *significant at $p \leq 0.05$ level*

Page 263, Section 8.5, Table 31 :

Replace the present page with the page overleaf:

Table 31 (contd.)

			Exp. category	PR* for glucosuria (CI)		
Cross-sectional	1595 people from 4 villages in Bangladesh; 1481 exposed to As and 114 non exposed controls	As in drinking-water. For 39, 36, 18, and 7%, the exposure was < 0.5, 0.5–1, and > 1 mg/litre, respectively.	mg/litre · year			Study limited to the 1595 individuals out of 1794 eligible who were at home at the time of the interview. 114 persons were con-sidered unexposed and were used as the refer-ence group. Used existing As water measurements (measured by flow-injection hydride generation AAS. *PR, Mantel–Haenszel prevalence ratio adjusted for age and sex Urinary glucose analysed using glucose strip	Rahman et al. (1999b)
			Patients without skin lesions:			
			< 1.0	0.4 (0.1–1.0)		
			1.0–5.0	0.9 (0.5–1.7)		
			5.0–10	1.2 (0.6–2.2)		
			> 10	1.7 (1.0–2.9)		
			Patients with skin lesions:			
			< 1.0	0.8 (0.3–1.9)		
			1.0–5.0	1.7 (0.9–2.9)		
			5.0–10	2.1 (1.0–4.0)		
			> 10	2.9 (1.6–5.2)		

Occupational exposure

			Exp. Category (mg/m³)	OR, diabetes	95% CI	
Case–referent	12 cases and 31 referents in the Swedish smelter cohort	exposure to airborne As	<< 0.5	2.0	0.1–2.7	Exposure assessment very crude and exposure categorization not well defined making the interpretation difficult	Rahman & Axelson (1995)
			≤ 0.5	4.2	0.3–54		
			> 0.5	7.0	0.7–79		

Page 311 Table 34:

In the Ecological study by Tsai et al (1999) [second item in the table], replace the third column by:

Skin cancer SMR for females and males combined	
SMR	CI
483	374-615 (local rates)
597	462-760 (national rates)